Lecture Notes in Control and Information Sciences 403

Editors: M. Thoma, F. Allgöwer, M. Morari

Roland Tóth

Modeling and Identification of Linear Parameter-Varying Systems

Springer

Author

Dr. Roland Tóth
Delft University of Technology
Faculty of Mechanical, Maritime and
Materials Engineering
Delft Center for Systems and Control
Mekelweg 2
2628 CD, Delft
The Netherlands

Co-Editors

Dr. Peter S.C. Heuberger
Delft University of Technology
Faculty of Mechanical, Maritime and
Materials Engineering
Delft Center for Systems and Control
Mekelweg 2
2628 CD, Delft
The Netherlands

Prof. Dr. Paul M.J. Van den Hof
Delft University of Technology
Faculty of Mechanical, Maritime and
Materials Engineering
Delft Center for Systems and Control
Mekelweg 2
2628 CD, Delft
The Netherlands

ISBN 978-3-642-13811-9 e-ISBN 978-3-642-13812-6

DOI 10.1007/978-3-642-13812-6

Lecture Notes in Control and Information Sciences ISSN 0170-8643

Library of Congress Control Number: 2010928272

Typeset & Cover Design: Scientific Publishing Services Pvt. Ltd., Chennai, India.

Printed in acid-free paper

5 4 3 2 1 0

springer.com

I dedicate this book to my beloved wife
Andrea, my "great" son Sándor
and my little daughter Lujza.

Preface

Through the past 20 years, the framework of *Linear Parameter-Varying* (LPV) systems has become a promising system theoretical approach to handle the control of mildly nonlinear and especially position dependent systems which are common in mechatronic applications and in the process industry. The birth of this system class was initiated by the need of engineers to achieve better performance for nonlinear and time-varying dynamics, common in many industrial applications, than what the classical framework of *Linear Time-Invariant* (LTI) control can provide. However, it was also a primary goal to preserve simplicity and "re-use" the powerful LTI results by extending them to the LPV case. The progress continued according to this philosophy and LPV control has become a well established field with many promising applications.

Unfortunately, modeling of LPV systems, especially based on measured data (which is called system identification) has seen a limited development since the birth of the framework. Currently this bottleneck of the LPV framework is halting the transfer of the LPV theory into industrial use. Without good models that fulfill the expectations of the users and without the understanding how these models correspond to the dynamics of the application, it is difficult to design high performance LPV control solutions. This book aims to bridge the gap between modeling and control by investigating the fundamental questions of LPV modeling and identification. It explores the missing details of the LPV system theory that have hindered the formulation of a well established identification framework. By proposing an unified LPV system theory that is based on a behavioral approach, the concepts of representations, equivalence transformations, and means to compare model structures are re-established, giving a solid basis for an identification theory. It is also explored when and how first-principle nonlinear models can be efficiently converted to LPV descriptions and what are the pitfalls that must be avoided. Building on well founded system theoretical concepts, the classical LTI prediction-error framework is extended to the LPV case via the use of series-expansion representations.

Beside completing the system theoretical aspects and founding of an LPV prediction-error framework, the book proposes a novel identification approach based on orthonormal basis functions, which provides an efficient and easy to use approach of LPV identification. It has been shown in the LTI case that decomposing dynamical systems in terms of orthogonal expansions enables the accurate approximation of the system with a finite length expansion. By tuning the basis functions to the underlying system characteristics, the rate of convergence can be drastically decreased. This leads to highly accurate models (small bias) being represented by a few parameters (small variance), which in fact can be estimated efficiently via simple linear regression. This philosophy gives the basic concept of the proposed identification approach, for which the applicability in practical scenarios is investigated and powerful algorithms are provided and analyzed.

The work presented here is the result of 5 years of research at the Delft University of Technology under the supervision of Prof. Paul M.J. Van den Hof and Peter S.C. Heuberger. Starting with the initial intention to apply simplicity of orthonormal basis function models to overcome the challenges of LPV system identification, a long road has led to the theory which is presented in this book. Walking on this road, many excellent people have contributed to this work. Especially Prof. Jan C. Willems, at the Catholic University of Leuven, whose advice and vision about mathematical modeling helped me to find the right track that lead to the LPV behavioral theory, that forms the basis of many concepts of this book. Prof. Carsten Scherer, at the Delft University of Technology, whose advice and extensive knowledge on LPV systems and control has always helped me to find the missing wheel. Federico Felici, whose earth-moving questions and discussions made me aware of missing key points of the LPV system theory. But the main thanks goes to Prof. Paul M.J. Van den Hof and Peter S.C. Heuberger whose excellent guidance led me through these years, culminating in my Ph.D. thesis [188], which served as the basis of this book. As co-editors, they also had a significant role in establishing this book in its current from.

The book is written as a research monograph with a broad scope, trying to cover the key issues from system theory to modeling and identification. It is meant to be interesting for both researchers and engineers but also for graduate students in systems and control who would like to learn about the LPV framework. It also offers an easy to use guide for engineers about the off-shelf solutions in LPV modeling and identification.

I hope that you as a reader will enjoy reading it as much as it was fun to write it.

Delft, December 2009 Roland Tóth

Contents

Acronyms

ACM	Adaptive cluster merging
ARMAX	Auto-regressive moving-average with exogenous variable (model)
ARX	Auto-regressive with exogenous variable (model)
BFR	Best fit rate
BIBO	Bounded-input bounded-output (stability)
BJ	Box-Jenkins (model)
CT	Continuous time
DT	Discrete time
FcM	Fuzzy c-means
FIR	Finite impulse response
FKcM	Fuzzy Kolmogorov c-max
GOBF	Generalized orthonormal basis function
H-LPV OBF	Hammerstein linear parameter-varying OBF (model)
HF-LPV OBF	Hammerstein feedback linear parameter-varying OBF (model)
IIR	Infinite impulse response
IO	Input-output
IRR	Impulse response representation
IV	Instrumental variable
KM	Kolmogorov metric
KnW	Kolmogorov n-width
KR	Kernel representation
LFR	Linear fractional representation
LMI	Linear matrix inequality
LPV	Linear parameter-varying
LS	Least squares
LSDP	Linear semi-definite programming
LTI	Linear time-invariant
LTV	Linear time-varying
LUT	Local unit truncation (error)
MIMO	Multiple-input multiple-output
MISO	Multiple-input single-output

MOESP	Multivariable output-error state-space (algorithm)
MSE	Mean squared error
NH	Nonlinear Hammerstein (model)
NL	Nonlinear
NW	Nonlinear Wiener (model)
OBF	Orthonormal basis function
ODE	Ordinary differential equation
OE	Output error (model)
PE	Persistency of excitation
pdf	Probability density function
PV	Parameter-varying
QC	Quadratic constraint
ROC	Region of convergence
SIMO	Single-input multiple-output
SISO	Single-input single-output
SNR	Signal-to-noise ratio
SoS	Sum of squares
SS	State-space
SVD	Singular value decomposition
TS	Takagi-Sugeno (model)
TV	Time-varying
VAF	Variance accounted for
W-LPV OBF	Wiener linear parameter-varying OBF (model)
WF-LPV OBF	Wiener feedback linear parameter-varying OBF (model)
ZOH	Zero-order-hold

List of Symbols

	Operators
q	Forward time shift
$\frac{\mathrm{d}}{\mathrm{d}t}$	Differentiation
δ_-	Shift and cut
δ_+	Inverse shift and cut
Σ_-	Shift map
$\diamond$	Evaluation of a coefficient function along a trajectory of p
$\circ$	Function concatenation
$\oplus$	Direct sum
$\odot$	Element by element product
$\circledast$	Evaluation of an extended solution
$< \bullet , \bullet >$	Inner product
$[\,\bullet\,]_{ij}$	Element of the i-th row and j-th column
$\bullet^\top$	Transposition
$\bullet^*$	Complex conjugate transpose
$\bar{\bullet}$	Closure
$\bullet^{-1}$	Inverse
$\succ, (\succeq)$	Positive (semi)-definite
$\succ, (\succeq)$	Negative (semi)-definite
Re	Real part
Im	Imaginary part
inf	Infimum
sup	Supremum
min	Minimum
max	Maximum
Var	Variance
$\mathcal{E}\{\bullet\}$	Mean/Expectation
$\bar{\mathcal{E}}\{\bullet\}$	Generalized expectation
Dim	Dimension
Ker	Kernel
Img	Image

Det	Determinant
Adj	Matrix adjoint
Diag	Diagonal matrix composition
Span	Algebraic span
$\mathrm{Span}_{\mathcal{R}}^{\mathrm{row}}$	Row span on the ring $\mathcal{R}[\xi]$, subspace in $\mathcal{R}[\xi]^{\cdot\times\cdot}$
$\mathrm{Span}_{\mathcal{R}}^{\mathrm{col}}$	Column span on the ring $\mathcal{R}[\xi]$, subspace in $\mathcal{R}[\xi]^{\cdot\times\cdot}$
Hom	Homeomorphism
Grad	Gradient
Card	Cardinality
Col	Column composition
Deg	Degree
Deg_*	Degree, minimal
Rank	Rank
$\mathrm{Module}_{\mathcal{R}[\xi]}$	Left module in $\mathcal{R}[\xi]^{\cdot\times\cdot}$
$\underset{t}{\wedge}$	Concatenation of signals at time instant t
$\mathscr{L}$	Laplace transformation
$\mathscr{Z}$	Z-transformation
	Geometrical Objects
E	Euclidian line
L	i-line
H	h-line
K	Euclidian circle
K_{h}	Hyperbolic circle
D	Euclidean disc
D_{h}	Hyperbolic disc
$\mathscr{D}$	Set of Euclidian discs
$\mathtt{e}$	Euclidian center
$\mathtt{r}$	Euclidian radius
$\mathtt{e}_{\mathrm{h}}$	Hyperbolic center
$\mathtt{r}_{\mathrm{h}}$	Hyperbolic radius
$\mathtt{x},\mathtt{y},\mathtt{z},\mathtt{u},\mathtt{v}$	Points
φ_{h}	Hyperbolic coefficient
γ_{h}	Hyperbolic angle
$\mathfrak{h}_H$	Hyperbolic inversion with respect to H
$\mathfrak{h}_{\mathtt{x}}$	Hyperbolic inversion s.t. $\mathfrak{h}_{\mathtt{x}}(\mathtt{x}) = 0$
$\mathcal{Z}$	Hyperbolic coverage
	Dynamical Systems
$\mathcal{G}$	Dynamical system
$\mathcal{G}_{\mathrm{NL}}$	Nonlinear dynamical system
$\mathcal{S}$	Parameter-varying dynamical system
$\mathcal{F}$	LTI system
$\mathscr{F}$	LTI system set
$\mathscr{F}_{\mathcal{S}}$	Frozen system set of an LPV system $\mathcal{S}$

	Behaviors
$\mathfrak{B}$	Behavior
$\mathfrak{B}_*$	Complete (extended) behavior
$\mathfrak{B}_{\mathrm{L}}$	Latent Behavior
$\mathfrak{B}_{\mathrm{SS}}$	State-space Behavior
$\mathfrak{B}_{\mathrm{p}}$	Frozen behavior for constant scheduling p
$\mathfrak{B}_{\mathbb{P}}$	Scheduling behavior (projected)
$\mathfrak{B}_{\mathbb{W}}$	Signal behavior (projected)
$\mathfrak{B}_{\mathbb{X}}$	Sate signal behavior (projected)
$\mathfrak{B}_p$	Projected behavior w.r.t. a scheduling trajectory p
	Representations and Models
$\mathfrak{R}_{\mathrm{SS}}(\cdot)$	State-space system representation
$\mathfrak{R}_{\mathrm{SS}}^{\top}(\cdot)$	Transpose of a state-space system representation
$\mathfrak{R}_{\mathrm{SS}}^{\mathcal{O}}(\cdot)$	State-space representation, observability
$\mathfrak{R}_{\mathrm{SS}}^{\mathcal{R}}(\cdot)$	State-space representation, reachability
$\mathfrak{R}_{\mathrm{SS}}^{\mathcal{O}_{\mathrm{c}}}(\cdot)$	State-space representation, companion-observability
$\mathfrak{R}_{\mathrm{SS}}^{\mathcal{R}_{\mathrm{c}}}(\cdot)$	State-space representation, companion-reachability
$\mathfrak{R}_{\mathrm{SS}}^{\mathcal{O}_{\mathrm{LTI}}}(\cdot)$	State-space representation, observability, generated via LTI rules
$\mathfrak{R}_{\mathrm{SS}}^{\mathcal{R}_{\mathrm{LTI}}}(\cdot)$	State-space representation, reachability, generated via LTI rules
$\mathfrak{R}_{\mathrm{IO}}(\cdot)$	Input-output representation
$\mathfrak{R}_{\mathrm{K}}(\cdot)$	Kernel representation
$\mathfrak{R}_{\mathrm{IM}}(\cdot)$	Impulse response representation
$\mathfrak{R}_{\mathrm{OBF}}(\cdot)$	OBF expansion representation
$\mathfrak{M}_{\mathrm{W}}(\cdot)$	Wiener LPV OBF model
$\mathfrak{M}_{\mathrm{H}}(\cdot)$	Hammerstein LPV OBF model
$\mathfrak{M}_{\mathrm{WF}}(\cdot)$	Wiener feedback LPV OBF model
$\mathfrak{M}_{\mathrm{HF}}(\cdot)$	Hammerstein feedback LPV OBF model
	Equivalence Classes, Transformations
$\mathscr{E}$	Equivalence class, LTI
$\mathscr{E}^{n_{\mathbb{P}}}$	Equivalence class, LPV
$\mathscr{E}_{\mathrm{can}}^{n_{\mathbb{P}}}$	Set of canonical representations, LPV with $\mathrm{Dim}(\mathbb{P}) = n_{\mathbb{P}}$
$\overset{n_{\mathbb{P}}}{\sim}$	Equivalence relation, LPV with $\mathrm{Dim}(\mathbb{P}) = n_{\mathbb{P}}$
T	State-transformation
T_{o}	State-transformation, observability
T_{co}	State-transformation, companion-observability
T_{r}	State-transformation, reachability
T_{cr}	State-transformation, companion- reachability
	Signals and Variables
w	System signal/variable
w_{L}	Latent variable
$\mathscr{w}$	Extended solution
u	Input signal

y	Output signal
x	State signal
p	Scheduling variable
x_{o}	State signal (observability canonical)
x_{r}	State signal (reachability canonical)
$u_{\mathrm{d}}, y_{\mathrm{d}}, x_{\mathrm{d}}, p_{\mathrm{d}}$	Discretized signals
h	Impulse response
p	Point of the scheduling space
ξ	Indeterminate variable, time operator
ζ	Indeterminate variable, scheduling dependence
ω	Frequency
Δ	Lagrangian
δ	Lagrangian variable
ε	Noise/error/residual
v, e	Noise, stochastic process
$\mathcal{R}_{\mathrm{y}}, \mathcal{R}_{\mathrm{yu}}$	(Cross)-covariance
$\Phi_{\mathrm{y}}(e^{\mathrm{i}\omega}), \Phi_{\mathrm{yu}}(e^{\mathrm{i}\omega})$	(Cross)-power spectral densities
ε_k	Local unit truncation error (k^{th}-interval)
η_k	Global error (k^{th}-interval)
ε_*	Maximal local unit truncation error
η_*	Maximal global error
$\varepsilon_{\max}$	Acceptable maximal local unit truncation error
$\eta_{\max}$	Acceptable maximal global error
	Functions
R, P, X, Q	Polynomial matrix functions
M	Polynomial matrix function, unimodular
R_{u}	Polynomial matrix function, input side
R_{y}	Polynomial matrix function, output side
R_{L}	Polynomial matrix function, latent side
R_{com}	Common divisor of polynomial matrix functions
$R_{\mathrm{A}}, \ldots, R_{\mathrm{F}}$	Polynomial functions, classical model parametrization
F, G	Transfer function, IRR form
G_0	Transfer function / IRR form of the nominal model
H	Transfer function / IRR associated with the noise
H_0	Transfer function / IRR of the noise part of the nominal model
G_{b}	Inner function
M	Transfer function vector
$\mathfrak{F}$	Transfer function set
$\mathfrak{F}_{\mathcal{S}}$	Transfer function set of frozen behaviors
$\mathfrak{A}(\cdot,\cdot)$	Transition matrix
$\mathfrak{C}(t)$	Observability map
$\mathcal{V}$	Lyapunov function
J	Cost-function
ϕ	Orthonormal basis function

Φ_n	Orthonormal basis function set with n elements
$\Phi_{n_g}^{n_e}$	Hambo functions generated by a G_b with n_g poles and n_e extensions
μ	Membership function
$\mathcal{W}$	Criterion function
ψ	Function defining coefficient dependency in a model parametrization
Ψ	Set of functions defining coefficient dependency
$\mathfrak{g}$	Scheduling function
$1(.)$	Unit step (Heaviside) function
	Spaces and Fields
$\mathbb{T}$	Time axis
$\mathbb{V}$	Clustering space
$\mathbb{W}$	Signal space
$\mathbb{W}_L$	Latent signal space
$\mathbb{R}$	Set of real numbers
$\mathbb{R}^+, \mathbb{R}^-, \mathbb{R}_0^+, \mathbb{R}_0^-$	Positive (negative) real numbers with or without zero
$\mathbb{Z}$	Set of integers
$\mathbb{Z}^+, \mathbb{Z}^-, \mathbb{Z}_0^+, \mathbb{Z}_0^-$	Positive (negative) integers with or without zero
$\mathbb{C}$	Set of complex numbers
$\mathbb{N}$	Set of natural numbers
$\mathbb{D}$	Unit disk
$\mathbb{J}$	Unit circle
$\mathbb{E}$	Exterior of the unit disk
$\mathbb{X}$	State-space
$\mathbb{U}$	Input-space
$\mathbb{Y}$	Output-space
$\mathbb{I}_{n_1}^{n_2}$	Index set $\{n_1, n_1+1, \ldots, n_2\}$
$\mathbb{P}$	Scheduling domain
$\mathbb{W}^{\mathbb{T}}$	Collection of all maps from $\mathbb{T}$ to $\mathbb{W}$
$\mathbb{P}^{\mathbb{T}}$	Collection of all maps from $\mathbb{T}$ to $\mathbb{P}$
$\mathcal{M}$	Subspace
$\breve{\mathcal{M}}$	Optimal Subspace
$\mathscr{M}$	Set of subspaces with the same dimension
$\mathcal{U}$	Membership space
$\mathbb{S}, \mathbb{K}$	Singularity set
$\mathcal{I}$	Set of indices
$\mathcal{D}$	Hyperbolic group
$\mathcal{H}_2$	Hardy space of proper complex functions square integrable on $\mathbb{J}$
$\mathcal{H}_{2-}$	Hardy space of strictly proper complex functions square integrable on $\mathbb{J}$
$\mathcal{H}_2^{\perp}$	Complement of $\mathcal{H}_2$
$\mathcal{H}_2(\mathbb{E})$	Space of all $\mathcal{H}_2$ functions that are analytic in $\mathbb{E}$

$\mathcal{H}_2(\mathbb{C}^+)$	Space of all $\mathcal{H}_2$ functions that are analytic in $\mathbb{C}^+$
$\mathcal{H}_{2-}(\mathbb{E})$	Space of all $\mathcal{H}_{2-}$ functions that are analytic in $\mathbb{E}$
$\mathcal{RH}_{2-}(\mathbb{E})$	Space of all real $\mathcal{H}_{2-}$ functions that are analytic in $\mathbb{E}$
$\mathcal{R}_n$	Field of meromorphic functions with n variables
$\bar{\mathcal{R}}_n$	Subset of $\mathcal{R}_n$, non-eliminatable n^{th} variable
$\mathcal{R}$	Field of meromorphic functions with finite many variables
$\mathcal{R}\vert_{n_{\mathbb{P}}}$	Field of meromorphic functions with maximal $n_{\mathbb{P}}$ variables
$\mathbb{R}[\xi]$	Ring of polynomials (in the indeterminant ξ) over $\mathbb{R}$
$\mathcal{R}[\xi]$	Ring of polynomials over $\mathcal{R}$
$\mathcal{Q}_n$	Space of extended solutions for KR's with coefficients in $\mathcal{R}_n$
$\bar{\mathcal{Q}}_n$	Subset of $\mathcal{Q}_n$, non-eliminatable n^{th} variable
$\mathcal{Q}$	Additive group of extended solutions
$\mathcal{C}^\infty$	Space of infinitely differentiable functions
$\mathcal{L}_1^{\text{loc}}$	Space of locally integrable functions
$\mathcal{U}(n_1, n_2)$	Uniform distribution on interval $[n_1, n_2]$
$\mathcal{N}(n_1, n_2)$	Normal distribution with mean n_1 and variance n_2

Poles and Eigenvalues

σ	Singular value
$\bar{\sigma}$	Maximal singular value
$\underline{\sigma}$	Minimal singular value
λ	System pole
Ω	Pole region
$\Omega_{\mathbb{P}}$	Pole manifest region
$\mathcal{P}$	Pole uncertainty region
Λ	Set of poles
$\breve{R}$	characteristic polynomial

Measures and Norms

$\Vert\cdot\Vert_n$	n^{th} vector norm
$\ell_n(\cdot)$	n^{th}-signal norm
S_{s}	Similarity measure
S_{e}	Normalized entropy
χ	Xie-Beni validity measure
π	Kolmogorov distance
κ_1	Kolmogorov measure
κ_n	Kolmogorov cost n-width
d	Dissimilarity measure
$d_{\mathcal{H}_2}$	Distance on $\mathcal{H}_2$

Coefficients, Constants, and Rates

A, B, C, D	State-space matrices
$A_{\text{o}}, B_{\text{o}}, C_{\text{o}}, D_{\text{o}}$	State-space matrices, observability canonical
$A_{\text{r}}, B_{\text{r}}, C_{\text{r}}, D_{\text{r}}$	State-space matrices, reachability canonical
$A_{\text{co}}, B_{\text{co}}, C_{\text{co}}, D_{\text{co}}$	State-space matrices, companion-observability canonical
$A_{\text{cr}}, B_{\text{cr}}, C_{\text{cr}}, D_{\text{cr}}$	State-space matrices, companion-reachability canonical
$A_{\text{d}}, B_{\text{d}}, C_{\text{d}}, D_{\text{d}}$	State-space matrices of a discretized representation

α, β	State-space matrix elements
$\alpha^{\mathrm{o}}, \beta^{\mathrm{o}}$	State-space matrix elements, observability canonical
$\alpha^{\mathrm{r}}, \beta^{\mathrm{r}}$	State-space matrix elements, reachability canonical
$\alpha^{\mathrm{co}}, \beta^{\mathrm{co}}$	State-space matrix elements, companion-observability canonical
$\alpha^{\mathrm{cr}}, \beta^{\mathrm{cr}}$	State-space matrix elements, companion-reachability canonical
a, b	IO representation coefficients
R	Reachability gramian
O	Observability gramian
$\mathcal{O}_n$	Observability matrix, n-step
$\mathcal{R}_n$	Reachability matrix, n-step
o	Observability matrix element
$\mathtt{r}$	Reachability matrix element
θ	Parameter vector
θ_0	Parameter vector associated with the true system
θ^*	Asymptotic parameter vector estimate
$\hat{\theta}_{N_{\mathrm{d}}}$	Parameter vector estimate, based on N_{d} samples
Θ	Parameter vector space
$\mathcal{E}$	Ellipsoidal parameter uncertainty region
g, h	Markov parameters (of the noise model)
H	Hankel/Toeplitz matrix
$\mathtt{T}_{\mathrm{d}}$	Discretization time
$\check{\mathtt{T}}_{\mathrm{d}}$	Stability upperbound of $\mathtt{T}_{\mathrm{d}}$
$\hat{\mathtt{T}}_{\mathrm{d}}$	Performance upperbound of $\mathtt{T}_{\mathrm{d}}$
m	Fuzzyness parameter
$\mathcal{O}$	Rate of convergence
ρ	Convergence rate of the series expansion
$\check{\rho}$	Worst-case convergence rate of the series expansion
γ	Regressor vector
Γ	Regressor matrix
i	Imaginary unit
$M^{(n)}$	N-sensitivity constant, n-th order
$M_{\mathrm{x}}^{\max}$	Maximal amplitude of x over all trajectories
k_{s}	Spring constant
$S_{\mathrm{u}}, S_{\mathrm{y}}, S_{\mathrm{w}}, S_{\mathrm{x}}, S_{\mathrm{p}}$	Selector matrices
Z	Data for clustering
υ	Cluster center
V	Vector of cluster centers
U	Membership matrix
w	Coefficient in OBF models / expansion coefficient
v	Coefficient in OBF models (feedback)
W, V	Coefficient matrices in OBF models
Q	Set of grid points
Q_+	Set of grid points inside a region

Q_-	Set of grid points outside a region
ε	Threshold value
ε_s	Similarity threshold
ε_a	Adaptive threshold
ε_t	Termination threshold
α	Confidence level

Dimensions

$n_\mathbb{W}$	Signal dimension
$n_\mathbb{U}$	Input dimension
$n_\mathbb{Y}$	Output dimension
$n_\mathbb{X}$	State dimension
$n_\mathbb{P}$	Scheduling dimension
n_r	Dimension of a differential equation (number of equations)
n_L	Dimension of the latent variables
n_ξ	Maximal power of ξ
n_ζ	Maximal power of ζ
n_g	Inner function dimension
n_e	Number of bases extension
n_a	Order of the denominator polynomial
n_b	Order of the nominator polynomial
n_c	Number of clusters
n_θ	Parameter vector dimension
n_loc	Number of local models
n_ψ	Number of used functions in dependency parametrization
n_w	Dimension of w
n_v	Dimension of v
N_z	Number of data points for clustering
N_av	Average number of iterations
N_d	Data record length
N_p	Number of linearization/scheduling points (basis selection)
N_loc	Number of linearization / scheduling points (identification)

Data Sets

P	Linearization points/ identification points
$\mathcal{D}$	Data record

Chapter 1
Introduction

Abstract. In this introductory chapter we give an overview on the origin of *linear parameter-varying* (LPV) systems and their role in the control of today's industrial applications. Beside the history of this framework it is investigated what the state-of-the-art of LPV systems offers engineers to meet with the increasing expectations towards modern control solutions. We will see that the main bottleneck of this promising framework is the lack of a well established LPV modeling and identification theory. To overcome this, the current book sets the goal to explore the lacking details of the LPV system theory and to establish a solid framework for the modeling and identification of such systems.

1.1 New Challenges for System Identification

Today, the need to optimize efficiency of plants in terms of performance or energy consumption and to improve reliability of automatization results in increasing expectations towards automatic control applications. Engineers working in the control field have to face challenges in terms of operating industrial process in a more accurate way but at the same time with a lower cost in terms of used energy. For example in lithography, moving stages of wafer scanners today require fast and accurate position tracking with servo error specified in the nanometer scale. In the emerging alternative energy field, the coupled nonlinear nature of wind turbines and the rapidly changing conditions of wind and grid load require more efficient and also easily reconfigurable control solutions. On the other hand, economical efficiency also drives the control field to replace existing control designs with solutions that require less sensors or actuators, but still provide the same performance. Like in the case of induction motors, high performance control based on less built-in sensors, like speed-sensorless drives, has great economical importance. To cope with these challenges, well applicable theoretical solutions have been developed like optimal, robust, and *nonlinear* (NL) control approaches, trying to refine and extend the results of *linear time-invariant* (LTI) control theory, widely used in automatization. However to achieve the aimed objectives by these approaches, it is vital that

R. Tóth: Modeling and Identification of Linear Parameter-Varying Systems, LNCIS 403, pp. 1–20.
springerlink.com

an accurate, compact, and reliable mathematical description of the actual physical phenomenon is available.

First principle laws of physics, chemistry, biology etc. are commonly used to construct a dynamic model of the system of interest. However, such a procedure requires detailed process knowledge from specialists. Often it is a challenge to assemble the existing knowledge into a coherent and compact mathematical description. Usually, this results in a very complex model of the system dynamics, as it is hard to decide which effects are relevant and must be included in the final model and which are negligible. Such an approach is also often found to be very laborious and expensive. If the specialist's knowledge is lacking, like in case of poorly understood systems, the derivation of a model from first principles is even impossible. Moreover, certain quantities, like coefficients, rates, etc. required to build the model are often unknown, and have to be estimated by performing dedicated experiments.

Descriptions of systems can alternatively be derived by system identification, where the estimation of a dynamical model is accomplished directly from measured input-output data. The expert's knowledge still has a major role, as it gives the basic source of information that is used in the decision on parametrization, model-structure selection, experiment design, and the actual way of deriving the estimate. This knowledge also helps in judging the quality and applicability of the obtained models. Even if system identification requires human intervention and expert knowledge to arrive at appropriate models, it also gives a general framework in which most of the steps can be automated, providing a less laborious and cost intensive modeling process.

Starting from frequency domain approaches in the early 1940s, over the years considerable attention has been given to the identification of LTI systems, which have proven their usefulness in many engineering applications. Today LTI system identification has become a strongly founded framework considering issues of uncertainty and closed-loop identification with a vast theory on experiment design (for an overview see [144]). But the need to operate processes with higher accuracy/efficiency, has soon resulted in the realization that the commonly NL and *time-varying* (TV) nature of many physical systems must be handled by the control designs. This required better models, which initiated a significant research effort spent on identification and modeling of NL and *linear time-varying* (LTV) systems (see [59, 126] about the developed approaches). Despite many theoretical solutions, dealing with NL models without any structure has been found infeasible in practice both in terms of identification and control.

Today, engineers working in the industry still prefer the application of LTI control-designs, due to the attractive approaches of optimal and robust control. These approaches are preferred as they guarantee high performance and reliability, have easy and quick design schemes, and engineers have a vast experience in their application. Additionally, it has also been observed in practice that many NL systems can be well approximated by multiple LTI models that describe the behavior of the plant around some operation points. The recognition manifested in the 1980s, that instead jumping into the deep-space of NL and TV systems, a model class is needed which can serve as an extension of the existing LTI control approaches, but

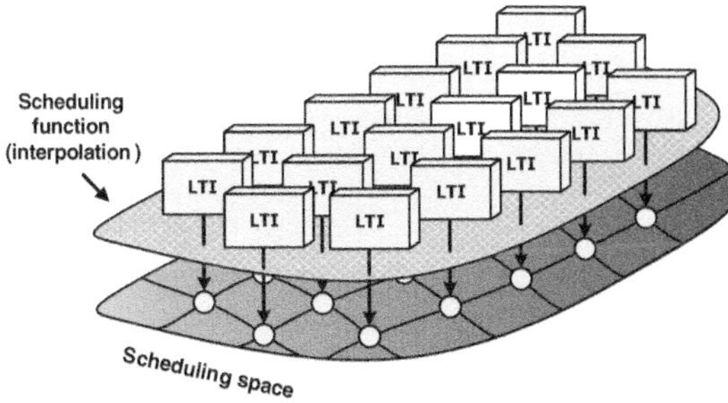

Fig. 1.1 The mechanism of gain-scheduling: interpolation of local LTI models/controllers of the plant to approximate the global behavior on the entire operation regime, i.e. scheduling space.

is still able to incorporate NL and TV dynamical aspects. This has led to the birth of *linear parameter-varying* (LPV) systems through the idea of gain-scheduling [166].

1.2 The Birth of LPV Systems

In gain-scheduling, the basic concept is to linearize the NL system model at different operating points resulting in a collection of local LTI descriptions of the plant (see Fig. 1.1). Then, subsequently, LTI controllers are designed for each local aspect. These controllers are interpolated to give a global control solution to the entire operation regime [157]. The used interpolation function is called the *scheduling function* in this framework and it is dependent on the current operating point of the plant. To describe the changes of the operating point, a signal is introduced, which is called the *scheduling signal* and often denoted by p. In this way, the parameters of the resulting controller are dependent on the varying signal p, hence the name *parameter-varying*, while the dynamic relation between the system signals is still *linear*. Due to many successful applications of this design methodology [177, 228, 179], gain-scheduling has become popular in industrial applications, even if guarantees for overall stability of the designed LPV controllers have not been available and the possibility of malfunction has existed. After 20 years, this was resolved by the introduction of interpolation based methods that guarantee global stability [182, 87]. In the mean time it was realized that in general many NL systems can be converted into an LPV form. Approaches have appeared that provided direct LPV models for gain-scheduling without the laborious process of NL system modeling or identification [120, 124, 36, 185]. LPV control has gained momentum during the 1990s, when the first results about the extension of $\mathcal{H}_\infty$ and $\mathcal{H}_2$ optimal control through *linear matrix inequalities* (LMIs) based optimization appeared (see [160, 6, 133]) together with μ-synthesis approaches, originating from robust LTI control [240]. Contrary to the former gain-scheduling methods, these approaches guarantee stability, optimal performance, and robustness over the entire operating regime of LPV models. Since then, the LPV field has evolved rapidly in the last 15 years and has become a promising framework for modern industrial control with a growing number of applications like aircrafts [112, 216], re-entry vehicles [212], automobiles [15], wind turbines [23, 103], induction motors [191, 147, 203], servo systems [229], wafer steppers [224, 180], internet web servers [152, 187], CD-players [45] and environmental modeling [18]. Unfortunately, LPV system identification and modeling could barely keep up with the advances of the control field. Only very recently, initiatives have been taken to explore many open problems and questions in this area.

1.3 The Present State of LPV Identification

In the following, a general picture about the state-of-the-art in LPV system identification is presented. Before going into details, we establish the key steps of the classical identification framework where the specifics of the LPV identification methods can be positioned and categorized.

1.3.1 The Identification Cycle

Identification of dynamical systems on the basis of experimentally measured data consists of several design steps, which need careful treatment in order to produce an acceptable model of the system. These steps are summarized in the so-called identification cycle presented in Table 1.1. This set of steps is referred as a cycle, due to the fact that several iterations might take place, using the knowledge gathered in the previous attempts, till the desired model is delivered. In the following a brief overview of each step is given, based on [105].

1.3.1.1 Experiment Design and Data Preprocessing

Experiment design focuses on the choice of the excitation of the system to be identified in order to maximize the information content in the measured signals. The design procedure is commonly accomplished with respect to a selected model class to minimize the variance, i.e. the estimation error, of the resulting model estimate. One of the most important problems is how to choose input signals that are *persistently exciting* (PE), i.e. they result in output signals which have enough information content to describe the relevant dynamical relations of the system for estimation with a given model structure. An equally important notion is the use of *adequately exciting* inputs that result in *informative* data sets which have enough information content to distinguish between different models in the considered model class. White noise inputs are in general considered to be optimal, as they excite all frequencies of an LTI system. As physical actuation by such signals is often infeasible, in practice random binary noises, frequency sweeps, and multisines are considered. Often it is also desirable to design excitation signals for a given frequency range related to the intended application of the identified model. Beside experiment design, data preprocessing is focusing on the attenuation of disturbances, aliasing or other defects in the measured data.

Table 1.1 The identification cycle

Step 1.	Experiment design, data collection and data manipulation.
Step 2.	Selection of model structure (including parametrization).
Step 3.	Choice of the identification criterion.
Step 4.	Estimation of a model that is optimal with respect to the criterion.
Step 5.	Validation of the resulting model estimate.

1.3.1.2 Choice of the Model Structure

The choice of an appropriate model structure is the most crucial part of the identification cycle. It determines the set in which a suitable description of the system is searched for. General questions considered in this step are the selection of the model structure in terms of the representation form (*state-space* (SS), *input-output* (IO), series expansion, etc.), parametrization, and the type of noise modeling. In terms of the well known bias/variance trade-off, the size of the model set is also important, like the number of parameters or order of the model structure. To obtain an adequate choice, the complexity of the algorithm delivering the model estimate or undesired local solutions of the estimation, non-uniqueness of the optimum, etc., also have to be considered.

1.3.1.3 Choice of the Identification Criterion

Selection of the identification criterion, the mathematical formulation of the performance measure of the model estimates, defines the user's purpose or expectation towards the model of the plant. In the literature, many identification criteria are presented, but the most commonly applied is the mean-squared error of the output prediction of the model estimate.

1.3.1.4 Model Estimation

The model estimation phase is the consequence of the previous choices of the identification cycle. Commonly the algorithmic solution of the estimation problem, defined in terms of the model structure and the identification criterion, is considered here.

1.3.1.5 Model (In)validation

A crucial question in identification is wether the obtained model is "good enough" for the intended purpose. Prior knowledge and experimental data are both used to confront the estimated model for answering this question. By using experimental data, validation is often accomplished by comparing simulation results of the model estimates to the measurements or by analyzing the model as a predictor of future outputs of the system, based on measured past data.

1.3.2 General Picture of LPV Identification

Next, the LPV identification problem is discussed, defining the notion of LPV systems and formulating the LPV model structures that are currently used in the literature. This sets the stage for the introduction of the state-of-the-art identification approaches.

1.3.2.1 LPV Systems and the Task of Identification

Based on the original gain-scheduling principle, LPV systems are often viewed as a *linear dynamical relation* between input signals u and output signals y, where the relation itself is dependent on an external variable, the so-called *scheduling signal* p. This provides the schematic view presented in Fig. 1.2. The relation can be formalized as a convolution in terms of u and p, which reads in *discrete time* (DT) as

$$y = \sum_{i=0}^{\infty} \mathrm{g}_i(p)\mathrm{q}^{-i}u, \tag{1.1}$$

where q denotes the *forward time shift operator*, i.e. $\mathrm{q}^i u(k) = u(k+i)$ and $\mathrm{q}^{-i}u(k) = u(k-i)$, $u : \mathbb{Z} \to \mathbb{R}^{n_\mathbb{U}}$ is the DT input, $y : \mathbb{Z} \to \mathbb{R}^{n_\mathbb{Y}}$ is the DT output, and $p : \mathbb{Z} \to \mathbb{P}$ is the DT scheduling signal of the system with a scheduling space $\mathbb{P} \subseteq \mathbb{R}^{n_\mathbb{P}}$ often considered to be compact. The coefficients g_i of (1.1) are functions of the scheduling variable and they define the varying linear dynamical relation between u and y. If the functions g_i are considered to be dependent only on the instantaneous value of the scheduling signal, i.e. $\mathrm{g}_i : \mathbb{P} \to \mathbb{R}^{n_\mathbb{Y} \times n_\mathbb{U}}$, then their functional dependence is called *static*. This means that (1.1) has the form

$$y(k) = \mathrm{g}_0(p(k))u(k) + \mathrm{g}_1(p(k))u(k-1) + \ldots \tag{1.2}$$

If the coefficients g_i also depend on the time-shifted versions of p, for instance if (1.1) is

$$y(k) = \mathrm{g}_0(p(k), p(k-1))u(k) + \mathrm{g}_1(p(k), p(k-1), p(k-2))u(k-1) + \ldots \tag{1.3}$$

then the functional dependence of g_i is called *dynamic*. An important property of LPV systems is that for a constant scheduling signal, i.e. $p(k) = \mathrm{p}$ for all $k \in \mathbb{Z}$, (1.1) is equal to a convolution describing an LTI system as each $\mathrm{g}_i(p)$ is constant. Thus, LPV systems can be seen to be similar to LTI systems, but their signal behavior is different due to the variation of the g_i parameters. Note that in the literature there are many formal definitions of LPV systems, commonly based on particular model structures with specific parameterizations. The convolution form (1.1) can be seen as their generalization. In identification, we aim to estimate a dynamical model of the system based on measured data, which corresponds to the estimation of the coefficients g_i in (1.1). Contrary to the LTI case where $\mathrm{g}_i \in \mathbb{R}$, here g_i are functions that depend on p, which makes their estimation to be a challenging task. This

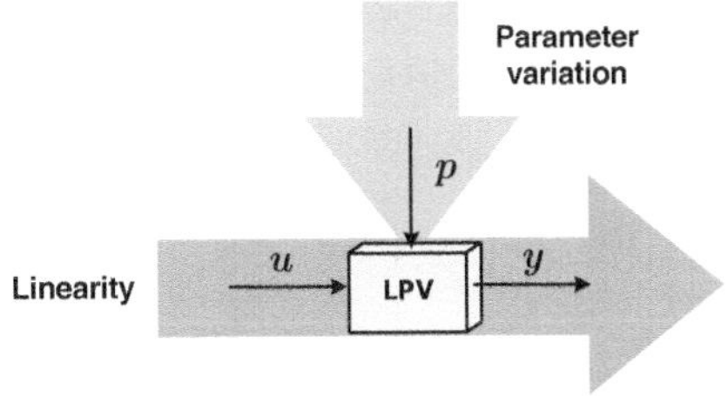

Fig. 1.2 Input-output signal flow of LPV systems.

estimation is formalized in terms of a model structure, an abstraction of (1.1), and an identification criterion.

1.3.2.2 Input-Output Representations

One particular type of model structure, which is used in some LPV identification approaches, originates from the IO type of representation of the data generating system in the LTI prediction-error setting. These LPV-IO representations are commonly defined in a filter form

$$y = -\sum_{i=1}^{n_a} a_i(p)q^{-i}y + \sum_{j=0}^{n_b} b_j(p)q^{-j}u + e, \tag{1.4}$$

where e is a noise process, $n_a \geq n_b$, and the coefficients $\{a_i\}_{i=1}^{n_a}$ and $\{b_i\}_{j=0}^{n_b}$ are functions of p with static dependence. It is commonly assumed that e is a zero-mean white noise. Alternatively (1.4) can be separated to a process and a noise part by $y = \breve{y} + v$ where

$$\breve{y} = -\sum_{i=1}^{n_a} a_i(p)q^{-i}\breve{y} + \sum_{j=0}^{n_b} b_j(p)q^{-j}u, \tag{1.5a}$$

$$v = -\sum_{i=1}^{n_a} a_i(p)q^{-i}v + e. \tag{1.5b}$$

LPV-IO model structures are formulated based on (1.5a-b) by parametrization of the coefficient functions in the process and noise part respectively (e.g. see (1.9)).

1.3.2.3 State-Space Representations

Another type of model structure is inspired by the classical SS representation based LTI models. The so-called LPV-SS representations of the data generating system are often given with an "innovation" type of noise model:

$$qx = A(p)x + B(p)u + E_1(p)e, \tag{1.6a}$$

$$y = C(p)x + D(p)u + E_2(p)e, \tag{1.6b}$$

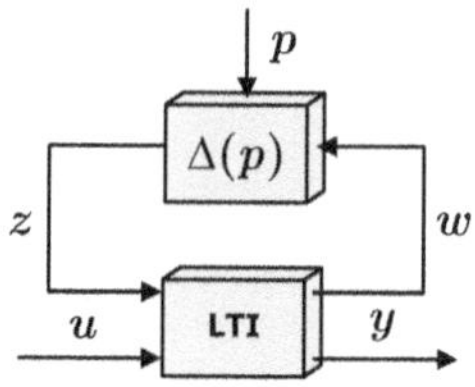

Fig. 1.3 Input-output signal flow of LFR models.

where $x:\mathbb{Z}\rightarrow\mathbb{R}^{n_\mathbb{X}}$ is the state-variable, e is a vector of independent zero-mean white noise processes, and (A,B,C,D,E_1,E_2) are matrix functions with static dependence on p. It is commonly assumed that the noise part is not dependent on p, i.e. E_1 and E_2 are constants, or $E_1 = I$ and $E_2 = I$. However there are recent initiatives with p-dependent noise representations. Additionally, the matrices are often considered with linear dependence. In case of A, such a dependence is defined as

$$A(p) = A_0 + \sum_{l=1}^{n_\mathbb{P}} A_l p_l, \tag{1.7}$$

where $A_l \in \mathbb{R}^{n_\mathbb{X}\times n_\mathbb{X}}$ and $p = [\, p_1 \ \dots \ p_{n_\mathbb{P}}\,]^\top$. This type of dependence is called *affine*, and used as a core assumption in many LPV control-design approaches.

Alternatively, the process part of (1.6a-b) is considered in an equivalent *linear fractional representation* (LFR), which originates from the robust control inspired μ-synthesis approaches. In this formalization, the scheduling dependence is extracted into a feedback gain, while the remaining part of the system is formulated as a LTI system (see Fig. 1.3 for an upper LFR form). This formulation is described mathematically as:

$$\begin{bmatrix} \mathrm{q}x \\ w \\ y \end{bmatrix} = \begin{bmatrix} A_1 & B_1 & B_2 \\ C_1 & D_{11} & D_{12} \\ C_2 & D_{21} & D_{22} \end{bmatrix} \begin{bmatrix} x \\ z \\ u \end{bmatrix}, \quad z = \Delta(p) w, \tag{1.8}$$

where w and z are auxiliary variables and $\{A_1,\dots,D_{22}\}$ are constant matrices, while Δ is a function of p with linear static dependence. An equivalent SS realization of (1.8) is defined as

$$\begin{aligned} A(p) &= A_1 + B_1\Delta(p)\,(I - D_{11}\Delta(p))^{-1} C_1, \\ B(p) &= B_2 + B_1\Delta(p)\,(I - D_{11}\Delta(p))^{-1} D_{12}, \\ C(p) &= C_2 + D_{21}\Delta(p)\,(I - D_{11}\Delta(p))^{-1} C_1, \\ D(p) &= D_{22} + D_{21}\Delta(p)\,(I - D_{11}\Delta(p))^{-1} D_{12}, \end{aligned}$$

if the matrix function $I - D_{11}\Delta(p)$ is invertible. If $D_{11} = 0$ then the resulting state-space matrices have linear dependence on p otherwise they can be polynomial or rational functions of p.

1.3.3 LPV-IO Representation Based Methods

The existing LPV identification approaches are almost exclusively formulated in discrete-time, they assume static dependence, and they are mainly characterized by the type of the used model structures. As a consequence, the identification methods of the LPV field are typically categorized as either LPV-IO or LPV-SS representations based methods. The resulting categories together with their main properties are investigated in the following sections. First we treat LPV-IO approaches which extend the results of LTI prediction-error identification. A common feature of these methods is that the approaches are derived by focusing on the *single-input single-output* (SISO) case, i.e. when $n_{\mathbb{Y}} = n_{\mathbb{U}} = 1$. The following subcategories are distinguished:

1.3.3.1 Interpolation Approaches

Methods that fall into this category apply the classical gain-scheduling concept: identification of the system for constant scheduling trajectories and interpolation of the resulting, so-called "frozen" models. These methods build on the well-worked out LTI prediction-error framework to achieve high quality estimates of the frozen aspects of the LPV system even in a closed-loop identification setting. Interpolation of the frozen models in an IO from is accomplished in various ways, like on the outputs of the local models [242], or on the inputs [241], or directly on the coefficients [115, 27]. All these interpolation forms result in different global models. A common feature of these approaches is that the interpolation is applied in an ad-hoc manner, most commonly using a polynomial or a spline approach, and that the methods show close relation with the local-linear-modeling framework [121] and with various multiple model approaches as well (see Sec. 1.3.4.3).

1.3.3.2 Linear Regression Methods

These methods use the LPV-IO model structure resulting from a specific parametrization of the coefficients of (1.4). The corresponding structure is an *auto-regressive* model with *exogenous* input (ARX), well known in the LTI identification framework. However, in the LPV case, the coefficients are functions of a varying p. Additionally, the approaches use a linear parametrization of $\{a_i\}_{i=1}^{n_a}$ and $\{b_j\}_{j=0}^{n_b}$ with polynomial scheduling dependence. In case of $n_{\mathbb{P}} = 1$, this reads for a_i as

$$a_i(p) = a_{i0} + \sum_{l=1}^{n} a_{il} p^l, \tag{1.9}$$

where $a_{il} \in \mathbb{R}$. As the resulting model is linear-in-the-parameters, the estimation of $\{a_{il}\}$ and $\{b_{jl}\}$ can be obtained by linear regression (see [226, 225, 12, 11]). In analogy, recursive least-squares or instrumental-variable methods can be applied to refine the estimate, building on the concepts of the LTI prediction-error identification framework (see [57, 13]). In [28] a closed-loop method is proposed for the estimation of LPV-ARX models by trying to directly apply the so-called *parameter adaptation algorithm* originating form the LTI case [93]. Extending the LTI concepts, some conditions of persistency of excitation for LPV-ARX models are derived in [226] and [13]. In [200] a statistical coefficient shrinkage method has been applied as well to assist selection of model order and required scheduling dependence in the parametrization of $a_i(p)$ and $b_i(p)$.

Due to an interest in specific applications, like environmental modeling, to identify LPV-IO models under heavy noise conditions, instrumental variable methods have been recently introduced for the estimation of LPV-IO models with *output-error* (OE) [34] and *Box-Jenkins* (BJ) [94] type of noise models. However, the classical nonlinear optimization approaches used in the LTI case to identify OE, BJ, etc. models have not been extended to the LPV case yet.

1.3.3.3 Set Membership Approaches

Set membership methods have been studied for the identification of LPV-IO models as well. Particular feature of these approaches is that the noise in the measured data is treated as deterministic uncertainty and, instead of a direct estimate of the parameters, a *feasible parameter set* is calculated. This feasible set represents the set of possible parameter values which satisfy the data equation (1.4) under a prior assumed error bound δ_e, where $\|e\| \leq \delta_e$ in terms of a given measure $\|\bullet\|$. A direct parameter estimate is often obtained by calculating the mean of the feasible set. In [19], identification of LPV *finite impulse response* (FIR) models, finite truncation of (1.1), has been studied by this methodology, where the expansion coefficients g_i were assumed to be nonlinear functions of a multi-dimensional p. Optimal worst-case experiment design has also been addressed in this context. In [38], this concept has been extended to general LPV-IO models in the form of (1.4). Opposite to [19], the calculation of the feasible parameter set in this case results in a non-convex optimization which is overcome by a polytopic outer approximation.

1.3.3.4 Nonlinear Optimization Methods

In this branch of IO approaches, the coefficients $\{a_i\}$ and $\{b_j\}$ of the LPV-IO model (1.4) are estimated by using nonlinear optimization to minimize the mean-squared prediction error. The aim is to give better estimates than the linear regression methods. In some approaches this is achieved by using a nonlinear parametrization

$$a_i(p) = a_{i0} + a_{i1}z, \tag{1.10}$$

where $a_{i0}, a_{i1} \in \mathbb{R}$ and z is the output of a feed-forward hidden layer neural network with inputs $\{y, q^{-1}y, \ldots\}$, $\{u, q^{-1}u, \ldots\}$, and $\{p, q^{-1}p, \ldots\}$. The estimation is accomplished via either a mixed linear/nonlinear procedure or by a separable least-squares approach. In the former case, the functional dependencies of the coefficients are identified through a neural-network approach while the linear part of the model is estimated by linear regression [150, 149, 148]. The approach is developed further in [151], using a separable least-squares strategy.

In [75], the estimation of LPV-ARX models is reformulated by applying a non-parametric method for the consistent estimation of the coefficient functions $a_i(p)$ and $b_i(p)$ resulting in a convex optimization. This method uses a *dispersion function approach*, originating from the machine-learning field, in order to learn the functional dependence of the coefficients in terms of piece-wise linear functions.

1.3.4 LPV-SS Representation Based Methods

Other type of approaches use the LPV-SS representation (1.6a–b) or its LFR equivalent (1.8) as a model structure by considering specific parametrization of the system matrices. In applications, these approaches are generally more appreciated, as LPV control theory requires a state-space representation and most of the LPV-SS identification methods can also easily handle MIMO plants. The methods of LPV-SS approaches fall into the following sub-categories:

1.3.4.1 Gradient Methods

These approaches formulate the estimation of the parameter-varying SS matrices as a NL optimization problem solved via gradient-search-based algorithms. Due to the nature of gradient-search optimization, the resulting estimate is often a locally optimal solution of the involved cost function. In [96] and [97], an LFR type of SS model structure is used where the scheduling dependence is extracted as $\Delta(p) = \mathrm{Diag}(Ip_1, \ldots, Ip_{n_\mathbb{P}})$. The estimation of the linear part is formulated as a NL optimization, which is solved in an iterative scheme based on the gradients of the mean-squared output error. In every step of the estimation, the system matrices can be estimated in a different state basis, i.e. a family of matrix estimates can be given which are all related by state transformations. To eliminate the non-uniqueness of the estimation, in each iteration step, the matrix estimates are restricted to a specific structure. Identifiability issues of such LFR structures are investigated in [98]. Other methods have been developed which apply the same methodology, but on the LPV-SS form (1.6a–b) with radial-basis functions based scheduling dependence [215, 214]. In this parametrization, the matrices are formulated as

$$A(p) = A_0 + \sum_{l=0}^{n} \mathfrak{g}_l(p) A_l \tag{1.11}$$

where $A_l \in \mathbb{R}^{n_\mathbb{X} \times n_\mathbb{X}}$ and each $\mathfrak{g}_l : \mathbb{P} \to [0,1]$ is a radial basis function.

In [91], the LPV state-space model (1.6a–b) with affine dependence is formulated as recurrent neural-network model. Estimation of the associated activation functions and system matrices is performed by training the network on measured data sets, similar to the neural network approaches in NL system identification [58].

1.3.4.2 Full-Measurement Approaches

These methods assume that the state x of the LPV-SS model, considered in the LFR form (1.8), is measurable. In this setting, under the assumption of linear dependence: $\Delta(p) = \text{Diag}(Ip_1, \ldots, Ip_{n_\mathbb{P}})$ and $D_{11} = 0$, the estimation problem reduces to a linear regression. In [124], the one-dimensional case of this approach has been treated, assuming a white noise scheduling signal and using recursive least-squares to obtain the estimate. Conservative conditions for persistency of excitation have also been derived. In [118] the robust extension of the method has been worked out, while in [109] and [107] the approach has been generalized to LFR structures with more complicated scheduling dependencies. In the latter case, instead of the state, the w and z signals in (1.8) are assumed to be measurable.

1.3.4.3 Multiple-Model Methods

These methods apply the classical gain-scheduling concept of identification similar to the interpolation approaches of the IO case (see Sec. 1.3.3.1). However in this case, interpolation is accomplished via a state-space representation usually in a rather intuitive manner. In some methods, the identified frozen LTI models are transformed into canonical SS forms [224, 204, 237, 180] or internally balanced modal form [108] to perform interpolation on $\mathbb{P}$. Some of the methods also apply model reduction independently on the obtained LTI models before interpolation [108, 180] or they use LTI discretization [224, 180]. In other approaches, interpolation is accomplished via pole locations [136]. As issues of noise and parametrization are dealt with in a local sense, i.e. by the applied LTI identification, these approaches focus on the question: how to accomplish interpolation in a more efficient sense. These approaches closely relate to the local-linear-modeling framework [121].

1.3.4.4 Set-Membership Approaches

Set-membership based identification of LPV-SS models has been first considered in [184] using a LFR form with a linear dependence: $\Delta(p) = \text{Diag}(Ip_1, \ldots, Ip_{n_\mathbb{P}})$ and $D_{11} = 0$. The noise/disturbance v of the system is assumed to be output additive with a moving average structure

$$v = \sum_{i=0}^{n} E_i \mathrm{q}^{-i} e, \tag{1.12}$$

where $E_i \in \mathbb{R}^{n_\mathbb{Y} \times n_\mathbb{Y}}$ and e is a ℓ_∞ sequence. This noise model, i.e. each E_i, is assumed to be known. In order to describe *inconsistency* with the measured data (error which cannot be explained by the LPV model nor the assumed noise model), an uncertainty block (extra Δ-block) is introduced. Under these assumptions and given bounds on $\|e\|_\infty$ and the uncertainty, the estimation problem of the LTI part is formulated as a LMIs-based optimization to derive a feasible set of parameter estimates. Based on a similar mechanism, validation of a LFR model with a linearly parameterized dependence can be formulated as a LMI feasibility problem if norm bounds and structural properties of the noise are known. In [24] this approach is reformulated to also compute directly the required bounds on $\|v\|_2$ and on the uncertainty block in order to satisfy the data equations. Under structural simplifications of the LFR form, the resulting optimization is formulated as a bilinear matrix inequality, loosing the computationally attractive property of the previous approach. It is common to these approaches that it is not understood how the conservatism of the used assumptions effect the validity of the calculated models.

1.3.4.5 Global Subspace Techniques

The family of these methods builds strongly on the concepts of the LTI subspace identification (see [105]), especially using the formulation of the *multivariable output-error state-space* (MOESP) algorithm (see [221, 220]). During the estimation process, a generalized data equation of the LPV-SS model (1.6a) is formulated to obtain both an estimate of the state evolution and the state-space matrices. In most methods, E_1 and E_2 are considered to be constant. The estimation is based on a discrete-time state-observability matrix of (1.6a-b) that is reconstructed from the measured data using a similar mechanism as in the MOESP algorithm. This identification strategy also enables the estimation of the system order, similar to the LTI case. Other similarities of the LPV subspace approaches are the assumption of affine scheduling dependence (see (1.7)) and the resemblance with bilinear system-identification methods.

- The early approaches used certain approximations during the reconstruction of the state, possibly leading to biased estimates [219, 213, 217]. The computational load of these methods turned out to be rather demanding in practice, as matrix dimensions quickly explode with an increasing number of scheduling variables and block dimensions. This has been the reason why the kernel-method-based modification has been proposed in [219] to regulate the computational load.
- To overcome the resulting bias, non-approximative methods have been derived by restricting the variation of the scheduling signals in the measured data to be periodic [53, 52] or piece-wise constant [207]. While the former method is similar to periodic LTV identification like in [222] and [104], the latter approach extends the approximative method of [218]. For these special scheduling signals, it is possible to find parameter-varying state-transformations such that the states in the subspace calculation are at the same basis at every time instant. Thus, no approximation is needed for the state-reconstruction in the noiseless case. However, the identifiability of SS models is not well understood with such a restricted class of scheduling signals and as a consequence numerical problems may result during the estimation.
- In the recent generation of LPV subspace methods, worked out in [210, 206, 208, 209], the predictor-based subspace identification approach, proposed originally in the LTI case [39, 40], is used to tackle the estimation problem of (1.6a–b). First the IO behavior of the LPV system is estimated in terms of Markov

parameters by using a high-order LPV-ARX model structure. Based on the fact that this estimated sequence of Markov parameters can be written as a product of the state sequence and the state-observability matrix, a *singular value decomposition* (SVD) is applied to estimate the state sequence. Consequently, the system matrices related to the affine parametrization are calculated via linear regression. In order to handle the high matrix dimensions in a computationally feasible manner, a kernel-based regularization approach is used [210, 206]. In case of small scale systems, this results in an attractive estimation scheme with negligible restrictions compared to earlier subspace approaches. An interesting property of the approach is that the so-called past window (the length of the estimated Markov sequence) plays as a trade-off between bias and variance of the estimates. Due to the quickly exploding number of estimated parameters, commonly a very small past window is preferred, contrary to the original subspace concept [206].
- In [47] LPV-SS models (1.6a–b) with affine dependence are formulated as LTI models by assuming white noise p, independent white noise u, and independent noise signals e_1 and e_2 in (1.6a–b). The estimation is solved as a bilinear identification problem via a Picard type of iterative scheme. In this approach, the state is reconstructed by a Kalman filter at each time instant, where the filter is based on the model obtained in the previous time steps. The disadvantage of the method is that it only provides a meaningful estimate in case the white noise assumptions of u and p are satisfied, which is hard to verify in practice. In [48] this approach has been modified to handle non-white noise scheduling sequences by extending the data equation used in the first iteration. Even if this modification does not provide an unbiased estimate in case of a colored p, it attenuates the possible bias or convergence problems.

1.3.4.6 Observer-Based Grey-Box Techniques

These approaches like [55, 54, 3], formulate the LPV identification problem as a parameter estimation of a known NL model structure. In that case, it is possible to use an adaptive observer to find the unknown parameters of the model based on measured data. The used observer is commonly an extended Kalman filter, which is applied on the augmented form of the NL model, where the state is extended with unknown parameters as variables. If the estimation has converged, then the obtained NL model is processed further, using the gain-scheduling approach to derive a LPV-SS form with affine coefficient dependence [3].

1.3.5 Similarity to Other System Classes

LPV systems are often considered to be similar in some aspects to other system classes. It has already been mentioned that LPV subspace techniques were inspired by the identification approaches of bilinear systems. These systems can be considered as LPV systems where p is equal to the input of the system and the system dynamics depend linearly on p.

LPV systems can also be seen as the extension of LTV systems. By restricting the coefficients of an LPV system to depend on a fixed linear trajectory of time, instead on a priori unknown trajectory of the scheduling variable, an LTV system results. Due to the structural similarity, many LPV approaches have been inspired by ideas of the LTV framework (IO methods, periodic identification) which do not exploit the priori known linear variation of the time trajectory.

In the fuzzy framework, *Takagi-Sugeno* (TS) dynamic fuzzy models with linear signal relations are often considered as LPV systems [86]. However, due to the if-then structure of the fuzzy rules, commonly LPV control cannot be applied on such systems and due to other structural differences such an equality of TS and LPV systems is dubious in the general sense. Thus, in the following, TS dynamic fuzzy models are treated as non-LPV systems.

1.4 Challenges and Open Problems

In the previous part we have seen that a wide variety of identification methods is available, approaching the underlying LPV identification problem (see Sect. 1.3.2) from different viewpoints. Many of these approaches are built around an assumed model structure and focusing only on the estimation task, which is just one step of the classical identification cycle. Due to these and many other issues, several challenges and open problems exist, that deserve further investigation. The most crucial questions are collected into the following list:

- The current methods use different identification settings, model structures, and even different views about what an LPV system is. So it is an obvious question how these concepts and ideas can be brought to a common ground, where they can be analyzed, compared, and refined.
- Often, the validity of the used identification concepts is not investigated. For example in prediction-error methods, like the least-squares LPV-IO methods, the formulation of the predictor, or even the description and analysis of the assumed noise structure is often omitted. Except for some recent contributions like [94, 34, 200], many approaches use linear regression as an optimization tool instead of estimation in a stochastic sense. How to formulate the prediction-error setting in the LPV case and how to derive well-founded identification approaches that dwell on the concepts of the classical framework remain questions to be investigated.
- In the SS variants of identification methods, based on the gain-scheduling principle, state-transformations are applied independently on each LTI-SS model estimate. For example, the frozen models are transformed to a canonical or balanced form and then they are interpolated. However, such transformed models do not have a common state, making the results of interpolation unpredictable [189]. Similarly, the problem of local transformations applies when model reduction is used independently on the frozen models or if the interpolation is based only on their pole locations. Unexpected problems can also occur in the interpolation process if the *McMillan degree* of the local models changes at some interpolation

points. Thus it is also important to investigate how the gain-scheduling principle can be used in identification such that the interpolation is well structured and issues of local transformations do not apply.

- Many LPV identification approaches, especially sub-space and gradient-search methods, have a significant computational load which renders their practical use to be quite limited in case of large scale systems. Therefore, it is important to develop approaches that solve the estimation problem efficiently and which are less effected by the curse of dimensionality.
- Except for some recent preliminary approaches like [75, 200], the selection of model structure, including parametrization, is often entirely skipped. The way how possible first-principle knowledge about the data-generating system is transformed to a discrete-time LPV form to assist at least the order selection is generally ad-hoc. The main reasons are the absence of sound results on LPV discretization theory or on the conversion of NL differential equations to LPV representations.
- The cardinal question, concerning the choice of the scheduling variable for a given physical system is commonly not investigated. As the entire dynamics of LPV systems depends on this variable, its choice should be part of model structure selection. This also implies the question when a given NL system can be efficiently described by a LPV model.
- In LTV system theory, it has been shown in discrete-time that equivalence transformations between SS and IO models result in coefficients that are constructed from time-shifted versions of the original coefficients [65]. If this is true for LTV systems, which can be considered as a special cases of LPV systems, may the same phenomenon hold in the LPV case? This would suggest that equivalent models in different representation forms depend on the time-shifted versions of the scheduling, which is called *dynamic dependence*.
- In LPV identification, many approaches build upon the assumption that LPV-SS and IO models with static dependence are equivalent representations of the same system. If in terms of the previous phenomenon, equivalence relations, like state-transformations and IO realization, result in dynamic dependence, then it becomes a cardinal question how this phenomenon effects the basic view of LPV identification and the validity of the used approaches and results.
- In view of the previous observation, further questions arise about how to define LPV systems, what kind of representations these systems have, what the equivalence relations are between these representations and how they correspond to the previously used concepts of LPV system theory. Answers to these questions are required to understand what LPV models correspond to, how they are related, what is the restriction of specific parameterizations and how the identification approaches can be compared.
- As LPV control is based on LPV-SS models it is also a question how to convert the model estimates to equivalent forms on which LPV control can be applied directly.
- Beside these issues, it is a general feature of the LPV field that many approaches try to build intuitively on the concepts of the LTI theory, often assuming that

time-varying filters commute. However, is it true that relations of the LTI framework apply directly to the LPV case? If not, then what are the merits of using the LTI concepts?

Observing these questions and problems we can conclude that, even though many LPV identification approaches have been considered in the literature, there is still a need for a mature modeling framework which can support LPV control synthesis and practical use of the developed models. By building on the previous results, it is necessary to establish a well-posed identification setting of LPV systems in terms of addressing the LPV identification problem in a general well-founded sense, paying attention to restrictions of the applied model structures and noise considerations, exploring the yet unknown relations of the LPV system theory, and investigating the identification cycle in full scope.

1.5 Perspectives of Orthonormal Basis Function Models

In view of the previous observations, a central problem of the LPV field is the absence of a model structure which has good representation capabilities with a limited number of parameters, useful for control, and its identification represents a low complexity problem. The latter means for example that the model structure is either well applicable in a gain-scheduling type of approach, i.e. it is easily interpolatable without the need of local transformations, or its estimation is available through linear regression.

1.5.1 *The Gain-Scheduling Perspective*

From the gain-scheduling perspective, *orthonormal basis functions* (OBF)s-based model representations offer an easily interpolatable structure with a well worked-out theory in the context of LTI system approximation and identification [74]. The basis functions, that provide bases for the space $\mathcal{H}_2$ (Hilbert space of complex functions that are squared integrable on the unit circle), are generated by a cascaded network of stable all-pass filters, whose pole locations represent the prior knowledge about the system at hand. This approach characterizes the transfer function of a proper SISO LTI system as

$$F(z) = \sum_{i=0}^{\infty} w_i \phi_i(z), \tag{1.13}$$

where $\{w_i\}_{i=0}^{\infty}$ is the set of constant coefficients and $\Phi_\infty = \{\phi_i\}_{i=0}^{\infty}$ with $\phi_0 = 1$ represents the sequence of OBFs. This implies that every transfer function $F_{\mathrm{p}}(z)$ that corresponds to (1.1) for constant scheduling $p(k) = \mathrm{p}$ can be represented as a linear combination of a given Φ_∞. In LTI identification, only a finite number of terms in (1.13) is used, like in FIR models. In contrast with FIR structures, the OBF parametrization can achieve almost zero modeling error with a relatively small number of parameters, due to the infinite impulse-response characteristics of the basis functions. In this way, it is generally possible to find a finite $\Phi_n \subset \Phi_\infty$, with a relatively small number of functions $n \in \mathbb{N}$, such that the representation error for all $F_{\mathrm{p}}(z)$ is negligible. Then, based on experiments with constant p, frozen aspects of (1.1) can be identified in the form (1.13) with finitely many terms, resulting in a set of local basis coefficients. Due to the linearity of (1.13) in these coefficients, model interpolation can easily be accomplished on the scheduling space $\mathbb{P}$ without the need of any local transformations, resulting in the LPV model:

$$y = \sum_{i=0}^{n} w_i(p)\phi_i(\mathrm{q})\,u. \tag{1.14}$$

This type of interpolation would also be well structured against local changes of the McMillan degree as the coefficients in (1.13) are not related directly to the order of the system. By using SS realizations of the basis functions $\{\phi_i\}_{i=0}^{n}$ in (1.14) a direct LPV-SS or LFR realization of this model is available, which means that model estimates can be directly used for control.

1.5.2 The Global Identification Perspective

OBF-based model structures have many attractive properties in the LTI case. In the prediction-error framework, they can be considered with an output-error type of noise model. The resulting model structure has a linear-in-the-parameter property, which implies that its estimation is available through linear regression. OBF models generally need less parameters than FIR models with similar properties. Non-asymptotic variance and bias bounds of the estimates are also available. These fruitful properties imply that direct identification of LPV systems may be beneficial in terms of the model structure (1.14) used in a prediction-error setting. In such a setting, parameterizing the coefficients as

$$w_i(p) = \sum_{l=0}^{n_\psi} \theta_{il}\,\psi_{il}(p), \tag{1.15}$$

where $\theta_{il} \in \mathbb{R}$ and $\{\psi_{il}\}$ are prior chosen functions of p, would yield that estimation of $\{\theta_{il}\}_{i=0,\dots,n}^{l=0,\dots,n_\psi}$ is possible via linear regression using experimental data with varying p. However, for the investigation of this approach, first an LPV prediction-error framework needs to be established.

1.5.3 Approximation via OBF Structures

In [30] it has been proved that model structures composed from a OBF filter bank followed by a static nonlinearity are general approximators of nonlinear systems with fading memory (NL dynamic systems with convolution representation). This means that a wide class of nonlinear systems can be identified with such OBF model structures with arbitrary precision. It is obvious that (1.14) is similar to these general approximators, which means that, if the same property can be shown for the LPV case, then these models have a wide representation capability of LPV systems. Based on the general approximator property and the attractive identification properties of OBF models, successful identification approaches based on these structures have been introduced in the NL and the fuzzy field (see [63, 159]). These methods provide low complexity and reliable estimates for the considered classes of systems, giving the hope that similar mechanisms could also be successfully applied in the LPV case.

It is well known in the LTI case that the approximation error, i.e. the resulting bias of the model estimate directly depends on how well the chosen basis functions Φ_n can represent the dynamics of the system. In terms of gain-scheduling, this refers to the size of the representation error of Φ_n with respect to each F_{p}. This refers back to the observation that model structure selection has a prime importance in the LPV case, and as for other type of LPV models, this choice influences the achievable maximal accuracy of the estimates.

In conclusion, the above discussed perspectives yield the observation that identification of LPV systems with OBF models like (1.14) could provide answers to some current challenges of the LPV identification field.

1.6 The Goal of the Book

In this chapter it has been shown that there are many open issues and unresolved problems/questions in the area of LPV modeling and identification. Based on this observation, we aim to develop an identification framework in this book where relations of model structures and the concept of estimation can be understood. In fact we intend to extend the prediction-error framework to the LPV case and investigate how we cab benefit from the classical results. Additionally, we also intend to develop an effective LPV identification mechanism, based on the promising properties of OBF expansion models, that overcomes the drawbacks of the existing state-of-the-art solutions (see Sect. 1.4).

However, it has been revealed in the previous section that the main obstacle in the formulation of such an identification framework originates from the gaps of LPV system theory. To fill these gaps and give a unified system theoretical framework in the LPV case, which enables the investigation and comparison of model structures and also provides an algebraic tool for manipulations on them, a new system theoretical framework, which is based on the extension of the LTI behavioral approach [146], is worked out in the first part of the book.

In the sequel we present a theory with new ideas of LPV behaviors, modeling, and identification methods, that have appealing properties and can be applied successfully to fulfill the aimed objectives. Due to the vast number of problems this book addresses, we also leave many open ends and questions for future research. The exploration of all issues of the LPV identification cycle of general physical systems remains a problem, but the use of orthonormal basis functions, the behavioral theory, and the proposed prediction-error identification framework open interesting and promising perspectives to approach the identification of this system class efficiently.

1.7 Overview of Contents

First we start our investigations by introducing in Chap. 2 some basic concepts of system identification with important mathematical and system theoretical tools which are used in the remainder.

In Chap. 3 we introduce an LPV behavioral approach that establishes a unified system theoretical framework. In this framework it becomes possible to understand relations of LPV models and the behavioral approach also gives the basic tool to introduce OBFs-based LPV model parameterizations later on.

In Chap. 4 we explore LPV-SS canonical forms and equivalence transformations between different representations of LPV systems. The discovered relations provide the tools to analyze and compare LPV model structures.

In Chap. 5 the representation of LPV systems by OBF-based series-expansions is investigated. It is shown that finite truncations of these representations can be used as model structures for LPV system identification. Such structures also have wide approximation capabilities.

Discretization of LPV systems is reviewed in Chap. 6. New discretization approaches are introduced together with criteria to choose the discretization step size.

In Chap. 7 the modeling of NL systems in a LPV form is investigated and the available solutions for this problem are studied. Using the framework of the LPV behavioral approach, a new mechanism is introduced that solves the LPV modeling issue of such systems. This approach together with the discretization methods of Chap. 6 are developed with the intention to assist the model-structure selection phase of the identification cycle based on first principle knowledge.

In Chap. 8 the basis-selection problem of OBFs-based LPV model structures is considered. The method is based on the clustering of sample poles that result from identification of the system with constant scheduling signals. The effect of noise on identification is also considered and a robust basis-selection procedure is developed based on hyperbolic-geometry results.

In Chap. 9 the extension of the LTI OBFs-based identification approach to the LPV case is developed, relying heavily on the tools derived in the previous chapters. The prediction-error framework for the LPV case is established and the model structures of the current approaches are analyzed. Two identification approaches, a global and a local one, are formulated with static dependence of the coefficients. The former approach utilizes the gain-scheduling-based interpolation concept while the latter yields a global estimate based on linear regression. To overcome the limiting assumption of static dependence needed for the parametrization of these approaches, two alternative OBFs-based model structures are worked out. These structures enable the approximation of dynamic dependence of the coefficients through a feedback with only static dependence. An identification approach of such feedback structures is derived through a separable least-squares strategy.

Chapter 2
LTI System Identification and the Role of OBFs

Abstract. This chapter is devoted to the introduction of basic concepts that are fundamental for the theory developed in the subsequent chapters. The notion of *orthonormal basis functions* (OBFs) is introduced with a brief overview of the most important relations and properties. Next, the basic concepts and theory of LTI system identification are reviewed focusing on prediction-error methods. OBFs related parameterizations and the associated identification approaches are also introduced together with the optimality concept of OBF model structures in terms of the Kolmogorov n-width theory.

2.1 The Concept of Orthonormal Basis Functions

In LTI system theory, it is common to represent transfer functions of dynamical systems in a series-expansion form. The most simplest of these expansions is the Laurent expansion. Consider a discrete time LTI system $\mathcal{F}$ with input $u : \mathbb{Z} \to \mathbb{R}^{n_\mathrm{U}}$ and output signal $y : \mathbb{Z} \to \mathbb{R}^{n_\mathrm{Y}}$. Denote the transfer function of $\mathcal{F}$ as $F(z) : \mathbb{C} \to \mathbb{C}^{n_\mathrm{Y} \times n_\mathrm{U}}$. Let (u, y) be valid signal trajectories of $\mathcal{F}$ with left compact support and denote the *Z-transform* of u and y by $Y(z) = \mathscr{Z}\{y\}$ and $U(z) = \mathscr{Z}\{u\}$ defined on their appropriate *region of convergence*[1] (ROC) with $z \in \mathbb{C}$ called the *Z-variable*. Then F satisfies that

$$Y(z) = F(z)U(z), \tag{2.1}$$

for any z in the intersection of the ROC of $Y(z)$ and $U(z)$. Substitution of z in $F(z)$ by $e^{\mathrm{i}\omega}$ gives the frequency response of the discrete-time system for $\omega \in (-\pi, \pi)$. Assume that $\mathcal{F}$ is stable, so the domain of $F(z)$ is the exterior of the unit circle. Then, by applying a Laurent series-expansion of $F(z)$ around $z = \infty$, it can be shown, that there exists a unique sequence of constants $\{\mathrm{g}_i\}_{i=0}^{\infty} \subset \mathbb{R}^{n_\mathrm{Y} \times n_\mathrm{U}}$, so-called Markov parameters, such that

$$F(z) = \sum_{i=0}^{\infty} \mathrm{g}_i z^{-i}. \tag{2.2}$$

[1] The region of convergence is the set of points in $\mathbb{C}$ for which the summation associated with the Z-transform converges.

R. Tóth: Modeling and Identification of Linear Parameter-Varying Systems, LNCIS 403, pp. 21–44.
springerlink.com

Moreover the signal

$$h = \mathscr{Z}^{-1}\left\{\sum_{i=0}^{\infty} g_i z^{-i}\right\}, \tag{2.3}$$

is called the *impulse response* of $\mathcal{F}$ ($\mathscr{Z}^{-1}$ is the standard notation of the *inverse Z-transformation*). Such a signal corresponds to the response of $\mathcal{F}$ for a pulse input at $k = 0$ and it uniquely represents the IO behavior of $\mathcal{F}$. Note that in case of unstable systems, a Laurent expansion of $F(z)$ is available but around $z = 0$, which results in an expression in the positive powers of z. By substituting z^{-1} in (2.2) with the backward-shift operator q^{-1}, it holds that

$$y = \sum_{i=0}^{\infty} g_i \mathrm{q}^{-i} u, \tag{2.4}$$

for all (u, y) valid signal trajectories of $\mathcal{F}$ with left compact support. Thus (2.4) can be considered as a representation of the system itself and it is called the *impulse response representation* (IRR). Note that such a representation is available for unstable systems as well, but in terms of the forward-shift operator q.

Finite truncations of the IRR are known as *finite impulse response* (FIR) models, which have proven their usefulness in many areas of engineering, ranging from signal processing to control design and also in approximative system identification. Despite numerous attractive properties of FIR models, these structures often require a large number of expansion terms to adequately approximate the original system dynamics. As an alternative, series-expansion models using general basis functions have been introduced:

$$y \approx \sum_{i=1}^{n} w_i \phi_i(\mathrm{q}) u. \tag{2.5}$$

for a predefined set of rational basis functions $\{\phi_i\}_{i=1}^{n}$ with $\phi_0 = 1$ and constant coefficients $\{w_i\}_{i=0}^{\infty}$. Such models preserve all the advantages of FIR structures, but in general they require much less expansion terms for adequate approximation due to the *infinite impulse response* (IIR) characteristics of the basis functions. In the following, the basic properties and the theory of *orthonormal basis functions* (OBF) are introduced briefly with the intention to derive OBFs based representations/models of LPV systems later on. To simplify the following discussion, which is based on [73, 129] and [74], we restrict our attention to *discrete time* (DT) stable SISO systems, however some hints are also given later how the theory extends to MIMO or *continuous time* (CT) systems.

2.2 Signal Spaces and Inner Functions

Before defining OBFs and their properties in system approximation, some important concepts of signal spaces and function properties are defined. We denote by $\mathbb{D} = \{z \in \mathbb{C} \mid |z| < 1\}$ the open unit disk in the complex plane and by

$\mathbb{J} = \{z \in \mathbb{C} \mid |z| = 1\}$ the unit circle. Also introduce $\mathbb{E}$ to represent the exterior of $\mathbb{J}$. The function space we frequently use in the sequel is the following:

Definition 2.1 (Hardy space on $\mathbb{E}$). Denote by $\mathcal{H}_2(\mathbb{E})$ the *Hardy space* of complex functions (transfer functions) $F : \mathbb{C} \to \mathbb{C}$, which are analytic (holomorphic) on $\mathbb{E}$, and squared integrable on $\mathbb{J}$:

$$\|F\|_{\mathcal{H}_2} := \sup_{1<r} \sqrt{\frac{1}{2\pi} \int_0^{2\pi} |F(re^{i\omega})|^2 \, d\omega} < \infty, \tag{2.6}$$

where $\| \,.\, \|_{\mathcal{H}_2}$ is a norm on $\mathcal{H}_2(\mathbb{E})$. □

$\mathcal{H}_2(\mathbb{E})$ can be interpreted as the space of stable proper transfer functions. Additionally we denote by $\mathcal{H}_{2-}(\mathbb{E})$ the subspace of strictly proper functions in $\mathcal{H}_2(\mathbb{E})$. Introduce also $\mathcal{RH}_{2-}(\mathbb{E})$ as the subspace of transfer functions in $\mathcal{H}_{2-}(\mathbb{E})$ with real valued impulse response. $\mathcal{H}_2(\mathbb{E})$ is equipped with the following inner product:

Definition 2.2 (Inner product of $\mathcal{H}_2(\mathbb{E})$). The inner product of $F_1, F_2 \in \mathcal{H}_2(\mathbb{E})$ is defined as:

$$\langle F_1, F_2 \rangle := \frac{1}{2\pi} \int_{-\pi}^{\pi} F_1(e^{i\omega}) F_2^*(e^{i\omega}) \, d\omega = \frac{1}{2i\pi} \oint_{\mathbb{J}} F_1(z) F_2^*(1/z^*) \, \frac{dz}{z}, \tag{2.7}$$

where * denotes complex conjugation. □

The norm of $F \in \mathcal{H}_2(\mathbb{E})$ satisfies $\|F\|_{\mathcal{H}_2} = \sqrt{\langle F, F \rangle}$. Two transfer functions $F_1, F_2 \in \mathcal{H}_2(\mathbb{E})$ are called orthonormal if the following conditions hold:

$$\langle F_1, F_2 \rangle = 0, \quad \|F_1\|_{\mathcal{H}_2} = \|F_2\|_{\mathcal{H}_2} = 1. \tag{2.8}$$

In this sense, the functions $\{z^{-k}\}_{k=1}^n$ in $\mathcal{H}_2(\mathbb{E})$ are orthonormal, as they trivially satisfy (2.8). Moreover, for a given transfer function $F \in \mathcal{H}_2(\mathbb{E})$, the Markov parameters can be computed as $g_i = \langle F, z^{-i} \rangle$, yielding the Laurent series-expansion (2.2). This series-expansion is called convergent if

$$\|F\|_{\mathcal{H}_2} = \sqrt{\sum_{i=0}^{\infty} |g_i|^2} < \infty. \tag{2.9}$$

Relation (2.9) implies that $\{z^{-i}\}_{i=1}^{\infty}$ are complete in $\mathcal{H}_{2-}(\mathbb{E})$. These functions, often referred as the *pulse basis*, are the most simple OBFs in $\mathcal{H}_{2-}(\mathbb{E})$ and they generate the IIR of stable dynamical systems.

We distinguish a special set of functions in $\mathcal{H}_2(\mathbb{E})$, the so-called *inner* functions. A function $G_b \in \mathcal{H}_2(\mathbb{E})$ is called inner, if $G_b(z) G_b^*(1/z^*) = 1$. Such a function in $\mathcal{RH}_{2-}(\mathbb{E})$ is completely determined, modulo the sign, by its poles $\Lambda_n = [\lambda_1 \, \dots \, \lambda_n] \in \mathbb{D}^n$:

$$G_b(z) = \pm \prod_{i=1}^{n} \frac{1 - \lambda_i^* z}{z - \lambda_i}, \tag{2.10}$$

often called a *Blaschke product.*

2.3 General Class of Orthonormal Basis Functions

First the class of DT stable basis functions is considered. Let $G_{b,0} = 1$ and $\{G_{b,i}\}_{i=1}^{\infty}$ be a sequence of DT inner functions with McMillan degrees $\{n_i\}_{i=1}^{\infty}$. Let (A_i, B_i, C_i, D_i) be a minimal balanced DT *state-space* (SS) representation[2] of the transfer function $G_{b,i}(z)$. Let $\{\lambda_1, \lambda_2, \ldots\}$ denote the collection of all poles of the inner functions $\{G_{b,i}\}_{i=1}^{\infty}$. Under the completeness (Szász) condition that $\sum_{i=1}^{\infty}(1 - |\lambda_i|) = \infty$, the scalar elements of the sequence of vector functions

$$M_i(z) := (zI - A_i)^{-1} B_i \prod_{l=0}^{i-1} G_{b,l}(z), \quad i > 0, \tag{2.11}$$

constitute a basis for $\mathcal{H}_{2-}(\mathbb{E})$, and each element $\phi_{ij} = [M_i]_j$ is orthonormal in $\mathcal{H}_{2-}(\mathbb{E})$ with respect to the entire sequence. An important aspect of these basis functions $\{\phi_{ij}\}_{i=1,j=1}^{\infty,n_i}$ is that they are uniquely determined, modulo the sign, by the poles of the generating inner functions (see (2.10) and (2.11)). However, note that (2.11) is only a particular way to construct OBFs and hence the generated basis sequence is not unique with respect to $\{\lambda_1, \lambda_2, \ldots\}$ [73].

Any $F \in \mathcal{H}_2(\mathbb{E})$ can be written as

$$F(z) = W_0 + \sum_{i=1}^{\infty} W_i M_i(z), \tag{2.12}$$

where $W_i^\top \in \mathbb{C}^{n_i}$ (if $F, G_{b,i} \in \mathcal{RH}_2(\mathbb{E})$, then $W_i^\top \in \mathbb{R}^{n_i}$) and it can be shown that the rate of convergence of this series is bounded. The IO relation of the OBF parametrization (2.12) is illustrated by Fig. 2.1. Note that in this figure, the state signals $\{x_i\}_{i=1}^{\infty}$ are the state variables of the balanced SS realizations of $\{G_{b,i}\}_{i=1}^{\infty}$. Additionally, $x_i = M_i(\mathrm{q})u$, i.e. the states are equal to the output of the generated basis functions.

Basically, the class of OBFs generalized by (2.11) can be classified into five function sets. These categories, which hierarchically contain the smaller classes from pulse basis to the Takenaka-Malmquist class, are defined as follows (see [73] for a detailed overview on these classes):

2.3.1 *Takenaka-Malmquist Basis*

The functions (2.11) are often referred to as the *Takenaka-Malmquist functions*. For a particular[3] balanced realization of the inner functions, it can be shown that the basis functions generated via (2.11) have the form of

[2] Note, that only inner functions in $\mathcal{RH}_{2-}(\mathbb{E})$ have SS representations with real matrices. We allow here SS representations, i.e. their matrices to generalize to the complex case.

[3] Note that a balanced state-space realization of a inner function is non-unique.

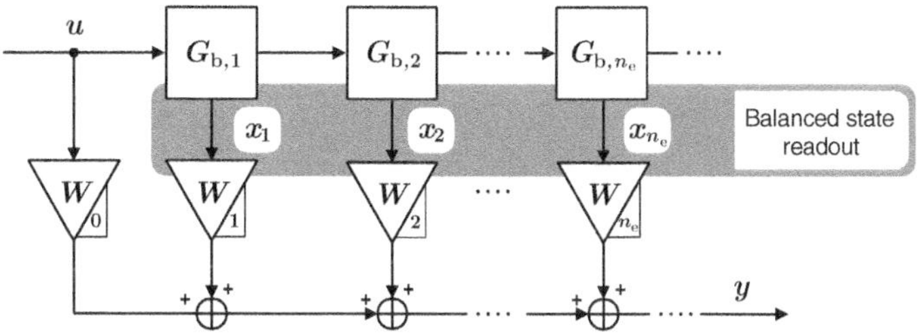

Fig. 2.1 IO signal flow of OBFs based series-expansion models.

$$\phi_{ij}(z) = \frac{\sqrt{1-|\lambda_{ij}|}}{z-\lambda_{ij}} \left(\prod_{l=1}^{j-1} \frac{1-z\lambda_{il}^*}{z-\lambda_{il}} \right) \left(\prod_{k=0}^{i-1} \prod_{l=1}^{n_k} \frac{1-z\lambda_{kl}^*}{z-\lambda_{kl}} \right), \tag{2.13}$$

where $\{\lambda_{ij}\}_{j=1}^{n_i} \subset \mathbb{D}$ are the poles of the inner function $G_{b,i}$.

2.3.2 *Hambo Basis*

The special cases when all $G_{b,i}(z)$ are equal, i.e. $G_{b,i} = G_b$, $\forall i > 0$, where $G_b(z)$ is an inner function with McMillan degree $n_g > 0$, are known as *Hambo functions* or *generalized orthonormal basis functions* (GOBFs). Let (A,B,C,D) be a minimal balanced SS representation of $G_b(z)$. Define

$$M_i^{AB}(z) := (zI-A)^{-1}BG_b^{i-1}(z), \tag{2.14a}$$
$$M_i^{AC}(z) := (zI-A^\top)^{-1}C^\top G_b^{i-1}(z). \tag{2.14b}$$

As system transposition is an equivalence transformation for minimal state-space representations (see [116]), there exists a unitary[4] $T \in \mathbb{C}^{n_g \times n_g}$ such that

$$M_1^{AB} = TM_1^{AC}. \tag{2.15}$$

Now by using $\phi_j = [M_1^{AB}]_j$, the Hambo basis consists of the functions

$$\Phi_{n_g}^{\infty} := \{\phi_j G_b^i\}_{j=1,\dots,n_g}^{i=0,\dots,\infty}. \tag{2.16}$$

We also introduce

$$\Phi_{n_g}^{n_e} := \{\phi_j G_b^i\}_{j=1,\dots,n_g}^{i=0,\dots,n_e}, \quad n_e \geq 0, \tag{2.17}$$

to denote the case when we talk only about a set of orthonormal functions generated by a finite extension of $G_b(z)$. Note that by using M_1^{AC} to define ϕ_j, a different basis sequence $\tilde{\Phi}_{n_g}^{\infty}$ results. However, due to (2.15), these basis sequences can be

[4] A square matrix $T \in \mathbb{C}^{n\times n}$ is called unitary if $TT^* = T^*T = I$, where * denotes conjugate transpose, also called Hermitian transpose.

considered equivalent. In the following we use the M_1^{AB} based construction if not indicated otherwise.

Similar to the previous generation method, $G_\mathrm{b}(z)$ and hence $\Phi_{n_\mathrm{g}}^{\infty}$ are completely determined by the poles of $G_\mathrm{b}(z)$, $\Lambda_{n_\mathrm{g}} = [\,\lambda_1 \; \dots \; \lambda_{n_\mathrm{g}}\,]$ and any $F \in \mathcal{H}_2(\mathbb{E})$ can be written as

$$F(z) = w_{00} + \sum_{i=0}^{\infty} \sum_{j=1}^{n_\mathrm{g}} w_{ij} \phi_j(z) G_\mathrm{b}^i(z), \tag{2.18}$$

where w_{00} is associated with the unit gain $\phi_{00} = 1$. It can be shown, that the rate of convergence of this series-expansion is bounded by

$$\rho = \max_k |G_\mathrm{b}(1/\lambda_k^{(0)})|, \tag{2.19}$$

where $\{\lambda_k^{(0)}\}$ are the poles of $F(z)$. ρ is often called the *convergence rate* of the Hambo series-expansion and it is an essential measure of the approximation quality of the basis function set with respect to $F(z)$. In the best case, where the poles of $F(z)$ are the same (with multiplicity) as the poles of $G_\mathrm{b}(z)$, only the terms with $i = 0$ in (2.18) are non-zero. Such a basis is commonly considered to be optimal for F.

2.3.3 *Kautz Basis*

When $G_{\mathrm{b},i} = G_\mathrm{b}$, $\forall i > 0$ with $n_\mathrm{g} = 2$, the resulting OBFs are called *2-parameter Kautz functions*. Such basis sequences can be considered to be adequate for the expansion of systems with dominating second order modes.

2.3.4 *Laguerre Basis*

The case when $G_{\mathrm{b},i} = G_\mathrm{b}$, $\forall i > 0$ with $n_\mathrm{g} = 1$ are called *Laguerre functions*. As this type of basis sequence in $\mathcal{RH}_{2-}(\mathbb{E})$ has only a real pole λ, it can provide an adequate basis for a $F \in \mathcal{RH}_{2-}(\mathbb{E})$ with a dominating first-order mode.

2.3.5 *Pulse Basis*

The case when $G_{\mathrm{b},i}(z) = z^{-1}$, $\forall i > 0$ are called *pulse functions*. Based on the expansion of the transfer functions of LTI dynamical systems in terms of the pulse basis, the impulse response representation of LTI systems is available.

2.3.6 *Orthonormal Basis Functions of MIMO Systems*

In the MIMO case, two approaches have been introduced for the construction of OBF functions in $\mathcal{H}_{2-}^{n_\mathrm{Y} \times n_\mathrm{U}}(\mathbb{E})$. The generated functions provide a series-expansion of any $F \in \mathcal{H}_{2-}(\mathbb{E})^{n_\mathrm{Y} \times n_\mathrm{U}}$. In one of the approaches, the key idea is to use MIMO functions that are composed from scalar basis sequences:

$$\breve{\phi}_l(z) := \begin{bmatrix} \phi_{l11}(z) & \dots & \phi_{l1n_\mathrm{U}}(z) \\ \vdots & \ddots & \vdots \\ \phi_{ln_\mathrm{Y}1}(z) & \dots & \phi_{ln_\mathrm{Y}n_\mathrm{U}}(z) \end{bmatrix} \tag{2.20}$$

where each $\{\phi_{lij}\}_{l=1}^{\infty}$ corresponds to a basis of $\mathcal{H}_{2-}(\mathbb{E})$. Then any $F \in \mathcal{H}_2^{n_\mathrm{Y} \times n_\mathrm{Y}}(\mathbb{E})$ can be represented as

$$F(z) = W_0 + \sum_{i=0}^{\infty} W_i \odot \breve{\phi}_i(z), \tag{2.21}$$

where $W_i \in \mathbb{C}^{n_\mathrm{Y} \times n_\mathrm{U}}$ and $\odot$ denotes the element-by-element matrix product. Similar to the SISO case, with different basis sequences $\{\phi_{lij}\}_{l=1}^{\infty}$, different convergence rates of the series-expansion can be achieved. However, the degree of freedom in the basis selection is much higher in the MIMO case. Note, that it is possible to use different basis sequences in the generation of (2.20), which gives the possibility of several structural classifications of this type of MIMO bases (see [205]).

Another formulation of MIMO orthonormal basis functions follows by using a multivariable, specifically square, inner function $G_\mathrm{b} \in \mathcal{H}_2^{n_\mathrm{U} \times n_\mathrm{U}}(\mathbb{E})$. The derivation is the same as the state-space construction approach presented by (2.11), but in this case $\phi_i = M_i$ constitutes a basis for $\mathcal{H}_{2-}^{n_\mathrm{Y} \times n_\mathrm{U}}(\mathbb{E})$ in the sense that any $F \in \mathcal{H}_2^{n_\mathrm{Y} \times n_\mathrm{U}}(\mathbb{E})$ can be written as

$$F(z) = W_0 + \sum_{i=1}^{\infty} W_i M_i(z), \tag{2.22}$$

where $W_i \in \mathbb{C}^{n_\mathrm{Y} \times n_\mathrm{U}}$. An important issue here is the construction of the square all-pass function $G_\mathrm{b}(z)$. Where in the scalar case, $G_\mathrm{b}(z)$ can be written as a Blaschke product and thus modulo the sign it is determined by the poles of the product, the multivariable-case is more involved and inhibits more freedom. Here also the structure and/or the dynamic directions are important [73]. See [205] and [72] for more on MIMO-OBFs of this form.

2.3.7 *Basis Functions in Continuous-Time*

All the results introduced so far are presented for DT systems as this time domain is used to formulate OBFs based identification of LPV systems later on. There is however a completely analogous theory for CT systems, where the Laplace transform is used instead of the Z-transform. The exterior of the unit disk $\mathbb{E}$ is replaced by the right half plane $\mathbb{C}^+$ and the inner product (2.7) is defined as integration over the imaginary axis:

$$\langle F_1, F_2 \rangle := \frac{1}{2i\pi} \int_{i\mathbb{R}} F_1(s) F_2^*(-s^*)\, \mathrm{d}s. \tag{2.23}$$

It is possible to construct an explicit isomorphism between $\mathcal{H}_2(\mathbb{E})$ and $\mathcal{H}_2(\mathbb{C}^+)$ using the bilinear transformation

$$z \to s = \gamma \frac{z-1}{z+1}, \quad \gamma > 0. \tag{2.24}$$

In this way, most of the introduced concepts extend trivially.

2.4 Modeling and Identification of LTI Systems

When identifying a dynamical system on the basis of experimentally measured data records, the outline of the procedure is summarized in the so-called identification cycle (see Table 1.1). The two most important steps of this cycle are the choices of an appropriate model set and the identification criterion. While the previous describes the set in which the suitable description of the system is sought, the latter defines the aimed performance of the model. The choice of the model set is crucial as it directly influences the maximum achievable accuracy or quality of the identified model in terms of the user-defined criterion. The model set should be as large as possible in order to contain as many candidate models as possible, which reduces the structural or bias error of the optimal model in the set. On the other hand, the number of parameters of the model should be kept as small as possible, because the variability of the identified models increases with increasing number of parameters. The conflict between these issues is the well-known bias/variance trade-off that is present in many estimation problems.

Model structures induced by orthonormal basis functions have attractive properties in terms of the variance/bias trade-off. When appropriately chosen, they require only a limited number of parameters to represent models that can accurately describe the dynamics of the considered system. The choice of basis functions then becomes a principle design issue. In this section, a brief coverage of DT prediction-error system identification is given, based on [105] and [73]. We focus on OBFs based model structures and concepts required for the derivation of the identification approaches of this book.

2.4.1 *The Identification Setting*

As a framework, the black-box setting of [105] is adopted. In this setting, identification of an unknown system is aimed at without the use of prior structural information. We assume that the underlying unknown system, so-called *data generating system*, is an LTI discrete-time SISO process:

$$y = G_0(\mathrm{q})u + v, \tag{2.25}$$

where $G \in \mathcal{H}_2(\mathbb{E})$, u is a *quasi-stationary* signal, and v is a stationary stochastic process (see [105] for a definition of these properties). Furthermore v satisfies

$$v = H_0(\mathrm{q})e, \tag{2.26}$$

with monic transfer function $H_0(\mathrm{q})$ such that $H_0, H_0^{-1} \in \mathcal{H}_2(\mathbb{E})$ and e is a zero-mean white noise process with variance σ_e^2. Assume furthermore that data sequences $\mathcal{D}_{N_\mathrm{d}} = \{u(k), y(k)\}_{k=0}^{N_\mathrm{d}-1}$, generated by (2.25), are available. Under the given assumptions, the so-called *one-step ahead prediction* of $y(k)$ based on $\{y(k-1), y(k-2), \ldots\}$ and $\{u(k), u(k-1), \ldots\}$ is

$$\hat{y} := (1 - H_0(\mathrm{q})^{-1})y + H_0(\mathrm{q})^{-1}G_0(\mathrm{q})u. \tag{2.27}$$

In prediction-error identification, a parameterized model $(G(\mathrm{q}, \theta), H(\mathrm{q}, \theta))$ is hypothesized where $\theta \subset \Theta$ represents the parameter vector, the coefficients of the model, and $\Theta \in \mathbb{R}^n$ is the allowed parameter space. This model structure leads to the one-step ahead predictor:

$$\hat{y}_\theta := (1 - H(\mathrm{q}, \theta)^{-1})y + H(\mathrm{q}, \theta)^{-1}G(\mathrm{q}, \theta)u. \tag{2.28}$$

Then in the prediction-error setting, we would like to choose θ such that the resulting $\hat{y}_\theta$ is a good approximation of y, i.e. the so-called *prediction error*

$$\varepsilon(k, \theta) := y(k) - \hat{y}_\theta(k), \tag{2.29}$$

is minimized. This is commonly performed by the minimization of the scalar valued *least-squares* (LS) identification criterion

$$\mathcal{W}_{N_\mathrm{d}}\left(\theta, \mathcal{D}_{N_\mathrm{d}}\right) = \frac{1}{N_\mathrm{d}} \sum_{k=0}^{N_\mathrm{d}-1} \varepsilon^2(k, \theta), \tag{2.30}$$

resulting in

$$\hat{\theta}_{N_\mathrm{d}} = \arg\min_{\theta \in \Theta} \mathcal{W}_{N_\mathrm{d}}\left(\theta, \mathcal{D}_{N_\mathrm{d}}\right), \tag{2.31}$$

based on the available data record $\mathcal{D}_{N_\mathrm{d}}$. Other criteria using different signal norms of ε can also be used or prefiltering can be applied on ε to deliver optimal estimates of θ based on certain considerations (see [105]). Optimization of the identification criterion according to (2.31) is generally a non-convex optimization problem for which iterative (gradient) algorithms have to be applied. This also implies that convergence to a global optimum can not be easily guaranteed. However, in specific cases of parametrization, the optimization reduces to a convex problem with an analytic solution.

For dealing with quasi-stationary signals, we introduce the generalized expectation operator $\bar{\mathcal{E}}$ defined as

$$\bar{\mathcal{E}}\{y\} = \lim_{N \to \infty} \frac{1}{N} \sum_{k=0}^{N-1} \mathcal{E}\{y(k)\}, \tag{2.32}$$

where $\mathcal{E}$ represents the mean value operator. Furthermore we define

Table 2.1 Black-box model structures

	ARX	ARMAX	OE	FIR	BJ
$G(q,\theta)$	$\frac{R_B(q^{-1},\theta)}{R_A(q^{-1},\theta)}$	$\frac{R_B(q^{-1},\theta)}{R_A(q^{-1},\theta)}$	$\frac{R_B(q^{-1},\theta)}{R_F(q^{-1},\theta)}$	$R_B(q^{-1},\theta)$	$\frac{R_B(q^{-1},\theta)}{R_F(q^{-1},\theta)}$
$H(q,\theta)$	$\frac{1}{R_A(q^{-1},\theta)}$	$\frac{R_C(q^{-1},\theta)}{R_A(q^{-1},\theta)}$	1	1	$\frac{R_C(q^{-1},\theta)}{R_D(q^{-1},\theta)}$

$$\mathrm{Var}\{y\} = \bar{\mathcal{E}}\left\{(y - \bar{\mathcal{E}}\{y\})^2\right\}, \tag{2.33}$$

as the variance operator. The related (cross)-covariance functions are

$$\mathcal{R}_y(\tau) := \bar{\mathcal{E}}\{y(t)y(t-\tau)\}, \quad \mathcal{R}_{yu}(\tau) := \bar{\mathcal{E}}\{y(t)u(t-\tau)\}. \tag{2.34}$$

Additionally, the (cross)-power spectral densities are given as

$$\Phi_y(e^{i\omega}) := \sum_{\tau=-\infty}^{\infty} \mathcal{R}_y(\tau)e^{-i\omega\tau}, \quad \Phi_{yu}(e^{i\omega}) := \sum_{\tau=-\infty}^{\infty} \mathcal{R}_{yu}(\tau)e^{-i\omega\tau}. \tag{2.35}$$

2.4.2 *Model Structures*

There are numerous different black-box model structures available for the parametrization of $G(q,\theta)$ and $H(q,\theta)$. Most of them, collected in Table 2.1, parameterize the two transfer functions in terms of ratio's of polynomials $R_A,\ldots,R_F \in \mathbb{R}[\xi]$ in the backward time-shift operator q^{-1}. These structures are known under the acronyms given in the table. The parameter vector θ of these model structures contains the collection of the coefficients of the polynomials. Commonly, the denominator polynomials are assumed to be monic to ensure uniqueness of the parametrization. Every model structure or parametrization induces a set of predictor models, commonly called the *model set*:

$$\{(G(q,\theta),H(q,\theta)) \in \mathcal{H}_2(\mathbb{E}) \times \mathcal{H}_2(\mathbb{E}) \mid \theta \in \Theta \subset \mathbb{R}^n\}. \tag{2.36}$$

This concept allows us to distinguish the following situations:

- The data generating system $(G_0(q),H_0(q))$ is in the model set, i.e. an exact representation of the data generating system can be found by the applied model structure.
- $(G_0(q),H_0(q))$ is not in the model set, i.e. no exact representation of the system exists by the model structure.

When the main attention is given to the IO dynamics of the system, i.e. the underlying deterministic behavior described by $G_0(q)$, it is attractive to deal with the set of so-called IO models:

$$\{G(q,\theta) \in \mathcal{H}_2(\mathbb{E}) \mid \theta \in \Theta \subset \mathbb{R}^n\}. \tag{2.37}$$

This leads to situations when the IO relation of the plant $G_0(q)$ can be or can not be captured within the chosen model set. Two important properties of the introduced model structures are the following

- For some model structures, the expression of the output predictor (2.28) is linear in the unknown parameters θ, i.e. both the terms $(1 - H(q,\theta))^{-1}$ and $H(q,\theta)^{-1}G(q,\theta)$ are polynomials. This property holds for the ARX and FIR structures and has the major benefit that the LS criterion can be minimized by solving a set of linear equations.
- If $G(q,\theta)$ and $H(q,\theta)$ are independently parameterized, the two transfer functions can be estimated independently. This property holds for the FIR, OE, and BJ model structures.

From these viewpoints it is particulary attractive to consider the FIR model, where both these properties are satisfied.

2.4.3 *Properties*

Next we investigate the statistical properties of the considered model structures in terms of prediction-error identification.

2.4.3.1 Consistency and Convergence

When applying the quadratic identification criterion (2.30), the asymptotic properties of the resulting parameter estimate can be derived in the situation when $N_d \to \infty$. If the noise in the measured data is normally distributed, the LS estimator is equivalent with a maximum likelihood (statistically optimal in an asymptotic sense) estimator [105]. With other noise distributions, attractive properties also hold (under weak conditions on the noise):

- *Convergence result*: For $N_d \to \infty$, the parameter estimate $\hat{\theta}_{N_d}$ converges, i.e. $\hat{\theta}_{N_d} \to \theta^*$, with probability 1, where $\theta^* = \arg\min_{\theta\in\Theta} \bar{\mathcal{E}}\{\varepsilon^2(\theta)\}$. This implies that the asymptotic parameter estimate is independent from the particular noise realization in the data sequence.
- *Consistency result*: If u is *persistently exciting* (PE) of a sufficient order, then the asymptotic parameter estimate θ^* has the following properties:
  - If the data generating system is in the model set, then $G_0(q) = G(q,\theta^*)$ and $H_0(q) = H(q,\theta^*)$.
  - If $G_0(q)$ is in the IO model set and additionally $G(q,\theta)$ and $H(q,\theta)$ are independently parameterized, then $G_0(q) = G(q,\theta^*)$. This means that consistency of the estimate $G(q,\theta)$ is also obtained if $H(q,\theta)$ is misspecified.

Persistency of excitation (PE) for an order n means in this context that the quasi-stationary u, used for excitation during the experiment, satisfy that

$$\mathrm{Det}\begin{bmatrix} \mathcal{R}_u(0) & \mathcal{R}_u(1) & \dots & \mathcal{R}_u(n-1) \\ \mathcal{R}_u(1) & \mathcal{R}_u(0) & \dots & \mathcal{R}_u(n-2) \\ \vdots & \ddots & \ddots & \vdots \\ \mathcal{R}_u(n-1) & \dots & \mathcal{R}_u(1) & \mathcal{R}_u(0) \end{bmatrix} \neq 0. \tag{2.38}$$

This condition guarantees that enough information on the dynamics of $G_0(q)$ is present in the measured y to approximate n parameters of a model. In the LTI case it is sufficient to require that $\Phi_u(e^{i\omega})$ has a non-zero contribution in the frequency range $-\pi < \omega \leq \pi$ in at least as many points as there are parameters to be estimated in $G(q,\theta)$.

2.4.3.2 Asymptotic Bias and Variance

In system identification, one also has to deal with estimation errors. This is due to the fact that information on the system to be estimated is only partially available: finite data records, effect of noise, etc. A well-accepted approach is to decompose the estimation error for $G(q,\hat{\theta}_{N_d})$ as:

$$G_0(q) - G(q,\hat{\theta}_{N_d}) = \underbrace{G_0(q) - G(q,\theta^*)}_{\text{bias}} + \underbrace{G(q,\theta^*) - G(q,\hat{\theta}_{N_d})}_{\text{variance}}. \tag{2.39}$$

In this decomposition, the first part is the *structural* or *bias error*, usually induced by the fact that the model set is not rich enough to exactly represent the plant. The second part is the *noise induced* or *variance error* which is due to noise contribution on the measured data. The bias can be characterized in terms of integral formulas over the frequency domain. Powerful formulas also exist to express variance error if both N_d and the model order tend to infinity [105].

One of the most basic results on variance error is formulated in terms of the variability of asymptotic parameter estimates. In the most general form, the characterization follows from the central limit theorem, proving that

$$\sqrt{N_d}(\hat{\theta}_{N_d} - \theta^*) \to \mathcal{N}(0,\mathcal{Q}_\theta) \quad \text{as } N_d \to \infty, \tag{2.40}$$

i.e. the random variable $\sqrt{N_d}(\hat{\theta}_{N_d} - \theta^*)$ converges in distribution to a Gaussian probability density function with zero mean and covariance matrix $\mathcal{Q}_\theta$. Note that $\mathcal{Q}_\theta$ can be calculated only in a limited number of situations. One of these is when the data generating system is in the model set, leading to the consistent estimate $\theta^* = \theta_0$. In this case:

$$\mathcal{Q}_\theta = \sigma_e^2 \left(\bar{\mathcal{E}}\left\{ \varphi(k,\theta_0)\varphi^\top(k,\theta_0) \right\} \right)^{-1} \text{ with } \varphi(k,\theta_0) := -\frac{\partial}{\partial\theta}\varepsilon(k,\theta)\Big|_{\theta=\theta_0}.$$

2.4.4 Linear Regression

If the model structure has the property of being linear-in-the-parameters, then the LS problem (2.31) becomes a convex optimization problem with the analytic solution:

$$\hat{\theta}_{N_d} = \left(\frac{1}{N_d}\Gamma_{N_d}^\top \Gamma_{N_d}\right)^{-1} \cdot \left(\frac{1}{N_d}\Gamma_{N_d}^\top Y_{N_d}\right). \tag{2.41}$$

where $Y_{N_d} = [y(0), \ldots, y(N_d - 1)]^\top$ is the collection of the measured output samples and $\Gamma_{N_d} = [\gamma(0), \ldots, \gamma(N_d - 1)]^\top$ contains the regressor vector γ that describes the data relation according to the one-step-ahead predictor: $\hat{y}_\theta(k) = \gamma^\top(k)\theta$. For the ARX case with $\mathrm{Deg}(R_A) = n_a$ and $\mathrm{Deg}(R_B) = n_b$, the regressor vector is

$$\gamma^\top(k) = \begin{bmatrix} y(k-1) & \ldots & y(k-n_a) & u(k) & \ldots & u(k-n_b) \end{bmatrix},$$

while in the FIR case with $\mathrm{Deg}(R_B) = n_b$, the regressor vector becomes

$$\gamma^\top(k) = \begin{bmatrix} u(k) & \ldots & u(k-n_b) \end{bmatrix}.$$

Note that on the basis of numerical considerations regarding matrix inversion, the solution (2.41) is not computed directly, but via a QR-algorithm. Moreover, statistical analysis of this estimator results in non-asymptotic expressions of the bias and variance error, which provide important advantages of linear-in-the-parameter model structures over other model parameterizations.

2.4.5 *Identification with OBFs*

Considering the classical identification results described in the previous part, it appears that there are two attractive properties of model structures: linear-in-the-parameter property and independent parametrization of the process and noise models. Among the presented classical structures a combination of these two properties can only be found in the FIR structure. However, the main disadvantage of this structure is that it generally requires a large number of parameters to capture the dynamics of the physical system, which implies a relatively large variance of the estimate. Using OBFs instead of the pulse basis like in (2.5) can significantly decreases the number of required parameters and preserve all the attractive properties. In this section we focus on the model structures:

$$G(\mathrm{q}, \theta) = \sum_{i=0}^{n} w_i \phi_i(\mathrm{q}), \quad H(\mathrm{q}, \theta) = 1, \tag{2.42}$$

where $\{\phi_i\}_{i=1}^n$ with $\phi_0 = 1$ are orthonormal basis functions in $\mathcal{RH}_2(\mathbb{E})$ with pole locations $\Lambda_n = \begin{bmatrix} \lambda_1 & \ldots & \lambda_n \end{bmatrix}$. The unknown series-expansion coefficients of (2.42) are collected into the parameter vector $\theta = \begin{bmatrix} w_0 & \ldots & w_n \end{bmatrix} \subset \mathbb{R}^{n+1}$. In a general sense, different identification criteria and settings can be applied for OBFs based model structures, like for example frequency domain identification in $\mathcal{H}_2/\mathcal{H}_\infty$, resulting in attractive alternatives of LTI system identification [73, 50]. In the following we explore the LS prediction-error setting to compare the properties of this model structure to the classical results of other structures. Later, these properties also form the basic motivation why this model structure with the LS prediction-error setting yields an attractive candidate for LPV identification.

2.4.5.1 Least-Squares Identification

Due to the linear-in-the-parameter property of the OBF parametrization, estimation of the parameters in the LS setting similarly follows as in the FIR or ARX cases. The only difference with respect to these model structures is that the regression vector is

$$\gamma^\top(k) = \begin{bmatrix} u(k) & (\phi_1(\mathrm{q})u)(k) & \dots & (\phi_n(\mathrm{q})u)(k) \end{bmatrix},$$

containing filtered versions of the input signal rather than delayed version of u or y. This form of the regressor implies that the LS parameter estimate of (2.42) can be obtained via (2.41). In terms of consistency/bias properties of the estimates all the classical results about FIR structures hold, if u is a white noise signal, i.e. $\Phi_\mathrm{u}(e^{\mathrm{i}\omega})$ is constant. This means, that if $G_0(\mathrm{q})$ is in the IO model set and the input is white noise, then the model estimate $G(\mathrm{q}, \hat{\theta}_{N_\mathrm{d}})$ is unbiased and consistent. This is the case when the true system has a finite series-expansion in terms of the used basis functions. In all other cases, the expansion coefficients of $G_0(\mathrm{q})$ in terms of the finite basis functions are estimated consistently, but a bias results due to the truncated tail of the required infinite expansion. Therefore there is a primal emphasis on the selection of appropriate basis functions, to reduce the bias by ensuring a fast convergence rate of the series-expansion.

A particularly interesting aspect results if the estimation of the parameters is formulated in state-space. As indicated in (2.14a–b), a basis functions based series-expansion model $G(\mathrm{q}, \theta)$ can be realized efficiently in two SS forms:

$$\left[\begin{array}{c|c} A & B \\ \hline W & w_0 \end{array}\right] \quad \text{and} \quad \left[\begin{array}{c|c} A & (WT)^\top \\ \hline C & w_0 \end{array}\right], \tag{2.43}$$

where (A,B,C,D) is the minimal balanced SS realization of the inner function G_b generating the basis functions $\{\phi_i\}_{i=1}^n$, $W = [w_1 \ \dots \ w_n]$, and $T \in \mathbb{R}^{n\times n}$ is a unitary matrix such that (2.15) is satisfied. The SS realization in the left is often called the AB-invariant while the representation in the right is recognized as the AC-invariant form. Due to property (2.15), estimation in a AB-invariant form results in a parameter estimate $\hat{\theta}_{N_\mathrm{d}}$, whose elements are the linear combinations of a parameter estimate based on the AC-invariant form. An additional property is that the initial condition of the AC-invariant form can be easily formulated as the part of the parameter vector, due to the different formulation of the regressors in that case. Thus, estimation of the initial condition is available by linear regression in the AC-invariant case. This property is important when, because of various reasons, the experiment providing the data record can not be accomplished on the system starting from zero initial condition.

2.4.5.2 Asymptotic Bias and Variance

As indicated before, the classical results in terms of the FIR structure trivially extend to OBF model structures if the input sequence is a white noise signal. However, due to the fact that finite series expansions often result in the case where $G_0(\mathrm{q})$ is not in the IO model set, it is important to investigate how the estimated coefficients relate to the expansion coefficients of $G_0(\mathrm{q})$ in terms of the finite basis function set. It can be shown that, if u is white, then the expansion coefficients of $G_0(\mathrm{q})$ are identified consistently in the LS setting. This means that even if $G_0(\mathrm{q})$ can not be identified consistently due to the truncated tail of the expansion, the modeled dynamical part (the considered finite expansion) is consistently identified. Moreover, in this case θ^* obeys:

$$\theta^* = \arg\min_{\theta\in\Theta} \frac{1}{2\pi}\int_{-\pi}^{\pi} \left|G_0(e^{\mathrm{i}\omega}) - G(e^{\mathrm{i}\omega},\theta)\right|^2 \frac{\Phi_\mathrm{u}(e^{\mathrm{i}\omega})\sum_{i=0}^{n}|\phi_i(e^{\mathrm{i}\omega})|^2}{|H_0(e^{\mathrm{i}\omega})|^2}\,\mathrm{d}\omega, \tag{2.44}$$

which shows that the basis functions act as data prefilters, emphasizing the fit of the estimated model on the frequency domain where the gain of the basis functions is significant. The noise spectrum also does not appear in the expression due to the fixed noise model, thus the convergence of θ is not influenced by the noise just like in the case of FIR models. The bias, due to the undermodeling, can be directly computed via the theory of reproducing kernels [73]. Furthermore, for specific transfer functions, like

$$G_0(z) = \sum_{j=1}^{n_0} \frac{b_j}{z-\lambda_j^{(0)}} \tag{2.45}$$

an upper bound on the approximation error can be given with respect to $G(\mathrm{q},\theta^*)$ using the basis functions $\{\phi_i\}_{i=1}^{n}$:

$$\begin{aligned}\left|G_0(e^{\mathrm{i}\omega}) - G(e^{\mathrm{i}\omega},\theta^*)\right| &\leq \sum_{j=1}^{n_0}\left|\frac{b_j}{e^{\mathrm{i}\omega}-\lambda_j^{(0)}}\right|\cdot\prod_{i=1}^{n}\left|\frac{\lambda_j^{(0)}-\lambda_i}{1-\lambda_i^*\lambda_j^{(0)}}\right| \\ &\leq \underbrace{\max_{j\in\mathbb{I}_1^{n_0}}\prod_{i=1}^{n}\left|\frac{\lambda_j^{(0)}-\lambda_i}{1-\lambda_i^*\lambda_j^{(0)}}\right|}_{\rho}\cdot\sum_{l=1}^{n_0}\left|\frac{b_l}{e^{\mathrm{i}\omega}-\lambda_l^{(0)}}\right|,\end{aligned} \tag{2.46}$$

where $\{\lambda_i\}_{i=1}^{n}$ are the poles of $\{\phi_i\}_{i=1}^{n}$ and $\mathbb{I}_{\tau_1}^{\tau_2} = \{\tau\in\mathbb{Z} \mid \tau_1\leq\tau\leq\tau_2\}$ is the index set. This expression shows a tight bound on the approximation error in terms of the bias. If for each pole of the system, there exists a matching pole of the basis $\lambda_j^{(0)} = \lambda_i$, then the upperbound is zero (G_0 has a finite series-expansion in terms of the basis). In the general case, the truncation error of the series-expansion is directly influenced by the convergence rate ρ (see (2.46)), which expresses the natural distance between the basis and the system poles. This implies that to minimize the bias, the poles of the basis should be as close to the poles of the system to be identified as possible.

The variance of the model estimate also obeys the classical results of FIR structures. In the frequency domain, if both $N_\mathrm{d}, n\to\infty$, $n\ll N_\mathrm{d}$, the variance of the estimate $G(\mathrm{q},\hat{\theta}_{N_\mathrm{d}})$ reads

$$\mathrm{Var}\{G(\mathrm{q},\hat{\theta}_{N_\mathrm{d}})\} \sim \frac{1}{N_\mathrm{d}} \frac{\Phi_\mathrm{e}(e^{\mathrm{i}\omega})}{\Phi_\mathrm{u}(e^{\mathrm{i}\omega})} \sum_{i=1}^{n} |\phi_i(e^{\mathrm{i}\omega})|^2, \tag{2.47}$$

where $\Phi_\mathrm{e}(e^{\mathrm{i}\omega})$ is the noise spectrum. This expression shows the classical result of FIR structures when the input is prefiltered. Note that by choosing the poles $\{\lambda_i\}_{i=1}^n$ of $\{\phi_i\}_{i=1}^n$ close to the poles of G_0, the variance expression (2.47) peaks if $e^{\mathrm{i}\omega}$ comes to the close neighborhood of λ_i. This corresponds to the variance/bias trade-off.

As discussed in this section, OBFs-based parametrization can be effectively used for LTI system representation with many fruitful properties, however it is required that the basis function set is "well chosen" with respect to system to be identified. In the next section, the concept of optimality of an OBF set with respect to a set of LTI systems is established, giving the key theorem to solve the basis selection problem of the identification scheme both in the LTI and in the LPV case later on. Before that, some additional aspects of identification are reviewed.

2.4.5.3 Identification in the MIMO Case

In case of MIMO systems, identification follows similar guidelines as in the SISO case, except that the model structure (2.42) is formulated with MIMO basis functions. Due to the FIR structure of OBFs models, the analytic solution of (2.31) is still obtained via a linear regression, however with a more extensive book keeping.

In case the MIMO basis are constructed from scalar basis functions (method 1), all properties in terms of identification trivially extend to MIMO case [128]. However, this model structure has a major disadvantage, namely that specific elements of $\{W_i\}$ can be insignificantly small for every $i > 0$, which can result in an over-parametrization, ergo in a significant bias of the estimate. MIMO basis sequences generated by square inner functions via (2.11) (method 2) are not affected by the previous disadvantage, however bias and variance properties of the estimates are not yet clearly understood. See [205] and [72] for more on identification properties of model structures based on this type of MIMO-OBFs.

2.4.6 *Pole Uncertainty of Model Estimates*

In practical situations, identification is unavoidably effected by noise, resulting in uncertainty of the model estimates. The resulting model uncertainties can be characterized based on numerous concepts of uncertainty in the parameter or frequency domain. See [67, 105, 127] and [49] for an overview on the available approaches. A well-known fact in the LTI case is that pole locations of a model estimate are generally sensitive to parameter uncertainties, see e.g. [66]. In the following, some basic concepts of pole uncertainty regions are briefly introduced for DT model estimates in the LS setting. The developed concepts are essential for the formulation of the robust basis selection approach in Chap. 8.

In the literature, many approaches have been developed for the calculation of confidence bounds for the estimated parameters (for a survey see [144]). However, commonly the first-order Taylor approximation based ellipsoidal bounds are used, like in the MATLAB *System Identification Toolbox* [106]. Consider the LS setting with the model structures given in Table 2.1. In this case, if the data generating system is in the model set, then this implies that the parameter estimate $\hat{\theta}_{N_\mathrm{d}}$ is consistent and has an asymptotically normal distribution with a covariance matrix $\mathcal{Q}_\theta$ (see Sect. 2.4.3). The previous properties provide that

$$(\hat{\theta}_{N_\mathrm{d}} - \theta_0)^\top \mathcal{Q}_\theta^{-1} (\hat{\theta}_{N_\mathrm{d}} - \theta_0) \to \chi^2(n), \quad \text{as } N_\mathrm{d} \to \infty, \tag{2.48}$$

where $\chi^2(n)$ is a χ^2-distribution with n-degrees of freedom and n is the number of parameters in θ. Let $[\,a_1\ \dots\ a_{n_\mathrm{a}}\,]$ be the parameters in θ characterizing the denominator part of the process model and let $[\,b_0\ \dots\ b_{n_\mathrm{b}}\,]$ be the nominator parameters. Denote $\triangle\theta := (\theta - \theta_0)$ for every $\theta \in \Theta$. Then for a given confidence level $\alpha \in [0,1]$, the parameter uncertainty of $G(\mathrm{q}, \hat{\theta}_{N_\mathrm{d}})$ can be defined as an ellipsoid

$$\mathcal{E}_\theta(\mathcal{Q}_\theta, \alpha) := \left\{ \theta \in \mathbb{R}^n \mid \triangle\theta^\top \mathcal{Q}_\theta^{-1} \triangle\theta \leq \chi^2_\alpha(n_\theta) \right\}, \tag{2.49}$$

where $\chi^2_\alpha(n)$ denotes the α-percentile of $\chi^2(n)$. This means that the probability of $\hat{\theta}_{N_\mathrm{d}} \in \mathcal{E}_\theta(\mathcal{Q}_\theta, \alpha)$ is equal to α as $N_\mathrm{d} \to \infty$. Often $\mathcal{E}_\theta(\mathcal{Q}_\theta, \alpha)$ is restricted to Θ.

In order to establish an uncertainty region of poles associated with each configuration of the parameters inside the derived ellipsoidal bound, a nonlinear transformation of the parameter confidence region $\mathcal{E}_\theta(\mathcal{Q}_\theta, \alpha)$ is needed. This transformation can be accomplished through the method of [223], which gives a hypothesis test to decide wether a $\lambda \in \mathbb{C}$ is a pole location of a model with $\theta \in \mathcal{E}_\theta(\mathcal{Q}_\theta, \alpha)$.

Denote by $\Lambda_0 \in \mathbb{C}^{n_\mathrm{a}}$ the poles associated with $G(\mathrm{q}, \theta_0)$ and define $\triangle\Lambda := \Lambda - \Lambda_0$ for every $\Lambda \in \mathbb{C}^{n_\mathrm{a}}$. Let λ_0 be a real valued pole in Λ_0 with $n_a > 1$. Then, define the perturbation of this pole as $\lambda = \lambda_0 + \triangle\lambda$ such that a parameter vector $\theta \in \Theta$ exists whose associated pole vector contains λ. Note that θ is not unique because λ only determines the denominator parameters. If $\theta \in \Theta$ exists, then it can be chosen such that the numerator parameters $[\,b_0\ \dots\ b_{n_\mathrm{b}}\,]$ of θ are equal to numerator parameters of θ_0. Write $G(\mathrm{q}, \theta)$ as

$$G(\mathrm{q}, \theta) = G_1(\mathrm{q}, \gamma) G_2(\mathrm{q}, \breve{\theta}), \tag{2.50}$$

with

$$G_1(\mathrm{q}, \gamma) := \frac{1}{1 + \gamma \mathrm{q}^{-1}} = \frac{1}{1 - \lambda \mathrm{q}^{-1}} \quad \text{and} \quad G_2(\mathrm{q}, \breve{\theta}) := \frac{\sum_{j=0}^{n_\mathrm{b}} b_j \mathrm{q}^{-j}}{1 + \sum_{i=1}^{n_\mathrm{a}-1} \breve{a}_i \mathrm{q}^{-i}},$$

where $\breve{\theta}$ contains the parameters of G_2. This factorization implies the existence of the transformation

$$\theta = T_1(\lambda)\breve{\theta} + T_2(\lambda), \tag{2.51}$$

where $\gamma = -\lambda$ and

$$T_1(\lambda) = \left[\begin{array}{ccccc|c} 1 & 0 & \dots & 0 & & \\ \gamma & 1 & \ddots & \vdots & & \\ 0 & \ddots & \ddots & 0 & & 0_{n_a \times (n-n_a)} \\ \vdots & \ddots & \ddots & 1 & & \\ 0 & \dots & 0 & \gamma & & \\ \hline \multicolumn{5}{c|}{0_{(n-n_a)\times(n_a-1)}} & I_{(n-n_a)\times(n-n_a)} \end{array}\right]_{n\times(n-1)} \qquad T_2(\lambda) = \begin{bmatrix} \gamma \\ 0 \\ \vdots \\ 0 \end{bmatrix}_{n\times 1}$$

In case of λ_0 is complex valued, define the perturbation of the complex pole pair λ_0, λ_0^* as $\lambda = \lambda_0 + \triangle\lambda$ and $\lambda^* = \lambda_0^* + \triangle\lambda^*$. Using the same mechanism, write $G(\mathrm{q}, \theta)$ as

$$G(\mathrm{q}, \tilde{\theta}) = G_1(\mathrm{q}, \gamma_1, \gamma_2) G_2(\mathrm{q}, \check{\theta}),$$

with

$$G_1(\mathrm{q}, \gamma_1, \gamma_2) := \frac{1}{1 + \gamma_1 \mathrm{q}^{-1} + \gamma_2 \mathrm{q}^{-2}} = \frac{1}{1 - 2\mathrm{Re}(\lambda)\mathrm{q}^{-1} + |\lambda|^2 \mathrm{q}^{-2}},$$

$$G_2(\mathrm{q}, \check{\theta}) := \frac{\sum_{j=0}^{n_b} b_j \mathrm{q}^{-j}}{1 + \sum_{i=1}^{n_a - 2} \check{a}_i \mathrm{q}^{-i}}.$$

Again, this factorization implies the existence of the transformation (2.51) with

$$T_1(\lambda) = \left[\begin{array}{ccccc|c} 1 & 0 & \dots & 0 & & \\ \gamma_1 & 1 & \ddots & 0 & & \\ \gamma_2 & \gamma_1 & \ddots & \vdots & & \\ 0 & \gamma_2 & \ddots & 1 & & 0_{n_a \times (n-n_a)} \\ \vdots & \ddots & \ddots & \gamma_1 & & \\ 0 & \dots & 0 & \gamma_2 & & \\ \hline \multicolumn{5}{c|}{0_{(n-n_a)\times(n_a-2)}} & I_{(n-n_a)\times(n-n_a)} \end{array}\right]_{n\times(n-2)} \qquad T_2(\lambda) = \begin{bmatrix} \gamma_1 \\ \gamma_2 \\ 0 \\ \vdots \\ 0 \end{bmatrix}_{n\times 1}$$

where $\gamma_1 = -2\mathrm{Re}(\lambda)$ and $\gamma_2 = |\lambda|^2$.

The above derived transformations qualify as a projection of a single pole or complex pole pair perturbation to the parameter domain Θ through the free parameter $\check{\theta} \in \mathbb{R}^{n-1}$ or $\check{\theta} \in \mathbb{R}^{n-2}$. In order to test that the parameter variation induced by $\triangle\lambda$ is inside the parameter uncertainty region $\mathcal{E}_\theta(\mathfrak{Q}_\theta, \alpha)$, it is sufficient to show that there exists a $\check{\theta} \in \mathbb{R}^{n-1}$ (or $\check{\theta} \in \mathbb{R}^{n-2}$) which minimizes

$$\left(T_1(\lambda)\check{\theta} + T_2(\lambda)\right)^\top \mathfrak{Q}_\theta^{-1} \left(T_1(\lambda)\check{\theta} + T_2(\lambda)\right), \tag{2.52}$$

and the minimum is smaller or equal than $\chi^2_\alpha(n_\theta)$ (see (2.49)). If this condition is not satisfied, then this proves the hypothesis that the pole perturbation cannot be associated with a parameter vector in $\mathcal{E}_\theta(\mathfrak{Q}_\theta, \alpha)$. The minimization problem of (2.52) has an analytic solution:

$$\check{\theta} = \frac{T_1^\top(\lambda)\mathcal{Q}_\theta^{-1}\theta_0 - T_1^\top(\lambda)\mathcal{Q}_\theta^{-1}T_2(\lambda)}{T_1^\top(\lambda)\mathcal{Q}_\theta^{-1}T_1(\lambda)}. \tag{2.53}$$

Thus for a given pole perturbation $\Delta\lambda$, if $\check{\theta}$ resulting from (2.53) satisfies that (2.52) is smaller or equal than $\chi^2_\alpha(n)$, then λ can be the pole of the asymptotic model estimate with probability α.

Based on the derived hypothesis test, it is possible to calculate the pole uncertainty region

$$\mathcal{P}(\mathcal{Q}_\theta,\alpha) := \{\lambda \in \mathbb{C} \mid \exists\theta \in \mathcal{E}_\theta(\mathcal{Q}_\theta,\alpha) \text{ s.t. } \lambda \text{ is a pole of } G(\mathrm{q},\theta)\}, \tag{2.54}$$

associated with $\mathcal{E}_\theta(\mathcal{Q}_\theta,\alpha)$. Note that $\mathcal{P}(\mathcal{Q}_\theta,\alpha) \subset \mathbb{C}$ which means that the hyper ellipsoid $\mathcal{E}_\theta(\mathcal{Q}_\theta,\alpha)$ is projected to a lower dimensional space. This implies that for a pole locations Λ, associated with $\theta \in \mathcal{E}_\theta(\mathcal{Q}_\theta,\alpha)$, $\Lambda \in (\mathcal{P}(\mathcal{Q}_\theta,\alpha))^{n_\mathrm{a}}$ holds. However, it is not true that any $\Lambda \in (\mathcal{P}(\mathcal{Q}_\theta,\alpha))^{n_\mathrm{a}}$ can be associated with a $\theta \in \mathcal{E}_\theta(\mathcal{Q}_\theta,\alpha)$. So the projection is surjective. This is an advantage in the sense that with the given hypothesis test, $\mathcal{P}(\mathcal{Q}_\theta,\alpha)$ can be efficiently computed and visualized. However, the surjective property is also a disadvantage. Instead of representing all configuration of poles associated with $\mathcal{E}_\theta(\mathcal{Q}_\theta,\alpha)$, the complex region $\mathcal{P}(\mathcal{Q}_\theta,\alpha)$ characterizes the set of pole locations that can occur with the given probability level in the model estimates. In this way, the perimeter bounds of $\mathcal{P}(\mathcal{Q}_\theta,\alpha)$ describe the uncertain pole locations in a worst-case sense.

In practical situations, θ_0 can be substituted by $\hat{\theta}_{N_\mathrm{d}}$ if N_d is large enough. Alternatively, using $\hat{\theta}_{N_\mathrm{d}}$ instead of θ_0 results in the hypothesis test that the true poles of the system are in the uncertainty region of the estimated poles with a given confidence level. Note that the above given mechanism can be also used for the calculation of the uncertainty region associated with the zeros of the model, by varying the numerator part in (2.50) instead of the denominator.

Using this pole uncertainty concept, the uncertainty regions can take various shapes in $\mathbb{C}$, ranging from real segments (real pole) and ellipsoidal or banana shaped forms to butterfly figures (complex pole pairs) as illustrated in Example 2.1. Therefore, it is not guaranteed that they constitute convex regions in $\mathbb{C}$. The regions can be all connected or separated into small sets due to the fact that they are the nonlinear projection of a hyper dimensional ellipsoid. Increasing α often results in the merging of previously separated regions. For an increasing α, regions can also popup unexpectedly in $\mathbb{C}$, due to the higher possibility of parameter variation. This yields the need of special algorithms to ensure correct calculation of the regions. For this purpose, an algorithm is developed in [198] that ensures calculation of possibly separated regions, unlike the existing solutions in the literature.

Note that other concepts of pole uncertainty regions have been developed in the literature as well. The commonly used ellipsoidal pole regions, also implemented in the identification toolbox of MATLAB, are calculated using a first order Taylor approximation of the nonlinear projection of $\mathcal{E}_\theta(\mathcal{Q}_\theta,\alpha)$ to $\mathbb{C}$. This type of approximation can introduce significant errors in the calculation of $\mathcal{P}(\mathcal{Q}_\theta,\alpha)$, unlike the previously presented approach. Other approaches, like discussed in [114], focus

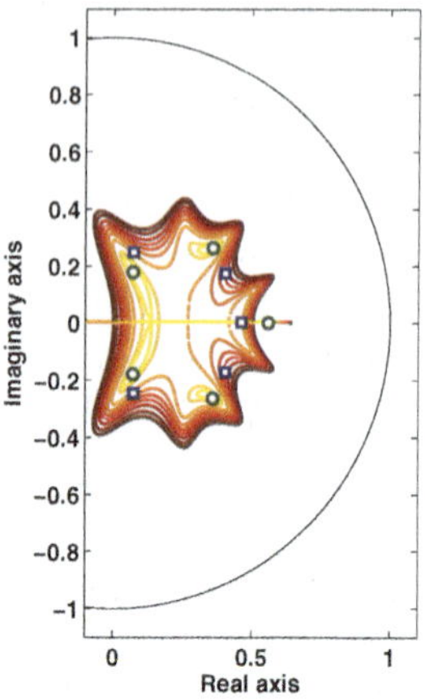

Fig. 2.2 Pole-uncertainty regions $\mathcal{P}(\mathcal{Q}_\theta, \alpha)$ of estimated poles (green ∘) with different confidence levels α (deep red 0.99 ↔ bright yellow 0.01). The true pole locations Λ_0 are denoted by blue squares.

more on the quantification of the variance of poles or zeros rather than on the actual calculation of their uncertainty regions. Based on these, the previously presented approach provides the state-of-the-art technique to calculate pole uncertainty regions associated with model estimates.

Example 2.1 (Pole uncertainty regions). Let the transfer function G_0 of a discrete-time asymptotically stable LTI system $\mathcal{F}$ with IO partition (u, y) be given as

$$G_0(z) = \frac{1.4}{1 - 1.5z^{-1} + 0.83z^{-2} - 0.26z^{-3} + 0.05z^{-4} - 0.006z^{-5}} z^{-1}. \tag{2.55}$$

For this system $\mathcal{F}$, a 500 sample long data record has been collected with a white u based on a uniform distribution $\mathcal{U}(-1, 1)$ and an additive white output noise ε with normal distribution $\mathcal{N}(0, 0.15)$. The LS prediction-error identification of G_0 has been accomplished via OE parametrization with correct denominator and nominator orders. The resulting pole uncertainty regions $\mathcal{P}(\mathcal{Q}_\theta, \alpha)$ of the estimate $G(\mathrm{q}, \hat{\theta}_{N_\mathrm{d}})$ have been calculated with the approach of Sect. 2.4.6 using confidence levels (0.99 ↔ 0.01). The perimeter lines of these regions, obtained with the algorithm of [198], are presented in Fig. 2.2. □

2.4.7 *Validation in the Prediction-Error Setting*

In the prediction-error setting, commonly either simulation or prediction by the model estimate based on a measured data record is used for (in)validation. One approach is to investigate the correlation of the residual, i.e. the error of the prediction, with respect to the input or itself. In other cases, error measures of the difference between the measured y and the simulated output $\hat{y}$ are calculated. These measures are used to decide on the validity of the model estimate. Some popular measures are the following:

Definition 2.3 (Mean squared error, [105]). The *mean squared error* (MSE) is the expected value of the squared estimation error :

$$\text{MSE} := \bar{\mathcal{E}}\{(y-\hat{y})^2\}. \tag{2.56}$$

□

Definition 2.4 (Best fit rate, [106]). The *best fit rate* (BFR) percentage or *fit score* is defined as

$$\text{BFR} := 100\% \cdot \max\left(1 - \frac{\|y-\hat{y}\|_2}{\|y-\bar{y}\|_2}, 0\right) = 100\% \cdot \max\left(1 - \sqrt{\frac{\bar{\mathcal{E}}\{(y-\hat{y})^2\}}{\bar{\mathcal{E}}\{(y-\bar{y})^2\}}}, 0\right),$$

where $\bar{y}$ is the mean of y, i.e. $\bar{y} = \bar{\mathcal{E}}\{y\}$. □

Definition 2.5 (Variance accounted for). The *variance accounted for* (VAF) percentage is the percentage of the output variation that is explained by the model:

$$\text{VAF} := 100\% \cdot \max\left(1 - \frac{\text{Var}\{y-\hat{y}\}}{\text{Var}\{y\}}, 0\right) = 100\% \cdot \max\left(1 - \frac{\bar{\mathcal{E}}\{(y-\hat{y}-\bar{e})^2\}}{\bar{\mathcal{E}}\{(y-\bar{y})^2\}}, 0\right),$$

where $\bar{e} = \bar{\mathcal{E}}\{y-\hat{y}\}$. □

Note that the MSE is equal to the LS criterion (2.30) evaluated for the simulated $\hat{y}$, instead of the predicted output signal. In this way, a high value indicates invalidity of the model. The BFR percentage is a relative measure, often used in the identification toolbox of MATLAB, and a low value of this measure indicates invalidity of the model. The VAF measure describes how much of the output variation is explained by the model, disregarding possible bias of the estimates.

2.5 The Kolmogorov n-Width Theory

In the identification cycle, one of the key steps is the choice of an adequate model structure, i.e. the model set, which can represent the system to be identified with a relatively small number of statistically meaningful parameters. In the identification approach based on OBF model structures, finding an appropriate model set translates to the search for a set of basis functions $\Phi_{n_b}^{n_e}$, that gives a series-expansion of the system with a fast convergence rate ρ. In LTI system identification, one approach to find appropriate model sets is based on the n-width concept [143], which has been shown to result in appropriate model sets for robust modeling of linear systems [111]. Using this concept, it has been shown in [131] that OBF model structures are optimal in the n-width sense for specific subsets of systems and finding the optimal OBF set for a given system set can be formulated as an optimization problem. In the following, the basic ingredients of this approach for discrete-time, stable, SISO systems are described. Later, this theory is used as the backbone of OBFs selection for the identification of LPV systems.

Let $\mathscr{F}$ denote a set of LTI SISO systems with transfer functions $\{F(z)\} = \mathfrak{F} \subseteq \mathcal{H}_{2-}(\mathbb{E})$ that we want to approximate with a linear combination of n elements of $\mathcal{H}_{2-}(\mathbb{E})$. Let $\Phi_n = \{\phi_i\}_{i=1}^{n}$ be a sequence of n linearly independent elements of $\mathcal{H}_{2-}(\mathbb{E})$, and let $\mathcal{M}_n = \mathrm{Span}(\Phi_n)$. Note that $\mathcal{M}_n$ describes all the possible linear combinations of Φ_n that can be used for the approximation of the elements of $\mathfrak{F}$. The distance $d_{\mathcal{H}_2}(F, \mathcal{M}_n)$ between a $F \in \mathcal{H}_{2-}(\mathbb{E})$ and $\mathcal{M}_n$ is defined as

$$d_{\mathcal{H}_2}(F, \mathcal{M}_n) = \inf_{F' \in \mathcal{M}_n} \left\| F - F' \right\|_{\mathcal{H}_2}. \tag{2.57}$$

This distance describes the best possible approximation error of a given $F \in \mathcal{H}_{2-}(\mathbb{E})$ in terms of the $\mathcal{H}_2$ norm if the linear combination of Φ_n is used as an approximation. With respect to the transfer function set $\mathfrak{F}$, we can define the worst-case approximation error by Φ_n as the maximum of (2.57) on $\mathfrak{F}$. Now we can use this concept to look for a set Φ_n that has a minimal worst-case approximation error for $\mathfrak{F}$. This minimum is called the *Kolmogorov n-width* (KnW) of $\mathfrak{F}$.

Definition 2.6 (Kolmogorov n-width, [143]). Let $\mathscr{M}_n$ be the collection of all n-dimensional subspaces of $\mathcal{H}_{2-}(\mathbb{E})$. The Kolmogorov n-width of a function set $\mathfrak{F}$ in $\mathcal{H}_{2-}(\mathbb{E})$ is

$$\pi_n(\mathfrak{F}, \mathcal{H}_{2-}(\mathbb{E})) = \inf_{\mathcal{M}_n \in \mathscr{M}_n} \sup_{F \in \mathfrak{F}} d_{\mathcal{H}_2}(F, \mathcal{M}_n). \tag{2.58}$$

□

In this way, $\pi_n(\mathfrak{F}, \mathcal{H}_{2-}(\mathbb{E}))$ describes the smallest possible $d_{\mathcal{H}_2}$ that can be achieved for all F in $\mathfrak{F}$ by the linear combination of n independent elements of $\mathcal{H}_{2-}(\mathbb{E})$. The subspace $\mathcal{M}_n \in \mathscr{M}_n$, for which π_n is minimal, is called the optimal subspace in the KnW sense. This optimal subspace describes a Φ_n that can approximate $\mathfrak{F}$ best in the worst-case sense. Now we can formulate this concept for OBFs.

Proposition 2.1 (n-width optimal OBFs, [131]). *Let $G_\mathrm{b} \in \mathcal{H}_{2-}(\mathbb{E})$ be an inner function with McMillan degree $n_\mathrm{g} > 0$, with poles Λ_{n_g}, and let $n_\mathrm{e} \geq 0$. Consider the subspace*

$$\mathcal{M}_n = \mathrm{Span}\left\{\phi_j(z)\, G_\mathrm{b}^i(z)\right\}_{j=1,\ldots,n_\mathrm{g}}^{i=0,\ldots,n_\mathrm{e}} \tag{2.59}$$

where $\phi_j = [M_1]_j$ and $M_1(z)$ is defined by (2.11) with respect to $G_\mathrm{b}(z)$. Then the subspace $\mathcal{M}_n$ is optimal in the Kolmogorov n-width sense with $n = (n_\mathrm{e}+1)n_\mathrm{g}$ for the set of systems with transfer functions $\mathfrak{F}$ analytic in the complement of the region

$$\Omega\left(\Lambda_{n_\mathrm{g}}, \rho\right) := \{z \in \mathbb{C} \mid |G_\mathrm{b}(z^{-1})| \leq \rho\}, \tag{2.60}$$

and are square integrable on its boundary. The worst-case approximation error, i.e. $\sup_{F \in \mathfrak{F}} d_{\mathcal{H}_2}(F, \mathcal{M}_n)$, is proportional to $\rho^{n_\mathrm{e}+1}$.

This remarkable result means that the set of OBFs $\Phi_{n_\mathrm{g}}^{n_\mathrm{e}} = \{\phi_j(z) G_\mathrm{b}^i(z)\}_{j=1,\ldots,n_\mathrm{g}}^{i=0,\ldots,n_\mathrm{e}}$ is the best in the worst-case sense to approximate transfer functions with all pole locations in $\Omega(\Lambda_{n_\mathrm{g}}, \rho)$. So if in an identification scenario it is known that the system poles lie in $\Omega(\Lambda_{n_\mathrm{g}}, \rho)$, then the optimal choice of basis functions is the set of OBFs associated

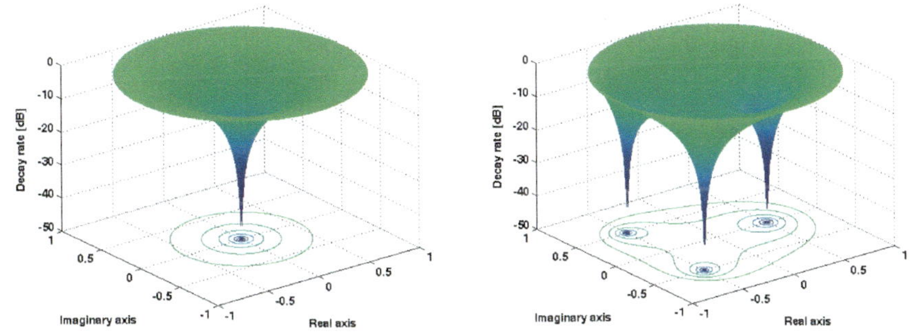

(a) $G_b(z) = z^{-1}$, pole at the origin (b) $G_b(z)$ with poles 0.5 and $-0.5 \pm 0.5i$

Fig. 2.3 The plot of the function $\left|G_b\left(z^{-1}\right)\right|$ for different choices of the inner function $G_b(z)$ and the convergence rate ρ (in dB). Level sets of $\left|G_b\left(z^{-1}\right)\right|$ give the boundaries of the regions $\{z \in \mathbb{C}, \left|G_b\left(z^{-1}\right)\right| \leq \rho\}$. Optimality of the $G_b(z)$ generated basis is ensured with a worst-case convergence rate ρ^{n_e+1} for systems with pole locations inside the regions defined by the level set boundaries.

with the poles Λ_{n_g}. Additionally, Proposition 2.1 shows that for the specified region one can not improve on the worst-case error by adding new poles to the n_g basis poles. It also generalizes the well-know fact that the set of *pulse functions* $\{z^{-i}\}_{i=1}^{n}$ is optimal for the class of stable systems analytic outside the region $\Omega(\Lambda_{n_g}, \rho) = \{|z| \leq \rho\}, \rho > 0$. The boundary of $\Omega(\Lambda_{n_g}, \rho)$ is displayed in Fig. 2.3a as a function of the convergence rate ρ. For a given $\rho > 0$, the boundary of the region results as the level set of this function, like the contour lines at the bottom of the figure. The worst-case approximation error in this case is proportional to ρ^n. This implies the optimality of FIR model structures with respect to the identification of such systems, which is a well-known result [143]. However, in case of arbitrary regions, like the regions in Fig. 2.3b, the level sets are commonly non-circular, containing separate regions that merge for increasing values of ρ. For these regions, the optimal choice of a basis has to be found among general basis functions (OBF model structures).

In an OBFs-based identification scenario we are dealing with the opposite problem. We are given a rough idea about the possible pole locations of the system described as a region $\Omega_0 \subset \mathbb{D}$ and we want to find a fixed n number of OBFs that are optimal with respect to Ω_0 in the KnW sense. This problem is referred to as the *inverse Kolmogorov problem*, where we want to find an inner function G_b to describe/approximate a given region of non-analyticity Ω_0 in the form $\Omega(\Lambda_{n_g}, \rho)$ with ρ as small as possible. The reason is that in terms of Proposition 2.1, the inner function G_b, associated with the best fitting $\Omega(\Lambda_{n_g}, \rho)$, generates the n_g-width optimal basis functions with respect to Ω_0. Let $G_{b,\Lambda_{n_g}}(z)$ be an inner function, associated with the poles Λ_{n_g} and define

$$\kappa_{n_g}(z, \Lambda_{n_g}) := |G_{b,\Lambda_{n_g}}(z^{-1})| = \prod_{j=1}^{n_g} \left| \frac{z - \lambda_j}{1 - z\lambda_j^*} \right|. \tag{2.61}$$

Then the solution of the inverse Kolmogorov problem for a given number of poles n_g, comes down to the min-max optimization problem:

$$\min_{\Lambda_{n_g} \subset \mathbb{D}} \max_{z \in \Omega} \kappa_{n_g}(z, \Lambda_{n_g}). \tag{2.62}$$

See [73] for details on this non-linear optimization problem and solution methods.

2.6 Conclusions

In this introductory chapter basic definitions and concepts of orthonormal basis functions and the prediction-error identification framework have been introduced with the aim to establish a solid background for our upcoming investigation of modeling and identification of LPV systems. Through this chapter we have shown that truncated series-expansion models like OBFs based model structures have attractive properties in LTI system identification. Such model structures not only provide consistent model estimation via simple linear regression (in case of an OE noise model), but need significantly less parameters than FIR models for an adequate approximation of the system and have a direct characterization of the truncation error. We will use these attractive properties to develop a low complexity and efficient identification approach of LPV systems in Chap. 9 using similar truncated series expansion models. Furthermore, we have discussed in Sect. 2.5 that in the context of OBF model structures, the optimal choice of the model set can be characterized by using the Kolmogorov n-width concept. As we will see, this theory will serve as the backbone of the later developed OBF-model-structure-based model set selection approach for LPV systems together with the improved concept of pole uncertainty regions discussed in Sect. 2.4.6. However, before such theories and methods can be introduced first we will study LPV systems and establish the basic system theoretical tools which will enable understanding of model structures and estimation in the LPV context.

Chapter 3
LPV Systems and Representations

Abstract. In this chapter, a behavioral framework of LPV systems is introduced as an extension of the LTI behavioral approach. This is done with the intention to give a unified view on LPV system theory, that enables to approach LPV system identification in a well-founded system theoretic sense. First we define LPV dynamical systems from the behavioral point of view. Then we introduce the algebraic structure in which we formulate kernel, state-space, and input-output representations of LPV systems. We also analyze the properties of LPV systems in terms of state-observability, state-reachability, and dynamic stability.

3.1 General Class of LPV Systems

In this section, we establish the basics of a behavioral framework for *linear parameter-varying* (LPV) systems where a representation-free definition of such systems can be given and the previously considered concepts of LPV representations and corresponding theories can be re-established. Our main motivation is to set this framework as a tool for the analysis of LPV system identification in a well-founded sense.

One of the key concepts that is required to establish the LPV behavioral framework is an algebraic structure with elements describing *differential equations*, like $\mathbb{R}[\xi]$ (the ring of polynomials with real constant coefficients) used in the LTI case. As we will see, the required structure in the LPV case is based on polynomials with coefficients that are functions of the scheduling variable p and its derivatives (continuous-time) or its time-shifts (discrete-time). The construction of this structure enables us to apply the results of the *linear time-varying* (LTV) behavioral approach, worked out by [239] and [78]. We use these results to establish three key theorems: the existence of kernel representations, the existence of state-kernel forms, and later in Sect. 3.2 the concept of left/right unimodular transformations. These theorems give the basic building blocks for the derivation of equivalence transformations between LPV representations, treated in Chap. 4, which have paramount significance for system identification.

R. Tóth: Modeling and Identification of Linear Parameter-Varying Systems, LNCIS 403, pp. 45–100.
springerlink.com

First, let's investigate what we call an LPV system from the behavioral point of view and what kind of physical phenomena are represented by this modeling concept.

3.1.1 Parameter Varying Dynamical Systems

In aerospace engineering, it is well-known that many airplanes, like the F-16 Fighting Falcon presented in Fig. 3.1, are nonlinear dynamical systems, but at a constant altitude they can be well approximated as an LTI system [181, 42]. Then, by viewing the aircraft as a collection of LTI behaviors corresponding to different altitude levels and using the altitude variable as a *scheduling* between them, we can arrive at an approximation of the global behavior. In this context, the concept of scheduling means the selection of the LTI behavior associated with a specific altitude level. This behavior describes the possible continuation of signal trajectories during the time interval in which the aircraft remains at the same altitude. Thus, the resulting representation of the global behavior involves coefficients that are functions of the scheduling. Such a modeling approach, that was introduced in Chap. 1 as the *gain-scheduling* principle, defines a parameter-varying (due to scheduling) and linear (in signal relation) system. Such systems are referred as LPV. However, it is important that an LPV system is more than just an array of LTI systems, because the governing scheduling rules or functions also define the dynamical behavior between each scheduling point, i.e. altitude points of this example. The concept of scheduling functions and "frozen" LTI behaviors provides an essential viewpoint on LPV systems which will be frequently used in the development of the identification approaches of Chap. 9.

In the general *parameter-varying* (PV) framework, the scheduling variable, commonly denoted by p, is an external[1], so-called free signal of the system, that governs the dynamical behavior. From this aspect, the role of p can be understood as an other "time-variable" that determines the change of signal relations. However, the trajectory of p is unknown in advance which property distinguishes LPV systems from the LTV system class, where the variations of signal relations is directly associated with time. Based on this, the class of PV systems can be defined as follows:

Definition 3.1 (Parameter-varying dynamical system). A parameter-varying dynamical system $\mathcal{S}$ is defined as a quadruple

$$\mathcal{S} = (\mathbb{T}, \mathbb{P}, \mathbb{W}, \mathfrak{B}), \tag{3.1}$$

with $\mathbb{T}$ the time-axis, $\mathbb{P}$ the scheduling space with dimension $n_{\mathbb{P}}$, $\mathbb{W}$ the signal space with dimension $n_{\mathbb{W}}$, and $\mathfrak{B} \subseteq (\mathbb{W} \times \mathbb{P})^{\mathbb{T}}$ the behavior ($\mathbb{X}^{\mathbb{T}}$ is the standard notation for the collection of all maps from $\mathbb{T}$ to $\mathbb{X}$). □

[1] Note that systems where p is an internal variable (like output, input, or state) are called *quasi parameter-varying systems*. Still, such systems are commonly treated as a PV system with external scheduling variable, therefore in the upcoming analysis, p is assumed to be an independent variable. For more on quasi-PV systems, see Chap. 7.

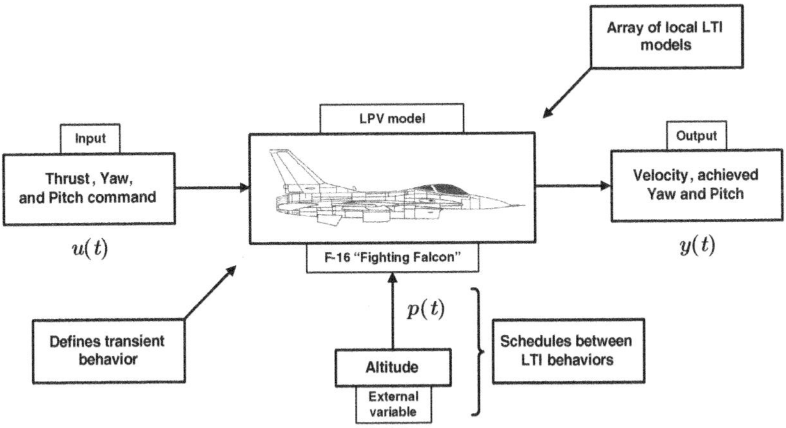

Fig. 3.1 The LPV modeling concept of the F-16 Fighting Falcon.

The set $\mathbb{T}$ defines the time-axis of the system, describing *continuous-time* (CT), $\mathbb{T} = \mathbb{R}$, and *discrete-time* (DT), $\mathbb{T} = \mathbb{Z}$, systems alike, while $\mathbb{W}$ gives the range of the system signals. The behavior $\mathfrak{B} \subseteq (\mathbb{W} \times \mathbb{P})^{\mathbb{T}}$ is the space of all signal and scheduling trajectories that are compatible with the system. Note that there is no prior distinction between inputs and outputs in this setting. The scheduling space $\mathbb{P}$ is usually a closed subset of a vector space on which the scheduling variable p varies: $p \in \mathbb{P}^{\mathbb{T}}$. In our example, $\mathbb{P}$ refers to the altitude range of the aircraft, which is positive and bounded by a maximum height of operation. Often, the admissible trajectories of p are further restricted to a subset of $\mathbb{P}^{\mathbb{T}}$ to bound their variations, e.g. it is not possible for an aircraft to have discontinuous jumps in altitude. This set of admissible scheduling trajectories is defined as the *projected scheduling behavior*:

$$\mathfrak{B}_{\mathbb{P}} = \pi_p \mathfrak{B} = \left\{ p \in \mathbb{P}^{\mathbb{T}} \mid \exists w \in \mathbb{W}^{\mathbb{T}} \text{ s.t. } (w, p) \in \mathfrak{B} \right\}, \tag{3.2}$$

where π_p denotes projection onto $\mathbb{P}^{\mathbb{T}}$. In other words, $\mathfrak{B}_{\mathbb{P}}$ describes all possible scheduling trajectories of $\mathcal{S}$. Similarly we can define the projected signal behavior $\mathfrak{B}_{\mathbb{W}}$. Additionally, for a given fixed scheduling trajectory $p \in \mathfrak{B}_{\mathbb{P}}$, the projected behavior

$$\mathfrak{B}_p = \left\{ w \in \mathbb{W}^{\mathbb{T}} \mid (w, p) \in \mathfrak{B} \right\}, \tag{3.3}$$

defines all the signal trajectories compatible with the fixed scheduling trajectory p. The projected behavior gives the possible course of actions or maneuvers that the aircraft in our example can take to follow a fixed altitude trajectory. In case of a constant scheduling trajectory, $p \in \mathfrak{B}_p$ with $p(t) = \mathrm{p}$ for all $t \in \mathbb{T}$ where $\mathrm{p} \in \mathbb{P}$, the projected behavior $\mathfrak{B}_p$ is called a *frozen behavior* and denoted as

$$\mathfrak{B}_{\mathrm{p}} = \left\{ w \in \mathbb{W}^{\mathbb{T}} \mid (w, p) \in \mathfrak{B} \text{ with } p(t) = \mathrm{p}, \ \forall t \in \mathbb{T} \right\}. \tag{3.4}$$

Definition 3.2 (Frozen system). Let $\mathcal{S} = (\mathbb{T}, \mathbb{P}, \mathbb{W}, \mathfrak{B})$ be a PV system and consider $\mathfrak{B}_\mathrm{p}$ for a given $\mathrm{p} \in \mathbb{P}$. The dynamical system

$$\mathcal{F}_\mathrm{p} = (\mathbb{T}, \mathbb{W}, \mathfrak{B}_\mathrm{p}) \tag{3.5}$$

is called a frozen system of $\mathcal{S}$. □

With the previously introduced concepts, we can define LPV systems as follows:

Definition 3.3 (LPV system). The parameter-varying system $\mathcal{S}$ is called LPV, if the following conditions are satisfied:

- $\mathbb{W}$ is a vector-space and $\mathfrak{B}_p$ is a linear subspace of $\mathbb{W}^\mathbb{T}$ for all $p \in \mathfrak{B}_\mathbb{P}$ (linearity).
- $\mathbb{T}$ is closed under addition.
- For any $(w, p) \in \mathfrak{B}$ (a signal trajectory associated with a scheduling trajectory) and any $\tau \in \mathbb{T}$, it holds that $(w(\cdot + \tau), p(\cdot + \tau)) \in \mathfrak{B}$, in other words $\mathrm{q}^\tau \mathfrak{B} = \mathfrak{B}$ (time-invariance). □

In terms of Def. 3.3, for a constant scheduling trajectory $p(k) \equiv \mathrm{p}$, time-invariance of $\mathcal{S}$ implies time-invariance of $\mathcal{F}_\mathrm{p}$. Based on this and the linearity condition of $\mathfrak{B}_p$, it holds for an LPV system that for each $\mathrm{p} \in \mathbb{P}$ the associated frozen system $\mathcal{F}_\mathrm{p}$ is an LTI system, which is in accordance with previous definitions of LPV systems (see [156]). In this way, the projected behaviors of a given LPV system $\mathcal{S}$ with respect to constant scheduling trajectories define a set of LTI systems:

Definition 3.4 (Frozen system set). Let $\mathcal{S} = (\mathbb{T}, \mathbb{P}, \mathbb{W}, \mathfrak{B})$ be an LPV system. The set of LTI systems

$$\mathscr{F}_\mathcal{S} = \{\mathcal{F} = (\mathbb{T}, \mathbb{W}, \mathfrak{B}') \mid \exists \mathrm{p} \in \mathbb{P} \;\; \text{s.t.} \;\; \mathfrak{B}' = \mathfrak{B}_\mathrm{p}\} \tag{3.6}$$

is called the frozen system set of $\mathcal{S}$. □

This set refers in our example to the LTI behaviors of the aircraft for constant altitude levels. We have already motivated that the LPV system concept is advantageous compared to nonlinear systems, as the relation of the signals is linear. Definition 3.3 also reveals the advantage of this system class over LTV systems: the variation of the system dynamics is not associated directly with time, but with the variation of a free signal. Thus, the LPV modeling concept, compared to LTV systems, is more suitable for non-stationary/coordinate dependent physical systems as it describes the underlying phenomena directly (see Example 3.1).

Example 3.1 (Varying mass on a spring). To emphasize the advantage of LPV systems, let's investigate the modeling of the motion of a varying mass connected to a spring (see Fig. 3.2). This problem is one of the typical phenomena occurring in systems with time-varying masses like in motion control (robotics, rotating crankshafts, rockets, etc.), biomechanics, and in fluid-structure interaction problems. Denote by w_x the position of the varying mass m. Let $\mathrm{k_s} > 0$ be the spring constant, introduce w_F as the force acting on the mass, and assume that there is no damping. By Newton's second law of motion, the following equation holds:

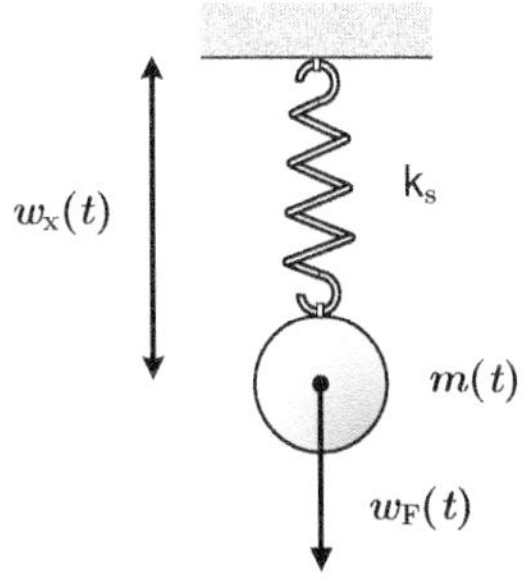

Fig. 3.2 Varying-mass connected to a spring.

$$\frac{d}{dt}\left(m\frac{d}{dt}w_X\right) = w_F - k_s w_X, \tag{3.7}$$

or equivalently

$$k_s w_X + \left(\frac{d}{dt}m\right)\frac{d}{dt}w_X + m\frac{d^2}{dt^2}w_X = w_F. \tag{3.8}$$

It is immediate that by taking m as a scheduling variable, the behavior of this plant can be described as an LPV system, preserving the physical insight of Newton's second law. On the other hand, viewing m as a time-varying parameter, whose trajectory is fixed in time, results in a LTV system. Such a system would explain the behavior of the plant for only a fixed trajectory of the mass.

It also holds that for any constant scheduling trajectory $p(t) = \mathrm{p}$, i.e. constant m for all $t \in \mathbb{R}$, the set of admissible signal trajectories of (3.8) is defined as the solutions of

$$k_s w_X - w_F + \mathrm{p}\frac{d^2}{dt^2}w_X = 0. \tag{3.9}$$

This yields that the frozen system set is the collection of LTI systems represented by (3.9) for all $\mathrm{p} \in \mathbb{P}$, i.e. all possible constant m. □

In the sequel, we restrict our attention to continuous-time ($\mathbb{T} = \mathbb{R}$) and discrete-time ($\mathbb{T} = \mathbb{Z}$) systems with finite dimensional and real signal space ($\mathbb{W} = \mathbb{R}^{n_W}$) and with finite dimensional, real, and closed scheduling space ($\mathbb{P} \subseteq \mathbb{R}^{n_P}$). In fact, we consider LPV systems described by finite order linear differential or difference equations with parameter-varying effects in the coefficients, and we call such systems *linear parameter-varying differential/difference* systems and denote them with $\mathcal{S}$. A basic property of such systems is that their behaviors $\mathfrak{B}$ are complete ($(w,p) \in \mathfrak{B} \Leftrightarrow (w,p)|_{[t_0,t_1]} \in \mathfrak{B}|_{[t_0,t_1]}, \forall [t_0,t_1] \subset \mathbb{T}$). In the sequel, if we refer to LPV systems, we refer to this system class.

3.1.2 Representations of Continuous-Time LPV Systems

In order to re-establish the concept of LPV representations, we introduce *kernel* (KR), *state-space* (SS) and *input-output* (IO) representations of *continuous-time* (CT) LPV systems. However, to define these representations, we first have to introduce differential equations with varying-coefficients as the representations of LPV behaviors. These differential equations are described by polynomials of an algebraic ring where equivalence of representations and other system theoretic concepts can be characterized by simple algebraic manipulations.

3.1.2.1 Coefficient Functions

As stated, the coefficients of the representations of LPV systems are functions of the possibly multidimensional scheduling variable p. Thus, first we define the class of functional dependencies that we consider in the sequel. In fact, we establish an algebraic field of a wide class of multivariable functions. We show later that if the variables of these functions are assigned to the elements of p and their derivatives, then equivalence transformations between different representation domains become possible. To formulate the class of multivariable functions that we consider, introduce the following definition:

Definition 3.5 (Real-meromorphic function, [88]). A real-meromorphic function $f:\mathbb{R}^n \to \mathbb{R}$ is a function in the form

$$f = \frac{g}{h}, \tag{3.10}$$

where $g, h : \mathbb{R}^n \to \mathbb{R}$ are holomorphic (analytic) functions and $h \neq 0$. □

Meromorphic functions consist of all rational, polynomial functions, trigonometric expressions, rational exponential functions etc., however functions like $f(\mathrm{x}) = \sin^{-1}(1/\mathrm{x})$ are not meromorphic ($\mathrm{x} = 0$ is an accumulation point of the singularities). Therefore, this class contains the common functional dependencies that result during LPV modeling of physical systems (see Chap. 1 and 7). Next we formulate an algebraic field over all multivariable real-meromorphic functions.

Let $\mathcal{R}_n$ denote the field of real-meromorphic functions with n variables. Denote the variables of a $r \in \mathcal{R}_n$ as $\zeta_1, \ldots \zeta_n$. Also define an operator $\mho_j$ on $\mathcal{R}_n$ with $1 \leq j \leq n$ such that

$$\mho_j(r(\zeta_1, \ldots, \zeta_n)) := r(\zeta_1, \ldots, \zeta_j, 0, \ldots, 0). \tag{3.11}$$

Note that $\mho_j$ projects a meromorphic function to a lower dimensional domain. Introduce $\bar{\mathcal{R}}_n$, defined as

$$\bar{\mathcal{R}}_n = \{ r \in \mathcal{R}_n \mid \mho_{n-1}(r) \neq r \}. \tag{3.12}$$

It is clear that $\bar{\mathcal{R}}_n$ consist of all functions $\mathcal{R}_n$ in which the variable ζ_n has a nonzero contribution, i.e. it plays a role in the function. Also define the operator $\mho_* : (\cup_{i\geq 0} \mathcal{R}_i) \to (\cup_{i\geq 0} \bar{\mathcal{R}}_i)$, which associates a given $r \in \mathcal{R}_n$ with a $r' \in \bar{\mathcal{R}}_{n'}$, $n \geq n'$, i.e. $\mho_*(r) = r'$, such that $r'(\zeta_1, \ldots, \zeta_{n'}) = r(\zeta_1, \ldots, \zeta_{n'}, 0, \ldots, 0)$ for all $\zeta_1, \ldots, \zeta_{n'} \in \mathbb{R}$, $\mho_{n'}(r) = r$ and n' is minimal. In this way, $\mho_*$ reduces the variables of a function till $\zeta_{n'}$ can not be left out from the expression because it has a nonzero contribution to the value of the function. Now define the collection of all real-meromorphic functions with finite many variables as

$$\mathcal{R} = \bigcup_{i \geq 0} \bar{\mathcal{R}}_i, \quad \text{with } \bar{\mathcal{R}}_0 = \mathbb{R}. \tag{3.13}$$

The function class $\mathcal{R}$ will be used as the collection of coefficient functions (like $\{A,\dots,D\}$ and $\{a_i, b_j\}$ in (1.6a-b) and (1.4)) for the representations, giving the basic building block of PV differential equations. These functions are not only used to express dependence over multidimensional p but also to enable a distinction between dynamic scheduling dependence of the coefficients and the dynamic relation between the signals of the system. The following lemma is important:

Lemma 3.1 (Field property of $\mathcal{R}$). *The set $\mathcal{R}$ is a field.*

The proof of this lemma is straightforward and is given in [188]. Denote by $\mathcal{R}^{\cdot\times\cdot}$ the matrices with elements in $\mathcal{R}$. It is immediate that $\mathcal{R}^{\cdot\times\cdot}$ is also a field.

3.1.2.2 Representing Scheduling Dependence

The next step towards the formal definition of PV differential equations and the associated kernel representations, is to associate the variables of the coefficient functions with elements of p and its derivatives. This is required to handle multiplication of coefficients by time operators. Given the scheduling dimension $n_{\mathbb{P}}$, denote the variables of $r \in \bar{\mathcal{R}}_n$ (n-dimensional function in $\bar{\mathcal{R}}$) as:

$$r(\{\zeta_{ij}\}_{n_\zeta,n_{\mathbb{P}},\tau}) := r\left(\zeta_{01}, \zeta_{02}, \dots, \zeta_{0n_{\mathbb{P}}}, \zeta_{11}, \zeta_{12}, \dots, \zeta_{n_\zeta 1}, \dots, \zeta_{n_\zeta \tau}\right), \tag{3.14}$$

where $n_\zeta \geq 0$ and $0 < \tau \leq n_{\mathbb{P}}$ such that $n_\zeta n_{\mathbb{P}} + \tau = n$. In this way, the first variable of r is denoted by ζ_{01}, the second is denoted by ζ_{02}, etc. Thus, (3.14) gives a unique labeling of the variables for each $\bar{\mathcal{R}}_n$ with $n \geq 1$. In continuous-time, associate each variable ζ_{ij} as

$$\zeta_{ij} = \frac{\mathrm{d}^i}{\mathrm{d}t^i} p_j \tag{3.15}$$

where p_j is the jth element of the scheduling signal p. This association provides the description of parameter-varying coefficients in $\mathcal{R}$, where each coefficient is a meromorphic function of the elements of p and their finite order derivatives (see Example 3.2). We call such a coefficient dependence *dynamic*. To compare this dependency class to class of dependencies used in the state-of-the-art of LPV identification (see Chap. 1), define the subset of $\mathcal{R}$ for a given $n_{\mathbb{P}}$ as $\mathcal{R}|_{n_{\mathbb{P}}} = \bigcup_{i=0}^{n_{\mathbb{P}}} \bar{\mathcal{R}}_i$. It is easy to show that $\mathcal{R}|_{n_{\mathbb{P}}}$ is a field and it consists of meromorphic coefficient functions dependent on the elements of p (without derivatives). This dependence type is called *static*. It has already been discussed in Chap. 1, that LPV representations based on coefficients in $\mathcal{R}|_{n_{\mathbb{P}}}$ are inequivalent (see Chap. 4 and [189, 202] for a detailed explanation). This is the main motivation to use coefficients in $\mathcal{R}$ as the building blocks of representations, because with these, equivalency of representations can be re-established.

To express the association (3.15), we introduce the operator $\diamond : (\mathcal{R}, \mathfrak{B}_{\mathbb{P}}) \to (\mathbb{R}^{\mathbb{R}})$ that associates $r \in \mathcal{R}$ and $p \in \mathfrak{B}_{\mathbb{P}}$ with a time function defined as:

$$r \diamond p = r\left(\left\{\tfrac{\mathrm{d}^i}{\mathrm{d}t^i} p_j\right\}_{n_\zeta, n_{\mathbb{P}}, \tau}\right). \tag{3.16}$$

This operator expresses the evaluation of the coefficient function r along a scheduling trajectory p and its derivatives.

Example 3.2 (Coefficient function). Let $\mathbb{P} = \mathbb{R}^{n_\mathbb{P}}$ with $n_\mathbb{P} = 2$. Consider the real-meromorphic coefficient function $r : \mathbb{R} \to \mathbb{R}^3$ defined as

$$r(\mathrm{x}_1, \mathrm{x}_2, \mathrm{x}_3) = \frac{\cos(\mathrm{x}_1)}{\sin(\mathrm{x}_3)}. \tag{3.17}$$

Then for a scheduling signal $p : \mathbb{R} \to \mathbb{R}^2$:

$$r \diamond p = r\left(p_1, p_2, \tfrac{\mathrm{d}}{\mathrm{d}t} p_1\right) = \frac{\cos(p_1)}{\sin(\frac{\mathrm{d}}{\mathrm{d}t} p_1)}. \tag{3.18}$$

On the other hand, if $n_\mathbb{P} = 3$, then

$$r \diamond p = r(p_1, p_2, p_3) = \frac{\cos(p_1)}{\sin(p_3)}. \tag{3.19}$$

□

In the sequel the (time-varying) coefficient sequence $(r \diamond p)$ will be used to operate on a signal w (like $a_i(p)$ in (1.4)), giving the varying coefficient sequence of the representations. In this respect an important property is that multiplication of the $\diamond$ operation with the operator $\frac{\mathrm{d}}{\mathrm{d}t}$ is not commutative, in other words $\frac{\mathrm{d}}{\mathrm{d}t}(r \diamond p) \neq (r \diamond p)\frac{\mathrm{d}}{\mathrm{d}t}$. To handle this multiplication, for $r \in \mathcal{R}$ we define the *dot* operator on $\mathcal{R}$ to describe differentiation of parameter-varying coefficient functions.

Definition 3.6 (Dot operator). Let $r \in \bar{\mathcal{R}}_n$ be a meromorphic function in $\mathcal{R}$. For a given scheduling dimension $n_\mathbb{P}$, denote the variables of r as $\{\zeta_{ij}\}_{n_\zeta, n_\mathbb{P}, \tau}$ based on (3.14). Then, in continuous-time, the dot operator on $\mathcal{R}$ is introduced as

$$\dot{r} := \mho_*(\check{r}), \tag{3.20}$$

where $\check{r} \in \bar{\mathcal{R}}_{n+n_\mathbb{P}}$ and is given by

$$\begin{aligned} \check{r}(\{\zeta_{ij}\}_{n_\zeta+1, n_\mathbb{P}, \tau}) = {} & \sum_{k=0}^{n_\zeta} \sum_{l=1}^{n_\mathbb{P}} \frac{\partial}{\partial \zeta_{kl}} r(\{\zeta_{ij}\}_{n_\zeta, n_\mathbb{P}, \tau}) \zeta_{(k+1)l} \\ & + \sum_{l=1}^{\tau} \frac{\partial}{\partial \zeta_{n_\zeta l}} r(\{\zeta_{ij}\}_{n_\zeta, n_\mathbb{P}, \tau}) \zeta_{(n_\zeta+1)l}. \end{aligned} \tag{3.21}$$

□

Note that the chain rules of differentiation imply that the dot operator fulfills the following rules for any $r_1, r_2 \in \mathcal{R}$:

$$\begin{cases} \text{if } r = r_1 \pm r_2 & \text{then } \dot{r} = \dot{r}_1 \pm \dot{r}_2, \\ \text{if } r = r_1 r_2 & \text{then } \dot{r} = \dot{r}_1 r_2 + r_1 \dot{r}_2, \\ \text{if } r = \frac{r_1}{r_2} & \text{then } \dot{r} = \frac{\dot{r}_1 r_2 - r_1 \dot{r}_2}{r_2^2}. \end{cases}$$

Due to the differentiation to which (3.21) corresponds, the number of variables in the coefficient functions can grow, representing an increase in the order of derivatives of p in the functional dependence.

Example 3.3 (Non-commutativity and the dot operation). Consider again Example 3.1 and rewrite the differential equation (3.7) into the form:

$$\mathrm{k_s} w_\mathrm{X} - w_\mathrm{F} + \xi(r \diamond m)\xi w_\mathrm{X} = 0, \tag{3.22}$$

with $\xi = \frac{\mathrm{d}}{\mathrm{d}t}$ and $r \in \mathcal{R}$ is the identity function: $r \diamond m = m$. Then, due to the non-commutative multiplication rule (3.32), equation (3.22) is equivalent with

$$\mathrm{k_s} w_\mathrm{X} - w_\mathrm{F} + (\dot{r} \diamond m)\xi w_\mathrm{X} + (r \diamond m)\xi^2 w_\mathrm{X} = 0, \tag{3.23}$$

where $\dot{r} \diamond m = \frac{\mathrm{d}}{\mathrm{d}t} m$. Note that (3.23) is identical to (3.8) by using the signal substitution $p = m$ and $w^\top = [w_\mathrm{X} \quad w_\mathrm{F}]$. □

3.1.2.3 Polynomials over $\mathcal{R}$

In order to introduce representations of LPV systems, an other key ingredient is needed, namely the formulation of polynomials with meromorphic coefficient functions that have a finite number of variables. Polynomials of this type are used to define PV differential equations describing the behavior of CT-LPV systems. First we define the ring of such polynomials and we show that multiplication, which corresponds to differentiation, is non-commutative over this ring.

Introduce $\mathcal{R}[\xi]$ as the collection of all polynomials in the indeterminant ξ and with coefficients in $\mathcal{R}$. It is a general property of polynomial spaces over a field, that they define a ring. Also introduce $\mathcal{R}[\xi]^{\cdot \times \cdot}$, the set of matrix polynomial functions with elements in $\mathcal{R}[\xi]$. Using $\mathcal{R}[\xi]$ and the operator $\diamond$, we are now able to define a parameter-varying differential equation:

Definition 3.7 (PV differential equation). Consider $R(\xi) = \sum_{i=0}^{n_\xi} r_i \xi^i \in \mathcal{R}[\xi]^{n_\mathrm{r} \times n_\mathbb{W}}$ and $(w, p) \in (\mathbb{R}^{n_\mathbb{W}} \times \mathbb{R}^{n_\mathbb{P}})^\mathbb{R}$.

$$(R(\frac{\mathrm{d}}{\mathrm{d}t}) \diamond p)w := \sum_{i=0}^{n_\xi} (r_i \diamond p)\frac{\mathrm{d}^i}{\mathrm{d}t^i} w = 0, \tag{3.24}$$

is called a PV differential equation with order $n_\xi = \mathrm{Deg}(R)$. □

PV differential equations in the form of (3.24) are used to define the class of CT-LPV systems we consider in the sequel. It will be shown, that this class contains all the popular definitions of LPV state-space and IO models. Furthermore this mathematical structure also enables the transformation of a wide class of nonlinear systems to an LPV form by preserving physical insight (see Example 3.4 and Chap. 7).

Example 3.4 (PV differential equation). Consider the mass-spring system of Example 3.1. Let $p = m$ with a scheduling space $\mathbb{P} = [1, 2]$ and let $w = [w_\mathrm{X} \quad w_\mathrm{F}]^\top$ with $\mathbb{W} = \mathbb{R}^2$. Then the possible signal trajectories are defined as the solutions of

$$\mathrm{k_s}w_1 - w_2 + \left(\frac{\mathrm{d}}{\mathrm{d}t}p\right)\frac{\mathrm{d}}{\mathrm{d}t}w_1 + p\frac{\mathrm{d}^2}{\mathrm{d}t^2}w_1 = 0. \tag{3.25}$$

for all smooth $p : \mathbb{R} \to \mathbb{P}$. Such a system equation can be written in the form (3.24) with $n_\mathbb{W} = 2$, $n_\xi = 1$, $n_\mathbb{P} = 1$, and

$$r_0 \diamond p = \begin{bmatrix} \mathrm{k_s} & -1 \end{bmatrix}, \quad r_1 \diamond p = \begin{bmatrix} \frac{\mathrm{d}}{\mathrm{d}t}p & 0 \end{bmatrix}, \quad r_2 \diamond p = \begin{bmatrix} p & 0 \end{bmatrix}. \qquad \square$$

So far we have defined PV differential equations in the form of

$$(R(\frac{\mathrm{d}}{\mathrm{d}t}) \diamond p)w = 0, \tag{3.26}$$

which for a given p corresponds to an ordinary differential equation in the variable w. In the behavioral framework, a PV system is defined as the union of signal and scheduling trajectories (w, p) that are compatible with the system. Thus to define a representation of a system in terms of a PV differential equation (3.26) which describes these signals, we require the concept of admissible signal trajectories of PV differential equations. Define $\mathcal{L}_1^{\mathrm{loc}}(\mathbb{R}, \mathbb{R}^{n_\mathbb{W}})$, the space of locally integrable functions $w : \mathbb{R} \to \mathbb{R}^{n_\mathbb{W}}$ satisfying:

$$\int_{\tau_1}^{\tau_2} \|w(t)\|_2 < \infty, \tag{3.27}$$

with $[\tau_1, \tau_2] \subset \mathbb{R}$ and $\|\bullet\|_2$ denoting the Euclidian norm on $\mathbb{R}^{n_\mathbb{W}}$. Also define $\mathcal{C}^\infty(\mathbb{R}, \mathbb{R}^\cdot)$ as the space of infinitely differentiable functions $w : \mathbb{R} \to \mathbb{R}^\cdot$. Now we can formulate the definition of solutions we consider for (3.26). Restricting ourselves to $\mathcal{C}^\infty$ would leave out important functions like steps. On the other hand, the space of distributions is too large to have the solution at every time instant well-defined. A vital alternative is $\mathcal{L}_1^{\mathrm{loc}}$, which is large enough to accommodate steps, ramps, etc. and still concrete enough to avoid problems with distributions. Thus, the concept of solution is introduced in the following sense:

Definition 3.8 (Weak solution). We call $w \in \mathcal{L}_1^{\mathrm{loc}}(\mathbb{R}, \mathbb{R}^{n_\mathbb{W}})$ a weak solution of (3.24) for a given smooth scheduling trajectory $p \in \mathfrak{B}_\mathbb{P} \subseteq \mathcal{L}_1^{\mathrm{loc}}(\mathbb{R}, \mathbb{R}^{n_\mathbb{P}})$, if

$$\langle w, (R^\dagger(\frac{\mathrm{d}}{\mathrm{d}t}) \diamond p)\varphi \rangle := \int_\mathbb{R} w^\top (R^\dagger(\frac{\mathrm{d}}{\mathrm{d}t}) \diamond p)\varphi \, dt = 0 \tag{3.28}$$

holds for every smooth, so-called test function, $\varphi : \mathbb{R} \to \mathbb{R}^{n_\mathrm{r}}$ with compact support, where $R^\dagger$ is

$$R^\dagger(\xi) = \sum_{i=0}^{n_\xi} (-1)^i \xi^i r_i^\top. \tag{3.29}$$

$\square$

In the following we only consider scheduling trajectories for which the coefficients of $R(\xi) \diamond p$ are bounded, so the set of solutions associated with R is well defined in terms of (3.28). Additionally, $R^\dagger$ (also called the adjoint of R) results in the form of (3.29), due to integrations by parts of (3.24) to transfer all differential operators acting on w to differential operators acting on φ, similar to the LTI case [51]. Each integration by parts entails a multiplication by -1. To compute $R^{\dagger}$ in a form where the coefficients are right-side multiplied by ξ, repeated use of the multiplication rule (3.32) is required in (3.29) (see Example 3.5).

Example 3.5 (Weak solution). Consider the parameter-varying differential equation (3.25). Then

$$R^{\dagger}(\xi) = r_0^{\top} - \xi r_1^{\top} + \xi^2 r_2^{\top} = \left(r_0^{\top} - \dot{r}_1^{\top} + \ddot{r}_2^{\top}\right) + \left(2\dot{r}_2^{\top} - r_1^{\top}\right)\xi + r_2^{\top}\xi,$$

$$R^{\dagger}(\xi)\diamond p = \begin{bmatrix} \mathrm{k_s} \\ -1 \end{bmatrix} + \begin{bmatrix} \frac{\mathrm{d}}{dt}p \\ 0 \end{bmatrix}\xi + \begin{bmatrix} p \\ 0 \end{bmatrix}\xi^2.$$

Choose a particular scheduling trajectory $p(t) = \cos(t)$ and a test function $\varphi(t) = \cos(t)$. Then

$$(R^{\dagger}(\frac{\mathrm{d}}{dt})\diamond p)(t)\cdot\varphi(t) = \begin{bmatrix} \mathrm{k_s}\cos(t) + \sin^2(t) - \cos^2(t) \\ -\cos(t) \end{bmatrix}, \tag{3.30}$$

which means that taking $w(t) = [\cos(t) \quad \mathrm{k_s}\cos(t) + \sin^2(t) - \cos^2(t)]^{\top}$ gives 0 for (3.28). It can be shown, that this holds for every φ, yielding that w is a weak solution. Substitution of w into (3.25) satisfies the differential equation with $p(t) = \cos(t)$ for all $t \in \mathbb{R}$. This implies that w is a strong solution (satisfies (3.25) for all $t \in \mathbb{R}$) of the varying-mass and spring system. However, taking $w(t) = [\frac{1}{\mathrm{k_s}\cos(t) - \cos^2(t) + \sin^2(t)} \quad \frac{1}{\cos(t)}]^{\top}$ also gives 0 for (3.28) with $\varphi(t) = \cos(t)$, but such a solution does not satisfies (3.25) for $\{k\cdot\pi\}_{k\in\mathbb{Z}}$ and as a result it can only be a weak solution. By considering other test functions, it can be shown that (3.28) is not satisfied in all cases, which proves that this choice of w is not a weak solution of (3.25). □

Due to its algebraic structure, it is possible to show that $\mathcal{R}[\xi]$ is a domain, i.e. for all $R_1, R_2 \in \mathcal{R}[\xi]$ it follows that

$$R_1(\xi)R_2 = 0(\xi) \quad \Rightarrow \quad R_1(\xi) = 0 \text{ or } R_2(\xi) = 0. \tag{3.31}$$

Since the indeterminant ξ is associated with $\frac{\mathrm{d}}{dt}$, multiplication with ξ on $\mathcal{R}[\xi]$ is non-commutative due to the chain rule of differentiation. Multiplication with ξ can be algebraically defined through the dot operator using the non-commutative rule:

$$\xi r = \dot{r} + r\xi, \tag{3.32}$$

where $r \in \mathcal{R}$ (see Example 3.3). Additionally, the ring $\mathcal{R}[\xi]$ is simple (i.e. the only ideals that are both right and left ideals are the trivial ones: 0 and $\mathcal{R}[\xi]$ itself) and it is a left and right principle domain (i.e. every left and right ideal can be generated by a single element). To show these properties, the argument similarly follows as in the case of polynomial rings with rational coefficient functions [64]. In fact this ring is even a right and left Euclidian domain, which means that there exist a right and left division with remainder [41]. Based on these properties and with the non-commutative multiplication rule (3.32), $\mathcal{R}[\xi]$ defines an Ore algebra [132].

Due to the fact that $\mathcal{R}[\xi]$ is a right and left Euclidean domain, there exist left and right division by remainder. This means, that if $R_1, R_2 \in \mathcal{R}[\xi]$ with $\mathrm{Deg}(R_1) \geq \mathrm{Deg}(R_2)$ and $R_2 \neq 0$, then there exist unique polynomials $R', R'' \in \mathcal{R}[\xi]$ such that

$$R_1(\xi) = R_2(\xi)R'(\xi) + R''(\xi), \tag{3.33}$$

where $\mathrm{Deg}(R_2) > \mathrm{Deg}(R'')$. Here we call R'' the right-remainder. Furthermore, as $\mathcal{R}[\xi]$ is simple, the rank of a matrix polynomial $R \in \mathcal{R}[\xi]^{n_\mathrm{r} \times n_\mathrm{W}}$ is well-defined [92]. Denote by $\mathrm{Span}^{\mathrm{row}}_{\mathcal{R}}(R)$ and $\mathrm{Span}^{\mathrm{col}}_{\mathcal{R}}(R)$ the subspace spanned by the rows (columns) of $R \in \mathcal{R}[\xi]^{\cdot \times \cdot}$, viewed as a linear space of polynomial vector functions with coefficients in $\mathcal{R}^{\cdot \times \cdot}$. Then

$$\mathrm{Rank}(R) = \mathrm{Dim}(\mathrm{Span}^{\mathrm{row}}_{\mathcal{R}}(R)) = \mathrm{Dim}(\mathrm{Span}^{\mathrm{col}}_{\mathcal{R}}(R)). \tag{3.34}$$

The notion of unimodular matrices, essential to characterize equivalent representations in Sec. 3.2, is also introduced:

Definition 3.9 (Unimodular polynomial matrix function). Let $M \in \mathcal{R}[\xi]^{n \times n}$. Then M is called unimodular, if there exists a $M^\dagger \in \mathcal{R}[\xi]^{n \times n}$, such that $M^\dagger(\xi)M(\xi) = I$ and $M(\xi)M^\dagger(\xi) = I$. □

Any unimodular matrix operator in $\mathcal{R}[\xi]^{\cdot \times \cdot}$ is equivalent to the product of finite many elementary row and column operations [41]:

1. Interchange row (column) i and row (column) j.
2. Multiply on the left (right) a row (column) i by a $r \in \mathcal{R}$, $r \neq 0$.
3. For $i \neq j$, add to row (column) i row (column) j multiplied by ξ^n, $n \geq 0$.
4. Replace the first element in each of two columns by their highest common left factor and 0 respectively.

Example 3.6 (Unimodular polynomial matrix function). The matrix polynomials $M, M^\dagger \in \mathcal{R}[\xi]^{2 \times 2}$, defined as

$$M(\xi) = \begin{bmatrix} r_2 & r_2\xi \\ r_1\xi & r_1\xi^2 + r_1 \end{bmatrix}, \quad M^\dagger(\xi) = \begin{bmatrix} r_1 + \xi^2 r_1 & -\xi r_2 \\ -\xi r_1 & r_2 \end{bmatrix} \frac{1}{r_1 r_2},$$

are unimodular as $M(\xi)M^\dagger(\xi) = M^\dagger(\xi)M(\xi) = I$. Note that $\xi r_1 \neq r_1 \xi$ due to the non-commutativity of the multiplication by ξ on $\mathcal{R}[\xi]$. □

Another important property of $\mathcal{R}[\xi]^{\cdot \times \cdot}$ is the existence of a Jacobson or so-called Smith form:

Theorem 3.1 (Jacobson form, [41]). *Let $R \in \mathcal{R}[\xi]^{n_\mathrm{r} \times n_\mathrm{W}}$ with $R \neq 0$ and $n =$* $\mathrm{Rank}(R)$. *Then there exist unimodular matrices $M_1 \in \mathcal{R}[\xi]^{n_\mathrm{r} \times n_\mathrm{r}}$ and $M_2 \in \mathcal{R}[\xi]^{n_\mathrm{W} \times n_\mathrm{W}}$ such that*

$$M_1(\xi)R(\xi)M_2(\xi) = \begin{bmatrix} Q(\xi) & 0 \\ 0 & 0 \end{bmatrix}, \tag{3.35}$$

where $Q = \mathrm{Diag}(1, \dots, 1, r) \in \mathcal{R}[\xi]^{n \times n}$ and $0 \neq r \in \mathcal{R}[\xi]$.

Due to the algebraic structure of $\mathcal{R}[\xi]^{\cdot\times\cdot}$, the proof of Th. 3.1 similarly follows as in [41, Sect. 8.1]. In terms of the Jacobson form (3.35) of a given $R(\xi)$, introduce $\mathrm{Deg}_*(R) := \mathrm{Deg}(r)$, describing the minimal degree of the polynomial $r(\xi)$ which generates $R(\xi)$.

Example 3.7 (Jacobson form, [239]). Consider

$$R(\xi) = \begin{bmatrix} r+\xi & -1 & -1 \\ 1-\frac{\mathrm{d}}{\mathrm{d}t}r & -\frac{1}{r}+\xi & -r \end{bmatrix} \in \mathcal{R}[\xi]^{2\times 3},$$

where r is a meromorphic function and $\xi = \frac{\mathrm{d}}{\mathrm{d}t}$. Then the Jacobson form of R is

$$M_1(\xi)R(\xi)M_2(\xi) = \begin{bmatrix} 1 & 0 & 0 \\ 0 & r-\frac{1}{r}+\xi & 0 \end{bmatrix},$$

with

$$M_1(\xi) = \begin{bmatrix} 1 & 0 \\ -r & 1 \end{bmatrix}, \quad M_2(\xi) = \begin{bmatrix} 0 & 0 & 1 \\ 0 & 1 & r \\ 1 & -1 & \xi \end{bmatrix}. \qquad \square$$

Now it is possible to show that there exists a duality between the solution spaces of PV differential equations and the polynomial modules in $\mathcal{R}[\xi]^{\cdot\times\cdot}$ associated with them, which is implied by a so-called *injective cogenerator* property. This property makes it possible to use the developed algebraic structure to characterize behaviors and manipulations on them. Originally the injective cogenerator property has been shown for the solution spaces of the polynomial ring over $\mathcal{R}_1$ in [239]. In Appendix A.1, this proof is extended to $\mathcal{R}[\xi]$.

3.1.2.4 Kernel Representations

As a next step in the foundation of the LPV behavioral approach, we develop the concept of KR representations of CT-LPV systems and investigate some relating properties. Based on the concept of weak solutions of a PV differential equation, we introduce continuous-time KR representations of LPV dynamic systems as follows:

Definition 3.10 (CT-KR-LPV representation). The parameter-varying differential equation (3.24) is called a continuous-time kernel representation, denoted by $\mathfrak{R}_{\mathrm{K}}(\mathcal{S})$, of the LPV dynamical system $\mathcal{S} = (\mathbb{R}, \mathbb{R}^{n_{\mathbb{P}}}, \mathbb{R}^{n_{\mathbb{W}}}, \mathfrak{B})$ with scheduling vector p and signals w, if

$$\mathfrak{B} = \left\{ (w,p) \in \mathcal{L}_1^{\mathrm{loc}}(\mathbb{R}, \mathbb{R}^{n_{\mathbb{W}}} \times \mathbb{R}^{n_{\mathbb{P}}}) \mid (R(\frac{\mathrm{d}}{\mathrm{d}t}) \diamond p)w = 0 \text{ holds weakly} \right\}. \qquad \square$$

It is obvious that the behavior $\mathfrak{B}$ associated with (3.24) always corresponds to a LPV system in terms of Def. 3.3. As stated previously, we only consider LPV systems with $\mathbb{T} = \mathbb{R}$ that have a KR representation, so existence of such a representation is explicitly assumed. It will be shown that this system class includes all LPV systems that can be described in the classical form of state-space and input-output models (see Sec. 1.3.2) that have been considered previously in the literature. It is also important that the allowed trajectories of p are not restricted by (3.24) (only those $p \in \mathcal{L}_1^{\mathrm{loc}}(\mathbb{R}, \mathbb{R}^{n_\mathbb{P}})$ are excluded from $\mathfrak{B}$ for which a coefficient $r_i \diamond p$ is unbounded). This is in accordance with the classical concept of p being an external variable of the system. One can also include further restrictions on $\mathfrak{B}_\mathbb{P} = \pi_p \mathfrak{B}$, like description of the admissible scheduling trajectories as solutions of a differential equation, etc. However, to preserve the generality of the developed framework, we do not consider the latter case.

Based on the concept of rank, the following theorem can be introduced:

Theorem 3.2 (Existence of full row rank KR representation). *Let $\mathfrak{B}$ be given with a KR representation (3.24). Then $\mathfrak{B}$ can also be represented by a $R' \in \mathcal{R}[\xi]^{\cdot \times n_\mathbb{W}}$ with full row rank.*

The proof of this theorem is given in Appendix A.1. The significance of Th. 3.2 is that it establishes the concept of minimality for KR representations (see Chap. 3.2).

3.1.2.5 Input-Output Representations

Another key representation form is the IO representation, which we define from the behavioral point of view in this subsection. Before establishing our definition, we need the concept of IO partition for LPV systems.

In many applications, in particular in control, it is necessary to group the signals of dynamical systems into sets of *input signals* $u : \mathbb{R} \to \mathbb{U}$ and *output signals* $y : \mathbb{R} \to \mathbb{Y}$. This is done to distinguish which of them we would like to (or can) actuate as inputs in order to drive the remaining so-called output signals to a desired trajectory of $\mathfrak{B}$. The definition of such a IO partition of $\mathcal{S}$, which is commonly non-unique, is as follows:

Definition 3.11 (IO partition of a LPV system). Let $\mathcal{S} = (\mathbb{T}, \mathbb{R}^{n_\mathbb{P}}, \mathbb{R}^{n_\mathbb{W}}, \mathfrak{B})$ be an LPV system. The partition of the signal space as $\mathbb{R}^{n_\mathbb{W}} = \mathbb{U} \times \mathbb{Y} = \mathbb{R}^{n_\mathbb{U}} \times \mathbb{R}^{n_\mathbb{Y}}$ and partition of $w \in \mathcal{L}_1^{\mathrm{loc}}(\mathbb{T}, \mathbb{R}^{n_\mathbb{W}})$ correspondingly with $u \in \mathcal{L}_1^{\mathrm{loc}}(\mathbb{T}, \mathbb{U})$ and $y \in \mathcal{L}_1^{\mathrm{loc}}(\mathbb{T}, \mathbb{Y})$ is called an IO partition of $\mathcal{S}$, if

1. u is free, i.e. for all $u \in \mathcal{L}_1^{\mathrm{loc}}(\mathbb{T}, \mathbb{U})$ and $p \in \mathfrak{B}_\mathbb{P}$, there exists a $y \in \mathcal{L}_1^{\mathrm{loc}}(\mathbb{T}, \mathbb{Y})$ such that $(\mathrm{Col}(u,y), p) \in \mathfrak{B}$.
2. y does not contain any further free component, i.e. given u, none of the components of y can be chosen freely for every $p \in \mathfrak{B}_\mathbb{P}$ (maximally free). □

An IO partition implies the existence of matrix-polynomial functions $R_\mathrm{y} \in \mathcal{R}[\xi]^{n_\mathbb{Y} \times n_\mathbb{Y}}$ and $R_\mathrm{u} \in \mathcal{R}[\xi]^{n_\mathbb{Y} \times n_\mathbb{U}}$ with R_y full row rank, such that (3.24) can be written as

$$(R_\mathrm{y}(\frac{\mathrm{d}}{\mathrm{d}t}) \diamond p) y = (R_\mathrm{u}(\frac{\mathrm{d}}{\mathrm{d}t}) \diamond p) u, \tag{3.36}$$

with $n_\mathbb{W} = n_\mathbb{U} + n_\mathbb{Y}$. The corresponding behavior $\mathfrak{B}$ is given by

$$\left\{(u,y,p)\in\mathcal{L}_1^{\mathrm{loc}}(\mathbb{R},\mathbb{U}\times\mathbb{Y}\times\mathbb{P})\mid(R_{\mathrm{y}}(\tfrac{\mathrm{d}}{\mathrm{d}t})\diamond p)y=(R_{\mathrm{u}}(\tfrac{\mathrm{d}}{\mathrm{d}t})\diamond p)u \text{ holds weakly}\right\},$$

with $\mathbb{U}=\mathbb{R}^{n_\mathbb{U}}$ and $\mathbb{Y}=\mathbb{R}^{n_\mathbb{Y}}$. For those scheduling trajectories p, for which the maximum freedom of the input signal u holds in $\mathfrak{B}_p$, an IO partition defines a causal mapping in case the solutions of (3.36) are restricted to have left compact support. Otherwise, initial conditions also matter. LPV systems with no IO partition are called autonomous. Note that for some systems it is possible that the freedom of components of w can change for specific scheduling trajectories. In other words, it can happen that some of the outputs become free for specific scheduling trajectories but not for all. In this case, the autonomous part of the behavior is related to the scheduling dependent nature of the system. In case $n_\mathbb{U}=n_\mathbb{Y}=1$, systems are referred to as *single-input single-output* (SISO), while systems with $n_\mathbb{U}>1, n_\mathbb{Y}>1$ are called *multiple-input multiple-output* (MIMO) systems.

Now it is possible to introduce IO representations of CT-LPV systems:

Definition 3.12 (CT-LPV-IO representation). The continuous-time IO representation of $\mathcal{S}=(\mathbb{R},\mathbb{P}\subseteq\mathbb{R}^{n_\mathbb{P}},\mathbb{R}^{n_\mathbb{U}+n_\mathbb{Y}},\mathfrak{B})$ with scheduling vector p and IO partition (u,y) is denoted by $\mathfrak{R}_{\mathrm{IO}}(\mathcal{S})$ and defined as a parameter-varying differential-equation system with order n_a:

$$\sum_{i=0}^{n_\mathrm{a}}(a_i\diamond p)\frac{\mathrm{d}^i}{\mathrm{d}t^i}y=\sum_{j=0}^{n_\mathrm{b}}(b_j\diamond p)\frac{d^j}{dt^j}u, \tag{3.37}$$

where $a_j\in\mathcal{R}^{n_\mathbb{Y}\times n_\mathbb{Y}}$ and $b_j\in\mathcal{R}^{n_\mathbb{Y}\times n_\mathbb{U}}$ with $a_{n_\mathrm{a}}\neq 0$ and $b_{n_\mathrm{b}}\neq 0$ are the meromorphic parameter-varying coefficients of the matrix polynomials $R_\mathrm{u}(\xi)=\sum_{j=0}^{n_\mathrm{b}}b_j\xi^j$ and full rank $R_\mathrm{y}(\xi)=\sum_{i=0}^{n_\mathrm{a}}a_i\xi^i$. As u is maximally free, such polynomials exist with $n_\mathrm{a}\geq n_\mathrm{b}\geq 0$ and $n_\mathrm{a}>0$. □

In terms of Th. 3.2, any KR representation has a full row-rank equivalent, thus the existence of IO representations is guaranteed for any valid IO partition. It is also apparent that (3.37) is the continuous-time "dynamic dependent" counterpart of (1.5a).

Example 3.8 (IO partition and representation). In Example 3.1, the force w_F is a free variable as it represents the inhomogeneous part of (3.7). Thus, the choice of $w=[y\ \ u]^\top=[w_\mathrm{F}\ \ w_\mathrm{x}]^\top$ yields a valid IO partition. With m chosen as the scheduling signal p the PV behavior can be represented in the form of (3.36) with polynomials

$$R_\mathrm{y}(\xi)=a_0+a_1\xi+a_2\xi^2,\quad R_\mathrm{u}(\xi)=b_0, \tag{3.38}$$

which have coefficients

$$a_0\diamond p=\mathrm{k_s},\quad a_1\diamond p=\frac{\mathrm{d}}{\mathrm{d}t}p,\quad a_2\diamond p=p,\quad b_0\diamond p=1.$$

Obviously, R_y has full rank. This implies that $R_\mathrm{y}(\xi)$ and $R_\mathrm{u}(\xi)$ define an LPV-IO representation of the system. □

In case of continuous-time LPV systems, the notion of transfer function or frequency response has no meaningful[2] interpretation. By using the approximative transfer function calculus of LTV systems [117], some interpretation of these notions can be given for LPV systems. However, these concepts do not satisfy the relations of the LTI case. On the other hand, each element of the frozen system set $\mathscr{F}_S$, is an LTI system. Therefore, each $\mathcal{F}_{\mathrm{p}} \in \mathscr{F}_S$ has a transfer function $F_{\mathrm{p}}(s)$, a frequency response $F_{\mathrm{p}}(\mathrm{i}\omega)$, and an impulse response $h_{\mathrm{p}}(t)$ with Markov parameters $\{\mathrm{g}_{i|\mathrm{p}}\}_{i=0}^{\infty}$. The notion of frozen transfer function set or impulse response set can be established for any LPV system. The set of frozen poles or zeros follows similarly.

3.1.2.6 State-Space Representations

Besides partitioning the signals of the system in an IO sense, we often need to introduce additional variables in most modeling exercises to express more conveniently the relationships of those signals we are particulary interested in. This is motivated by the first principle laws of physics where we can often meet physically non-existing (virtual) variables like the potential field in the well-known Maxwell's equations. It also can happen that we are just simply not interested in some "inner" variables of the system, like in the voltage drop on a resistor of a space shuttle, if we would like to control its motion around the planet. We call these other, auxiliary variables *latent variables*. The introduction of latent variables and their associated representations are essential for LPV systems as they give the mathematical concept of signals that corresponds to inner variables or states of the system. In the following we extend the definition of latent variables and the property of state to the LPV case. Then we introduce the state-kernel form, proving that all latent variable representations have a first order PV differential equation form. The latter property is used to define SS representations of LPV systems.

If the continuous-time LPV system contains n_{L} latent (eliminatable) and n_{W} manifest (non-latent) variables, then as a generalization of (3.24):

$$(R_{\mathrm{W}}(\frac{\mathrm{d}}{\mathrm{d}t}) \diamond p)w = (R_{\mathrm{L}}(\frac{\mathrm{d}}{\mathrm{d}t}) \diamond p)w_{\mathrm{L}} \tag{3.39}$$

holds, where $w : \mathbb{R} \to \mathbb{R}^{n_{\mathrm{W}}}$ is the manifest variable, $w_{\mathrm{L}} : \mathbb{R} \to \mathbb{R}^{n_{\mathrm{L}}}$ is the latent variable, $R_{\mathrm{W}} \in \mathcal{R}[\xi]^{n_{\mathrm{r}} \times n_{\mathrm{W}}}$ and $R_{\mathrm{L}} \in \mathcal{R}[\xi]^{n_{\mathrm{r}} \times n_{\mathrm{L}}}$ are polynomial matrices with meromorphic coefficients. The set of equations (3.39) is called a *latent variable representation* of the LPV *latent variable system* $(\mathbb{Z}, \mathbb{R}^{n_{\mathrm{P}}}, \mathbb{R}^{n_{\mathrm{W}}} \times \mathbb{R}^{n_{\mathrm{L}}}, \mathfrak{B}_{\mathrm{L}})$, where the so-called *full behavior* $\mathfrak{B}_{\mathrm{L}}$ of this system is defined as

$$\mathfrak{B}_{\mathrm{L}} = \left\{ (w, w_{\mathrm{L}}, p) \in \mathcal{L}_1^{\mathrm{loc}}(\mathbb{R}, \mathbb{R}^{n_{\mathrm{W}}} \times \mathbb{R}^{n_{\mathrm{L}}} \times \mathbb{R}^{n_{\mathrm{P}}}) \mid (3.39) \text{ holds weakly} \right\}.$$

Additionally, $\mathfrak{B} = \pi_{(w,p)} \mathfrak{B}_{\mathrm{L}}$ is introduced as the *manifest behavior* associated with $\mathfrak{B}_{\mathrm{L}}$:

[2] Some authors [225, 135, 125] introduce LPV transfer functions with varying parameters. As they commonly refer only to the collection of transfer functions associated with $\mathscr{F}_S$, this notion of the LPV transfer function is misleading.

$$\mathfrak{B} = \left\{ (w,p) \in \mathcal{L}_1^{\mathrm{loc}}(\mathbb{R}, \mathbb{R}^{n_\mathrm{W}} \times \mathbb{R}^{n_\mathbb{P}}) \mid \exists w_\mathrm{L} \in \mathcal{L}_1^{\mathrm{loc}}(\mathbb{R}, \mathbb{R}^{n_\mathrm{L}}) \text{ s.t. } (w, w_\mathrm{L}, p) \in \mathfrak{B}_\mathrm{L} \right\}.$$

Elimination of latent variables is always possible on $\mathcal{R}[\xi]^{\cdot \times \cdot}$.

Theorem 3.3 (Elimination property). *Given a LPV latent variable system* $(\mathbb{T}, \mathbb{R}^{n_\mathbb{P}}, \mathbb{R}^{n_\mathrm{W}} \times \mathbb{R}^{n_\mathrm{L}}, \mathfrak{B}_\mathrm{L})$ *with a signal variable* w*, a latent variable* w_L*, and scheduling variable* p*, then there exists a* $R' \in \mathcal{R}[\xi]^{\cdot \times n_\mathrm{W}}$ *which defines a LPV-KR representation of* $\mathfrak{B} = \pi_{(w,p)}\mathfrak{B}_\mathrm{L}$.

For a proof see Appendix A.1. Note that elimination of latent variables in CT, can result in a loss of smoothness constraints of w which have been imposed by the latent form (3.39) and can not be represented by the resulting KR representation (in detail see [145]). The most natural way to avoid such problems is to drop these constraints and consider the closure of $\pi_{(w,p)}\mathfrak{B}_\mathrm{L}$ in the topology of $\mathcal{L}_1^{\mathrm{loc}}$, i.e. $\overline{\pi_{(w,p)}\mathfrak{B}_\mathrm{L}}$. This allows both to keep $\mathcal{L}_1^{\mathrm{loc}}$ as the solution space and the proper elimination of latent variables. According to this, in Th. 3.3 the LPV-KR representation has an equal behavior with $\overline{\pi_{(w,p)}\mathfrak{B}_\mathrm{L}}$. Now it is possible to define the concept of state for LPV systems.

Definition 3.13 (Property of state for LPV systems). Let $\mathcal{S} = (\mathbb{T}, \mathbb{R}^{n_\mathbb{P}}, \mathbb{R}^{n_\mathrm{W}} \times \mathbb{R}^{n_\mathrm{L}}, \mathfrak{B}_\mathrm{L})$ be a LPV latent variable system with a latent variable w_L. Then w_L is a state if for every $t_0 \in \mathbb{T}$ and $(w_1, w_{\mathrm{L},1}, p), (w_2, w_{\mathrm{L},2}, p) \in \mathfrak{B}_\mathrm{L}$ with $w_{\mathrm{L},1}(t_0) = w_{\mathrm{L},2}(t_0)$ and in case of $\mathbb{T} = \mathbb{R}$, with $w_{\mathrm{L},1}$ and $w_{\mathrm{L},2}$ continuous on $\mathbb{R}$, it follows that the concatenation of these signals at t_0 satisfies

$$(w_1, w_{\mathrm{L},1}, p) \underset{t_0}{\wedge} (w_2, w_{\mathrm{L},2}, p) \in \mathfrak{B}_\mathrm{L}. \tag{3.40}$$

Then $\mathfrak{B}_\mathrm{L}$ is called a state-space behavior, and the latent variable w_L is called the state. □

In terms of Def. 3.13, w_L needs to qualify as a state for each scheduling trajectory of $\mathfrak{B}_\mathbb{P}$.

Example 3.9 (Latent variable representation). By considering the system in Example 3.1 with scheduling $p = m$ and $\mathbb{P} = [1,2]$, the following latent variable representation of the model has the same manifest behavior:

$$\begin{bmatrix} -k_\mathrm{s} & 1 \\ 0 & 0 \\ 1 & 0 \end{bmatrix} \begin{bmatrix} w_\mathrm{x} \\ w_\mathrm{F} \end{bmatrix} = \begin{bmatrix} \frac{\mathrm{d}}{dt} & 0 \\ -\frac{1}{p} & \frac{\mathrm{d}}{dt} \\ 0 & 1 \end{bmatrix} w_\mathrm{L}. \tag{3.41}$$

This can be proved by substituting the third row of (3.41) into the second row, giving

$$w_{\mathrm{L},1} = p \frac{\mathrm{d}}{\mathrm{d}t} w_\mathrm{F}. \tag{3.42}$$

Substitution of (3.42) into the first row of (3.41) gives a PV *differential equation* in the variables w_x and w_F, which is equal to (3.25). Any $w_\mathrm{L} \in \mathcal{L}_1^{\mathrm{loc}}(\mathbb{R}, \mathbb{R}^2)$ satisfying weakly the previous equations, trivially fulfills the property of state. □

To decide whether a latent variable is a state, the following theorem is important:

Theorem 3.4 (State-kernel form). *The latent variable w_{L} is a state, iff there exist matrices $r_{\mathrm{w}} \in \mathcal{R}^{n_{\mathrm{r}} \times n_{\mathrm{W}}}$ and $r_0, r_1 \in \mathcal{R}^{n_{\mathrm{r}} \times n_{\mathrm{L}}}$ such that the full behavior $\mathfrak{B}_{\mathrm{L}}$ has the kernel representation:*

$$r_{\mathrm{w}} w + r_0 w_{\mathrm{L}} + r_1 \xi w_{\mathrm{L}} = 0. \tag{3.43}$$

The proof is given in Appendix A.1. In this way, a CT-LPV *state-space* (SS) behavior is defined by a first-order meromorphic-coefficient differential equation. Now we can give the definition of SS representations of a continuous-time $\mathcal{S}$:

Definition 3.14 (CT-LPV-SS representation). The continuous-time state-space representation of $\mathcal{S} = (\mathbb{R}, \mathbb{P} \subseteq \mathbb{R}^{n_{\mathbb{P}}}, \mathbb{R}^{n_{\mathbb{U}}+n_{\mathbb{Y}}}, \mathfrak{B})$ with scheduling vector p is denoted by $\mathfrak{R}_{\mathrm{SS}}(\mathcal{S})$ and defined as a first-order parameter-varying differential equation system in the latent variable $x : \mathbb{R} \to \mathbb{X}$:

$$\frac{\mathrm{d}}{\mathrm{d}t} x = (A \diamond p)x + (B \diamond p)u, \tag{3.44a}$$

$$y = (C \diamond p)x + (D \diamond p)u, \tag{3.44b}$$

where (u, y) is the IO partition of $\mathcal{S}$, x is the state-vector, $\mathbb{X} = \mathbb{R}^{n_{\mathbb{X}}}$ is the state-space,

$$\mathfrak{B}_{\mathrm{SS}} = \left\{ (u, x, y, p) \in \mathcal{L}_1^{\mathrm{loc}}(\mathbb{R}, \mathbb{U} \times \mathbb{X} \times \mathbb{Y} \times \mathbb{P}) \mid \text{(3.44a–b) hold weakly} \right\},$$

is the full behavior of (3.44a–b), $\mathfrak{B}$ is equal to the closure of the manifest behavior of (3.44a–b), i.e. $\mathfrak{B} = \overline{\pi_{(u,y,p)} \mathfrak{B}_{\mathrm{SS}}}$, and

$$\left[\begin{array}{c|c} A & B \\ \hline C & D \end{array}\right] \in \left[\begin{array}{c|c} \mathcal{R}^{n_{\mathbb{X}} \times n_{\mathbb{X}}} & \mathcal{R}^{n_{\mathbb{X}} \times n_{\mathbb{U}}} \\ \hline \mathcal{R}^{n_{\mathbb{Y}} \times n_{\mathbb{X}}} & \mathcal{R}^{n_{\mathbb{Y}} \times n_{\mathbb{U}}} \end{array}\right].$$

□

Example 3.10 (SS representation). Continuing Example 3.9, the LPV state-space representation of the model follows by taking $[y \;\; u]^\top = [w_{\mathrm{x}} \;\; w_{\mathrm{F}}]^\top$ as the IO partition and $x = w_{\mathrm{L}}$ as the state:

$$\frac{\mathrm{d}}{\mathrm{d}t} x = \begin{bmatrix} 0 & 0 \\ \frac{1}{p} & 0 \end{bmatrix} x + \begin{bmatrix} -k_{\mathrm{s}} & 1 \\ 0 & 0 \end{bmatrix} \begin{bmatrix} y \\ u \end{bmatrix}, \tag{3.45a}$$

$$y = \begin{bmatrix} 0 & 1 \end{bmatrix} x. \tag{3.45b}$$

By substitution of (3.45b) into (3.45a), the state equation (3.44a) results, while (3.45b) gives the output equation (3.44b). Thus the corresponding SS representation is

$$\left[\begin{array}{c|c} A \diamond p & B \diamond p \\ \hline C \diamond p & D \diamond p \end{array}\right] = \left[\begin{array}{cc|c} 0 & -k_{\mathrm{s}} & 1 \\ \frac{1}{p} & 0 & 0 \\ \hline 0 & 1 & 0 \end{array}\right]. \tag{3.46}$$

□

Note that in the full behavior $\mathfrak{B}_{\mathrm{SS}}$, the latent variable x trivially fulfills the state property in terms of Th. 3.4. Additionally, this behavioral type of definition of CT-SS representations includes the definition of LPV state-space models used in the state-of-the-art of LPV control. In those models, the dependence of the matrix functions are assumed to be rational and static, which is trivially included in $\mathcal{R}$. Similar to LPV-IO representations the notions of transfer function, frequency response, and impulse response can only be defined in a frozen sense for LPV-SS representations.

3.1.3 Representations of Discrete-Time LPV Systems

The concept of *discrete-time* (DT) parameter-varying dynamical systems is also important for engineering applications. In the following we formulate the behavioral framework for DT-LPV systems. First we investigate the concept of DT parameter-varying systems and then we define their KR, SS and IO representations. To do so, we use the previously developed ring of polynomials $\mathcal{R}[\xi]$ with meromorphic coefficient functions. In the DT case, the time operator q is substituted as the indeterminate of these polynomials which results in a different non-commutative multiplication rule in $\mathcal{R}[\xi]$ than in CT case, but the algebraic structure remains the same. This property provides that the previously developed theories extend to the DT case.

3.1.3.1 The Discrete-Time Parameter-Varying Concept

In discrete-time, the time-axis is restricted to $\mathbb{T} = \mathbb{Z}$. Signals on this axis can be viewed (but not necessarily) as observations of continuous-time signal trajectories at equidistant time points. This concept is called periodic, equidistant sampling that defines the DT projection of a CT signal $w : \mathbb{R} \to \mathbb{W}$ as $w' : \mathbb{Z} \to \mathbb{W}$ satisfying

$$w'(k) = w(k\mathrm{T_d}), \quad \forall k \in \mathbb{Z}, \tag{3.47}$$

with discretization-step $\mathrm{T_d} \in \mathbb{R}^+$ where $\mathbb{R}^+$ is the set of positive real numbers. However in the parameter-varying case, the scheduling signal is also restricted to the DT time-axis as its observations are only available at the sampling instants. Thus, the DT projection of a CT scheduling signal $p : \mathbb{R} \to \mathbb{P}$ is defined as $p' : \mathbb{Z} \to \mathbb{P}$ satisfying

$$p'(k) = p(k\mathrm{T_d}), \quad \forall k \in \mathbb{Z}. \tag{3.48}$$

In this way, we call $\mathcal{S}' = (\mathbb{Z}, \mathbb{W}, \mathbb{P}, \mathfrak{B}')$ the DT equivalent of $\mathcal{S} = (\mathbb{R}, \mathbb{W}, \mathbb{P}, \mathfrak{B})$ under the sampling time $\mathrm{T_d}$ if

$$\mathfrak{B}' = \left\{ (w', p') \in (\mathbb{W} \times \mathbb{P})^{\mathbb{Z}} \mid \exists (w, p) \in \mathfrak{B} \text{ s.t. (3.47) \& (3.48) hold} \right\}. \tag{3.49}$$

Note that for arbitrary $\mathrm{T_d}$ it is not guaranteed that there exists a DT-LPV system such that (3.49) is satisfied. On the other hand, not every DT-LPV system is equivalent with a sampled CT-LPV system. Thus sampling provides only a particular viewpoint for understanding DT systems and it must be emphasized that DT-LPV systems are a stand alone mathematical concept of modeling just like in the LTI case. Additionally, the time projection is defined on the signals and not on their derivatives. This means that finding the DT equivalent of a $\mathcal{S} = (\mathbb{R}, \mathbb{W}, \mathbb{P}, \mathfrak{B})$, described by a differential equation with meromorphic coefficients dependent on p and its derivatives, is a non-trivial problem (see Sect. 6.2).

3.1.3.2 Polynomials over $\mathcal{R}$

In the following, the DT analog of the concepts introduced in the CT-LPV case is developed. Similar to the LTI case, the time operator that we use in the discrete-time PV case to define *difference equations* is the forward-time shift operator q. Such difference equations are used to describe the behavior of DT-LPV systems. However, the operator q has different properties than $\frac{\mathrm{d}}{\mathrm{d}t}$, used in the CT case. Additionally, q being the time operator also implies that the coefficients in the DT case are dependent on p and its time-shifted versions. Due to these differences, in the following we reformulate concepts of coefficient dependence (association of coefficient variables with the scheduling) and commutation rules of multiplication in $\mathcal{R}[\xi]$ in order to define the analog of the concepts of the CT case.

As a first step, we define the variable association of the coefficient functions $r \in \mathcal{R}$ with elements of p and their forward and backward time-shifts. Similar to the CT case, such a dynamic dependence is needed to establish equivalence transformations (state elimination/construction) between discrete-time SS and IO representations (see Chap. 4 and [189, 202] for the details). Contrary to the CT case, in DT the coefficient functions are required to depend on both forward and backward time-shifts of p. Thus, labeling of the variables must contain positive and negative indexes as well.

For a fixed scheduling dimension $n_{\mathbb{P}}$, denote the variables of a $r \in \bar{\mathcal{R}}_n$ (n-dimensional function in $\mathcal{R}$) as:

$$r\left(\{\zeta_{ij}\}_{n_\zeta,n_{\mathbb{P}},\tau}\right) := r\left(\zeta_{0,1},\ldots,\zeta_{0,n_{\mathbb{P}}},\zeta_{1,1},\ldots,\zeta_{1,n_{\mathbb{P}}},\zeta_{-1,1},\ldots,\zeta_{-1,n_{\mathbb{P}}},\zeta_{2,1},\ldots,\zeta_{\tau_1,\tau_2}\right)$$

where $n_\zeta = 0$ and $\tau = n$ if $n \leq n_{\mathbb{P}}$ otherwise $n = (2n_\zeta - 1)n_{\mathbb{P}} + \tau$ with $n_\zeta > 0$ and $0 < \tau \leq 2n_{\mathbb{P}}$ and

$$(\tau_1, \tau_2) = \begin{cases} (n_\zeta, \tau) & \text{if } \tau \leq n_{\mathbb{P}}; \\ (-n_\zeta, \tau - n_{\mathbb{P}}) & \text{if } \tau > n_{\mathbb{P}}. \end{cases} \tag{3.50}$$

The basic mechanism of this variable labeling scheme is presented in Fig. 3.3 for $n_{\mathbb{P}} = 2$ and $n = 1,\ldots,6$. In this figure in each row, the yellow dots represent the labels of the variables for a n-variable coefficient function. Note that for all finite dimensions, this labeling sequence is unique.

Now we can associate the variables $\{\zeta_{ij}\}_{n_{\mathbb{P}},n_\zeta,\tau}$ of a $r \in \mathcal{R}$ as

$$\zeta_{ij} = \mathrm{q}^i p_j, \tag{3.51}$$

where p_j is the jth element of p, $n_{\mathbb{P}}$ is the dimension of p, $n_\zeta \in \mathbb{N}$ is the maximal order of the shifted versions of p on which r is dependent. For this association, we define the operator $\diamond : (\mathcal{R}, \mathfrak{B}_{\mathbb{P}}) \to (\mathbb{R}^{\mathbb{Z}})$ in DT as:

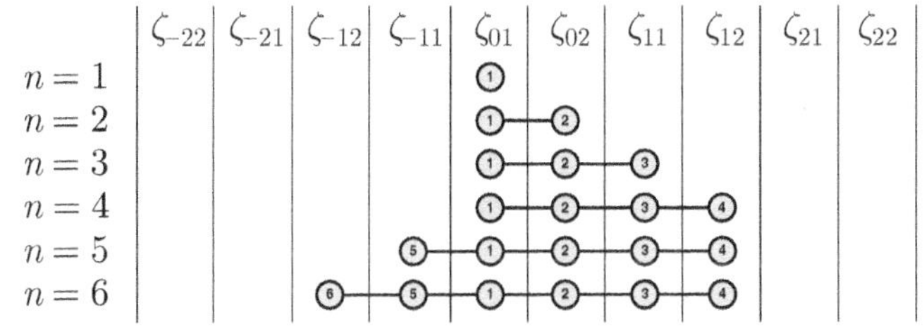

Fig. 3.3 Labeling scheme for the variables of meromorphic coefficient functions $r : \bar{\mathcal{R}}_n$ for $n_\mathbb{P} = 2$ and $n = 1, \ldots, 6$.

$$r \diamond p = r\left(\{\mathrm{q}^i p_j\}_{n_\varsigma, n_\mathbb{P}, \tau}\right) = r(p, \mathrm{q}p, \mathrm{q}^{-1}p, \ldots). \tag{3.52}$$

This operator evaluates a given coefficient function $r \in \mathcal{R}$ for a given scheduling trajectory $p \in \mathfrak{B}_\mathbb{P}$ according to the association rule (3.51). For an example of this variable association and the associated coefficient dependence see Example 3.11.

Example 3.11 (Coefficient function). Let $\mathbb{P} = \mathbb{R}^{n_\mathbb{P}}$ with $n_\mathbb{P} = 2$. Consider the real-meromorphic coefficient function $r : \mathbb{R}^3 \to \mathbb{R}$ defined as

$$r(\mathrm{x}_1, \mathrm{x}_2, \mathrm{x}_3) = \frac{1 + \mathrm{x}_3}{1 - \mathrm{x}_2}.$$

Then for a scheduling trajectory $p : \mathbb{R} \to \mathbb{R}^2$:

$$(r \diamond p)(k) = r(p_1, p_2, \mathrm{q}p_1)(k) = \frac{1 + p_1(k+1)}{1 - p_2(k)}.$$

On the other hand, if $n_\mathbb{P} = 1$, then

$$(r \diamond p)(k) = r(p_1, \mathrm{q}p_1, \mathrm{q}^{-1}p_1)(k) = \frac{1 + p_1(k-1)}{1 - p_1(k+1)}.$$ □

With the association rule (3.51) and evaluation operator (3.52) in mind, we can use polynomial matrices in $\mathcal{R}[\xi]^{n_\mathrm{r} \times n_\mathbb{W}}$ to define a parameter-varying difference equation:

Definition 3.15 (PV difference equation). Consider $R(\xi) = \sum_{i=0}^{n_\xi} r_i \xi^i \in \mathcal{R}[\xi]^{n_\mathrm{r} \times n_\mathbb{W}}$ and $(w, p) \in (\mathbb{R}^{n_\mathbb{W}} \times \mathbb{R}^{n_\mathbb{P}})^\mathbb{Z}$.

$$(R(\mathrm{q}) \diamond p)w = \sum_{i=0}^{n_\xi} (r_i \diamond p)\mathrm{q}^i w = 0, \tag{3.53}$$

is called a PV difference equation with order $n_\xi = \mathrm{Deg}(R)$. □

In discrete-time all trajectories in $(\mathbb{W} \times \mathbb{P})^\mathbb{Z}$ that satisfy (3.53) are considered as solutions. PV differential equations in the form (3.53) are used to define the class of

DT-LPV systems we consider in the sequel. It will be shown that this class contains all popular definitions of LPV-SS and IO models used in LPV system identification.

Example 3.12 (PV difference equation). Consider again Example 3.1. Let $0 < \mathrm{T_d} \ll 1$ and develop an approximation of the CT behavior of this system through the Euler (forward) type of approximation:

$$\frac{\mathrm{d}}{\mathrm{d}t} w \approx \frac{\mathrm{q}w' - w'}{\mathrm{T_d}}, \tag{3.54}$$

where $w'(k) = w(k\mathrm{T_d})$ for all $k \in \mathbb{Z}$. Repeated substitution of (3.54) for the derivatives of m, w_x, and w_F in (3.7) yields[3]

$$(\mathrm{T_d^2 k_s} + m(k))w_\mathrm{x}(k) - (m(k+1) + m(k))w_\mathrm{x}(k+1) \\ + m(k+1)w_\mathrm{x}(k+2) = \mathrm{T_d^2} w_\mathrm{F}(k), \tag{3.55}$$

where the time index k denotes the values of the signals at $k\mathrm{T_d}$ on $\mathbb{R}$. We consider these signals as DT signals in the following. Let $p = m$ with a scheduling space $\mathbb{P} = [1,2]$ and let $w = [\, w_\mathrm{x} \;\; w_\mathrm{F} \,]^\top$. Then the difference equation (3.55), which defines the possible signal trajectories of the DT approximation of the mass-spring system, can be written in the form of (3.53) with $n_\mathbb{W} = 2$, $n_\xi = 1$, $n_\mathbb{P} = 1$:

$$(R(q) \diamond p)w = (r_0 \diamond p)w + (r_1 \diamond p)\mathrm{q}w + (r_2 \diamond p)\mathrm{q}^2 w = 0 \tag{3.56}$$

where

$$r_0 \diamond p = \left[\, \mathrm{T_d^2 k_s} + p \;\; -\mathrm{T_d^2} \,\right], \quad r_1 \diamond p = \left[\, -\mathrm{q}p - p \;\; 0 \,\right], \quad r_2 \diamond p = \left[\, \mathrm{q}p \;\; 0 \,\right]. \quad \square$$

So far we have used the polynomials of the ring $\mathcal{R}[\xi]$ to define parameter-varying difference equations. Thus, by using the algebraic structure of this ring we can develop the analog of the CT results using similar arguments. However, an important difference with the CT case is that multiplication in $\mathcal{R}[\xi]$ obeys a different non-commutative rule in the DT case. To explore the non-commutative multiplication rule by $\xi = \mathrm{q}$, first introduce the *shift* operators on $\mathcal{R}$, to describe time-shifts of parameter-varying coefficient functions:

Definition 3.16 (Shift operators). Let $r \in \bar{\mathcal{R}}_n$. For a given scheduling dimension $n_\mathbb{P}$, denote the variables of r as $\{\zeta_{ij}\}_{n_\zeta, n_\mathbb{P}, \tau}$ based on the previously introduced labeling. The forward-shift and backward-shift operators on $\mathcal{R}$ are defined as

$$\overrightarrow{r} = \mho_*(r \circ m_1), \tag{3.57a}$$

$$\overleftarrow{r} = \mho_*(r \circ m_2), \tag{3.57b}$$

where $\circ$ denotes the function concatenation, $m_1, m_2 \in \mathcal{R}^n_{n+2n_\mathbb{P}}$, and m_1 assigns each variable ζ_{ij} to $\zeta_{(i+1)j}$, while m_2 assigns each variable ζ_{ij} to $\zeta_{(i-1)j}$ as depicted in Fig. 3.4. $\square$

[3] Note that applying the Euler derivative approximation on different representations (SS, IO, etc.) can result in inequivalent DT descriptions.

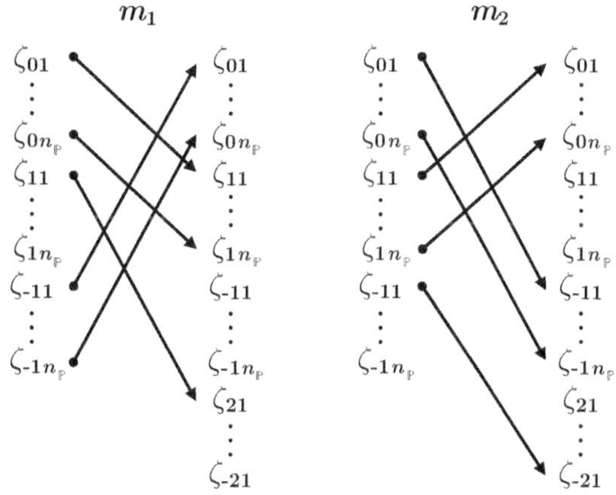

Fig. 3.4 Variable assignment by the functions m_1 and m_2 in Def. 3.16.

In other words, if $r \diamond p$ is dependent on for example p and $\mathrm{q}p$, then $\overrightarrow{r}$ is the "same" function (disregarding the number of variables) except it is now dependent on $\mathrm{q}p$ and $\mathrm{q}^2 p$. Similarly, $\overleftarrow{r}$ is also the "same" as r except it is now dependent on $\mathrm{q}^{-1}p$ and p. In this way, the shift operators describe time-shifts on the parameter-varying coefficient functions just like the dot operator describe differentiation in CT (see Example 3.13). The shift operators fulfill the following rules for any $r_1, r_2 \in \mathcal{R}$:

$$\begin{cases} \text{if } r = r_1 \pm r_2 & \text{then } \overrightarrow{r} = \overrightarrow{r_1} \pm \overrightarrow{r_2} \text{ and } \overleftarrow{r} = \overleftarrow{r_1} \pm \overleftarrow{r_2}, \\ \text{if } r = r_1 r_2 & \text{then } \overrightarrow{r} = \overrightarrow{r_1}\,\overrightarrow{r_2} \text{ and } \overleftarrow{r} = \overleftarrow{r_1}\,\overleftarrow{r_2}, \\ \text{if } r = \frac{r_1}{r_2} & \text{then } \overrightarrow{r} = \frac{\overrightarrow{r_1}}{\overrightarrow{r_2}} \text{ and } \overleftarrow{r} = \frac{\overleftarrow{r_1}}{\overleftarrow{r_2}}. \end{cases}$$

Example 3.13 (Shift operators). Consider the coefficient function r given in Example 3.11. Then

$$\overrightarrow{r}(\{\zeta_{ij}\}_{2,2,1}) = \frac{1+\zeta_{21}}{1-\zeta_{12}}, \quad \overleftarrow{r}(\{\zeta_{ij}\}_{1,2,4}) = \frac{1+\zeta_{01}}{1-\zeta_{-12}}.$$

For a scheduling trajectory $p : \mathbb{R} \to \mathbb{R}^2$, it holds that

$$(\overrightarrow{r} \diamond p)(k) = \frac{1+p_1(k+2)}{1-p_2(k+1)}, \quad (\overleftarrow{r} \diamond p)(k) = \frac{1+p_1(k)}{1-p_2(k-1)}. \qquad \square$$

Multiplication on $\mathcal{R}[\xi]$ with ξ now can be defined through the forward-shift operator using the non-commutative rule:

$$\xi r = \overrightarrow{r}\xi, \tag{3.58}$$

where $r \in \mathcal{R}$. With (3.58) it is possible to show that $\mathcal{R}[\xi]$ still defines an Ore algebra [132]. Contrary to the CT case, $\mathcal{R}[\xi]$, with the multiplication rule (3.58), is a principle domain but it does not have the *simple* property (see Sect. 3.1.2.3). These imply that the general algebraic structure of $\mathcal{R}[\xi]$ remains the same as in the CT case, except the Jacobson form becomes more involved

Theorem 3.5 (Jacobson form [41]). *Let $R \in \mathcal{R}[\xi]^{n_r \times n_w}$ with $R \neq 0$ and $n = \mathrm{Rank}(R)$. Then there exist unimodular matrices $M_1 \in \mathcal{R}[\xi]^{n_r \times n_r}$ and $M_2 \in \mathcal{R}[\xi]^{n_w \times n_w}$ such that*

$$M_1(\xi)R(\xi)M_2(\xi) = \begin{bmatrix} Q(\xi) & 0 \\ 0 & 0 \end{bmatrix}, \tag{3.59}$$

where $Q = \mathrm{Diag}(r'_1, \ldots, r'_n) \in \mathcal{R}[\xi]^{n \times n}$ with monic non-zero $r'_i \in \mathcal{R}[\xi]$. Furthermore, there exist $g'_i \in \mathcal{R}[\xi]$ such that $r'_{i+1} = g'_i r'_i$ for $i = 1, \ldots, n-1$.

Due to the algebraic structure of $\mathcal{R}[\xi]^{\cdot \times \cdot}$, the proof of Th. 3.5 similarly follows as in [41, Chap. 8.1.]. In this case, $\mathrm{Deg}_*(R) := \mathrm{Deg}(r'_n)$ in terms of the Jacobson form (3.59) of $R(\xi)$.

Example 3.14 (Jacobson form). Consider

$$R(\xi) = \begin{bmatrix} r + \xi & -1 & -1 \\ -r & 1 + \xi & -\overrightarrow{r} \end{bmatrix} \in \mathcal{R}[\xi]^{2 \times 3},$$

where r is a meromorphic function and $\xi = q$. Then the Jacobson form of R is

$$M_1(\xi)R(\xi)M_2(\xi) = \begin{bmatrix} 1 & 0 & 0 \\ 0 & 1 + \overrightarrow{r} + \xi & 0 \end{bmatrix},$$

with

$$M_1(\xi) = \begin{bmatrix} 1 & 0 \\ -\overrightarrow{r} & 1 \end{bmatrix}, \quad M_2(\xi) = \begin{bmatrix} 0 & 0 & 1 \\ 0 & 1 & r \\ -1 & -1 & \xi \end{bmatrix}. \qquad \square$$

Due to the fact that all required algebraic properties are still satisfied for $\mathcal{R}[\xi]$ the proof of the injective cogenerator property similarly follows in DT (see Appendix A.1). The latter property implies that the theorems, introduced and used in Sect. 3.1.2, also hold in the DT case.

3.1.3.3 Kernel Representations

As a next step, we develop the concept of KR representations of DT-LPV systems following the same line of discussion as in the CT case.

Definition 3.17 (DT-KR-LPV representation). The parameter-varying difference equation (3.53) is called a discrete-time kernel representation, denoted by $\mathfrak{R}_K(\mathcal{S})$, of the LPV dynamical system $\mathcal{S} = (\mathbb{Z}, \mathbb{R}^{n_P}, \mathbb{R}^{n_W}, \mathfrak{B})$ with scheduling vector p and signals w, if

$$\mathfrak{B} = \left\{ (w, p) \in (\mathbb{R}^{n_W} \times \mathbb{R}^{n_P})^{\mathbb{Z}} \mid (R(\mathrm{q}) \diamond p)w = 0 \right\}. \qquad \square$$

Obviously the behavior, i.e. all solutions associated with (3.53) corresponds to a LPV system in terms of Def. 3.3. On the other hand, in the LPV system class we consider LPV systems with $\mathbb{T} = \mathbb{Z}$ that have a KR representation, so existence of such a representation is explicitly assumed in the following. Additionally, we denote the DT-KR representation of a CT-LPV dynamical system $\mathcal{S} = (\mathbb{R}, \mathbb{P}, \mathbb{R}^{n_\mathrm{W}}, \mathfrak{B})$ by $\mathfrak{R}_\mathrm{K}(\mathcal{S}, \mathrm{T_d})$, if the DT behavior $\mathfrak{B}'$ of this representation is equivalent with $\mathfrak{B}$ in terms of (3.49) for $\mathrm{T_d}$.

The existence of full row rank representations follows directly:

Theorem 3.6 (Existence of full row rank KR representation). *Let $\mathfrak{B}$ be given with a KR representation (3.53). Then $\mathfrak{B}$ can also be represented by a $R \in \mathcal{R}[\xi]^{\cdot \times n_\mathrm{W}}$ with full row rank.*

The proof is given in Appendix A.1. This theorem is crucial as the concept of minimality for KR representations is also based on the full row rank of the associated matrix polynomials R. (see Sec. 3.2).

3.1.3.4 Input-Output Representations

IO representations are also important for the DT behavioral framework of LPV systems. Beside system theoretical aspects, this type of representations connects the developed behavioral theory to IO models used in the-state-of-the-art of LPV system identification. In this way, it enables comparison and analysis of IO identification methods. In the following, we define DT-IO representations from the behavioral point of view, based on the same line of discussion as in the CT case.

For DT-LPV systems, IO partitions are also characterized by Def. 3.11. The existence of IO partition (u, y) implies the existence of matrix polynomials $R_\mathrm{y} \in \mathcal{R}[\xi]^{n_\mathrm{Y} \times n_\mathrm{Y}}$ and $R_\mathrm{u} \in \mathcal{R}[\xi]^{n_\mathrm{Y} \times n_\mathrm{U}}$ with R_y full rank, such that (3.53) can be written in a similar form as (3.36). Based on this, it is possible to introduce IO representations of DT-LPV systems as follows:

Definition 3.18 (DT-LPV-IO representation). The discrete-time IO representation of $\mathcal{S} = (\mathbb{Z}, \mathbb{P} \subseteq \mathbb{R}^{n_\mathrm{P}}, \mathbb{R}^{n_\mathrm{U}+n_\mathrm{Y}}, \mathfrak{B})$ with IO partition (u, y) and scheduling vector p is denoted by $\mathfrak{R}_\mathrm{IO}(\mathcal{S})$ and defined as a parameter-varying difference-equation system with order n_a:

$$\sum_{i=0}^{n_\mathrm{a}} (a_i \diamond p)\mathrm{q}^i y = \sum_{j=0}^{n_\mathrm{b}} (b_j \diamond p)\mathrm{q}^j u, \tag{3.60}$$

where $a_j \in \mathcal{R}^{n_\mathrm{Y} \times n_\mathrm{Y}}$ and $b_j \in \mathcal{R}^{n_\mathrm{Y} \times n_\mathrm{U}}$ with $a_{n_\mathrm{a}} \neq 0$ and $b_{n_\mathrm{b}} \neq 0$ are the meromorphic parameter-varying coefficients of the matrix polynomials $R_\mathrm{u}(\xi) = \sum_{j=0}^{n_\mathrm{b}} b_j \xi^j$ and full rank $R_\mathrm{y}(\xi) = \sum_{i=0}^{n_\mathrm{a}} a_i \xi^i$ with $n_\mathrm{a} \geq n_\mathrm{b} \geq 0$ and $n_\mathrm{a} > 0$. □

Note that the coefficient dependencies in Def. 3.18 can be polynomial functions of p. Thus the behavioral definition of IO representations also contains the deterministic part of IO models (1.5a) used in LPV system identification (see Sect. 1.3.2.2). By defining $u(k) := u(k\mathrm{T_d})$ and $y(k) := y(k\mathrm{T_d})$, the DT-IO representation, denoted by $\mathfrak{R}_\mathrm{IO}(\mathcal{S}, \mathrm{T_d})$, of a CT-LPV system $\mathcal{S}$ with scheduling p and IO partition (u, y) can similarly be given with equivalent behaviors in terms of (3.49) under the sampling-time $\mathrm{T_d}$. Note that the discrete-time projection with arbitrary $\mathrm{T_d}$ can change the validity of the IO partition as dynamic components of the system can be lost due to slow sampling. Thus, it is important to emphasize that existence of $\mathfrak{R}_{\mathrm{IO}}(\mathcal{S},\mathrm{T_d})$ is not guaranteed.

Example 3.15 (IO partition and representation). In Example 3.12, the sampled force variable w_{x} is still a free variable as it represents the inhomogen part of difference equation (3.55). Thus the choice of $w = [y \;\; u]^\top = [w_{\mathrm{x}} \;\; w_{\mathrm{F}}]^\top$ yields a valid IO partition. With m being the scheduling signal, the discrete-time PV behavior can be represented in the form of (3.60) with polynomials

$$R_{\mathrm{y}}(\xi) = a_0 + a_1\xi + a_2\xi^2, \quad R_{\mathrm{u}}(\xi) = b_0,$$

which have coefficients:

$$a_0 \diamond p = \mathrm{T_d^2 k_s} + p, \quad a_1 \diamond p = -p - \mathrm{q}p, \quad a_2 \diamond p = \mathrm{q}p, \quad b_0 \diamond p = \mathrm{T_d^2}.$$

Obviously, $R_{\mathrm{y}}(\xi)$ have full rank. This implies that $R_{\mathrm{y}}(\xi)$ and $R_{\mathrm{u}}(\xi)$ define an IO representation of the model with coefficients as above. □

Similar to the CT-LPV case, the notion of transfer functions for DT-LPV systems is not well-defined. The introduction of a viable formulation can be tackled via the formal series approach of [81], constructed for DT-SS systems of the LTV case. However, the extension of this approximative transfer function calculus to the class of systems considered here is not available. The notion of frozen transfer functions, frequency responses, impulse responses, poles, and zeros can be similarly defined for the frozen system set as in the CT case.

3.1.3.5 State-Space Representations

For the discrete-time LPV behavior $\mathfrak{B}$, we can also introduce latent variables with the property of state (see Def. 3.13). Existence of latent variable representations with equivalent LPV manifest behaviors is guaranteed in the DT case as Th. 3.4 holds regardless wether ξ is associated with q or $\frac{d}{dt}$ (see Example 3.16). Similarly, Def. 3.13 of the property of state also applies in the DT case with the exception that continuity of $w_{\mathrm{L},1}$ and $w_{\mathrm{L},2}$ is not required. Furthermore, the elimination property also applies for the DT case.

Example 3.16 (Latent variable representation). By considering the DT system in Example 3.12 with scheduling $p = m$ and $\mathbb{P} = [1,2]$, the following latent variable representation of the model has the same manifest behavior:

$$\begin{bmatrix} \mathrm{T_d^2 k_s} + p & -\mathrm{T_d^2} \\ (-p - \mathrm{q}^{-1}p) & 0 \\ (-\mathrm{q}^{-1}p) & 0 \end{bmatrix} \begin{bmatrix} w_{\mathrm{x}} \\ w_{\mathrm{F}} \end{bmatrix} = \begin{bmatrix} \mathrm{q} & 0 \\ -1 & \mathrm{q} \\ 0 & 1 \end{bmatrix} w_{\mathrm{L}}. \tag{3.61}$$

This can be proved by substituting the third row of (3.61) into the second row, giving

$$w_{\mathrm{L},1} = (p + \mathrm{q}^{-1}p)w_{\mathrm{X}} - p\mathrm{q}w_{\mathrm{X}}.$$

Substitution of the previous equation into the first row of (3.61) gives a PV difference equation in the variables w_{X} and w_{F}, which is equal to (3.55). Note, that any $w_{\mathrm{L}} \in (\mathbb{R}^2)^{\mathbb{Z}}$ satisfying (3.61) trivially fulfills the property of state. □

Now we can introduce SS representations of DT-LPV systems as follows:

Definition 3.19 (DT-LPV-SS representation). The discrete-time state-space representation of $\mathcal{S} = (\mathbb{Z}, \mathbb{P} \subseteq \mathbb{R}^{n_{\mathbb{P}}}, \mathbb{R}^{n_{\mathbb{U}}+n_{\mathbb{Y}}}, \mathfrak{B})$ with scheduling vector p is denoted by $\mathfrak{R}_{\mathrm{SS}}(\mathcal{S})$ and defined as a first-order parameter-varying difference equation system in the latent variable $x : \mathbb{Z} \to \mathbb{X}$:

$$\mathrm{q}x = (A \diamond p)x + (B \diamond p)u, \tag{3.62a}$$

$$y = (C \diamond p)x + (D \diamond p)u, \tag{3.62b}$$

where (u, y) is the IO partition of $\mathcal{S}$, x is the state-vector, $\mathbb{X} = \mathbb{R}^{n_{\mathbb{X}}}$ is the state-space,

$$\mathfrak{B}_{\mathrm{SS}} = \left\{ (u, x, y, p) \in (\mathbb{U} \times \mathbb{X} \times \mathbb{Y} \times \mathbb{P})^{\mathbb{Z}} \mid \text{(3.62a–b) holds} \right\},$$

is the full behavior of (3.62a–b), $\mathfrak{B}$ is equal to the manifest behavior of (3.62a–b), i.e. $\mathfrak{B} = \pi_{(u,y,p}\mathfrak{B}_{\mathrm{SS}}$, and

$$\left[\begin{array}{c|c} A & B \\ \hline C & D \end{array}\right] \in \left[\begin{array}{c|c} \mathcal{R}^{n_{\mathbb{X}} \times n_{\mathbb{X}}} & \mathcal{R}^{n_{\mathbb{X}} \times n_{\mathbb{U}}} \\ \hline \mathcal{R}^{n_{\mathbb{Y}} \times n_{\mathbb{X}}} & \mathcal{R}^{n_{\mathbb{Y}} \times n_{\mathbb{U}}} \end{array}\right].$$ □

Again, in the full behavior $\mathfrak{B}_{\mathrm{SS}}$ the latent variable x trivially fulfills the state property. Also the class of SS representations formulated by Def. 3.19 contains the SS models used in LPV system identification as the coefficient dependencies in the behavioral definition can be linear functions of p. This enables the analysis and comparison of the results of LPV-SS system identification. The DT-SS representation of a CT-LPV dynamical system $\mathcal{S} = (\mathbb{R}, \mathbb{P}, \mathbb{R}^{n_{\mathbb{U}}+n_{\mathbb{Y}}}, \mathfrak{B})$ is denoted as $\mathfrak{R}_{\mathrm{SS}}(\mathcal{S}, \mathrm{T_d})$, where the manifest behavior of $\mathfrak{R}_{\mathrm{SS}}(\mathcal{S}, \mathrm{T_d})$ is equivalent in terms of (3.49) with $\mathfrak{B}$ under the sampling-time $\mathrm{T_d}$. As before, existence of such representations is not guaranteed for arbitrary $\mathrm{T_d}$.

Example 3.17 (SS representation). Continuing Example 3.16, the LPV state-space representation of the model follows by taking $[\,y \quad u\,]^\top = [\,w_{\mathrm{X}} \; w_{\mathrm{F}}\,]^\top$ as the IO partition and $x = w_{\mathrm{L}}$ as the state:

$$\mathrm{q}x = \begin{bmatrix} 0 & 0 \\ 1 & 0 \end{bmatrix} x + \begin{bmatrix} \mathrm{T_d^2 k_s} + p & -\mathrm{T_d^2} \\ -p - \mathrm{q}^{-1}p & 0 \end{bmatrix} \begin{bmatrix} y \\ u \end{bmatrix}, \tag{3.63}$$

$$y = \begin{bmatrix} 0 & \frac{1}{-\mathrm{q}^{-1}p} \end{bmatrix} x. \tag{3.64}$$

By substitution of the second equation into the first one, the state equation in the form of (3.62a) results, while the second equation gives the output equation in the form of (3.62b). Thus, the corresponding SS representation is

$$\left[\begin{array}{c|c} A\diamond p & B\diamond p \\ \hline C\diamond p & D\diamond p \end{array}\right] = \left[\begin{array}{cc|c} 0 & -\frac{p+\mathrm{T}_\mathrm{d}^2\mathrm{k}_\mathrm{s}}{\mathrm{q}^{-1}p} & -\mathrm{T}_\mathrm{d}^2 \\ 1 & 1+\frac{p}{\mathrm{q}^{-1}p} & 0 \\ \hline 0 & \frac{-1}{\mathrm{q}^{-1}p} & 0 \end{array}\right]. \qquad \square$$

Similar to LPV-IO representations, the notion of transfer function, frequency response, and impulse response can only be defined in a frozen sense for LPV-SS representations.

3.2 Equivalence Classes and Relations

In this section, we continue the introduction of the LPV behavioral framework by defining equivalence relations and classes for the introduced representation forms. These are essential aspects of the theory as they characterize which representations describe the same system, providing tools to compare and analyze representation capabilities/validity of LPV models for identification. First we define equality of behaviors and investigate why this equality needs to be understood in an almost everywhere. Then, we define equivalency, equivalence relations and equivalence classes of all representation forms together with the concept of minimality and canonical forms. To do so, we introduce key theorems of left/right side unimodular transformations.

The most important concept to begin with is to define equality with respect to behaviors of PV systems.

Definition 3.20 (Equal behaviors of PV systems). Let $\mathfrak{B}_1, \mathfrak{B}_2 \subseteq (\mathbb{W}\times\mathbb{P})^{\mathbb{T}}$ with $\mathbb{W} = \mathbb{R}^{n_\mathrm{W}}$, $\mathbb{P} \subseteq \mathbb{R}^{n_\mathrm{P}}$, and with $\mathbb{T}$ equal to either $\mathbb{R}$ or $\mathbb{Z}$. We call $\mathfrak{B}_1$ and $\mathfrak{B}_2$ equal, if in case of $\mathbb{T} = \mathbb{R}$:

$$(w,p) \in \mathfrak{B}_1 \cap \mathcal{C}^\infty(\mathbb{R}, \mathbb{W}\times\mathbb{P}) \Leftrightarrow (w,p) \in \mathfrak{B}_2 \cap \mathcal{C}^\infty(\mathbb{R}, \mathbb{W}\times\mathbb{P}),$$

or in the discrete-time case ($\mathbb{T} = \mathbb{Z}$):

$$(w,p) \in \mathfrak{B}_1 \Leftrightarrow (w,p) \in \mathfrak{B}_2. \qquad \square$$

Two systems are called equal if they have equal behaviors. However, it is important to remark that Def. 3.20 establishes the concept of equal behaviors in terms of internal, and hence not external equality. This means, that even if two LPV systems have the same behavior in terms of IO signals for a particular IO partition, they are not necessarily equal as their scheduling signals or latent variables can differ. In case of systems with latent variables, equality can be defined in terms of the equality of the manifest behaviors (see Def. 3.29). On the other hand, there also exist systems with equal signal behavior and isomorphic projected scheduling behavior (see Example 3.18). However, due to technical reasons, we do not consider such systems to be equal in the sequel.

Example 3.18 (Equal PV behaviors). Consider the varying-mass and spring system of Example 3.1 with m as the scheduling function. We have seen in Example 3.4, that the PV differential equation (KR representation)

$$\mathrm{k_s} w_1 - w_2 + \left(\frac{\mathrm{d}}{\mathrm{d}t} p\right) \frac{\mathrm{d}}{\mathrm{d}t} w_1 + p \frac{\mathrm{d}^2}{\mathrm{d}t^2} w_1 = 0. \tag{3.65}$$

is basically the same as the differential equation (3.7) describing the behavior of the varying-mass and spring system. Based on the equivalence of the differential equations, the associated LPV behaviors are also equal. Now consider the latent variable representation in Example 3.9. The behavior associated with (3.65) is not equal with the associated behavior of (3.41), as in the latter case the trajectories of the latent variable are part of the behavior. Another example of inequality is

$$\mathrm{k_s} w_1 - w_2 - \left(\frac{\mathrm{d}}{\mathrm{d}t} \breve{p}\right) \frac{\mathrm{d}}{\mathrm{d}t} w_1 + (1 - \breve{p}) \frac{\mathrm{d}^2}{\mathrm{d}t^2} w_1 = 0. \tag{3.66}$$

with a scheduling variable $\breve{p} = 1 - m$. In this case, the solution trajectories of w for this KR representation are the same as for (3.65), but the behaviors are not equal due to the different scheduling function. □

3.2.1 Equivalent Kernel Representations

According to the LTI case, $R_1, R_2 \in \mathcal{R}[\xi]$ are expected to define an equal behavior if they are equivalent up to multiplication by a $r \in \mathcal{R}$, $r \neq 0$. However, r can be a rational function for which $(r \diamond p)(t)$ is unbounded for some scheduling trajectories p and $t \in \mathbb{T}$. The behavior of a kernel representation is defined to contain only those trajectories of p for which a (weak) solution exists and is well defined. The latter is guaranteed by the boundedness of $r \diamond p$. In this way, the behavior (solutions) of R_1 is equal with the behavior of $R_2(\xi) = rR_1(\xi)$ except for those trajectories for which $r \diamond p$ is unbounded. To consider equality of LPV-KR representations with this phenomenon of singularity in mind, we define the restriction of $\mathfrak{B} \subseteq (\mathbb{W} \times \mathbb{P})^{\mathbb{T}}$ to $\bar{\mathfrak{B}}_{\mathbb{P}} \subseteq \mathfrak{B}_{\mathbb{P}}$ as

$$\mathfrak{B}\mid_{\bar{\mathfrak{B}}_{\mathbb{P}}} = \left\{(w,p) \in \mathfrak{B} \mid p \in \bar{\mathfrak{B}}_{\mathbb{P}}\right\}. \tag{3.67}$$

The equivalence of LPV-KR representations can now be introduced in an *almost-everywhere sense*[4]:

Definition 3.21 (Equivalent KR representations). Two kernel representations with polynomials $R, R' \in \mathcal{R}[\xi]^{\cdot \times n_{\mathbb{W}}}$, $\mathbb{P} = \mathbb{R}^{n_{\mathbb{P}}}$ and behaviors $\mathfrak{B}, \mathfrak{B}' \subseteq \mathcal{L}_1^{\mathrm{loc}}(\mathbb{R}, \mathbb{R}^{n_{\mathbb{W}}} \times \mathbb{R}^{n_{\mathbb{P}}})$ (or $\mathfrak{B}, \mathfrak{B}' \subseteq (\mathbb{R}^{n_{\mathbb{W}}} \times \mathbb{R}^{n_{\mathbb{P}}})^{\mathbb{Z}}$ in discrete-time) are called equivalent if $\mathfrak{B}\mid_{\mathfrak{B}_{\mathbb{P}} \cap \mathfrak{B}'_{\mathbb{P}}} = \mathfrak{B}'\mid_{\mathfrak{B}_{\mathbb{P}} \cap \mathfrak{B}'_{\mathbb{P}}}$, i.e. their behaviors are equal for all possible mutually valid trajectories

[4] Technically, equivalence of KR representations can be considered in a more complicated setting without the need of the almost-everywhere concept. However, this requires the use of a so-called extended solution set (see the proofs in Appendix A.1). To avoid the use of a too technical language, these details are only presented in the Appendix.

of p. If equality holds with $\mathfrak{B}'_{\mathbb{P}} = \mathfrak{B}_{\mathbb{P}}$, then the equivalence of the representations is called full equivalence. □

Example 3.19 (Almost everywhere equivalence). By continuing Example 3.18, the KR representation

$$\frac{\mathrm{k_s}}{p}w_1 - \frac{1}{p}w_2 - \frac{1}{p}\left(\frac{\mathrm{d}}{\mathrm{d}t}p\right)\frac{\mathrm{d}}{\mathrm{d}t}w_1 + \frac{\mathrm{d}^2}{\mathrm{d}t^2}w_1 = 0.$$

has the same weak solutions as (3.65) except for those trajectories of $p = m$, where $m(t) = 0$ for some $t \in \mathbb{T}$. Thus, this KR representation and (3.65) are equivalent in the almost everywhere sense. □

The existence of equivalent KR equations implies, similar to the LTI case, that representations of PV dynamical systems are non-unique. Thus, we need to characterize when two KR representations define the same behavior. This characterization follows trough left/right unimodular transformations just like in the LTI case (see [230]). To extend these theories to the LPV case, we use the concept of unimodular matrices in $\mathcal{R}[\xi]^{\cdot\times\cdot}$ and the Jacobson form introduced in Sec. 3.1.2. Based on these concepts it is possible to show that the following theorem holds in the LPV case:

Theorem 3.7 (Left-side unimodular transformation). *Let $R \in \mathcal{R}[\xi]^{n_\mathrm{r}\times n_\mathbb{W}}$ and $M \in \mathcal{R}[\xi]^{n_\mathrm{r}\times n_\mathrm{r}}$ with M unimodular. For a given scheduling dimension $n_\mathbb{P}$, define $R'(\xi) := M(\xi)R(\xi)$. Denote the behaviors corresponding to R and R' by $\mathfrak{B}$ and $\mathfrak{B}'$ with scheduling space $\mathbb{P} \subseteq \mathbb{R}^{n_\mathbb{P}}$ and signal space $\mathbb{W} = \mathbb{R}^{n_\mathbb{W}}$. Then $\mathfrak{B}$ and $\mathfrak{B}'$ are equal (almost everywhere).*

The proof is given in Appendix A.1. Furthermore, if $R \in \mathcal{R}[\xi]^{n_\mathrm{r}\times n_\mathbb{W}}$ is not full row rank, i.e. $\mathrm{Rank}(R) = n < n_\mathrm{r}$, then there exists an unimodular $M \in \mathcal{R}[\xi]^{n_\mathrm{r}\times n_\mathrm{r}}$ such that

$$M(\xi)R(\xi) = \begin{bmatrix} R'(\xi) \\ 0 \end{bmatrix}, \tag{3.68}$$

where $R' \in \mathcal{R}[\xi]^{n\times n_\mathbb{W}}$ is full row rank and the corresponding behaviors are equal in terms of Th. 3.7. Note that this theorem establishes the concept when two representations can be considered equivalent. However, to establish equivalency of SS representations or latent variable systems, we require the concept of right-side unimodular transformation. It can be shown that the following theorem holds:

Theorem 3.8 (Right-side unimodular transformation). *Let $R \in \mathcal{R}[\xi]^{n_\mathrm{r}\times n_\mathbb{W}}$ and $M \in \mathcal{R}[\xi]^{n_\mathbb{W}\times n_\mathbb{W}}$ with M unimodular. Denote the behaviors defined by R and $R'(\xi) := R(\xi)M(\xi)$ as $\mathfrak{B}$ and $\mathfrak{B}'$ with scheduling domain $\mathbb{P} \subseteq \mathbb{R}^{n_\mathbb{P}}$ and signal space $\mathbb{W} = \mathbb{R}^{n_\mathbb{W}}$. If $\mathbb{T} = \mathbb{R}$, then $\mathfrak{B} \cap \mathcal{C}^\infty(\mathbb{R}, \mathbb{W} \times \mathbb{P})$ and $\mathfrak{B}' \cap \mathcal{C}^\infty(\mathbb{R}, \mathbb{W} \times \mathbb{P})$ are isomorphic in the almost everywhere sense. If $\mathbb{T} = \mathbb{Z}$, then $\mathfrak{B}$ and $\mathfrak{B}'$ are isomorphic in the almost everywhere sense.*

The proof is given in Appendix A.1. It is important that right-side unimodular transformations do not change the underlying relation between the system signals nor the projected scheduling behavior, but they do change the system signals, the trajectories of the behavior.

With all the required tools established, it is now possible to introduce the notion of equivalence relation for LPV kernel representations. This relation is the key to characterize the set of equivalent representations of the same behavior, which we call an equivalence class. In the LPV case, the scheduling dimension $n_{\mathbb{P}}$ plays an important role in $\mathcal{R}[\xi]$ as it determines the exact coefficient dependence and also how the dot or shift operators behave on $\mathcal{R}$. Thus, an equivalence relation must be dependent on $n_{\mathbb{P}}$. Based on this, the following definition is given:

Definition 3.22 (LPV Equivalence relation). Introduce the symbol $\overset{n_{\mathbb{P}}}{\sim}$ to denote the equivalence relation on $\bigcup \mathcal{R}[\xi]^{\cdot\times\cdot}$ (all polynomial matrices with finite dimension) for an $n_{\mathbb{P}}$-dimensional scheduling space. $R_1 \in \mathcal{R}[\xi]^{n_1 \times n_{\mathbb{W}}}$ and $R_2 \in \mathcal{R}[\xi]^{n_2 \times n_{\mathbb{W}}}$ with $n_1 \geq n_2$ are called equivalent, i.e. $R_1 \overset{n_{\mathbb{P}}}{\sim} R_2$, if there exists a unimodular matrix function $M \in \mathcal{R}[\xi]^{n_1 \times n_1}$ such that

$$M(\xi)R_1(\xi) = \begin{bmatrix} R_2(\xi) \\ 0 \end{bmatrix} \begin{matrix} \updownarrow & n_2 \\ \updownarrow & n_1 - n_2 \end{matrix}. \tag{3.69}$$

□

This implies that if $R_1 \overset{n_{\mathbb{P}}}{\sim} R_2$, then the corresponding behaviors with $\mathbb{P} \subseteq \mathbb{R}^{n_{\mathbb{P}}}$ and $\mathbb{W} = \mathbb{R}^{n_{\mathbb{W}}}$ are equal (almost everywhere). Using $\overset{n_{\mathbb{P}}}{\sim}$ we can define equivalence classes as follows:

Definition 3.23 (LPV Equivalence class). For a given scheduling dimension $n_{\mathbb{P}}$, the set $\mathscr{E}^{n_{\mathbb{P}}} \subseteq \bigcup \mathcal{R}[\xi]^{\cdot\times\cdot}$ is called an equivalence class, if it is a maximal subset of $\bigcup \mathcal{R}[\xi]^{\cdot\times\cdot}$ such that for all $R_1, R_2 \in \mathscr{E}^{n_{\mathbb{P}}}$ it holds that $R_1 \overset{n_{\mathbb{P}}}{\sim} R_2$. □

An equivalence class defines the set of all KR representations which have equal behavior. An important subset of an equivalence class are the so-called minimal representations:

Definition 3.24 (Minimality). Let $R \in \mathcal{R}[\xi]^{n_{\mathrm{r}} \times n_{\mathbb{W}}}$. Then R is called minimal if it has full row rank, i.e. $\mathrm{Rank}(R) = n_{\mathrm{r}}$. □

Based on the concept of minimality with respect to KR representations and the Jacobson form, we can also define the order of the system, i.e. the required minimal number of state variables in a SS realization of $\mathfrak{R}_{\mathrm{K}}(\mathcal{S})$.

Definition 3.25 (Minimal degree of a KR representation). Consider a minimal $\mathfrak{R}_{\mathrm{K}}(\mathcal{S})$ described by a full row rank $R \in \mathcal{R}[\xi]^{n_{\mathrm{r}} \times n_{\mathbb{W}}}$. Let $R(\xi) = [R'(\xi) \;\; R''(\xi)]$ where $R' \in \mathcal{R}[\xi]^{n_{\mathrm{r}} \times n_{\mathrm{r}}}$ has full column rank. Note that such form can always be obtained by the permutation of the signal variables and it is not unique. Consider $n_{\mathrm{deg}} := \mathrm{Deg}_*(R')$ associated with the Jacobson form (see Th. 3.1 and 3.5) of R'. Assume that R' is chosen with respect to R such that n_{deg} is maximal. Then n_{deg} is called the minimal degree of $\mathfrak{R}_{\mathrm{K}}(\mathcal{S})$. □

It follows from Th. 3.7, that all KR representations in the equivalence class of $\mathfrak{R}_{\mathrm{K}}(\mathcal{S})$ have the same minimal degree n_{deg}. It can also be shown that this degree is equal to the required minimal number of state variables in a SS realization of $\mathfrak{R}_{\mathrm{K}}(\mathcal{S})$, hence $n_{\deg}$ can be considered as the order, i.e. *McMillan degree* of $\mathcal{S}$. It is also important to consider the subclass of representations, so-called canonical forms, that can uniquely characterize each equivalence class:

Definition 3.26 (Canonical forms). For a given scheduling dimension $n_{\mathbb{P}}$, $\mathscr{E}_{\mathrm{can}}^{n_{\mathbb{P}}} \subset \bigcup \mathcal{R}[\xi]^{\cdot\times\cdot}$ is called a set of canonical forms if each element of $\bigcup \mathcal{R}[\xi]^{\cdot\times\cdot}$ is equivalent under $\overset{n_{\mathbb{P}}}{\sim}$ with only one element of $\mathscr{E}_{\mathrm{can}}^{n_{\mathbb{P}}}$. ($\mathscr{E}_{\mathrm{can}}^{n_{\mathbb{P}}}$ is the class representative of $\bigcup \mathcal{R}[\xi]^{\cdot\times\cdot}$ under $\overset{n_{\mathbb{P}}}{\sim}$). □

Example 3.20 (LPV equivalence relation and minimality). Let the KR representation $\mathfrak{R}_{\mathrm{K}}(\mathcal{S})$ of an CT-LPV system $\mathcal{S}$ with $\mathbb{P} \subseteq \mathbb{R}$ given by

$$R(\xi)\diamond p = \begin{bmatrix} (\sin(p)p - \cos(p))\frac{\mathrm{d}}{\mathrm{d}t}p & 0 \\ p & \frac{1}{\cos(p)} \end{bmatrix} + \begin{bmatrix} -\cos(p)p & -1 \\ \frac{1}{\cos(p)} & 0 \end{bmatrix}\xi + \begin{bmatrix} -1 & 0 \\ 0 & 0 \end{bmatrix}\xi^2.$$

Then, there exists a unimodular matrix $M \in \mathcal{R}[\xi]^{2\times 2}$ with

$$M(\xi)\diamond p = \begin{bmatrix} 0 & 1 \\ 1 & \cos(p)\xi - \sin(p)\frac{\mathrm{d}}{\mathrm{d}t}p \end{bmatrix} \text{ s.t. } (M(\xi)R(\xi))\diamond p = \begin{bmatrix} p + \frac{1}{\cos(p)}\xi & \frac{1}{\cos(p)} \\ 0 & 0 \end{bmatrix}.$$

Note that this result is obtained by using (3.32), i.e $\xi r = \dot{r} + r\xi$, so

$$\xi\frac{1}{\cos(p)} = \frac{\sin(p)}{\cos^2(p)}\left(\frac{\mathrm{d}}{\mathrm{d}t}p\right) + \frac{1}{\cos(p)}\xi.$$

From Th. 3.7 it follows that $R \overset{1}{\sim} R'$ where $\left[\,(R')^\top(\xi) \;\; 0\,\right]^\top = M(\xi)R(\xi)$. Furthermore, $\mathrm{Rank}(R') = 1$ implies that $\mathrm{Rank}(R) = 1$, hence R' is minimal while R is not. By computing $n_{\deg}$ of R', it follows that the McMillan degree of $\mathcal{S}$ is 1. □

3.2.2 *Equivalent IO Representations*

The introduced concepts generalize to LPV-IO representations as well:

Definition 3.27 (Equivalence relation of LPV-IO representations). Let $(R_{\mathrm{y}}, R_{\mathrm{u}})$ and $(R'_{\mathrm{y}}, R'_{\mathrm{u}})$ be LPV-IO representations (see Def. 3.12 in CT and Def. 3.18 in DT) with the same input and output dimensions $(n_{\mathbb{Y}}, n_{\mathbb{U}})$. For a given scheduling dimension $n_{\mathbb{P}}$, we call $(R_{\mathrm{y}}, R_{\mathrm{u}})$ and $(R'_{\mathrm{y}}, R'_{\mathrm{u}})$ equivalent

$$(R_{\mathrm{y}}, R_{\mathrm{u}}) \overset{n_{\mathbb{P}}}{\sim} (R'_{\mathrm{y}}, R'_{\mathrm{u}}), \tag{3.70}$$

if there exists a unimodular matrix $M \in \mathcal{R}[\xi]^{n_{\mathbb{Y}}\times n_{\mathbb{Y}}}$ such that

$$R'_{\mathrm{y}}(\xi) = M(\xi)R_{\mathrm{y}}(\xi) \quad \text{and} \quad R'_{\mathrm{u}}(\xi) = M(\xi)R_{\mathrm{u}}(\xi). \tag{3.71}$$

□

Definition 3.28 (Minimal LPV-IO representation). An IO representation defined through $R_{\mathrm{y}} \in \mathcal{R}[\xi]^{n_{\mathbb{Y}} \times n_{\mathbb{Y}}}$ and $R_{\mathrm{u}} \in \mathcal{R}[\xi]^{n_{\mathbb{Y}} \times n_{\mathbb{U}}}$ is called minimal for a given scheduling dimension $n_{\mathbb{P}}$, if there are no polynomials $R'_{\mathrm{y}} \in \mathcal{R}[\xi]^{n_{\mathbb{Y}} \times n_{\mathbb{Y}}}$ and $R'_{\mathrm{u}} \in \mathcal{R}[\xi]^{n_{\mathbb{Y}} \times n_{\mathbb{U}}}$ with $\mathrm{Deg}(R_{\mathrm{y}}) < \mathrm{Deg}(R'_{\mathrm{y}})$ such that

$$(R_{\mathrm{y}}, R_{\mathrm{u}}) \overset{n_{\mathbb{P}}}{\sim} (R'_{\mathrm{y}}, R'_{\mathrm{u}}). \qquad \square$$

Using the IO equivalence relation and minimality, the definition of IO equivalence classes and canonical forms follows naturally. In the IO case, the McMillan degree is equal to $n_{\deg} = \mathrm{Deg}_*(R_{\mathrm{y}})$ associated with the Jacobson form of R_{y}.

Example 3.21 (LPV-IO equivalence relation and minimality). Let the IO representation $\mathfrak{R}_{\mathrm{IO}}(\mathcal{S})$ of an CT-LPV system $\mathcal{S}$ with $\mathbb{P} \subseteq \mathbb{R}$ be given by

$$R_{\mathrm{y}}(\xi) \diamond p = \begin{bmatrix} \sin^2(p)\xi & \sin^2(p)p \\ -2\cot(p)\frac{\mathrm{d}}{\mathrm{d}t}p\xi - \xi^2 & (1-p)\xi - \frac{\mathrm{d}}{\mathrm{d}t}p(1+2\cot(p)p) \end{bmatrix},$$

$$R_{\mathrm{u}}(\xi) \diamond p = \begin{bmatrix} \sin^2(p) \\ p - 2\cot(p)\frac{\mathrm{d}}{\mathrm{d}t}p - \xi \end{bmatrix}.$$

Then, there exists a unimodular matrix $M \in \mathcal{R}[\xi]^{2\times 2}$ given by

$$M(\xi) \diamond p = \begin{bmatrix} \frac{1}{\sin^2(p)} & 0 \\ \frac{1}{\sin^2(p)}\xi & 1 \end{bmatrix} \quad \text{s.t.} \quad \begin{cases} (M(\xi)R_{\mathrm{y}}(\xi)) \diamond p = \begin{bmatrix} \xi & p \\ 0 & \xi \end{bmatrix}, \\ (M(\xi)R_{\mathrm{u}}(\xi)) \diamond p = \begin{bmatrix} 1 \\ p \end{bmatrix}, \end{cases}$$

which can be verified by using

$$\frac{1}{\sin^2(p)}\xi \sin^2(p) = 2\cot(p)\frac{\mathrm{d}}{\mathrm{d}t}p + \xi.$$

This implies that $(R'_{\mathrm{y}}, R'_{\mathrm{u}}) = (MR_{\mathrm{y}}, MR_{\mathrm{u}})$ and $(R_{\mathrm{y}}, R_{\mathrm{u}})$ are equivalent for $n_{\mathbb{P}} = 1$ in terms of Th. 3.7. From Def. 3.28 it follows that $\mathfrak{R}_{\mathrm{IO}}(\mathcal{S})$ is not minimal as $\mathrm{Deg}(R_{\mathrm{y}}) = 2$ is greater than $\mathrm{Deg}(R'_{\mathrm{y}}) = 1$. On the other hand, it is trivial that $(R'_{\mathrm{y}}, R'_{\mathrm{u}})$ defines a minimal IO representation of $\mathcal{S}$. By computing the Jacobson form of R'_{y} it follows that the McMillan degree of $\mathcal{S}$ is 1. $\square$

3.2.3 *Equivalent State-space Representations*

We can also generalize the introduced concepts to LPV-SS representations. To do so, we first have to clarify state-transformation in the LPV case which is an important ingredient to formulate equivalence relations of SS representations.

By definition, the full behavior of a $\mathfrak{R}_{\mathrm{SS}}(\mathcal{S})$ is represented by a zero-order and a first-order polynomial matrix $R_{\mathrm{W}} \in \mathcal{R}[\xi]^{n_{\mathrm{r}} \times (n_{\mathbb{Y}} + n_{\mathbb{U}})}$ and $R_{\mathrm{L}} \in \mathcal{R}[\xi]^{n_{\mathrm{r}} \times n_{\mathbb{X}}}$ in the form of

$$(R_W(\xi) \diamond p)\mathrm{Col}(u,y) = (R_L(\xi) \diamond p)x \tag{3.72}$$

where ξ is either q or $\frac{d}{dt}$, (u,y) is an IO partition of $\mathcal{S}$, x is the state variable of $\mathfrak{R}_{SS}(\mathcal{S})$, and p is the scheduling signal. Similar to the LTI case, left and right side multiplication of R_W and R_L with unimodular $M_1 \in \mathcal{R}[\xi]^{n_r \times n_r}$ and $M_2 \in \mathcal{R}[\xi]^{n_\mathbb{X} \times n_\mathbb{X}}$ leads to

$$R'_W(\xi) = M_1(\xi)R(\xi), \quad R'_L(\xi) = M_1(\xi)R_L(\xi)M_2(\xi). \tag{3.73}$$

In terms of Th. 3.7 and 3.8, the resulting polynomials R' and R'_L define an equivalent latent variable representation of $\mathcal{S}$. The new latent variable, given by

$$x' = (M_2^\dagger(\xi) \diamond p)x, \tag{3.74}$$

fulfills the property of state as $M_2^\dagger$ is unimodular. To guarantee that the resulting latent variable representation qualifies as a SS representation, R'_L needs to be monic and $\mathrm{Deg}(R'_W) = 0$ with $\mathrm{Deg}(R'_L) = 1$ must be satisfied. This implies that the unimodular matrices must have zero order, i.e. $M_1 \in \mathcal{R}^{n_r \times n_r}$ and $M_2 \in \mathcal{R}^{n_\mathbb{X} \times n_\mathbb{X}}$, and M_1 has a special structure in order to guarantee that R'_W and R'_L correspond to an equivalent SS representation. In that case, (3.74) is called a *state-transformation* and $T = M_2^\dagger$ is called the *state-transformation* matrix resulting in

$$x' = (T \diamond p)x. \tag{3.75}$$

A major difference with respect to LTI state-transformations is that, in the LPV case, T is inherently dependent on p and this dependence is dynamic, i.e. $T \in \mathcal{R}^{n_\mathbb{X} \times n_\mathbb{X}}$. Additionally, it can be shown that an invertible $T \in \mathcal{R}^{n_\mathbb{X} \times n_\mathbb{X}}$ used as a state-transformation is always equivalent with a right and left-side multiplication by unimodular matrix functions yielding a valid SS representation of the LPV system. This is proved by the algebraic equivalency, see (3.75), of the original and the new state variables [172]. The converse, namely that the state variables of any two SS representations of the same LPV system are algebraically equivalent up to a state-transformation, follows directly from Th. 3.7 and 3.8. Based on this, two SS representations are equivalent if and only if their states can be related via an invertible state-transformation (3.75).

Similar to the LTI case, the SS representation, resulting from the state-transformation of $\mathfrak{R}_{SS}(\mathcal{S})$, can be analytically computed from the meromorphic matrices of $\mathfrak{R}_{SS}(\mathcal{S})$. However, different commutation rules in $\mathcal{R}[\xi]$ for the continuous and discrete-time cases yield different consequences with respect to the system matrices. Let $\mathfrak{R}_{SS}(\mathcal{S})$ be a given LPV-SS representation with $\mathbb{X} = \mathbb{R}^{n_\mathbb{X}}$ and state-equation

$$\xi x = (A \diamond p)x + (B \diamond p)u. \tag{3.76}$$

Let $T \in \mathcal{R}^{n_\mathbb{X} \times n_\mathbb{X}}$ be an invertible matrix function and consider x', given by (3.75), as a new state variable. It is immediate that substitution of (3.75) into (3.76) yields

$$\xi (T^{-1} \diamond p)x' = (A \diamond p)(T^{-1} \diamond p)x' + (B \diamond p)u. \tag{3.77}$$

In the continuous-time case, the multiplication by $\xi = \frac{\mathrm{d}}{\mathrm{d}t}$ in (3.77) yields:

$$\frac{\mathrm{d}}{\mathrm{d}t}x' = \left((TAT^{-1} + \dot{T}T^{-1}) \diamond p\right)x' + \left((TB) \diamond p\right)u. \tag{3.78}$$

Here we use the fact that $\dot{T}T^{-1} = -T\dot{T}^{-1}$. In the discrete-time case, the multiplication by $\xi = \mathrm{q}$ in (3.77) gives:

$$\mathrm{q}x' = \left(\left(\overrightarrow{T}AT^{-1}\right) \diamond p\right)x' + \left(\left(\overrightarrow{T}B\right) \diamond p\right)u. \tag{3.79}$$

In both cases the new state-equation defines a LPV-SS representation with state-vector x'. Note that, due to the different commutation rules of the time-operators, the transformation rules of the original system matrices are different in the CT and in the DT cases. Now we can give the following definition of equivalence classes:

Definition 3.29 (Equivalence relation of LPV-SS representations). Consider two LPV-SS representations with SS matrices (A_1, B_1, C_1, D_1) and (A_2, B_2, C_2, D_2) in $\mathcal{R}^{\cdot \times \cdot}$ where $A_1 \in \mathcal{R}^{n_1 \times n_1}$ and $A_2 \in \mathcal{R}^{n_2 \times n_2}$ and $n_1 \geq n_2$. For a given scheduling dimension $n_{\mathbb{P}}$, these representations are called equivalent,

$$\left[\begin{array}{c|c} A_1 & B_1 \\ \hline C_1 & D_1 \end{array}\right] \overset{n_{\mathbb{P}}}{\sim} \left[\begin{array}{c|c} A_2 & B_2 \\ \hline C_2 & D_2 \end{array}\right], \tag{3.80}$$

if there exists an invertible $T \in \mathcal{R}^{n_1 \times n_1}$ such that in case of $\mathbb{T} = \mathbb{R}$ the following holds:

$$\begin{aligned} TA_1T^{-1} + \dot{T}T^{-1} &= \begin{bmatrix} A_2 & 0 \\ * & * \end{bmatrix}, \ TB_1 = \begin{bmatrix} B_2 \\ * \end{bmatrix}, \ \begin{matrix} \updownarrow & n_2 \\ \updownarrow & n_1 - n_2 \end{matrix} \\ C_1T^{-1} &= \begin{bmatrix} C_2 & 0 \end{bmatrix}, \quad D_1 = D_2, \end{aligned} \tag{3.81a}$$

while in case of $\mathbb{T} = \mathbb{Z}$:

$$\begin{aligned} \overrightarrow{T}A_1T^{-1} &= \begin{bmatrix} A_2 & 0 \\ * & * \end{bmatrix}, \ \overrightarrow{T}B_1 = \begin{bmatrix} B_2 \\ * \end{bmatrix}, \ \begin{matrix} \updownarrow & n_2 \\ \updownarrow & n_1 - n_2 \end{matrix} \\ C_1T^{-1} &= \begin{bmatrix} C_2 & 0 \end{bmatrix}, \quad D_1 = D_2. \end{aligned} \tag{3.81b}$$

□

Based on the previous considerations, the existence of a state-transformation T between the matrices of two SS representations implies that they have the same manifest behavior. Furthermore, the states are related as

$$(T \diamond p)(t)x_1(t) = \begin{bmatrix} x_2(t) \\ * \end{bmatrix}, \ \begin{matrix} \updownarrow & n_2 \\ \updownarrow & n_1 - n_2 \end{matrix} \qquad \forall t \in \mathbb{T}. \tag{3.82}$$

From the concept of LPV-SS equivalence the concept of minimality directly follows:

Definition 3.30 (Minimal LPV-SS representation). For a given scheduling dimension $n_{\mathbb{P}}$, an SS representation defined through the matrix functions (A, B, C, D) is called minimal if there exist no (A', B', C', D') with $n'_{\mathbb{X}} < n_{\mathbb{X}}$ such that

$$\left[\begin{array}{c|c} A & B \\ \hline C & D \end{array}\right] \overset{n_{\mathbb{P}}}{\sim} \left[\begin{array}{c|c} A' & B' \\ \hline C' & D' \end{array}\right]. \qquad \square$$

Again, using the concept of SS equivalence relation and minimality, the definition of LPV-SS equivalence classes and canonical forms follows naturally. In addition, the state-dimension $n_{\mathbb{X}}$ of a minimal $\mathfrak{R}_{\mathrm{SS}}(\mathcal{S})$ is equal to the McMillan degree of $\mathcal{S}$.

Example 3.22 (LPV-SS equivalence relation and minimality). Consider the LPV-SS representation derived in Example 3.17. Let $T \in \mathcal{R}^{2\times 2}$ be an invertible state-transformation defined by

$$T \diamond p = \begin{bmatrix} -1 & -1 \\ 0 & -\frac{1}{q^{-1}p} \end{bmatrix}.$$

Then

$$T^{-1} \diamond p = \begin{bmatrix} -1 & q^{-1}p \\ 0 & -q^{-1}p \end{bmatrix}, \quad \overrightarrow{T} \diamond p = \begin{bmatrix} -1 & -1 \\ 0 & -\frac{1}{p} \end{bmatrix},$$

implying

$$\left[\begin{array}{c|c} \overrightarrow{T} A T^{-1} & \overrightarrow{T} B \\ \hline C T^{-1} & D \end{array}\right] \diamond p = \left[\begin{array}{cc|c} 1 & -\mathrm{T}_{\mathrm{d}}^2 \mathrm{k}_{\mathrm{s}} & \mathrm{T}_{\mathrm{d}}^2 \\ \frac{1}{p} & 1 & 0 \\ \hline 0 & 1 & 0 \end{array}\right].$$

The obtained SS representation is an equivalent minimal SS representation of $\mathcal{S}$ as it is in an equivalence relation with $\mathfrak{R}_{\mathrm{SS}}(\mathcal{S})$ and its state dimension is the same. Note that this realization has only static dependence. $\square$

3.3 Properties of LPV Systems and Representations

In the previous section we have developed the basics of a LPV behavioral framework. In order to use this framework as an analysis tool for LPV system identification, we also need to investigate key properties of systems and representations in terms of dynamic stability and state-observability/reachability. We analyze in this section these concepts using the results of LTV system theory. We also compare the developed results with their counterparts in the existing LPV theory. As the developed behavioral approach addresses a larger set of LPV systems than the state-of-the-art of LPV system theory, the existing LPV concepts follow as special cases of the theory presented in this section.

3.3.1 State-Observability and Reachability

The concepts of state-observability and reachability of LPV-SS system representations are important properties in the LPV case. They are not only key concepts for control and subspace-based identification, but they also provide the formulation of special SS canonical forms. These so-called observability or reachability canonical forms are strongly connected to the observability and reachability matrices of

SS representations and they are required to develop equivalence transformation between different representation domains (see Chap. 4).

In the following discussion, we explore complete state-observability and reachability[5] of LPV-SS representations both in CT and DT, based on concepts of the LTV system theory. We show that these properties are equivalent with the existence of invertible linear maps for all time instances and scheduling trajectories. Based on this, we define observability and reachability matrices of SS representations. We show that complete state-observability and reachability are very strong properties in the LPV case as only a rather restricted class of representations fulfills them. Moreover, they are not required for minimality nor the generation of observability or reachability canonical forms. We show that a weaker property, the so-called structural state-observability/reachability, which defines the state-observability/reachability concept in an almost everywhere sense, is a necessary and sufficient property to generate these canonical forms.

As a first step, we extended the behavioral definitions of complete state-observability and reachability (see [231]) to the LPV case:

Definition 3.31 (Complete LPV state-observability). $\mathfrak{R}_{\mathrm{SS}}(\mathcal{S})$ is called completely state-observable, if for all $(u,x,y,p),(u,x',y,p) \in \mathfrak{B}_{\mathrm{SS}}$ it holds that $x = x'$. □

Definition 3.32 (Complete LPV state-reachability). $\mathfrak{R}_{\mathrm{SS}}(\mathcal{S})$ is called completely state-reachable, if for any given two states $\mathrm{x}_1,\mathrm{x}_2 \in \mathbb{X}$ and any scheduling signal $p \in \mathfrak{B}_{\mathbb{P}}$, there exist an input signal u and an output signal y such that $(u,x,y,p) \in \mathfrak{B}_{\mathrm{SS}}$ with $x(t_1) = \mathrm{x}_1$ and $x(t_2) = \mathrm{x}_2$ for some $t_1,t_2 \in \mathbb{T}$. □

The main difference of these definitions with respect to the LTI case follows from the presence of the scheduling signal p that acts as an extra "time-axis" of the system. By freezing this axis, i.e. using a constant scheduling, these concepts coincide with the original LTI definitions (see [146]). To establish conditions when an LPV-SS representation is completely state-observable or reachable, as a next step, the concept of state-observability and reachability matrices is introduced. Appropriate conditions are derived to formulate when and in which sense the full rank of these matrices implies complete state-observability or reachability. However, the formulation of these matrices follows a different track than in the LTI case, due to the non-commutative multiplication on $\mathcal{R}[\xi]$.

3.3.1.1 State-Observability in Continuous-Time

First the CT observability case is investigated based on the results of LTV system theory by [174] and [35]. Using these approaches, we establish the concept of state-observability of $\mathfrak{R}_{\mathrm{SS}}(\mathcal{S})$ on a finite time interval and show that this property is equivalent with the existence of an invertible linear map. Then we define

[5] Note that in many text-books, reachability is also called controllability. As in DT, sometimes controllability is identified as state-stabilizability, usually much confusion rises around this term. To avoid this problem, here the terminology of reachability is used.

this linear map as the state-observability matrix of $\mathfrak{R}_{\mathrm{SS}}(\mathcal{S})$ and claim that state-observability (invertibility of the map) on any finite time interval implies complete state-observability.

For a given continuous-time LPV-SS representation $\mathfrak{R}_{\mathrm{SS}}(\mathcal{S})$ with matrix functions (A,B,C,D), define $\mathfrak{A}(t_1,t_0,p)$ as the *state transition matrix function* for $\dot{x} = (A\diamond p)x$. Then the state and output evolution of $\mathfrak{R}_{\mathrm{SS}}(\mathcal{S})$ satisfy that

$$x(t_1) = \mathfrak{A}(t_1,t_0,p)x(t_0) + \int_{t_0}^{t_1} \mathfrak{A}(t_1,\tau,p)(B\diamond p)(\tau)u(\tau)\,\mathrm{d}\tau, \tag{3.83a}$$

$$y(t_1) = (C\diamond p)(t_1)x(t_1), \tag{3.83b}$$

for all $t_1 \geq t_0$ along a given scheduling trajectory $p \in \mathfrak{B}_{\mathbb{P}}$. For LPV-SS representations in general, the explicit form of $\mathfrak{A}(t_1,t_0,p)$ is hard to derive as it involves integration over the scheduling trajectory, while in the LTI case, $\mathfrak{A}(t_1,t_0) = \mathrm{e}^{A(t_1-t_0)}$.

Complete state-observability of $\mathfrak{R}_{\mathrm{SS}}(\mathcal{S})$ can be described using the concept of reconstructibility. For a given $p \in \mathfrak{B}_{\mathbb{P}}$, define the following linear map from $\mathbb{X}$ to $\mathcal{L}_1^{\mathrm{loc}}([t_0,t_1],\mathbb{Y})$:

$$y(t) = \underbrace{(C\diamond p)(t))\mathfrak{A}(t,t_0,p)}_{\mathfrak{C}(t)}\mathrm{x}, \quad \mathrm{x}\in\mathbb{X},\ t\in[t_0,t_1]. \tag{3.84}$$

For the given $p \in \mathfrak{B}_{\mathbb{P}}$, the mapping $y(t) = \mathfrak{C}(t)\mathrm{x}$ defines the output evolution of $\mathfrak{R}_{\mathrm{SS}}(\mathcal{S})$ on the finite interval $[t_0,t_1] \subset \mathbb{R}$ with initial state $x(t_0) = \mathrm{x}$ and with zero input signal $u = 0$. The state $\mathrm{x} \in \mathbb{X}$ is said to be reconstructible on $[t_0,t_1]$, if x lies in the kernel of the linear map (3.84). This means that if (3.84) is injective, then every state is reconstructible on $[t_0,t_1]$. If (3.84) is injective for every $p \in \mathfrak{B}_{\mathbb{P}}$, then $\mathfrak{R}_{\mathrm{SS}}(\mathcal{S})$ is called state-observable on $[t_0,t_1]$. It is easy to show that $\mathfrak{R}_{\mathrm{SS}}(\mathcal{S})$ is completely state-observable, if it is state-observable for any finite interval of $\mathbb{R}$. Now we can introduce the following matrix function to describe the map (3.84):

Definition 3.33 (n-step CT state-observability matrix function, [175]). In continuous time, the n-step state-observability matrix function of $\mathfrak{R}_{\mathrm{SS}}(\mathcal{S})$ is defined as $\mathcal{O}_n \in \mathcal{R}^{(n n_{\mathbb{Y}})\times n_{\mathbb{X}}}$ with

$$\mathcal{O}_n = \begin{bmatrix} \mathrm{o}_1^\top & \mathrm{o}_2^\top & \dots & \mathrm{o}_n^\top \end{bmatrix}^\top, \tag{3.85}$$

where $\mathrm{o}_1 = C$ and

$$\mathrm{o}_{i+1} = \mathrm{o}_i A + \dot{\mathrm{o}}_i, \quad i > 1. \tag{3.86}$$

□

The n-step state-observability matrix function has a similar role as the state-observability matrix for LTI representations, as in case of complete state-observability, it provides an invertible map for the reconstruction of the state from the derivatives of y.

Example 3.23 (CT state-observability matrix function). Let the SS representation $\mathfrak{R}_{\mathrm{SS}}(\mathcal{S})$ of a CT-LPV system $\mathcal{S}$ with $\mathbb{P} \subseteq \mathbb{R}^+$ be given by

$$\left[\begin{array}{c|c} A & B \\ \hline C & D \end{array}\right] \diamond p = \left[\begin{array}{cc|c} p & 1 & 1 \\ 0 & \frac{1}{p} & p \\ \hline 1 & p & p \end{array}\right].$$

Then the observability matrices of $\mathfrak{R}_{\mathrm{SS}}(\mathcal{S})$ for $n = 1, 2, 3$ are as follows:

$$\mathcal{O}_1 \diamond p = \begin{bmatrix} 1 & p \end{bmatrix}, \ \mathcal{O}_2 \diamond p = \begin{bmatrix} 1 & p \\ p & 2 + \frac{\mathrm{d}}{\mathrm{d}t} p \end{bmatrix}, \ \mathcal{O}_3 \diamond p = \begin{bmatrix} 1 & p \\ p & 2 + \frac{\mathrm{d}}{\mathrm{d}t} p \\ \frac{\mathrm{d}}{\mathrm{d}t} p + p^2 & \frac{p^2 + 2 + \frac{\mathrm{d}}{\mathrm{d}t} p}{p} + \frac{\mathrm{d}^2}{\mathrm{d}t^2} p \end{bmatrix}.$$

These matrices were computed by using:

$$\dot{\mathrm{o}}_1 \diamond p = \begin{bmatrix} 0 & \frac{\mathrm{d}}{\mathrm{d}t} p \end{bmatrix}, \qquad \dot{\mathrm{o}}_2 \diamond p = \begin{bmatrix} \frac{\mathrm{d}}{\mathrm{d}t} p & \frac{\mathrm{d}^2}{\mathrm{d}t^2} p \end{bmatrix}.$$ □

Based on the LTI case, one could expect that full rank of $\mathcal{O}_{n_{\mathbb{X}}}$, i.e. $\mathrm{Rank}(\mathcal{O}_{n_{\mathbb{X}}}) = n_{\mathbb{X}}$ (full rank in a functional sense), is the necessary and sufficient condition for complete state-observability. However, this not true in the LPV case. It can also be shown that instead of the functional full rank, full rank of $\mathcal{O}_{n_{\mathbb{X}}}$ for every time instance along every scheduling trajectory, i.e. $\mathrm{Rank}((\mathcal{O}_{n_{\mathbb{X}}} \diamond p)(t)) = n_{\mathbb{X}}$ for all $t \in \mathbb{R}$ and $p \in \mathfrak{B}_{\mathbb{P}}$, is a sufficient but not a necessary condition for complete state-observability. To show this and to derive the sufficient and necessary condition for complete state-observability, first introduce a weaker notion of state-observability:

Definition 3.34 (Structural state-observability). $\mathfrak{R}_{\mathrm{SS}}(\mathcal{S})$ with state-dimension $n_{\mathbb{X}}$ is called structurally state-observable if its $n_{\mathbb{X}}$-step observability matrix $\mathcal{O}_{n_{\mathbb{X}}}$ is full (column) rank, i.e. $\mathrm{Rank}(\mathcal{O}_{n_{\mathbb{X}}}) = n_{\mathbb{X}}$. □

Note that full rank in a functional sense, does not guarantee that $\mathcal{O}_{n_{\mathbb{X}}}$ is invertible for all $t \in \mathbb{R}$ and $p \in \mathfrak{B}_{\mathbb{P}}$. Therefore, for specific scheduling trajectories and time instances, reconstructibility of state x by the linear map $\mathcal{O}_{n_{\mathbb{X}}}$ is not guaranteed. In this way, complete state-observability is not implied. Even if reconstructibility may fail for some scheduling trajectories, for the rest state-observability holds on $\mathbb{R}$. This gives that structural observability can be understood as complete state-observability in an almost everywhere sense. In the following we show that in fact structural state-observability is a necessary condition for complete state-observability.

To derive the appropriate conditions for complete state-observability the following lemma has a key importance:

Lemma 3.2 (Constant observability rank, [175]). *Let a representation $\mathfrak{R}_{\mathrm{SS}}(\mathcal{S})$ be given with a projected scheduling behavior $\mathfrak{B}_{\mathbb{P}}$ and observability matrices $\mathcal{O}_n$. For a scheduling trajectory $p \in \mathfrak{B}_{\mathbb{P}}$, it holds that, if there exists $n > 0$ such that*

$$\mathrm{Rank}((\mathcal{O}_n \diamond p)(\tau)) = \mathrm{Rank}((\mathcal{O}_{n+1} \diamond p)(\tau)) = \gamma, \tag{3.87}$$

for all $\tau \in \mathbb{T}$, then $\mathrm{Rank}((\mathcal{O}_l \diamond p)(\tau)) = \gamma$ for all $l \geq n$ and $\tau \in \mathbb{T}$.

The proof is given in [175]. The minimal $n > 0$, for which (3.87) holds, is called the observability radius of $\mathfrak{R}_{\mathrm{SS}}(\mathcal{S})$ with respect to the scheduling trajectory p. This lemma has the obvious consequence that for a given $p \in \mathfrak{B}_{\mathbb{P}}$, if the ranks of $\mathcal{O}_n \diamond p$ and $\mathcal{O}_{n+1} \diamond p$ are constant and equal along the entire trajectory of p, then the rank of $\mathcal{O}_l \diamond p$ remains constant for all $l \geq n$. Contrary to the LTI case, it is not guaranteed that the rank of $\mathcal{O}_{n_{\mathbb{X}}} \diamond p$ is equal to the rank of $\mathcal{O}_{n_{\mathbb{X}}+1} \diamond p$ for all t, i.e. the observability radius is smaller or equal than $n_{\mathbb{X}}$. As the observability radius can vary for each trajectory of p, the introduction of the notion of constant observability rank representation is required:

Definition 3.35 (Constant observability rank representation, [175]). A $\mathfrak{R}_{\mathrm{SS}}(\mathcal{S})$ representation has constant observability rank if there exist a $n \in \{1,\dots,n_{\mathbb{X}}\}$ and a $l > 0$ such that

$$\mathrm{Rank}((\mathcal{O}_n \diamond p)(\tau)) = \mathrm{Rank}((\mathcal{O}_{n+1} \diamond p)(\tau)) = l \leq n_{\mathbb{X}}, \tag{3.88}$$

for all $p \in \mathfrak{B}_{\mathbb{P}}$ and $\tau \in \mathbb{T}$. □

Example 3.24 (CT non-constant observability rank representation). For Example 3.23 it holds, that $\mathrm{Rank}((\mathcal{O}_2 \diamond p)(t)) = 2$ for all possible scheduling signals and time instances, except when for a $t \in \mathbb{R}$ the scheduling signal satisfies $p^2(t) = 2 + \frac{\mathrm{d}}{\mathrm{d}t} p(t)$. At that time instant, $\mathrm{Rank}((\mathcal{O}_2 \diamond p)(t)) = 1$. Hence it is not a constant observability rank representation, but its minimal observability radius is 1. For a constant observability rank representation see Example 3.25. □

Using the previously introduced concepts the following theorem holds:

Theorem 3.9 (Induced complete LPV state-observability in CT, [174]). *The CT-LPV-SS representation $\mathfrak{R}_{\mathrm{SS}}(\mathcal{S})$ is completely state-observable, iff for every $p \in \mathcal{C}^\infty(\mathbb{R},\mathbb{P}) \cap \mathfrak{B}_{\mathbb{P}}$ there exists a $0 < n < \infty$ such that* $\mathrm{Rank}((\mathcal{O}_n \diamond p)(t)) = n_{\mathbb{X}}$ *for all $t \in \mathbb{R}$. If $\mathfrak{R}_{\mathrm{SS}}(\mathcal{S})$ is a constant observability rank representation, then the condition is* $\mathrm{Rank}((\mathcal{O}_{n_{\mathbb{X}}} \diamond p)(t)) = n_{\mathbb{X}}$ *for all $t \in \mathbb{R}$.*

The proof follows similarly as in [174]. The clear interpretation of this result is important. If a LPV-SS representation is completely state-observable, then it is not guaranteed that the reconstruction of the state is available for every time instance through the linear map $(\mathcal{O}_{n_{\mathbb{X}}} \diamond p)(t)$, as it is not injective. It can happen that this property is only satisfied for $n > n_{\mathbb{X}}$. In case $\mathfrak{R}_{\mathrm{SS}}(\mathcal{S})$ is a constant observability rank representation, then, similar to the LTI case, full rank of $(\mathcal{O}_{n_{\mathbb{X}}} \diamond p)(t)$ along every $p \in \mathfrak{B}_{\mathbb{P}}$ guarantees complete state-observability. For an SS representation with complete state-observability, see Example 3.25.

It can be shown that the class of LPV-SS representations with constant observability rank includes the class of LTI-SS representations, similar to the LTV case [172]. In LPV control design, instead of observability, the so-called *detectability* of the state is investigated together with *stabilizibility* [96]. These concepts are formulated in a quadratic sense, similar to quadratic Lyapunov stability (see later in Sect. 3.3.2). It can be shown that if quadratic detectability is satisfied for a LPV-SS representation with static linear dependence, then it is a necessary condition in terms of Th. 3.9 for complete state-observability. In the existing LPV system theory the notion of complete-state observability is only considered in [10], where invariant observability subspaces of LPV-SS representations with static linear dependence are explored using the concepts of nonlinear system theory. However, the connection of these results to Th. 3.9 is currently not fully understood.

Example 3.25 (CT complete state-observability). Let the SS representation $\mathfrak{R}_{\mathrm{SS}}(\mathcal{S})$ of an CT-LPV system $\mathcal{S}$ with $\mathbb{P} = [-\frac{1}{2}, \frac{1}{2}]$ be given by

$$\left[\begin{array}{c|c} A & B \\ \hline C & D \end{array}\right] \diamond p = \left[\begin{array}{cc|c} \tan(p)\frac{\mathrm{d}}{\mathrm{d}t}p & 0 & 1 \\ \sin(p) & 2 & 1 \\ \hline \cos(p) & 1 & 0 \end{array}\right].$$

Then it follows that

$$\mathcal{O}_2 \diamond p = \begin{bmatrix} \cos(p) & 1 \\ \sin(p) & 2 \end{bmatrix}.$$

As $\cos(p) \neq 2\sin(p)$ on $\mathbb{P}$, $\mathcal{O}_2$ has a constant rank of 2 for all $p \in \mathfrak{B}_{\mathbb{P}}$ and $t \in \mathbb{R}$. This means that $\mathfrak{R}_{\mathrm{SS}}(\mathcal{S})$ is completely state observable on $\mathbb{P}$. □

In case $\mathrm{Rank}(\mathcal{O}_{n_{\mathbb{X}}}) = n_{\mathbb{X}}$ is satisfied, i.e. the SS representation is structurally state-observable, then it holds that $\mathrm{Rank}((\mathcal{O}_{n_{\mathbb{X}}} \diamond p)(t)) = n_{\mathbb{X}}$ for all $t \in \mathbb{R}$ except for some $p \in \mathcal{C}^{\infty}(\mathbb{R}, \mathbb{P}) \cap \mathfrak{B}_{\mathbb{P}}$. Then it is obvious that structural state-observability is a necessary condition for complete state-observability (see Example 3.26). This claim is also proved in [172].

Example 3.26 (Structural state-observability). The representation defined in Example 3.23 has been shown to be not completely state-observable as $\min_{t \in \mathbb{R}} \mathrm{Rank}((\mathcal{O}_2 \diamond p)(t)) = 1$ for scheduling trajectories that satisfy $p^2(t) = 2 + \frac{\mathrm{d}}{\mathrm{d}t}p(t)$ at a $t \in \mathbb{R}$. However, it is obvious that it is completely observable for all other scheduling trajectories, as $\mathrm{Rank}(\mathcal{O}_2) = 2$ in the functional sense. This implies complete state-observability of the representation in an almost everywhere sense, i.e. the representation defined in Example 3.23 is structurally state-observable. □

To check complete state-observability of a given $\mathfrak{R}_{\mathrm{SS}}(\mathcal{S})$, an iterative computation strategy must be applied in terms of Th. 3.9, checking the rank of $\mathcal{O}_l \diamond p$ for increasing $l > 0$. In step l of this iterative scheme, computation of the minimum of $\mathrm{Rank}((\mathcal{O}_l \diamond p)(t))$ for all $p \in \mathfrak{B}_{\mathbb{P}}$ and $t \in \mathbb{R}$ is required which is usually an infinite dimensional and hence unsolvable optimization problem. An approximative solution may follow through the parametrization of p like polynomial, piecewise continuous, periodical, etc. and then using this parameterized scheduling to set up a feasibility problem for the full rank of the matrix function on a large interval of $\mathbb{R}$. If in iteration step l it holds that $\mathrm{Rank}((\mathcal{O}_l \diamond p)(t)) = \mathrm{Rank}((\mathcal{O}_{l-1} \diamond p)(t))$ for all $p \in \mathfrak{B}_{\mathbb{P}}$ and $t \in \mathbb{R}$, then the observability rank of the representation is found. As this condition cannot be checked in a non-conservative sense, the iterative scheme can only prove complete state-observability if full rank of $\mathcal{O}_l \diamond p$ can be shown in a computationally feasible number of steps.

On the other hand, the rank test for structural state-observability can be accomplished in the SISO case based on symbolic computation of the determinant of $\mathcal{O}_{n_{\mathbb{X}}}$.

In case the result is a non-zero function, then structural state-observability is satisfied. In the MIMO case, full rank of $\mathcal{O}_{n_\mathbb{X}}$ can be checked by forming square matrices from the rows of $\mathcal{O}_{n_\mathbb{X}}$ using all possible combinations and checking if any of the determinants of these matrices is non-zero.

3.3.1.2 State-Reachability in Continuous-Time

The concepts introduced in the observability case can be similarly introduced in the reachability sense. The only difference is that to show the results, instead of state-reconstructibility, we need to introduce the so-called *reachability map*. By introducing this map, we develop the analog of the results of the observability case.

In terms of Def. 3.32, reachability of an LPV representation represents the ability to transfer an arbitrary initial state to an arbitrary target state of $\mathbb{X}$ in case of any scheduling trajectory. This can be explained using controllability maps between the space of input signals and the state-space $\mathbb{X}$. For a given $p \in \mathfrak{B}_\mathbb{P}$, define the following linear map from $\mathcal{L}_1^{\mathrm{loc}}([t_2,t_1],\mathbb{U})$ to $\mathbb{X}$:

$$\mathrm{x} = \int_{t_0}^{t_1} \mathfrak{A}(t_1,\tau,p)B(p(\tau))u(\tau)\mathrm{d}\tau, \quad \mathrm{x} \in \mathbb{X},\ t \in [t_0,t_1]. \tag{3.89}$$

For the given p, the mapping (3.89) defines the state evolution of $\mathfrak{R}_{\mathrm{SS}}(\mathcal{S})$ in the finite interval $[t_0,t_1]$ for an initial condition $x(t_0) = 0$ and input $u \in \mathcal{L}_1^{\mathrm{loc}}([t_0,t_1],\mathbb{U})$. The state $\mathrm{x} \in \mathbb{X}$ is said to be controllable on $[t_0,t_1]$, if x lies in the image of the linear map (3.89). Note that in case of initial condition $x(t_0) = \mathrm{x}_0$ and target state $x(t_1) = \mathrm{x}_1$, linearity of the signal behavior implies that the input $u \in \mathcal{L}_1^{\mathrm{loc}}([t_0,t_1],\mathbb{U})$ which satisfies (3.89) for $\mathrm{x} = \mathrm{x}_1 - \mathfrak{A}(t_1,t_0,p)\mathrm{x}_0$, transfers the state x_0 to x_1 in $[t_0,t_1]$. This means that if (3.89) is surjective, then every state can be reached from an arbitrary state in the time-interval $[t_0,t_1]$. If (3.89) is surjective for every $p \in \mathfrak{B}_\mathbb{P}$, then $\mathfrak{R}_{\mathrm{SS}}(\mathcal{S})$ is called state-reachable on $[t_0,t_1]$. It is easy to show that $\mathfrak{R}_{\mathrm{SS}}(\mathcal{S})$ is completely state-reachable, if it is state-observable for any finite interval in $\mathbb{R}$. Now we can introduce the following matrix function to describe the linear map (3.89):

Definition 3.36 (n-step CT state-reachability matrix function, [175]). In continuous-time, the n-step state-reachability matrix function of $\mathfrak{R}_{\mathrm{SS}}(\mathcal{S})$ is defined as $\mathcal{R}_n \in \mathcal{R}^{n_\mathbb{X} \times (n n_\mathbb{U})}$ with

$$\mathcal{R}_n = \begin{bmatrix} \mathrm{r}_1 & \mathrm{r}_2 & \dots & \mathrm{r}_n \end{bmatrix}, \tag{3.90}$$

where $\mathrm{r}_1 = B$ and

$$\mathrm{r}_{i+1} = -A\mathrm{r}_i + \dot{\mathrm{r}}_i, \quad i > 1. \tag{3.91}$$

□

The n-step state-reachability matrix function has similar role as the state-reachability matrix for LTI representations.

Example 3.27 (CT state-reachability matrix function). Consider the LPV-SS representation $\mathfrak{R}_{\mathrm{SS}}(\mathcal{S})$ defined in Example 3.23. The reachability matrices of $\mathfrak{R}_{\mathrm{SS}}(\mathcal{S})$ for $n = 1,2,3$ are as follows:

$$\mathcal{R}_1 \diamond p = \begin{bmatrix} 1 \\ p \end{bmatrix}, \quad \mathcal{R}_2 \diamond p = \begin{bmatrix} 1 & -2p \\ p & \frac{d}{dt}p - 1 \end{bmatrix}, \quad \mathcal{R}_3 \diamond p = \begin{bmatrix} 1 & -2p & 1 + 2p^2 - 3\frac{d}{dt}p \\ p & \frac{d}{dt}p - 1 & \frac{1-\frac{d}{dt}p}{p} + \frac{d^2}{dt^2}p \end{bmatrix}.$$

These matrices were computed by using:

$$\dot{\mathfrak{r}}_1 \diamond p = \begin{bmatrix} 0 & \frac{d}{dt}p \end{bmatrix}^\top, \qquad \dot{\mathfrak{r}}_2 \diamond p = \begin{bmatrix} -2\frac{d}{dt}p & \frac{d^2}{dt^2}p \end{bmatrix}^\top. \qquad \square$$

We can also introduce the notion of structural state-reachability:

Definition 3.37 (Structural state-reachability). $\mathfrak{R}_{SS}(\mathcal{S})$ with state-dimension $n_\mathbb{X}$ is called structurally state-reachable if its $n_\mathbb{X}$-step reachability matrix $\mathcal{R}_{n_\mathbb{X}}$ is full (row) rank, i.e. $\mathrm{Rank}(\mathcal{R}_{n_\mathbb{X}}) = n_\mathbb{X}$. □

Moreover, it can be shown that Lemma 3.2 holds also for $\mathcal{R}_n$, implying all the properties that have been noted in the observability case [175]. Similarly we can introduce the reachability radius of a given SS representation with respect to a scheduling signal $p \in \mathfrak{B}_\mathbb{P}$. As the reachability radius can vary for each trajectory of p, thus the notion of constant reachability rank representation is introduced:

Definition 3.38 (Constant reachability rank representation, [175]). A $\mathfrak{R}_{SS}(\mathcal{S})$ representation has constant reachability rank if there exist a $n \in \{1, \ldots, n_\mathbb{X}\}$ and a $l > 0$ such that

$$\mathrm{Rank}((\mathcal{R}_n \diamond p)(\tau)) = \mathrm{Rank}((\mathcal{R}_{n+1} \diamond p)(\tau)) = l \leq n_\mathbb{X}, \tag{3.92}$$

for all $p \in \mathfrak{B}_\mathbb{P}$ and $\tau \in \mathbb{T}$. □

Example 3.28 (CT non-constant reachability rank representation). In Example 3.27 it holds that $\mathrm{Rank}((\mathcal{R}_2 \diamond) p(t)) = 2$ for all possible scheduling signals and time instances, except when for a $t \in \mathbb{R}$ the scheduling signal satisfies $2p^2(t) = -\frac{d}{dt}p(t) + 1$. At that time instant, $\mathrm{Rank}((\mathcal{R}_2 \diamond p)(t)) = 1$. Hence it is not a constant reachability rank representation, but its minimal reachability radius is 1. For a constant reachability rank representation see Example 3.29. □

Using the previously introduced concepts, the analog of Th. 3.9 holds in the reachability case as well:

Theorem 3.10 (Induced complete LPV state-reachability in CT, [174]). *The CT-LPV-SS representation $\mathfrak{R}_{SS}(\mathcal{S})$ is completely state-reachable, iff for any $p \in \mathcal{C}^\infty(\mathbb{R}, \mathbb{P}) \cap \mathfrak{B}_\mathbb{P}$ it holds that there exists a $0 < n < \infty$ such that* $\mathrm{Rank}((\mathcal{R}_n \diamond p)(t)) = n_\mathbb{X}$ *for all $t \in \mathbb{R}$. If $\mathfrak{R}_{SS}(\mathcal{S})$ has constant reachability rank, then the condition is* $\mathrm{Rank}((\mathcal{R}_{n_\mathbb{X}} \diamond p)(t)) = n_\mathbb{X}$ *for all $t \in \mathbb{R}$.*

The proof follows similarly as in [174]. The interpretation of this theorem is the same as in the observability case: an LPV-SS can be completely state-reachable, even if the $n_\mathbb{X}$-step state-reachability matrix is not full rank along every scheduling trajectory. In case $\mathfrak{R}_{SS}(\mathcal{S})$ is a constant reachability rank representation, then, similar to the LTI case, full rank of $(\mathcal{R}_{n_\mathbb{X}} \diamond p)(t)$ along every $p \in \mathfrak{B}_\mathbb{P}$ guarantees complete state-reachability. For an SS representation with complete state-reachability, see Example 3.29.

It can also be shown that the class of LPV-SS representations with constant reachability rank includes the class of LTI-SS representations (see [172]). Additionally, quadratic stabilizibility (see [96]) of a LPV-SS representation with static linear dependence is a necessary condition in terms of Th. 3.10 for complete state-reachability.

Example 3.29 (CT complete state-reachability). Let the SS representation $\mathfrak{R}_{\mathrm{SS}}(\mathcal{S})$ of an CT-LPV system $\mathcal{S}$ with $\mathbb{P} = [-\frac{1}{2}, \frac{1}{2}]$ be given by

$$\left[\begin{array}{c|c} A & B \\ \hline C & D \end{array}\right] \diamond p = \left[\begin{array}{cc|c} -\tan(p)\frac{\mathrm{d}}{\mathrm{d}t}p & \sin(p) & \cos(p) \\ 0 & 2 & 1 \\ \hline 1 & 1 & 0 \end{array}\right].$$

Then it follows that

$$\mathcal{R}_2 \diamond p = \begin{bmatrix} \cos(p) & \sin(p) \\ 1 & 2 \end{bmatrix}.$$

As $2\cos(p) \neq \sin(p)$ on $\mathbb{P}$, $\mathcal{R}_2$ has rank 2 for all $p \in \mathfrak{B}_{\mathbb{P}}$ and $t \in \mathbb{R}$. This means that $\mathfrak{R}_{\mathrm{SS}}(\mathcal{S})$ is completely state reachable on $\mathbb{P}$. □

In case $\mathrm{Rank}(\mathcal{R}_{n_{\mathbb{X}}}) = n_{\mathbb{X}}$ is satisfied, i.e. the SS representation is structurally state-reachable, then it holds that $\mathrm{Rank}((\mathcal{R}_n \diamond p)(t)) = n_{\mathbb{X}}$ for all $t \in \mathbb{R}$ except some $p \in \mathcal{C}^{\infty}(\mathbb{R}, \mathbb{P}) \cap \mathfrak{B}_{\mathbb{P}}$. Then, similar to the previous case, it is obvious that structural state-reachability is a necessary condition for complete state-reachability. To check complete or structural state-reachability the computational considerations are the same as discussed for the observability case.

3.3.1.3 State-Observability in Discrete-Time

The concept of state-observability can be similarly investigated in DT for a given LPV-SS representation $\mathfrak{R}_{\mathrm{SS}}(\mathcal{S})$ with state and output equations (3.62a–b). By using a similar line of reasoning, we explore the previously introduced concepts based on the theory for DT-LTV systems in [61]. Define the state transition matrix as

$$\mathfrak{A}(k_1, k_0, p) = \begin{cases} \prod_{i=0}^{k_1-k_0-1} (A \diamond p)(k_1 - i), & \text{if } k_1 > k_0; \\ I, & \text{if } k_1 \leq k_0; \end{cases} \tag{3.93}$$

for $k_1, k_0 \in \mathbb{Z}$. Then, based on (3.62a–b), the state and output evolution of $\mathfrak{R}_{\mathrm{SS}}(\mathcal{S})$ in the finite time interval $[k_0, k_1] \subset \mathbb{Z}$ satisfies

$$x(k_1) = \mathfrak{A}(k_1, k_0, p)x(k_0) + \sum_{i=k_0}^{k_1} \mathfrak{A}(k_1, i, p)(B \diamond p)(i)u(i), \tag{3.94a}$$

$$y(k_1) = (C \diamond p)(k_1)x(k_1), \tag{3.94b}$$

for all $k_1 \geq k_0$ along a scheduling trajectory $p \in \mathfrak{B}_{\mathbb{P}}$. Again, complete state-observability of $\mathfrak{R}_{SS}(\mathcal{S})$ can be considered as a reconstruction problem of any $\mathrm{x}_0 = x(k_0)$ from a zero input response in the time interval $[k_0, k_1] \subset \mathbb{Z}$. Then state-observability on $[k_0, k_1]$ requires that the linear map

$$y(k) = (C \diamond p)(k)\mathfrak{A}(k, k_0, p)\mathrm{x}, \quad \mathrm{x} \in \mathbb{X},\ k \in [k_0, k_1], \tag{3.95}$$

from $\mathbb{X}$ to $\mathbb{Y}^{[k_0,k_1]}$ is injective for every $p \in \mathbb{P}^{[k_0,k_1]}$. It can be shown again, that $\mathfrak{R}_{SS}(\mathcal{S})$ is completely state-observable if it is state-observable for any finite interval of $\mathbb{Z}$ with $n_{\mathbb{X}} \leq k_1 - k_0 < \infty$. Now we can introduce the following observability matrix in DT to describe the linear map (3.95):

Definition 3.39 (n-step DT state-observability matrix function, [61]). In discrete-time, the n-step state-observability matrix function $\mathcal{O}_n \in \mathcal{R}^{n_{\mathbb{X}} \times n n_{\mathbb{U}}}$ of $\mathfrak{R}_{SS}(\mathcal{S})$ is defined as (3.85) with

$$\mathrm{o}_1 = C, \quad \mathrm{o}_{i+1} = \overrightarrow{\mathrm{o}_i} A, \quad \forall i > 1. \tag{3.96}$$

□

The n-step state-observability matrix function has a similar role as the state-observability matrix for LTI representations, as in case of complete state-observability, it provides an invertible map for the reconstruction of the state from the samples of y. Note that the difference in the structure of the discrete-time n-step state-observability matrix with respect to its continuous-time counterpart is due to the different commutation rules for the $\frac{\mathrm{d}}{\mathrm{d}t}$ and q operators on $\mathcal{R}[\xi]$.

Example 3.30 (DT state-observability matrix function). Consider the DT-LPV-SS representation $\mathfrak{R}_{SS}(\mathcal{S})$ given by

$$\left[\begin{array}{c|c} A & B \\ \hline C & D \end{array}\right] \diamond p = \left[\begin{array}{cc|c} p & 1 & 1 \\ 0 & p & p \\ \hline 1 & \frac{1}{p} & p \end{array}\right],$$

with $\mathbb{P} = [\frac{1}{4}, \frac{3}{4}]$. Then the observability matrices of $\mathfrak{R}_{SS}(\mathcal{S})$ for $n = 1, 2, 3$ are as follows:

$$\mathcal{O}_1 \diamond p = \begin{bmatrix} 1 & \frac{1}{p} \end{bmatrix}, \ \mathcal{O}_2 \diamond p = \begin{bmatrix} 1 & \frac{1}{p} \\ p & 1 + \frac{p}{\mathrm{q}p} \end{bmatrix}, \ \mathcal{O}_3 \diamond p = \begin{bmatrix} 1 & \frac{1}{p} \\ p & 1 + \frac{p}{\mathrm{q}p} \\ p(\mathrm{q}p) & p(1 + \frac{(\mathrm{q}p)}{(\mathrm{q}^2 p)}) + \mathrm{q}p \end{bmatrix}.$$

These matrices were computed by using:

$$\overrightarrow{\mathrm{o}}_1 \diamond p = \begin{bmatrix} 0 & \frac{1}{\mathrm{q}p} \end{bmatrix}, \qquad \overrightarrow{\mathrm{o}}_2 \diamond p = \begin{bmatrix} \mathrm{q}p & 1 + \frac{(\mathrm{q}p)}{(\mathrm{q}^2 p)} \end{bmatrix}.$$

□

Again we can introduce structural state-observability in terms of Def. 3.34. Similarly, Lemma 3.2 also holds in the DT case and by using the concepts of n-step state-observability matrix and constant observability-rank representation, given by Def. 3.39 and 3.35, the induced complete state-observability, Th. 3.9, can be shown to hold in DT as well:

Theorem 3.11 (Induced complete LPV state-observability in DT, [61]). *The LPV-SS representation $\mathfrak{R}_{SS}(\mathcal{S})$ is completely state-observable, iff for any $p \in \mathfrak{B}_{\mathbb{P}}$ it holds that there exists a $0 < n < \infty$ such that* $\mathrm{Rank}((\mathcal{O}_n \diamond p)(k)) = n_{\mathbb{X}}$ *for all $k \in \mathbb{Z}$. If $\mathfrak{R}_{SS}(\mathcal{S})$ has constant observability rank, then the condition is* $\mathrm{Rank}((\mathcal{O}_{n_{\mathbb{X}}} \diamond p)(k)) = n_{\mathbb{X}}$ *for all $k \in \mathbb{Z}$.*

The proof follows similarly as in [61]. The consequences of this theorem are similar as in the CT case, also implying that discrete-time LPV-SS representations with constant observability rank include the class of DT-LTI-SS representations (see [61]). Furthermore it can be shown, that structural state-observability is a necessary condition for complete state-observability also in the DT case. Again, basic DT results of the existing LPV system theory follow as special cases of Th. 3.11.

Example 3.31 (DT complete state-observability). Consider Example 3.30. It is trivial that $\mathrm{Det}(\mathcal{O}_2 \diamond p) = \frac{p}{\mathrm{q}p}$ can not be zero on $\mathbb{P} = [\frac{1}{4}, \frac{3}{4}]$, thus $\mathrm{Rank}((\mathcal{O}_2 \diamond p)(k)) = 2$ for all $p \in \mathfrak{B}_{\mathbb{P}}$ and $k \in \mathbb{Z}$. Naturally, the same holds for $\mathcal{O}_3$. Hence the SS representation of Example 3.30 is completely state-observable. □

Again an iterative test can be applied to check complete state-observability in terms of Th. 3.11. However, in the DT case, computation of the minimal rank of $(\mathcal{O}_n \diamond p)(\tau)$ can be accomplished by the generalization of the *Popov-Belevitch-Hautus* (PBH) spectral test, resulting in an *almost eigenvalue* problem [140]. As computation of the almost eigenvalue/eigenvectors is difficult in most general cases (see [16]), even for representations with linear dependence, therefore in practice an approximative approach is suggested. By this approach, the full rank condition of $(\mathcal{O}_n \diamond p)(\tau)$ is checked for finite sequences of p, like $\{\mathrm{p}_0, \ldots, \mathrm{p}_{N-1}\}$. Each shifted instance of p in $\mathcal{O}_n$, like $\mathrm{q}^l p$, is associated with the appropriate element of the sequence, i.e. $\mathrm{q}^l p = \mathrm{p}_l$. In this way, the full rank test of $(\mathcal{O}_n \diamond p)(\tau)$, can be formulated as a feasibility problem on $\mathbb{P}^N$, which can be solved via nonlinear optimization or by gridding. This mechanism corresponds to a conservative rank test, as the feasibility is checked for arbitrary variations of p. To compute the actual rank of $\mathcal{O}_n$, the previous method is applied iteratively, checking the full rank of $\mathcal{O}_l$ for increasing l. This yields an approach that is easily computable for small dimensions. To check structural state-observability, the same symbolic approach can be used as given in the CT case.

3.3.1.4 State-Reachability in Discrete-Time

In discrete-time, the concept of complete state-reachability can be similarly investigated as in CT, except that in case of state-reachability on a discrete interval $[k_0, k_1]$, it is required that the linear map

$$\mathrm{x} = \sum_{i=k_0}^{k} \mathfrak{A}(k, i, p)(B \diamond p)(i)u(i), \quad k \in [k_0, k_1], \tag{3.97}$$

from $\mathbb{U}^{[k_0,k_1]}$ to $\mathbb{X}$ is surjective for every $p \in \mathfrak{B}_{\mathbb{P}}$. Then $\mathfrak{R}_{SS}(\mathcal{S})$ is completely state-reachable if it is state-reachable for any finite interval of $\mathbb{Z}$. Similar to the CT case, we can introduce the following matrix function to describe the linear map (3.97):

Definition 3.40 (n-step DT state-reachability matrix function, [61]). In discrete-time, the n-step state-reachability matrix function $\mathcal{R}_n \in \mathcal{R}^{n_\mathbb{X} \times n n_\mathbb{U}}$ of $\mathfrak{R}_{SS}(\mathcal{S})$ is defined as (3.90) with

$$\mathtt{r}_1 = B, \quad \mathtt{r}_{i+1} = A\overleftarrow{\mathtt{r}_i}, \quad \forall i > 1. \tag{3.98}$$

□

Similar to the previous part, the difference in the structure of the discrete-time n-step state-reachability matrix with respect to its continuous-time counterpart is due to the different commutation rules of the $\frac{d}{dt}$ and q operators on $\mathcal{R}[\xi]$.

Example 3.32 (DT state-reachability matrix function). Consider the SS representation $\mathfrak{R}_{SS}(\mathcal{S})$ defined in Example 3.30. The observability matrices of $\mathfrak{R}_{SS}(\mathcal{S})$ for $n = 1, 2, 3$ are as follows:

$$\mathcal{R}_1 \diamond p = \begin{bmatrix} 1 \\ p \end{bmatrix}, \quad \mathcal{R}_2 \diamond p = \begin{bmatrix} 1 & p + \mathrm{q}^{-1}p \\ p & p(\mathrm{q}^{-1}p) \end{bmatrix},$$

$$\mathcal{R}_3 \diamond p = \begin{bmatrix} 1 & p + \mathrm{q}^{-1}p & p(\mathrm{q}^{-1}p) + (p + \mathrm{q}^{-1}p)\mathrm{q}^{-2}p \\ p & p(\mathrm{q}^{-1}p) & p(\mathrm{q}^{-1}p)(\mathrm{q}^{-2}p) \end{bmatrix}.$$

These matrices were computed by using:

$$\overleftarrow{\mathtt{r}}_1 = \begin{bmatrix} 0 & \mathrm{q}^{-1}p \end{bmatrix}^\top, \qquad \overleftarrow{\mathtt{r}}_2 = \begin{bmatrix} \mathrm{q}^{-1}p + \mathrm{q}^{-2}p & \mathrm{q}^{-1}p(\mathrm{q}^{-2}p) \end{bmatrix}^\top.$$

□

Again we can introduce structural state-reachability in terms of Def. 3.37. Aditionally, using the previously introduced concepts, the induced complete state-reachability, Th. 3.10, holds in DT as well:

Theorem 3.12 (Induced complete LPV state-reachability in DT, [61]). *The discrete time LPV-SS representation $\mathfrak{R}_{SS}(\mathcal{S})$ is completely state-reachable, iff for any $p \in \mathfrak{B}_\mathbb{P}$ it holds that there exists a $0 < n < \infty$ such that* $\mathrm{Rank}((\mathcal{R}_n \diamond p)(k)) = n_\mathbb{X}$ *for all $k \in \mathbb{Z}$. If $\mathfrak{R}_{SS}(\mathcal{S})$ has constant reachability rank, then the condition is* $\mathrm{Rank}((\mathcal{R}_{n_\mathbb{X}} \diamond p)(k)) = n_\mathbb{X}$ *for all $k \in \mathbb{Z}$.*

The proof is follows similarly as in [61]. The consequences of this theorem are the same as in the CT case, also implying that discrete-time LPV-SS representations with constant reachability rank include the class of DT-LTI-SS representations [61]. Furthermore, it can be shown that structural state-reachability is a necessary condition for complete state-reachability also in the DT case. Again, basic DT results of the existing LPV systems theory follow as special cases of Th. 3.11. To check complete or structural state-reachability in discrete-time, the computational considerations are the same as given in the observability case.

Example 3.33 (DT complete state-reachability). Consider Example 3.32. $\mathrm{Det}(\mathcal{R}_2 \diamond p) = -p^2$ implies that $\mathrm{Rank}((\mathcal{R}_2 \diamond p)(k)) = 2$ for all $p \in \mathfrak{B}_\mathbb{P}$ and $k \in \mathbb{Z}$. Trivially, the same holds for $\mathcal{R}_3$. Hence the SS representation of Example 3.30 is completely state-reachable. □

3.3.1.5 General Properties and Minimality

As a next step we show that structural state-observability/reachability are the necessary ingredients to develop observability and reachability canonical forms of SS representations which are similar to their LTI counterparts. Additionally, minimality of LPV-SS representations is implied by structural state-observability instead of the complete concept.

Based on the definition of structural state-observability/reachability, an important corollary is the following:

Corollary 3.1. *If $\mathfrak{R}_{\mathrm{SS}}(\mathcal{S})$ is structurally state-observable (reachable), i.e.* $\mathrm{Rank}(\mathcal{O}_{n_{\mathbb{X}}}) = n_{\mathbb{X}}$ *(*$\mathrm{Rank}(\mathcal{R}_{n_{\mathbb{X}}}) = n_{\mathbb{X}}$*), then at least $n_{\mathbb{X}}$ number of rows of $\mathcal{O}_{n_{\mathbb{X}}}$ (columns of $\mathcal{R}_{n_{\mathbb{X}}}$) are linearly independent in the functional sense. This implies that in the SISO case, $\mathcal{O}_{n_{\mathbb{X}}}$ ($\mathcal{R}_{n_{\mathbb{X}}}$) is invertible.*

Based on this property, $\mathcal{O}_n$ and $\mathcal{R}_n$ can be used to define state-transformations in the behavioral framework and, by using similar argumentation as in the LTI case, to develop canonical forms for LPV-SS representations. Note that, in case of complete state-observability/reachability, $\mathcal{O}_n$ or $\mathcal{R}_n$ are invertible for all scheduling trajectories and time instances, which is a much stronger property than the previous one. Thus by requiring this stronger property to generate canonical forms, we would exclude a large set of SS representations that have an equivalent (in the almost everywhere sense) observability/reachability canonical form (see Chap. 4).

A key theorem that enables construction of observability/reachability canonical forms in LPV case is the following:

Theorem 3.13 (Transformation of the state-observability/reachability structure, [173]). *If the matrices of two LPV-SS representations, with state dimensions $n_{\mathbb{X}}$ and with a common $n_{\mathbb{P}}$ dimensional scheduling space, fulfill the equivalence relation $\overset{n_{\mathbb{P}}}{\sim}$ via state-transformation $T \in \mathcal{R}^{n_{\mathbb{X}} \times n_{\mathbb{X}}}$, then for all $n \in \mathbb{N}$:*

$$\begin{cases} \textit{for } \mathbb{T} = \mathbb{R}, & \mathcal{O}'_n = \mathcal{O}_n T^{-1} \textit{ and } \mathcal{R}'_n = T\mathcal{R}_n \\ \textit{for } \mathbb{T} = \mathbb{Z}, & \mathcal{O}'_n = \mathcal{O}_n T^{-1} \textit{ and } \mathcal{R}'_n = \overrightarrow{T}\mathcal{R}_n \end{cases} \tag{3.99}$$

hold, where $\mathcal{O}_n$ and $\mathcal{O}'_n$, respectively $\mathcal{R}_n$ and $\mathcal{R}'_n$, are the corresponding n-step state-observability/reachability matrices of the representations.

The proof of this theorem similarly follows as in [173]. This means that the state-observability/reachability structure of equivalent representations is projected through the state-transformation that connects them, which is the required property to develop canonical forms with special structure of $\mathcal{O}_n$ or $\mathcal{R}_n$. The following theorem is the consequence of the minimality concept of Def. 3.30:

Theorem 3.14 (Induced PV-SS minimality). *The representation $\mathfrak{R}_{\mathrm{SS}}(\mathcal{S})$ is minimal iff it is structurally state-observable and it is state-trim, i.e. for all $\mathrm{x} \in \mathbb{X}$ there exists, in case of $\mathbb{T} = \mathbb{R}$, a $(u,x,y,p) \in \mathfrak{B}_{\mathrm{SS}} \cap \mathcal{C}^{\infty}(\mathbb{R}, \mathbb{W} \times \mathbb{X} \times \mathbb{P},)$ or $(u,x,y,p) \in \mathfrak{B}_{\mathrm{SS}}$ in case of $\mathbb{T} = \mathbb{Z}$, such that $x(0) = \mathrm{x}$.*

The proof of this theorem similarly follows as in the LTI case [153]. Note that, opposite to the orthodox linear system theory [80], minimality in this context does not require[6] state-reachability. However, when it is necessary to refer to the Kalman concept of minimality, we use the terminology of joint minimality:

Definition 3.41 (Joint minimality). If $\mathfrak{R}_{\mathrm{SS}}(\mathcal{S})$ is minimal and structurally state-reachable, then the representation is called jointly minimal. □

Example 3.34 (Induced minimality). Consider Example 3.30 and 3.32. Then the DT-SS representation is minimal as it is completely state-observable. Additionally it is completely state-reachable, hence it is jointly minimal. In case of Example 3.23, structural observability of the representation holds even if it is not completely state-observable. Thus, this representation is also minimal. □

3.3.2 *Stability of LPV Systems*

In the literature, there exist various stability concepts of LPV systems, originating either from the concepts of stability along frozen, i.e. constant scheduling trajectories (frozen stability) or stability along arbitrary varying p (global stability). While the first aspect defines stability in the LTI sense of the frozen behaviors, the latter establishes this concept on the full behavior. In many works, LPV stability issues are only discussed for the state-space case with static dependence, involving the notion of state equilibrium points and Lyapunov functions [160, 6], or mixing the concepts of frozen and global stability by considering slow variations of p [164, 176, 155]. Here we intend to define stability in the developed behavioral framework, investigating dynamic, IO, and Lyapunov stability both in a global and frozen sense. We also show the connection of the derived theory with the existing results of the LPV control synthesis framework.

3.3.2.1 Global Stability

Global stability is the natural concept of stability for LPV dynamic systems, as it means that small causes produce small effects for any scheduling trajectory. Historically, it originates from LTV system theory [68, 178] and robust control synthesis [240], where the problem of stability over the variations of the system has been first encountered. Let $\mathbb{R}_0^+$ denote the set of positive real numbers including 0. Then from the behavioral point of view, this concept extends the notion of dynamic stability for LPV systems as follows:

Definition 3.42 (Global dynamic stability). The autonomous LPV dynamical system $\mathcal{S} = (\mathbb{T}, \mathbb{R}^{n_{\mathrm{W}}}, \mathbb{P} \subseteq \mathbb{R}^{n_{\mathrm{P}}}, \mathfrak{B})$ is said to be globally dynamically stable, if $((w, p) \in \mathfrak{B}) \Rightarrow (\exists \varepsilon \in \mathbb{R}_0^+$ such that $\|w(t)\| \leq \varepsilon$ for all $t \geq 0)$ (in an arbitrary norm $\|\bullet\|$). It is

[6] Non-reachable systems are very common and they allow a state-minimal representation. Consider for instance autonomous systems. Such systems define a unique behavior, but their state representation is never state-reachable.

said to be globally unstable, if it is not stable; it is said to be globally dynamically asymptotically stable, if $((w,p) \in \mathfrak{B}) \Rightarrow (w(t) \to 0 \text{ as } t \to \infty)$. □

The definition strongly builds on the linearity (the only fixed point of the dynamic relation is 0) and time-invariance (stability on $t \geq 0$ implies stability on $t \geq t_0$ for all $t_0 \in \mathbb{R}$) of the LPV system class. Note that in terms of Def. 3.42 any signal trajectory, i.e. signal evolution of the system on the half-line, is bounded no matter the scheduling trajectory it is associated with, and the bound depends on the particular solution w. It is obvious that global stability includes stability with respect to frozen behaviors, as boundedness of w must hold for constant scheduling trajectories as well. This also emphasizes the difference between LPV and LTV systems, as in the latter case stability is defined with respect to only one, linear-trajectory of the scheduling ($p(t) = t$).

Contrary to the LTI case, conditions of global dynamic stability for LPV-KR representations can not be formulated in terms of eigenvalues or root conditions of polynomials in $\mathcal{R}[\xi]$. To show this, consider the following argument: In the CT-LTI case, $\frac{\mathrm{d}}{\mathrm{d}t}w = rw$ has solutions on the half-line in the form of $w(t) = \mathrm{e}^{r}w(0)$, thus the condition $r < 0$ guarantees boundedness of the solutions. However in the CT-LPV case $\frac{\mathrm{d}}{\mathrm{d}t}w = (r \diamond p)w$ with $r \in \mathcal{R}$, has solutions on the half-line in the form of $w(t) = \mathrm{e}^{(r \diamond p)(t)}w(0)$ only for constant p. Thus boundedness is not guaranteed by $(r \diamond p)(t) < 0$, $t \geq 0$. In fact, it is often possible to find a scheduling trajectory p such that the solution diverges even if $(r \diamond p)(t) < 0$, $t \geq 0$. In the sequel, unless indicated otherwise, we will call a LPV system *asymptotically stable* if it is globally dynamically asymptotically stable.

In case of an IO partition of $\mathcal{S}$, the concept of dynamic stability is formulated around the autonomous part of the behavior on the half line $[0, \infty)$, where $u = 0$. Similarly, the notion of global dynamic stability generalizes for systems with state-variables. Furthermore, global dynamic stability of LPV systems with IO partition also implies *bounded-input bounded-output* (BIBO) stability in the ℓ_∞ norm, and global asymptotic dynamic stability implies BIBO stability in the ℓ_τ norm, $1 \leq \tau < \infty$:

Definition 3.43 (BIBO stability). The LPV dynamical system $\mathcal{S} = (\mathbb{T}, \mathbb{R}^{n_\mathrm{W}}, \mathbb{P} \subseteq \mathbb{R}^{n_\mathbb{P}}, \mathfrak{B})$ with IO partition (u,y) is said to be BIBO stable in the ℓ_τ norm with $1 \leq \tau < \infty$, if for all $(u,y,p) \in \mathfrak{B}$ it holds that

$$\begin{cases} \text{for } \mathbb{T} = \mathbb{R}, & \displaystyle\int_0^\infty \|u(t)\|^\tau \mathrm{d}t < \infty \;\Rightarrow\; \int_0^\infty \|y(t)\|^\tau \mathrm{d}t < \infty; \\ \text{for } \mathbb{T} = \mathbb{Z}, & \displaystyle\sum_{k=0}^{\infty} \|u(k)\|^\tau < \infty \;\Rightarrow\; \sum_{k=0}^{\infty} \|y(k)\|^\tau < \infty. \end{cases}$$

It is said to be BIBO stable in the ℓ_∞ norm, if for all $(u,y,p) \in \mathfrak{B}$ it holds that

$$\sup_{t \geq 0} \|u(t)\| < \infty \;\Rightarrow\; \sup_{t \geq 0} \|y(t)\| < \infty. \qquad \square$$

Dynamic stability implies BIBO stability in the ℓ_∞ norm as all trajectories of y are bounded in case of dynamic stability. This boundedness holds due to the fact that the autonomous part of the behavior is bounded and $\mathfrak{B}$ fulfills the linearity and time-invariance properties in Def. 3.3. Similarly, asymptotic dynamic stability implies BIBO stability in an arbitrary ℓ_τ norm as all trajectories of y in the autonomous part of the behavior converge to zero. The concept of *bounded-input bounded-state* (BIBS) stability can be defined for LPV latent variable systems with both IO partition and state variables in a similar manner as BIBO stability. Also in the LPV case, BIBS stability always implies BIBO stability.

Another notion of stability leads through the approach of Lyapunov, which is widely used for stability analysis of linear and nonlinear systems, with both time-varying and time-invariant nature. While for LTI systems the Lyapunov method gives an informative alternative approach, for LPV systems it is the most applicable way to characterize or test stability of a given SS representation. First we introduce the intuitive idea in the context of first-order parameter-varying differential/difference equations:

$$\xi w = f(w, p), \tag{3.100}$$

where $f : \mathbb{R}^{n_\mathrm{W} \times n_\mathbb{P}} \to \mathbb{R}^{n_\mathrm{W}}$ is a Lipschitz continuous function and ξ is either $\frac{\mathrm{d}}{\mathrm{d}t}$ or q. For notational convenience, assume that $f(0, \centerdot) = 0$, so the equilibrium of (3.100) is 0 for which we investigate the concept of stability. Note, that any isolated non-zero equilibrium point $\mathtt{w} \in \mathbb{R}^{n_\mathrm{W}}$ of (3.100), i.e. $f(\mathtt{w}, \centerdot) = 0$, can be transferred to the origin by an exchange of variables $w' = w - \mathtt{w}$. We would like to find conditions ensuring that every solution $w : \mathbb{T} \to \mathbb{R}^{n_\mathrm{W}}$ of (3.100) goes to zero as $t \to \infty$. Suppose that $\mathbb{T} = \mathbb{R}$ and a continuously partially differentiable function $\mathcal{V} : \mathbb{R}^{n_\mathrm{W}} \to \mathbb{R}$ is given with $\mathcal{V}(0) = 0$ and $\mathcal{V}(\tau) > 0$ for $\tau \neq 0$. Assume that the derivative of $\mathcal{V}$ along every solution of (3.100) associated with a given $p \in \mathfrak{B}_\mathbb{P}$ is non-positive. Then $\mathcal{V}(w(t))$ is non-increasing for any w solution and under some additional requirements it implies that $w(t) \to 0$ for $t \to \infty$ along the scheduling trajectory p. To implement this idea, define the following:

Definition 3.44 (Definiteness of a function). A real valued function $g : \mathbb{R}^n \to \mathbb{R}$ is

- positive semi-definit (denoted by $g \succeq 0$) if $g(\tau) \geq 0, \forall \tau \in \mathbb{R}^n$,
- positive definit (denoted by $g \succ 0$) if $g \succeq 0$ and $(g(\tau) = 0) \Leftrightarrow (\tau = 0)$,
- negative semi-definit (denoted by $g \preceq 0$) if $g(\tau) \leq 0, \forall \tau \in \mathbb{R}^n$,
- negative definit (denoted by $g \prec 0$) if $g \preceq 0$ and $(g(\tau) = 0) \Leftrightarrow (\tau = 0)$. □

Denote $f_p = f(\cdot, p)$ for a given $p \in (\mathbb{R}^{n_\mathbb{P}})^\mathbb{T}$. Then based on the previously given considerations about $\mathcal{V}$, the following theorem holds:

Theorem 3.15 (Lyapunov stability, based on [146]). *The origin is an asymptotically stable equilibrium point of (3.100) in a global sense for a given scheduling trajectory $p \in \mathfrak{B}_\mathbb{P}$, if there exists a so-called Lyapunov function $\mathcal{V} : \mathbb{R}^{n_\mathrm{W}} \to \mathbb{R}$ such that the following conditions are satisfied:*

1. $\mathcal{V} \succ 0$ *(positive-definit),*
2. *If* $\mathbb{T} = \mathbb{R}$, *then* $\mathcal{V}$ *is continuously partially differentiable and* $\mathrm{Grad}[\mathcal{V}] f_p \prec 0$,
3. *If* $\mathbb{T} = \mathbb{Z}$, *then* $\mathcal{V}$ *is continuous at* 0 *and* $(\mathcal{V} \circ f_p) - \mathcal{V} \prec 0$,
4. $\mathcal{V}(\tau) \to \infty$ *as* $\|\tau\| \to \infty$.

Similar conditions can be given for stability of the equilibrium point by relaxing Th. 3.15 to require only semi-definiteness in the conditions. Instability condition of the equilibrium point can also be introduced by exchanging negative-definiteness of item 2 and 3 with positive-definiteness. However, Th. 3.15 is non-constructive in the determination of the Lyapunov function which can be laborious in practice.

In the following we focus on LPV systems in the context of the of the Lyapunov stability concept. Let $x \in \mathbb{X}^{\mathbb{T}}$ be the solution of the autonomous part of a SS representation $\mathfrak{R}_{\mathrm{SS}}(\mathcal{S})$ for a given scheduling trajectory $p \in \mathfrak{B}_{\mathbb{P}}$:

$$\xi x = (A \diamond p)x, \tag{3.101}$$

where ξ is either $\frac{\mathrm{d}}{\mathrm{d}t}$ or q. Consider the class of quadratic functions as Lyapunov functions $\mathcal{V}(\tau, p) = \tau^\top (P \diamond p)\tau$, where $\tau \in \mathbb{R}^{n_{\mathbb{X}}}$, $P \in \mathcal{R}^{n_{\mathbb{X}} \times n_{\mathbb{X}}}$, and $P = P^\top$ (symmetric). Then in continuous-time, using the chain rule of differentiation, it holds that

$$\frac{\mathrm{d}}{\mathrm{d}t}\mathcal{V}(x,p) = x^\top ((\underbrace{A^\top P + PA + \dot{P}}_{Q}) \diamond p)x, \tag{3.102}$$

where $Q \in \mathcal{R}^{n_{\mathbb{X}} \times n_{\mathbb{X}}}$ is symmetric. The term $Q = A^\top P + PA + \dot{P}$ is called the parameter-varying CT Lyapunov equation. In discrete-time, using a quadratic Lyapunov function yields

$$\mathcal{V}(\mathrm{q}x, \mathrm{q}p) - \mathcal{V}(x,p) = x^\top ((\underbrace{A^\top \overrightarrow{P} A - P}_{Q}) \diamond p)x, \tag{3.103}$$

where $Q \in \mathcal{R}^{n_{\mathbb{X}} \times n_{\mathbb{X}}}$ is also symmetric. Here the term $Q = A^\top \overrightarrow{P} A - P$ is the parameter-varying DT Lyapunov equation. Similar to Th. 3.15, the concept of stability is formulated around the "definiteness" property of (3.102) or (3.103) and the quadratic Lyapunov function. In case of a quadratic parameter-varying function $\mathcal{V}(\tau, p) = \tau^\top (P \diamond p)\tau$ with symmetric $P \in \mathcal{R}^{n_{\mathbb{X}} \times n_{\mathbb{X}}}$, we can define for a given $p \in \mathfrak{B}_{\mathbb{P}}$ the positive definiteness of P. We call P positive definite for p, i.e. $(P \diamond p) \succ 0$, if there exists a $\varepsilon > 0$ such that $(P \diamond p)(t) \succeq \varepsilon I$ for all $t \geq 0$ ($P \diamond p$ is bounded away from 0). The definition of negative definit, semi definit, etc. similarly follows.

Theorem 3.16 (LPV Quadratic stability). *Consider (3.101) and a projected scheduling behavior $\mathfrak{B}_{\mathbb{P}}$. Assume that $P \in \mathcal{R}^{n_{\mathbb{X}} \times n_{\mathbb{X}}}$, $P = P^\top$, and $Q = Q^\top$ satisfy the corresponding Lyapunov equation (see* (3.102) *and* (3.103)*) and $P \diamond p$ is bounded for all $p \in \mathfrak{B}_{\mathbb{P}}$. If for all $p \in \mathfrak{B}_{\mathbb{P}}$ it holds that*

- *$(P \diamond p) \succ 0$ and $(Q \diamond p) \preceq 0$, then (3.101) is globally dynamically stable.*
- *$(P \diamond p) \succ 0$, $(Q \diamond p) \preceq 0$, and (A, Q) is completely state-observable, then (3.101) is globally dynamically asymptotically stable.*
- *$(P \diamond p) \prec 0$, $(Q \diamond p) \preceq 0$, and (A, Q) is completely state-observable, then (3.101) is globally dynamically unstable.*

The proof of this theorem can be given according to [154] in CT and [2] in DT. However, there are two important facts to be noted. First, complete state-observability is required to ensure that the behavior of all state trajectories are characterized by the

Lyapunov function. In this way, as $\mathcal{V}$ is bounded away from zero with a positive ε and its derivative (difference) is bounded away from zero with a non-positive ε along every state and scheduling trajectory, convergence of the state to the origin is ensured. However, checking complete state-observability of (A,Q) is a computationally difficult problem (see Sect. 3.3.1). Second, due to the freedom of the functional dependence of P, the theorem is non-constructive in the general case.

Based on the previous considerations, in practice $\mathcal{R}^{n_\mathbb{X}\times n_\mathbb{X}}$ can be found to be too general for Lyapunov function construction. Especially in LPV control, the search for quadratic Lyapunov functions is commonly restricted to either a constant matrix $P \in \mathbb{R}^{n_\mathbb{X}\times n_\mathbb{X}}$ or to rational functions with static dependence on p. This restriction introduces conservative use of the Lyapunov theorem: If such a $P \in \mathbb{R}^{n_\mathbb{X}\times n_\mathbb{X}}$ can be found that either of the first two items of Th. 3.16 is satisfied, then stability of the system can be concluded, however, the lack of such a P does not necessarily imply instability. What we gain by the restriction of $P \in \mathbb{R}^{n_\mathbb{X}\times n_\mathbb{X}}$, is that the parameter-varying Lyapunov equations are modified as

$$\begin{cases} \text{for } \mathbb{T}=\mathbb{R}, & (A^\top \diamond p)P + P(A\diamond p) \prec 0, \\ \text{for } \mathbb{T}=\mathbb{Z}, & (A^\top \diamond p)P(A\diamond p) - P \prec 0. \end{cases} \tag{3.104}$$

By assuming that A has linear and static dependence on p, then the inequalities (3.104) become LMIs, defining an infinite dimensional *linear semi-definite programming* (LSDP) problem on $\mathbb{P}$ for the synthesis of P. If additionally it is assumed that $\mathbb{P}$ is a convex polytope[7] in $\mathbb{R}^{n_\mathbb{P}}$, then (3.104) reduces to a finite LSDP problem, where the feasibility of the Lyapunov equations is only checked at the vertices of the polytope $\mathbb{P}$ (see [160, 6]). This gives the foundation of the traditional $\mathcal{H}_2$ and $\mathcal{H}_\infty$ LPV control synthesis. Additionally, the use of full-block multipliers also enables to handle A with rational dependence through an LFR representation of the system [161]. In some works, parameter-varying Lyapunov functions with rational [233] or general [5], static dependence are also considered to overcome the restrictions of searching for a constant P. In the most simple case, assuming linear dependence of both P and A on p and restricting the derivative of p to a polytopic set, gives that the parameter-varying Lyapunov equations (3.102) and (3.103) can be formulated as LMIs by the use of relaxations, which again translates to a finite LSDP problem.

Example 3.35 (Global LPV stability). Consider the DT-SS representation $\mathfrak{R}_{\mathrm{SS}}(\mathcal{S})$, defined in Example 3.30. Choose a quadratic Lyapunov function

$$\mathcal{V}(\mathtt{x},p) = \mathtt{x}^\top \underbrace{\begin{bmatrix} 0.1 & 0 \\ 0 & 1 \end{bmatrix}}_{P} \mathtt{x}.$$

All eigenvalues of P are positive, thus $P \succ 0$. By computing the DT Lyapunov equation, it follows that

[7] A convex polytope is the convex hull of a finite set of points, i.e. it is the intersection of half-spaces.

$$Q \diamond p = (A^\top \overrightarrow{P} A - P) \diamond p = \frac{1}{10} \begin{bmatrix} p^2 - 1 & p \\ p & 10p^2 - 9 \end{bmatrix}.$$

Because $\mathbb{P} = [\frac{1}{4}, \frac{3}{4}]$, it holds that there exists a $\varepsilon < 0$ such that $(Q \diamond p)(k) \preceq \varepsilon I$ for all $k \in \mathbb{Z}$ and $p \in \mathfrak{B}_\mathbb{P}$. Furthermore (A, Q) is completely state-observable as the rank of Q is always 2 along any $p \in \mathfrak{B}_\mathbb{P}$. Thus the chosen Lyapunov function proves global asymptotic dynamic stability of $\mathcal{S}$. □

3.3.2.2 Frozen Stability

Another important aspect of LPV stability is the so-called frozen stability. Stability analysis of the frozen behaviors has been in the focus of LPV control during the gain-scheduling area, before the appearance of global LPV control synthesis techniques. At that time, researchers concluded global dynamic stability of the system based on the dynamic stability of the frozen behaviors by assuming appropriately slowly varying scheduling signals [235, 176, 155]. This view has been found misleading as the term "appropriate" was not well-defined (see the arguments of [157] and [166]). Today, frozen stability is still important in LPV analysis as it is a necessary ingredient for global stability.

Definition 3.45 (Frozen stability). Let $\mathscr{F}_\mathcal{S}$ be the frozen system set (see Def. 3.4) of the LPV system $\mathcal{S}$ with scheduling space $\mathbb{P} \subseteq \mathbb{R}^{n_\mathbb{P}}$. Then, in the frozen sense, $\mathcal{S}$ is

- Uniformly asymptotically stable, if for all $\mathrm{p} \in \mathbb{P}$, $\mathcal{F}_\mathrm{p} \in \mathscr{F}_\mathcal{S}$ is dynamically asymptotically stable.
- Uniformly stable, if for all $\mathrm{p} \in \mathbb{P}$, $\mathcal{F}_\mathrm{p}$ is dynamically stable.
- Non-uniformly stable, if it is not uniformly stable but there exists a $\mathrm{p} \in \mathbb{P}$ s.t. $\mathcal{F}_\mathrm{p}$ is dynamically stable.
- Uniformly unstable, if for all $\mathrm{p} \in \mathbb{P}$, $\mathcal{F}_\mathrm{p}$ is dynamically unstable. □

Note that uniform frozen stability of $\mathcal{S}$ does not imply global dynamic stability, though the converse is true. In terms of Def. 3.45, the stability of the frozen aspects of the system is checked separately for each frozen behavior. For example this means the construction of quadratic Lyapunov functions separately for each $\mathrm{p} \in \mathbb{P}$. If there exists such a common quadratic Lyapunov function which proves stability for all $\mathrm{p} \in \mathbb{P}$, where $\mathbb{P}$ is convex, then in case of linear and static dependence of A, it also implies global dynamic stability in terms of Th. 3.16. This makes an important distinction for systems with dynamic dependence, where the evaluation of A along a constant scheduling trajectory excludes the effect of the dependence on the derivatives/time shifts of p. For these systems, a common Lyapunov function of the frozen systems set does not imply global dynamic stability.

Example 3.36 (Frozen LPV stability). Consider again the DT-SS representation $\mathfrak{R}_\mathrm{SS}(\mathcal{S})$, defined in Example 3.30. For every constant scheduling trajectory $p \in \mathfrak{B}_\mathbb{P}$, where $p(k) = \mathrm{p}$, $\forall k \in \mathbb{Z}$, it holds that the eigenvalues of $(A \diamond p)$ are equal to p. As

$\mathrm{p} < 1$ for all $\mathrm{p} \in \mathbb{P}$, uniform frozen asymptotic stability of $\mathcal{S}$ holds. This is a not surprising discovery, as global asymptotic dynamic stability of $\mathcal{S}$, proved in Example 3.35, implies uniform frozen asymptotic stability of $\mathcal{S}$. Now consider the CT-SS representation defined in Example 3.25. For this representation, eigenvalues of $(A \diamond p)$ are $\{0,2\}$ for any constant scheduling trajectory which proves uniform frozen instability of the represented system. Uniform frozen instability also implies global dynamic instability. □

3.3.3 Gramians of LPV State-Space Representations

Gramians are also important concepts for LPV-SS representations, as they describe the complete state-observability and reachability properties and they can also characterize model reduction (see [232]). Using the previously developed linear input-map (3.84) and linear output-map (3.89) of the CT case, and their DT equivalents (3.95) and (3.97), the concept of PV gramians is introduced as follows:

Definition 3.46 (PV gramians). In CT, the observability gramian O and the reachability gramian R of a asymptotically stable $\mathfrak{R}_{\mathrm{SS}}(\mathcal{S})$ on the time interval $[t_0, t_1] \subset \mathbb{R}$ and along a scheduling trajectory $p \in \mathfrak{B}_{\mathbb{P}}$ are defined as:

$$\mathsf{O}(t_1, t_0, p) = \int_{t_0}^{t_1} \mathfrak{A}^T(\tau, t_0, p)(C^\top \diamond p)(\tau)(C \diamond p)(\tau)\mathfrak{A}(\tau, t_0, p)\, \mathrm{d}\tau, \quad (3.105a)$$

$$\mathsf{R}(t_1, t_0, p) = \int_{t_0}^{t_1} \mathfrak{A}(t_1, \tau, p)(B \diamond p)(\tau)(B^\top \diamond p)(\tau)\mathfrak{A}^T(t_1, \tau, p)\, \mathrm{d}\tau, \quad (3.105b)$$

while in DT, they are given on the time interval $[k_0, k_1] \subset \mathbb{Z}$ as:

$$\mathsf{O}(k_1, k_0, p) = \sum_{i=k_0}^{k_1} \mathfrak{A}^T(i, k_0, p)(C^\top \diamond p)(i)(C \diamond p)(i)\mathfrak{A}(i, k_0, p), \quad (3.106a)$$

$$\mathsf{R}(k_1, k_0, p) = \sum_{i=k_0}^{k_1} \mathfrak{A}(k_1, i, p)(B \diamond p)(i)(B^\top \diamond p)(i)\mathfrak{A}^T(k_1, i, p). \quad (3.106b)$$

□

Similar to the LTI case, the full rank property of gramians implies complete state-observability and reachability on the considered time interval and scheduling trajectory:

Theorem 3.17 (Induced PV observability/reachability, [174]). *The LPV-SS representation $\mathfrak{R}_{\mathrm{SS}}(\mathcal{S})$, is completely state-observable/reachable) iff its observability/reachability gramian is full rank for any finite time interval and each scheduling trajectory of $\mathfrak{B}_{\mathbb{P}}$.*

See the proofs in [174] for the CT case and in [61] for the DT case.

3.4 Conclusions

In this chapter, we have developed a behavioral framework of LPV systems as an extension of the LTI behavioral approach. The introduced theory has been established to give a unified view on LPV systems and their representations and to enable to approach LPV system identification in a well-founded system theoretic sense. We has shown that the behavioral approach provides a well established view on equivalence relations between LPV representations, gives a clear representation free definition of LPV systems and it is compatible with the existing system theoretical results. It was also shown that the use of dynamic scheduling dependence is necessary in order to establish equivalence relations and also in general to fill the gaps of the existing theory. In the next chapter, we continue by extending the concept of equivalence transformations between different representation domains to the LPV system class. This contribution gives the finishing details of the developed behavioral framework and enables the comparison of different LPV model structures.

Chapter 4
LPV Equivalence Transformations

Abstract. In this chapter, we continue the discussion of the LPV behavioral framework by establishing equivalence transformations between the state-space and the input-output representation domains. These equivalence transformations enable the comparison and analysis of LPV model structures and provide essential tools to formulate the identification approach of this thesis. First we define LPV canonical forms based on the concept of observability and reachability and we give an algorithmic scheme to construct them from an existing SS representation of the system. Then, transformations are introduced which provide an equivalent IO representation of a SS representation and vice versa. In both cases, the introduced LPV canonical forms are special cases of the transformation problem, serving as a simple gateway between the representation domains.

4.1 State-Space Canonical Forms

Specially structured canonical forms of state-space representations of LPV systems are essential ingredients to accommodate equivalence transformations between the *state-space* (SS) and the *input-output* (IO) representation domains. One set of these canonical forms are the so-called observability/reachability canonical forms which are also used in the state-of-the-art of LPV identification and control design.

In this section, LPV canonical forms in *continuous-time* (CT) and *discrete-time* (DT) are introduced through a transformation mechanism applied on a given SS representation of the LPV system. Using the concept of structural state-observability/reachability, i.e. the associated observability/reachability matrices, state-transformations are defined, that result in an equivalent SS representation with special structure in the system matrices. It is shown that this special structure implies complete state-observability/reachability of the resulting representation and hence it is called a LPV observability/reachability canonical form. Additionally, the introduced canonical forms give a unique representation of their associated equivalence

R. Tóth: Modeling and Identification of Linear Parameter-Varying Systems, LNCIS 403, pp. 101–130.
springerlink.com

class. The applied transformation mechanism is based on the results of LTV systems theory and it can be seen as extension of the LTI canonical form construction approach (see [80, 110]).

Besides the derivation of observability/reachability canonical forms, a construction approach for their companion counterparts is introduced. The concept of transposition of SS representations is also investigated, with the main conclusion that contrary to the LTI case, the transpose of a SS representation in the LPV framework does not have an equal manifest behavior. This means that such a transformation alters the dynamical relation.

It is also investigated how the developed concept of canonical forms relates to applied theories of the current LPV literature. It is shown that the common practice to use LTI theory to compute canonical forms for LPV systems results in SS representations that do not have an equal manifest behavior.

4.1.1 *The Observability Canonical Form*

At first, observability canonical forms are considered. It is assumed that a structurally state-observable state-space representation $\mathfrak{R}_{\mathrm{SS}}(\mathcal{S})$ is given for the SISO LPV system $\mathcal{S}$. Due to structural state-observability, i.e. full rank of $\mathcal{O}_{n_{\mathbb{X}}}$ associated with $\mathfrak{R}_{\mathrm{SS}}(\mathcal{S})$, it is possible to introduce a new state-basis for the representation with the parameter-varying transformation matrix, $T_{\mathrm{o}} \in \mathcal{R}^{n_{\mathbb{X}} \times n_{\mathbb{X}}}$:

$$T_{\mathrm{o}} := \mathcal{O}_{n_{\mathbb{X}}}. \tag{4.1}$$

This leads to a new state variable x_{o}, obtained as

$$x_{\mathrm{o}} := (T_{\mathrm{o}} \diamond p)x, \quad \forall p \in \mathfrak{B}_{\mathbb{P}}. \tag{4.2}$$

Due to the full rank property of $\mathcal{O}_{n_{\mathbb{X}}}$, T_{o} is invertible in $\mathcal{R}^{n_{\mathbb{X}} \times n_{\mathbb{X}}}$. If $\mathfrak{R}_{\mathrm{SS}}(\mathcal{S})$ is completely state-observable with an observability radius $n_{\mathbb{X}}$, then T_{o} is invertible for any scheduling trajectory and time instant. Thus, in that case, (4.2) implies algebraic equivalence between x_{o} and x. If only structural observability holds, then invertibility is guaranteed in a functional sense, which means that (4.2) implies algebraic equivalence almost everywhere. However, the latter is a sufficient property for T_{o} to be a PV state-transformation, which leads to an equivalent SS representation of $\mathcal{S}$ in terms of $\overset{n_{\mathbb{P}}}{\sim}$. Moreover, this transformation projects the observability structure in terms of Th. 3.13 to the identity matrix ($\mathcal{O}_{n_{\mathbb{X}}} \mathcal{O}_{n_{\mathbb{X}}}^{-1} = I$). Thus, similar to the LTI case, we call the equivalent SS representation, resulting by the state transformation T_{o}, the observability canonical form.

To obtain the equivalent representation of $\mathcal{S}$ in terms of the new state variable, T_{o} is applied to the system matrices in accordance with $\overset{n_{\mathbb{P}}}{\sim}$, resulting in a SS representation with the following special structure:

$$\left[\begin{array}{c|c} A_\mathrm{o} & B_\mathrm{o} \\ \hline C_\mathrm{o} & D_\mathrm{o} \end{array}\right] := \left[\begin{array}{cccc|c} 0 & 1 & \dots & 0 & \beta^\mathrm{o}_{n_\mathbb{X}-1} \\ \vdots & \vdots & \ddots & \vdots & \beta^\mathrm{o}_{n_\mathbb{X}-2} \\ 0 & 0 & \dots & 1 & \vdots \\ -\alpha^\mathrm{o}_0 & -\alpha^\mathrm{o}_1 & \dots & -\alpha^\mathrm{o}_{n_\mathbb{X}-1} & \beta^\mathrm{o}_0 \\ \hline 1 & 0 & \dots & 0 & \beta^\mathrm{o}_{n_\mathbb{X}} \end{array}\right].$$

Then

$$\mathfrak{R}^{\mathrm{O}}_{\mathrm{SS}}(\mathcal{S}) := \left[\begin{array}{c|c} A_\mathrm{o} & B_\mathrm{o} \\ \hline C_\mathrm{o} & D_\mathrm{o} \end{array}\right] \in \left[\begin{array}{c|c} \mathcal{R}^{n_\mathbb{X}\times n_\mathbb{X}} & \mathcal{R}^{n_\mathbb{X}\times 1} \\ \hline \mathcal{R}^{1\times n_\mathbb{X}} & \mathcal{R} \end{array}\right], \tag{4.3}$$

is called the *observability canonical state-space representation* of $\mathcal{S}$. Proof of that the invertible state-transformation based on (4.2) always results in the above given structure follows similarly as in the LTV case (see [172, 238] for the CT case and [113, 65] for the DT case).

Example 4.1 (CT-LPV observability canonical form, SISO). Consider the structurally state-observable CT-LPV-SS representation $\mathfrak{R}_{\mathrm{SS}}(\mathcal{S})$ of Example 3.23. By applying state-transformation (4.2) in terms of Def. 3.29, the resulting observability canonical form is as follows:

$$\mathfrak{R}^{\mathrm{O}}_{\mathrm{SS}}(\mathcal{S}) = \left[\begin{array}{cc|c} 0 & 1 & 1+p^2 \\ \frac{p\frac{\mathrm{d}^2}{\mathrm{d}t^2}p-\left(2+p^2+\frac{\mathrm{d}}{\mathrm{d}t}p\right)\frac{\mathrm{d}}{\mathrm{d}t}p}{p^2-2-\frac{\mathrm{d}}{\mathrm{d}t}p} - 1 & \frac{p^3-2p+p\frac{\mathrm{d}}{\mathrm{d}t}p-\frac{\mathrm{d}^2}{\mathrm{d}t^2}p}{p^2-2-\frac{\mathrm{d}}{\mathrm{d}t}p} - \frac{1}{p} & 3p+p\frac{\mathrm{d}}{\mathrm{d}t}p \\ \hline 1 & 0 & p \end{array}\right].$$

Because the original representation $\mathfrak{R}_{\mathrm{SS}}(\mathcal{S})$ is not completely state-observable, the resulting canonical representation $\mathfrak{R}^{\mathrm{O}}_{\mathrm{SS}}(\mathcal{S})$ is equivalent with $\mathfrak{R}_{\mathrm{SS}}(\mathcal{S})$ only in the almost everywhere sense. Furthermore, $\mathfrak{R}^{\mathrm{O}}_{\mathrm{SS}}(\mathcal{S})$ is completely state-observable (its observability matrix is an identity matrix) even if $\mathfrak{R}_{\mathrm{SS}}(\mathcal{S})$ is not. □

It is important to note that due to the state-transformation (4.2), the complexity of the dependence of the meromorphic coefficients of $\mathfrak{R}^{\mathrm{O}}_{\mathrm{SS}}(\mathcal{S})$ can increase considerably. If in the representation $\mathfrak{R}_{\mathrm{SS}}(\mathcal{S})$ all the coefficients/matrices are linear static functions of p, then the matrix functions defining $\mathfrak{R}^{\mathrm{O}}_{\mathrm{SS}}(\mathcal{S})$ can have rational dependence on p and its derivatives/forward time-shifts up to the order $n_\mathbb{X}$. This property has been one of the reasons to define coefficient dependence of LPV systems over the field of meromorphic functions $\mathcal{R}$ with variables associated with p and its derivatives/time-shifts in the introduced framework.

Example 4.2 (DT-LPV observability canonical form, SISO). As a DT example, we consider the DT-LPV-SS representation defined in Example 3.30. Again, by applying state-transformation (4.2) in terms of Def. 3.29, the resulting observability canonical form of this representation is the following:

$$\mathfrak{R}^{\mathrm{O}}_{\mathrm{SS}}(\mathcal{S}) = \left[\begin{array}{cc|c} 0 & 1 & 1+\frac{p}{\mathrm{q}p} \\ -\left(\mathrm{q}p^2\right)\frac{p}{\mathrm{q}^2p} & \mathrm{q}p+\frac{\mathrm{q}p^2}{\mathrm{q}^2p} & \frac{p(\mathrm{q}p)}{\mathrm{q}^2p}+p+\mathrm{q}p \\ \hline 1 & 0 & p \end{array}\right].$$

Due to the complete state-observability of the DT example, the resulting canonical form is fully equivalent with the original representation and it is also completely state-observable. Note that in the following, we restrict examples to the DT case to simplify the discussion. □

Next consider the MIMO case. According to theory given for LTV systems in [65], observability type of canonical forms with respect to a MIMO, structurally state-observable $\mathfrak{R}_{\mathrm{SS}}(\mathcal{S})$ can be realized by a mapping rule of three steps, similar to the LTI case:

1. Choose $n_{\mathbb{X}}$-independent rows of the full column-rank $\mathcal{O}_{n_{\mathbb{X}}}$ with a given ordering sequence.
2. Rearrange those $n_{\mathbb{X}}$-independent rows with a fixed ordering to form a nonsingular state transformation matrix $T_{\mathrm{o}} \in \mathcal{R}^{n_{\mathbb{X}} \times n_{\mathbb{X}}}$.
3. By applying the equivalence transformation defined via T_{o}, compute the canonical representation.

In the following, these steps form the line of reasoning for the introduction of a MIMO observability canonical form through its construction mechanism.

According to Step 1 of the previous algorithm, write $\mathcal{O}_{n_{\mathbb{X}}}$ as the sequence of row vectors:

$$\mathcal{O}_{n_{\mathbb{X}}} = \begin{bmatrix} \mathrm{o}_{11}^{\top} & \cdots & \mathrm{o}_{n_{\mathbb{Y}}1}^{\top} & \mathrm{o}_{12}^{\top} & \cdots & \mathrm{o}_{n_{\mathbb{Y}}2}^{\top} \cdots \end{bmatrix}^{\top}, \tag{4.4}$$

where $C = \begin{bmatrix} \mathrm{o}_{11}^{\top} & \cdots & \mathrm{o}_{n_{\mathbb{Y}}1}^{\top} \end{bmatrix}^{\top}$. Each $\mathrm{o}_j = [\mathrm{o}_{ij}]_{i=1}^{n_{\mathbb{Y}}}$, $j > 1$ is defined similarly as (3.86) in the CT case and as (3.96) in the DT case. To complete Step 1 of the algorithm, the selection of $n_{\mathbb{X}}$ linearly independent vector functions from the $n_{\mathbb{X}} \times n_{\mathbb{Y}}$ rows of (4.4) is needed, to form the new state-basis of the canonical representation. Due to the structural state-observability of the system, it is always possible to make such a selection, but in general it is not unique [65, 110]. Depending on the particular way of the selection procedure, different canonical forms can be obtained. In the following, the selection strategy that reproduces the structure of the previously introduced SISO LPV observability canonical form is used. According to this, select the vectors of (4.4) in terms of the following ordering:

$$\left\{\mathrm{o}_{11}, \mathrm{o}_{21}, \ldots, \mathrm{o}_{n_{\mathbb{Y}}1}, \mathrm{o}_{12}, \mathrm{o}_{22}, \ldots, \mathrm{o}_{n_{\mathbb{Y}}2}, \ldots\right\}, \tag{4.5}$$

which matches with the generalization of Young's selection scheme II (see [110]). For the sake of simplicity, temporally assume that $\mathrm{Rank}(C) = n_{\mathbb{Y}}$, meaning that $\{\mathrm{o}_{11}, \mathrm{o}_{21}, \ldots, \mathrm{o}_{n_{\mathbb{Y}}1}\}$ are linearly independent vector functions. Then, the linear dependence of every vector function from the ordered sequence (4.5) can be analyzed one after the other: if $\tau_i \in \mathbb{I}_1^{n_x}$ is the smallest number such that $\mathrm{o}_{i\tau_i}$ is linearly dependent on the previous vectors, then there exists a set of unique functions $\{\alpha_{ijl}^{\mathrm{o}} \in \mathcal{R}\}$, such that

$$\mathrm{o}_{i\tau_i} = \sum_{j=1}^{n_{\mathbb{Y}}} \sum_{l=0}^{\tau_{ij}-1} \alpha_{ijl}^{\mathrm{o}}\, \mathrm{o}_{jl}, \tag{4.6}$$

where the ordering of the vectors implies that τ_{ij} satisfies

$$\tau_{ij} = \begin{cases} \tau_i & \text{for } i = j, \\ \min(\tau_i + 1, \tau_j) & \text{for } i > j, \\ \min(\tau_i, \tau_j) & \text{for } i < j. \end{cases} \tag{4.7}$$

Once all dependent vector functions have been found, a total of $n_\mathbb{X} = \sum_{i=1}^{n_\mathbb{Y}} \tau_i$ independent vectors are selected due to the full rank assumption of $\mathcal{O}_{n_\mathbb{X}}$. Furthermore, as the first $n_\mathbb{Y}$ vectors $\{o_{11}, o_{21}, \ldots, o_{n_\mathbb{Y}1}\}$ are independent, they are automatically selected, implying that

$$\max_{i \in \mathbb{I}_1^{n_\mathbb{Y}}} \tau_i = \tau_{\max} \leq n_\mathbb{X} - n_\mathbb{Y} + 1. \tag{4.8}$$

Moreover, the remaining linearly dependent relations are described by $\sum_{i=1}^{n_\mathbb{Y}} \sum_{j=1}^{n_\mathbb{Y}} \tau_{ij}$ number of functions $\{\alpha_{ijl}^{\mathrm{o}}\}$. This accomplishes Step 1 of the algorithm.

Using the previously selected vectors, the new state-basis is defined by

$$T_\mathrm{o} := \begin{bmatrix} o_{11}^\top & \cdots & o_{1(\tau_i-1)}^\top & \cdots & o_{n_\mathbb{Y}1}^\top & \cdots & o_{n_\mathbb{Y}(\tau_{n_\mathbb{Y}}-1)}^\top \end{bmatrix}^\top. \tag{4.9}$$

Due to the linear independence of the rows, T_o is invertible and implies an equivalence relation in terms of Def. 3.29. This completes Step 2 of the algorithm.

As a final step, applying the previously constructed equivalence relation on $\mathfrak{R}_{\mathrm{SS}}(\mathcal{S})$, yields the transformed matrices in the following form:

$$\left[\begin{array}{c|c} A_\mathrm{o} & B_\mathrm{o} \\ \hline C_\mathrm{o} & D_\mathrm{o} \end{array}\right] := \left[\begin{array}{cccc|c} \multicolumn{4}{c|}{\left[A_{ij}^\mathrm{o}\right],\ i,j \in \mathbb{I}_1^{n_\mathbb{Y}}} & \begin{matrix} B_1^\mathrm{o} \\ \vdots \\ B_{n_\mathbb{Y}}^\mathrm{o} \end{matrix} \\ \hline e_1 \;\; 0_{n_\mathbb{Y}\times(\tau_1-1)} & \cdots & e_{n_\mathbb{Y}} \;\; 0_{n_\mathbb{Y}\times(\tau_{n_\mathbb{Y}}-1)} & & D \end{array}\right],$$

where $\{e_i\}_{i=1}^{n_\mathbb{Y}}$ is the standard basis of $\mathbb{R}^{n_\mathbb{Y}}$, and

$$A_{ii}^\mathrm{o} = \begin{bmatrix} 0 & \cdots & 0 & -\alpha_{ii0}^\mathrm{o} \\ 1 & \ddots & \vdots & -\alpha_{ii1}^\mathrm{o} \\ \vdots & \ddots & 0 & \vdots \\ 0 & \cdots & 1 & -\alpha_{ii(\tau_i-1)}^\mathrm{o} \end{bmatrix}_{(\tau_i\times\tau_i)}^\top \quad A_{ij}^\mathrm{o} = \begin{bmatrix} 0 & \cdots & 0 & -\alpha_{ij0}^\mathrm{o} \\ \vdots & & \vdots & \vdots \\ \vdots & & \vdots & -\alpha_{ij(\tau_{ij}-1)}^\mathrm{o} \\ \vdots & & \vdots & 0 \\ \vdots & & \vdots & \vdots \\ 0 & \cdots & 0 & 0 \end{bmatrix}_{(\tau_i\times\tau_j)}^\top$$

$$B_i^\mathrm{o} = \begin{bmatrix} \beta_{i1(\tau_i-1)}^\mathrm{o} & \cdots & \beta_{in_\mathbb{U}(\tau_i-1)}^\mathrm{o} \\ \vdots & & \vdots \\ \beta_{i10}^\mathrm{o} & \cdots & \beta_{in_\mathbb{U}0}^\mathrm{o} \end{bmatrix}_{(\tau_i\times n_\mathbb{U})} \quad D_\mathrm{o} = \begin{bmatrix} \beta_{11\tau_1}^\mathrm{o} & \cdots & \beta_{1n_\mathbb{U}\tau_1}^\mathrm{o} \\ \vdots & & \vdots \\ \beta_{n_\mathbb{Y}1\tau_{n_\mathbb{Y}}}^\mathrm{o} & \cdots & \beta_{n_\mathbb{Y}n_\mathbb{U}\tau_{n_\mathbb{Y}}}^\mathrm{o} \end{bmatrix}_{(n_\mathbb{Y}\times n_\mathbb{U})}$$

Proof of that the invertible state-transformation based on (4.9) always results in the above given structure follows similarly as in [65]. Based on this representation, the LTI system is separated to an interconnection of subsystems characterized by the A_{ii}^{o} and B_i^{o} matrices and the connection of these subsystems is defined through the A_{ij}^{o} matrices. In this way, using the constructed state-space transformation T_{o} applied on $\mathfrak{R}_{\mathrm{SS}}(\mathcal{S})$, we have constructed a canonical SS representation of $\mathcal{S}$. Due to the fact, that the projected $n_{\mathbb{X}}$-step observability matrix of $\mathfrak{R}_{\mathrm{SS}}^{\mathrm{O}}(\mathcal{S})$ is the identity matrix (SISO case) or composed from zero row vectors and the standard basis of $\mathbb{R}^{1\times n_{\mathbb{X}}}$ (MIMO case) such a canonical representation is always completely state-observable. Furthermore, such a structure of the observability matrix also obviously implies that $\mathfrak{R}_{\mathrm{SS}}^{\mathrm{O}}(\mathcal{S})$ is state trim. Thus the following corollary holds for all representations of $\mathcal{S}$ with the structural form of $(A_{\mathrm{o}}, B_{\mathrm{o}}, C_{\mathrm{o}}, D_{\mathrm{o}})$ both in the SISO and the MIMO cases:

Corollary 4.1. $\mathfrak{R}_{\mathrm{SS}}^{\mathrm{O}}(\mathcal{S})$ *is completely state-observable and state-trim hence it is minimal.*

In case $\mathrm{Rank}(C) \neq n_{\mathbb{Y}}$, the observability canonical form does not exist in the previously introduced structure as C_{o} cannot be a matrix composed from zero vectors and standard bases. In this case, T_{o} is constructed by considering the system only with output channels which are associated with the independent rows of C. Then, T_{o} is applied to the original matrix functions. The resulting C_{o} retains the structure of the conventional canonical form for the linearly independent output channels (containing only zero vectors and standard bases), however it also contains meromorphic coefficient functions (the weights of the linear combination of the independent channels) in the rows corresponding to the dependent output channels. If the LPV-SS representation is not structurally state-observable, then computation of an equivalent observability canonical form by the presented algorithm requires to search for a SS representation of the LPV system with structural state-observability. Then, such a representation can be converted to an observability canonical form. As we will see in Sect. 4.3, the construction of such a representation is always possible for LPV systems.

Example 4.3 (DT-LPV observability canonical form, MIMO). Let the SS representation $\mathfrak{R}_{\mathrm{SS}}(\mathcal{S})$ of an DT-LPV system $\mathcal{S}$ with $\mathbb{P} = [0.1, 0.3]$ be given by

$$\left[\begin{array}{c|c} A & B \\ \hline C & D \end{array}\right] \diamond p = \left[\begin{array}{cccc|cc} p & 0 & 0 & -0.5 & 1 & 1 \\ 0 & -p & 0 & 0 & p & p \\ 0 & 0 & -p & 0 & -1 & 0 \\ 0.5 & 0 & 0 & p & 0 & 1 \\ \hline 0 & p & 0 & 0 & 0 & 0 \\ 0 & 0 & 1 & p & 0 & 0 \\ p & 1 & 0 & 0 & 0 & 0 \end{array}\right].$$

By computing the 2-step observability matrix of $\mathfrak{R}_{\mathrm{SS}}(\mathcal{S})$,

$$\mathcal{O}_2 \diamond p = \begin{bmatrix} 0 & p & 0 & 0 \\ 0 & 0 & 1 & p \\ p & 1 & 0 & 0 \\ \hdashline 0 & -p(qp) & 0 & 0 \\ 0.5qp & 0 & -p & p(qp) \\ p(qp) & -p & 0 & -0.5qp \end{bmatrix}$$

results. It is clear, that the first four rows of $\mathcal{O}_2$ are independent in the functional sense, thus $\mathfrak{R}_{\mathrm{SS}}(\mathcal{S})$ is structurally state-observable. However, the first three rows and the sixth are independent along all possible scheduling trajectories on $\mathbb{P}$, thus $\mathfrak{R}_{\mathrm{SS}}(\mathcal{S})$ is also completely state-observable. Note that in this case, computation of $\mathcal{O}_4$ is not necessary to show these properties. By calculating the observability canonical form $\mathfrak{R}_{\mathrm{SS}}^{\mathcal{O}}(\mathcal{S})$ of $\mathfrak{R}_{\mathrm{SS}}(\mathcal{S})$ using the first three and the sixth rows of $\mathcal{O}_2$, the resulting matrices are the following (the sub-matrices are denoted by dashed lines):

$$A_{\mathrm{o}} \diamond p = \begin{bmatrix} \mathrm{q}p & 0 & 0 & 0 \\ \hdashline \frac{\mathrm{q}p^2+4p^2+8p(\mathrm{q}p)+4p^2(\mathrm{q}p^2)}{2p^2(\mathrm{q}p)} & p & -\frac{\mathrm{q}p+4p^3+4p^2(\mathrm{q}p)}{2p} & \frac{p(p+\mathrm{q}p)}{2\mathrm{q}p} \\ \hdashline 0 & 0 & 0 & 1 \\ -\frac{(\mathrm{q}p)(\mathrm{q}^2p)+4p^2(\mathrm{q}p^2)+4p^3(\mathrm{q}^3p)+8p^2(\mathrm{q}p)(\mathrm{q}^2p)}{4p^2(\mathrm{q}p)} & 0 & \frac{(1+4p^2)\mathrm{q}^2p}{4p} & -\frac{(p+\mathrm{q}p)\mathrm{q}^2p}{\mathrm{q}^2p} \end{bmatrix}$$

$$B_{\mathrm{o}} \diamond p = \begin{bmatrix} p(\mathrm{q}p) & p(\mathrm{q}p) \\ \hdashline -1 & \mathrm{q}p \\ \hdashline p+\mathrm{q}p & p+\mathrm{q}p \\ (\mathrm{q}^2p-p)\mathrm{q}p & \frac{2(\mathrm{q}^2p-2p)\mathrm{q}p-\mathrm{q}^2p}{2} \end{bmatrix}, \qquad C_{\mathrm{o}} \diamond p = \begin{bmatrix} 1 & 0 & 0 & 0 \\ 0 & 1 & 0 & 0 \\ 0 & 0 & 1 & 0 \end{bmatrix}.$$

Note that the resulting LPV-SS representation is the interconnection of 3 subsystems, each associated with a specific output channel. Furthermore, this observability canonical form is not generated using Young's selection scheme as in that case the first 4 rows of $\mathcal{O}_2$ would have been selected for the transformation. In opposite with the used state-transformation, the transformation of the state, based on the first 4 rows, only provides an equivalent representation in the almost everywhere sense, as independence of the rows only holds in the functional sense. This underlines that in the MIMO case there is a freedom in the construction of observability canonical forms and the provided selection scheme is only one from the available possibilities. □

4.1.2 Reachability Canonical Form

As a next step, we extend the previously introduced mechanism to the reachability case. Similar to the previous part, SISO systems are considered first. It is assumed that a structurally state-reachable state-space representation $\mathfrak{R}_{\mathrm{SS}}(\mathcal{S})$ is given for the

SISO LPV system $\mathcal{S}$. Due to the full rank of $\mathcal{R}_{n_\mathbb{X}}$ associated with $\mathfrak{R}_{\mathrm{SS}}(\mathcal{S})$, it is possible to introduce a new state-basis for the representation by using

$$T_\mathrm{r}^{-1} := \begin{cases} \mathcal{R}_{n_\mathbb{X}}, & \text{if } \mathbb{T} = \mathbb{R}, \\ \overleftarrow{\mathcal{R}}_{n_\mathbb{X}}, & \text{if } \mathbb{T} = \mathbb{Z}, \end{cases} \tag{4.10}$$

that leads to a new state variable x_r, obtained as

$$x_\mathrm{r} := (T_\mathrm{r} \diamond p)x, \quad \forall p \in \mathfrak{B}_\mathbb{P}. \tag{4.11}$$

Again, the full rank property of $\mathcal{R}_{n_\mathbb{X}}$ implies that T_r is invertible in $\mathcal{R}^{n_\mathbb{X} \times n_\mathbb{X}}$. Thus, (4.11) yields an equivalent SS representation of $\mathcal{S}$ in terms of $\overset{n_\mathbb{P}}{\sim}$. In case of complete state-reachability with reachability radius $n_\mathbb{X}$, (4.11) also implies algebraic equivalence of the original and the new states. Moreover, this state-transformation projects the reachability structure in terms of Th. 3.13 to the identity matrix. Therefore, we call this equivalent SS representation the reachability canonical form.

By applying the transformation associated with T_r on the matrices of $\mathfrak{R}_{\mathrm{SS}}(\mathcal{S})$ in accordance with $\overset{n_\mathbb{P}}{\sim}$, the transformed matrices are given by:

$$\left[\begin{array}{c|c} A_\mathrm{r} & B_\mathrm{r} \\ \hline C_\mathrm{r} & D_\mathrm{r} \end{array}\right] := \left[\begin{array}{cccc|c} 0 & \cdots & 0 & -\alpha_0^\mathrm{r} & 1 \\ 1 & \ddots & \vdots & -\alpha_1^\mathrm{r} & 0 \\ \vdots & \ddots & 0 & \vdots & \vdots \\ 0 & \cdots & 1 & -\alpha_{n_\mathbb{X}-1}^\mathrm{r} & 0 \\ \hline \beta_{n_\mathbb{X}-1}^\mathrm{r} & \beta_{n_\mathbb{X}-2}^\mathrm{r} & \cdots & \beta_0^\mathrm{r} & \beta_{n_\mathbb{X}}^\mathrm{r} \end{array}\right].$$

Then,

$$\mathfrak{R}_{\mathrm{SS}}^{\mathcal{R}}(\mathcal{S}) := \left[\begin{array}{c|c} A_\mathrm{r} & B_\mathrm{r} \\ \hline C_\mathrm{r} & D_\mathrm{r} \end{array}\right] \in \left[\begin{array}{c|c} \mathcal{R}^{n_\mathbb{X} \times n_\mathbb{X}} & \mathcal{R}^{n_\mathbb{X} \times 1} \\ \hline \mathcal{R}^{1 \times n_\mathbb{X}} & \mathcal{R} \end{array}\right], \tag{4.12}$$

is called the *reachability canonical state-space representation* of $\mathcal{S}$ and it is equivalent with $\mathfrak{R}_{\mathrm{SS}}(\mathcal{S})$. Proof of the above given matrix operations similarly follows as for LTV-SS representations (see [172, 238] for the CT case and [113, 139] for the DT case).

Example 4.4 (LPV reachability canonical form, SISO). Consider the completely state-reachable DT-LPV-SS representation defined in Example 3.30. By applying state-transformation (4.11) in terms of Def. 3.29, the resulting reachability canonical form of this representation is the following:

$$\mathfrak{R}_{\mathrm{SS}}^{\mathcal{R}}(\mathcal{S}) = \left[\begin{array}{cc|c} 0 & -\mathrm{q}^{-1}p^2 & 1 \\ 1 & \mathrm{q}^{-1}p + \mathrm{q}^{-2}p & 0 \\ \hline 1 + \frac{\mathrm{q}^{-1}p}{p} & \left(1 + \frac{\mathrm{q}^{-2}p}{p}\right)\mathrm{q}^{-1}p + \mathrm{q}^{-2}p & p \end{array}\right].$$

Due to the complete state-reachability, the resulting canonical form is fully equivalent with the original representation. □

Note that the previously introduced algorithm can also be applied to generate reachability-based MIMO canonical forms, but instead of the rows of $\mathcal{O}_{n_\mathbb{X}}$, the columns of $\mathcal{R}_{n_\mathbb{X}}$ are used in this case. According to Step 1 of this mechanism, $\mathcal{R}_{n_\mathbb{X}}$ is rewritten as a sequence of its column vectors:

$$\mathcal{R}_{n_\mathbb{X}} = \begin{bmatrix} \mathtt{r}_{11} & \dots & \mathtt{r}_{n_\mathbb{U}1} & \mathtt{r}_{12} & \dots & \mathtt{r}_{n_\mathbb{U}2} & \dots \end{bmatrix}, \tag{4.13}$$

where $B = \begin{bmatrix} \mathtt{r}_{11} & \dots & \mathtt{r}_{n_\mathbb{U}1} \end{bmatrix}$ and each $\mathtt{r}_j = [\mathtt{r}_{ij}]_{i=1}^{n_\mathbb{U}}$, $j > 1$ is defined similarly as (3.91) for CT and as (3.98) for DT. To accomplish Step 1, the selection of $n_\mathbb{X}$ linearly independent vectors from the $n_\mathbb{X} \times n_\mathbb{U}$ column vectors is required in order to determine the state-basis of the reachability canonical form. In the following, such a selection strategy is used that reproduces the structure of the previously introduced SISO LPV reachability canonical form. According to this, select the rows of $\mathcal{R}_{n_\mathbb{X}}$ as

$$\{\mathtt{r}_{11}, \mathtt{r}_{12}, \dots, \mathtt{r}_{1(n_\mathbb{X} n_\mathbb{U})}, \mathtt{r}_{21}, \mathtt{r}_{22}, \dots\},$$

which matches the extension of Young's selection scheme I (see [110]). Temporally assume that $\mathrm{Rank}(B) = n_\mathbb{U}$ which means that $\{\mathtt{r}_{11}, \mathtt{r}_{12}, \dots, \mathtt{r}_{1n_\mathbb{U}}\}$ are linearly independent vector functions. Then, the linear dependence of every vector function from the ordered sequence can be analyzed one after the other just like in the observability case. However, in Young's selection scheme I, the vectors $\{\mathtt{r}_{11}, \mathtt{r}_{21}, \dots, \mathtt{r}_{n_\mathbb{U}1}\}$ have to be selected to the state transformation even if the ordering would indicate it else. The explanation is that in the reachability canonical form, every row of B_r must be zero or a standard basis in $\mathbb{R}^{1\times n_\mathbb{U}}$, which needs that all $\{\mathtt{r}_1, \mathtt{r}_2, \dots, \mathtt{r}_{n_\mathbb{U}}\}$ must be the part of T_r. According to this selection scheme, if $\tau_i \in \mathbb{I}_1^{n_x}$ is the smallest number such that $\mathtt{r}_{i\tau_i}$ is linearly dependent on the previous vectors, then there exists a set of unique meromorphic functions $\{\alpha^\mathrm{r}_{ijl} \in \mathcal{R}\}$, such that

$$\mathtt{r}_{i\tau_i} = \sum_{j=1}^{i} \sum_{l=0}^{\tau_{ij}-1} \alpha^\mathrm{r}_{ijl}\, \mathtt{r}_{il}, \tag{4.14}$$

where $\{\tau_{ij}\}$ satisfies (4.7) because of the ordering of the vectors. Once that all dependent vector functions have been found, a total of $n_\mathbb{X} = \sum_{i=1}^{n_\mathbb{U}} \tau_i$ independent vector functions are selected due to the full rank assumption of $\mathcal{R}_{n_\mathbb{X}}$. Furthermore, by the selection scheme, the $n_\mathbb{U}$ number of vectors $\{\mathtt{r}_{11}, \mathtt{r}_{12}, \dots, \mathtt{r}_{1n_\mathbb{U}}\}$ are automatically selected implying

$$\max_{i\in\mathbb{I}_1^{n_\mathbb{U}}} \tau_i = \tau_{\max} \leq n_\mathbb{X} - n_\mathbb{U} + 1. \tag{4.15}$$

The remaining linearly dependent relations are described by $\sum_{i=1}^{n_\mathbb{U}} \sum_{j=1}^{i} \tau_{ij}$ rational functions $\{\alpha^\mathrm{r}_{ijl}\} \in \mathcal{R}$, and T_r is defined as:

$$T_\mathrm{r}^{-1} := \begin{cases} \begin{bmatrix} \mathtt{r}_{11} & \dots & \mathtt{r}_{1(\tau_1-1)} & \dots & \mathtt{r}_{n_\mathbb{U}1} & \dots & \mathtt{r}_{n_\mathbb{U}(\tau_{n_\mathbb{U}}-1)} \end{bmatrix}, & \text{if } \mathbb{T} = \mathbb{R}; \\ \begin{bmatrix} \overleftarrow{\mathtt{r}}_{11} & \dots & \overleftarrow{\mathtt{r}}_{1(\tau_1-1)} & \dots & \overleftarrow{\mathtt{r}}_{n_\mathbb{U}1} & \dots & \overleftarrow{\mathtt{r}}_{n_\mathbb{U}(\tau_{n_\mathbb{U}}-1)} \end{bmatrix}, & \text{if } \mathbb{T} = \mathbb{Z}. \end{cases}$$

which accomplishes Step 2. Again, linear independence of the selected vector functions assures the existence of the inverse, thus T_r implies an equivalence relation in terms of Def. 3.29. As a final step, applying the previously constructed equivalence relation on $\mathfrak{R}_{\mathrm{SS}}(\mathcal{S})$, yields the transformed matrices in the following form:

$$\left[\begin{array}{c|c} A_{\mathrm{r}} & B_{\mathrm{r}} \\ \hline C_{\mathrm{r}} & D_{\mathrm{r}} \end{array}\right] := \left[\begin{array}{c|c} \left[A^{\mathrm{r}}_{ij}\right],\ i,j \in \mathbb{I}_1^{n_{\mathbb{U}}} & \begin{array}{c} e_1^\top \\ 0_{(\tau_1-1)\times n_{\mathbb{U}}} \\ \vdots \\ e_{n_{\mathbb{U}}}^\top \\ 0_{(\tau_{n_{\mathbb{Y}}}-1)\times n_{\mathbb{U}}} \end{array} \\ \hline C^{\mathrm{r}}_1 \ \ldots \ C^{\mathrm{r}}_{n_{\mathbb{U}}} & D \end{array}\right],$$

where $\{e_i\}_{i=1}^{n_{\mathbb{U}}}$ is the standard basis of $\mathbb{R}^{n_{\mathbb{U}}}$ and

$$A^{\mathrm{r}}_{ii} = \begin{bmatrix} 0 & \ldots & 0 & -\alpha^{\mathrm{r}}_{ii0} \\ 1 & \ddots & \vdots & -\alpha^{\mathrm{r}}_{ii1} \\ \vdots & \ddots & 0 & \vdots \\ 0 & \ldots & 1 & -\alpha^{\mathrm{r}}_{ii(\tau_i-1)} \end{bmatrix}_{(\tau_i\times\tau_i)} \qquad A^{\mathrm{r}}_{ij} = \begin{bmatrix} 0 & \ldots & 0 & -\alpha^{\mathrm{r}}_{ij0} \\ \vdots & & \vdots & \vdots \\ \vdots & & \vdots & -\alpha^{\mathrm{r}}_{ij(\tau_{ij}-1)} \\ \vdots & & \vdots & 0 \\ \vdots & & \vdots & \vdots \\ 0 & \ldots & 0 & 0 \end{bmatrix}_{(\tau_i\times\tau_j)}$$

$$C^{\mathrm{r}}_i = \begin{bmatrix} \beta^{\mathrm{r}}_{1i(\tau_i-1)} & \cdots & \beta^{\mathrm{r}}_{1i0} \\ \vdots & & \vdots \\ \beta^{\mathrm{r}}_{n_{\mathbb{Y}}i(\tau_i-1)} & \cdots & \beta^{\mathrm{r}}_{n_{\mathbb{Y}}i0} \end{bmatrix}_{(n_{\mathbb{Y}}\times\tau_i)} \qquad D_{\mathrm{r}} = \begin{bmatrix} \beta^{\mathrm{r}}_{11\tau_1} & \cdots & \beta^{\mathrm{r}}_{1n_{\mathbb{U}}\tau_1} \\ \vdots & & \vdots \\ \beta^{\mathrm{r}}_{n_{\mathbb{Y}}1\tau_{n_{\mathbb{Y}}}} & \cdots & \beta^{\mathrm{r}}_{n_{\mathbb{Y}}n_{\mathbb{U}}\tau_{n_{\mathbb{Y}}}} \end{bmatrix}_{(n_{\mathbb{Y}}\times n_{\mathbb{U}})}$$

Proof of that the invertible state-transformation T_{r} always results in the above given structure follows similarly as in [139]. Again, $\mathfrak{R}^{\mathcal{R}}_{\mathrm{SS}}(\mathcal{S})$ is equivalent with $\mathfrak{R}_{\mathrm{SS}}(\mathcal{S})$ and characterizes the decomposition of $\mathfrak{R}^{\mathcal{R}}_{\mathrm{SS}}(\mathcal{S})$ into state-reachable subsystems associated with each output channels. Due to the fact that the projected $n_{\mathbb{X}}$-step reachability matrix of $\mathfrak{R}^{\mathcal{R}}_{\mathrm{SS}}(\mathcal{S})$ is the identity matrix (SISO case) or composed from zero column vectors and the standard basis of $\mathbb{R}^{n_{\mathbb{X}}}$ (MIMO case) the resulting canonical representation is always completely state-reachable. Thus the following corollary holds for all representations of $\mathcal{S}$ with the structural form of $(A_{\mathrm{r}}, B_{\mathrm{r}}, C_{\mathrm{r}}, D_{\mathrm{r}})$ both in the SISO and the MIMO cases:

Corollary 4.2. *$\mathfrak{R}^{\mathcal{R}}_{\mathrm{SS}}(\mathcal{S})$ is completely state-reachable.*

Furthermore it also holds that if $\mathfrak{R}_{\mathrm{SS}}(\mathcal{S})$ is minimal, then the resulting $\mathfrak{R}^{\mathcal{R}}_{\mathrm{SS}}(\mathcal{S})$ by the given construction procedure is also minimal. In the case of dependent columns of B, the state transformation is constructed based on the independent input channels. If the LPV-SS representation is not structurally state-reachable, then computation of an equivalent reachability canonical form is possible by finding an SS realization of the system which is structurally state-reachable. Similar to the observability case, such a realization always exists, if $\mathcal{S}$ has no autonomous dynamics.

4.1.3 Companion Canonical Forms

The LPV observability and reachability canonical forms can also be given in an other, so-called *companion* or *phase-variable* form. These representations $\mathfrak{R}_{\mathrm{SS}}^{\mathcal{O}_c}(\mathcal{S})$ and $\mathfrak{R}_{\mathrm{SS}}^{\mathcal{R}_c}(\mathcal{S})$ are defined in the SISO case as:

$$\mathfrak{R}_{\mathrm{SS}}^{\mathcal{O}_c}(\mathcal{S}) := \left[\begin{array}{c|c} A_{\mathrm{co}} & B_{\mathrm{co}} \\ \hline C_{\mathrm{co}} & D_{\mathrm{co}} \end{array}\right] = \left[\begin{array}{cccc|c} 0 & \dots & 0 & -\alpha_0^{\mathrm{co}} & \beta_0^{\mathrm{co}} \\ 1 & \ddots & \vdots & -\alpha_1^{\mathrm{co}} & \beta_1^{\mathrm{co}} \\ \vdots & \ddots & 0 & \vdots & \vdots \\ 0 & \dots & 1 & -\alpha_{n_{\mathbb{X}}-1}^{\mathrm{co}} & \beta_{n_{\mathbb{X}}-1}^{\mathrm{co}} \\ \hline 0 & \dots & 0 & 1 & \beta_{n_{\mathbb{X}}}^{\mathrm{co}} \end{array}\right],$$

$$\mathfrak{R}_{\mathrm{SS}}^{\mathcal{R}_c}(\mathcal{S}) := \left[\begin{array}{c|c} A_{\mathrm{cr}} & B_{\mathrm{cr}} \\ \hline C_{\mathrm{cr}} & D_{\mathrm{cr}} \end{array}\right] = \left[\begin{array}{cccc|c} 0 & 1 & \dots & 0 & 0 \\ \vdots & \vdots & \ddots & \vdots & \vdots \\ 0 & 0 & \dots & 1 & 0 \\ -\alpha_0^{\mathrm{cr}} & -\alpha_1^{\mathrm{cr}} & \dots & -\alpha_{n_{\mathbb{X}}-1}^{\mathrm{cr}} & 1 \\ \hline \beta_0^{\mathrm{cr}} & \beta_1^{\mathrm{cr}} & \dots & \beta_{n_{\mathbb{X}}-1}^{\mathrm{cr}} & \beta_{n_{\mathbb{X}}}^{\mathrm{cr}} \end{array}\right],$$

Again, it can be proved, based on [171] and [227], that every LPV system $\mathcal{S}$ admits a state-variable representation in these forms and they are equivalent with all SS representations of $\mathcal{S}$. The state-transformations that lead to these canonical forms can be constructed as:

$$T_{\mathrm{co}}^{-1} := \begin{cases} \left[\,\mathtt{r}_1,\ \mathtt{r}_2,\ \dots,\ \mathtt{r}_{n_{\mathbb{X}}}\,\right], & \text{if } \mathbb{T}=\mathbb{R}; \\ \left[\,\overleftarrow{\mathtt{r}}_1,\ \overleftarrow{\mathtt{r}}_2,\ \dots,\ \overleftarrow{\mathtt{r}}_{n_{\mathbb{X}}}\,\right], & \text{if } \mathbb{T}=\mathbb{Z}; \end{cases} \tag{4.16a}$$

$$T_{\mathrm{cr}}^{\top} := \left[\,\mathtt{o}_1^{\top},\ \mathtt{o}_2^{\top},\ \dots,\ \mathtt{o}_{n_{\mathbb{X}}}^{\top}\,\right], \tag{4.16b}$$

where $\mathtt{r}_1$ is the last column of $\mathcal{O}_{n_{\mathbb{X}}}^{-1}$ which is additionally shifted forward in time in case of $\mathbb{T}=\mathbb{Z}$, $\mathtt{o}_1$ is the last row of $\mathcal{R}_{n_{\mathbb{X}}}^{-1}$ which is additionally shifted backward in time in case of $\mathbb{T}=\mathbb{Z}$, and $\mathtt{r}_i$, $\mathtt{o}_i$ are generated recursively by (3.86) and (3.91) in continuous-time and by (3.96) and (3.98) in discrete-time. In the MIMO case, the companion forms are generated by selecting the linearly independent rows (columns) based on a different ordering. Similar to the LTI case, the ordering of the rows of $\mathcal{O}_{n_{\mathbb{X}}}$ by Young's selection scheme I results in the companion observability canonical form, while the ordering of the columns of $\mathcal{R}_{n_{\mathbb{X}}}$ by Young's selection scheme II results in the companion reachability canonical form.

Example 4.5 (Companion canonical forms). Consider again the DT-LPV-SS representation defined in Example 3.30. By constructing the state-transformations (4.16a–b) and applying them to the original SS representation, the following companion canonical forms result.

$$T_{\text{co}} \diamond p = \begin{bmatrix} -\frac{p^2}{\mathrm{q}p} & 0 \\ 1 & \frac{1}{p} \end{bmatrix} \quad \Rightarrow \quad \mathfrak{R}_{\text{SS}}^{\mathcal{O}_\text{c}}(\mathcal{S}) = \left[\begin{array}{cc|c} 0 & -p\frac{\mathrm{q}p^2}{\mathrm{q}^2 p} & -\frac{\mathrm{q}p^2}{\mathrm{q}^2 p} \\ 1 & p + \frac{p^2}{\mathrm{q}p} & 1 + \frac{p}{\mathrm{q}p} \\ \hline 0 & 1 & p \end{array}\right],$$

$$T_{\text{cr}} \diamond p = \begin{bmatrix} \frac{1}{\mathrm{q}^{-1}p} & -\frac{1}{\mathrm{q}^{-1}p^2} \\ 1 & 0 \end{bmatrix} \quad \Rightarrow \quad \mathfrak{R}_{\text{SS}}^{\mathcal{R}_\text{c}}(\mathcal{S}) = \left[\begin{array}{cc|c} 0 & 1 & 0 \\ -\mathrm{q}^{-1}p^2 & p + \mathrm{q}^{-1}p & 1 \\ \hline -\frac{\mathrm{q}^{-1}p^2}{p} & 1 + \frac{\mathrm{q}^{-1}p}{p} & p \end{array}\right].$$

□

4.1.4 *Transpose of SS Representations*

An important difference with respect to the LTI state-space representations is that transposed LPV-SS representations do not have an equal manifest behavior. In the LPV case, this hinders the use of important relations of the LTI system theory and identification (see [189]) which are based on the *transposition property* [116]. To show this, consider the following argument:

Let $\mathfrak{R}_{\text{SS}}(\mathcal{S})$ be a state-space representation of a given SISO LPV system. Then the transposed SS representation, defined as

$$\mathfrak{R}_{\text{SS}}^{\top}(\mathcal{S}) := \left[\begin{array}{c|c} A^\top & C^\top \\ \hline B^\top & D \end{array}\right] \quad \text{when} \quad \mathfrak{R}_{\text{SS}}(\mathcal{S}) = \left[\begin{array}{c|c} A & B \\ \hline C & D \end{array}\right], \tag{4.17}$$

is not a SS representation of $\mathcal{S}$, because the associated output trajectories of these representations are not equal in case of a varying scheduling signal (see Example 4.6 and [189]). In discrete time, this can be easily proved by computing the output responses of the representations for an impulsive input at $k = 0$, i.e $u(0) = 1$, and zero initial state $x(0) = 0$. Denote the resulting output sequences by y for $\mathfrak{R}_{\text{SS}}(\mathcal{S})$ and y' for $\mathfrak{R}_{\text{SS}}^{\top}(\mathcal{S})$. Then these sequences reads as

$$\begin{aligned} y(0) &= (D \diamond p)(0), & y'(0) &= (D \diamond p)(0), \\ y(1) &= (C \diamond p)(1)(B \diamond p)(0), & y'(1) &= (B^\top \diamond p)(1)(C^\top \diamond p)(0), \\ y(2) &= (C \diamond p)(2)(A \diamond p)(1)(B \diamond p)(0), & y'(2) &= (B^\top \diamond p)(2)(A^\top \diamond p)(1)(C^\top \diamond p)(0), \end{aligned}$$

It is obvious that these sequences are not equal if p is not a constant signal. The reason for this phenomenon is based on the non-commutativity of the multiplication on $\mathcal{R}[\xi]$. Similar arguments hold in the continuous time case.

Example 4.6 (LPV system transposition). In this example, the connection between LPV canonical forms and their transpose is investigated. Consider the canonical forms derived in Example 4.2 and 4.4 which are equivalent with the DT-SS representation of Example 3.30. The transpose of these canonical forms have been obtained according to (4.17) by computing the transpose of the matrices. The output response of these transposed canonical forms and the original SS representation have been calculated for

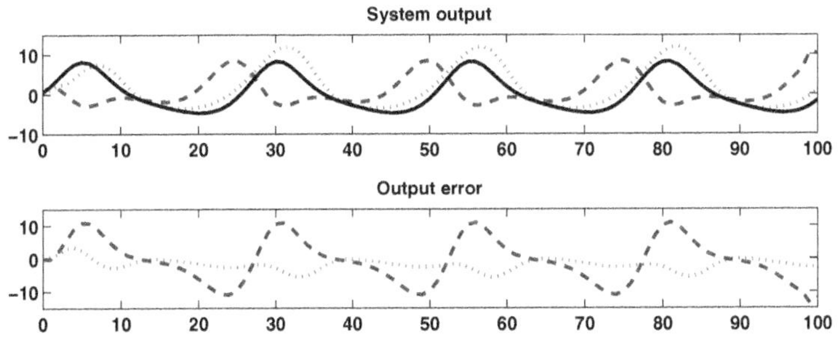

Fig. 4.1 Comparison of the transpose of canonical representations given in Examples 4.2 and 4.4 in terms of their output response, i.e. their output error with respect to the original representation. Transposed canonical observability form (dashed line), transposed canonical reachability form (dotted line), original representation (solid line).

$$u(k) = \sin\left(\frac{1}{4}k + \frac{\pi}{6}\right), \qquad p(k) = \frac{1}{2} + \frac{1}{4}\sin\left(\frac{1}{4}k + \frac{\pi}{2}\right),$$

and with zero initial conditions of the state variables at $k = 0$. The results are presented in Fig. 4.1. From these signals one can conclude that the transposed forms are not equal to the original SS representation. This proves that transposition of SS representations changes the manifest behavior of LPV-SS representations in general. □

4.1.5 LTI vs. LPV State Transformation

In the previous part, we have seen that through the developed behavioral framework, canonical forms of LPV systems can be similarly formulated as in the LTI case. It has also been emphasized how this framework extends the concepts of LTI system theory to LPV systems by proper handling of the time operators and their effect on scheduling-dependent coefficient functions. Comparing the introduced construction mechanism of LPV canonical forms to the state-of-the-art of the LPV literature, similar mechanisms can be found, which have been developed for CT-SS representations with linear dependence (see [89] as a notable approach). Thus the develop algorithm can be seen as a generalization of these approaches.

However in the general LPV literature, at many occasions the LTI theory is used intuitively, applying-state transformations in the form:

$$A' = T^{-1}AT, \tag{4.18}$$

where T is dependent on p. Such a state-transformation is equivalent to a state-transformation applied separately for every constant scheduling signal of $\mathfrak{B}_{\mathbb{P}}$. Based on Sect. 3.2, it is obvious that this transformation does not imply equivalence in any sense if T is not constant. It is also common that canonical forms are usually "achieved" by generating $\mathcal{O}_{n_{\mathbb{X}}}$ and $\mathcal{R}_{n_{\mathbb{X}}}$ similar to the LTI case (see Example

4.7). This corresponds to the observability/reachability matrices of the representation with respect to constant scheduling signals. Using independent rows (columns) of these matrices a state-transformation matrix T is formed. Then T is applied according to (4.18) to calculate the "canonical" form (see [224] and [180] as examples). It is not surprising that by this methodology the resulting structures resemble the observability/reachability canonical forms, however they are not equivalent in manifest behavior with the original system. To illustrate this see Example 4.7.

Example 4.7 (LTI vs LPV state transformation). Consider the canonical forms of the DT-SS representation derived in Example 4.2 and 4.4 for the DT-SS representation $\mathfrak{R}_{\mathrm{SS}}(\mathcal{S})$ defined in Example 3.30. Let $\mathcal{O}^{\dagger}_{n_{\mathbb{X}}}$ and $\mathcal{R}^{\dagger}_{n_{\mathbb{X}}}$ denote the "observability " and "reachability " matrices constructed for $\mathfrak{R}_{\mathrm{SS}}(\mathcal{S})$ in the LTI sense:

$$\mathcal{O}^{\dagger}_{n_{\mathbb{X}}} = \begin{bmatrix} C^{\top} & A^{\top}C^{\top} \end{bmatrix}^{\top}, \qquad \mathcal{R}^{\dagger}_{n_{\mathbb{X}}} = \begin{bmatrix} B & AB \end{bmatrix}$$

$$\mathcal{O}^{\dagger}_{n_{\mathbb{X}}} \diamond p = \begin{bmatrix} 1 & \frac{1}{p} \\ p & 2 \end{bmatrix}, \qquad \mathcal{R}^{\dagger}_{n_{\mathbb{X}}} \diamond p = \begin{bmatrix} 1 & 2p \\ p & p^2 \end{bmatrix}.$$

Now compute what would result by applying the state transformation (4.18) with $T = \mathcal{O}^{\dagger}_{n_{\mathbb{X}}}$ or $T^{-1} = \mathcal{R}^{\dagger}_{n_{\mathbb{X}}}$ just like in the LTI case. This intuitive approach produces the following so-called "frozen" canonical forms:

$$\mathfrak{R}_{\mathrm{SS}}^{\mathcal{O}_{\mathrm{LTI}}}(\mathcal{S}) = \left[\begin{array}{cc|c} 0 & 1 & p \\ -p & -p & -p^2 \\ \hline 1 & 0 & p \end{array}\right], \qquad \mathfrak{R}_{\mathrm{SS}}^{\mathcal{R}_{\mathrm{LTI}}}(\mathcal{S}) = \left(\mathfrak{R}_{\mathrm{SS}}^{\mathcal{O}_{\mathrm{LTI}}}(\mathcal{S})\right)^{\top}$$

Generally in the literature some follow this approach (see [224, 180]). It is important to note that the two sets of LPV representations derived here and in Example 4.2 and 4.4 are equivalent for constant scheduling trajectories, but they are unequal globally. To show this phenomenon, the output response of these frozen canonical representations, the global canonical forms, and the original SS representation have been calculated for the signals p and u indicated in Example 4.6. The results are presented in Fig. 4.2. As can be seen, the global canonical forms $\mathfrak{R}_{\mathrm{SS}}^{\mathcal{O}}(\mathcal{S})$ and $\mathfrak{R}_{\mathrm{SS}}^{\mathcal{R}}(\mathcal{S})$ completely reproduce the original output with zero error. However, $\mathfrak{R}_{\mathrm{SS}}^{\mathcal{O}_{\mathrm{LTI}}}(\mathcal{S})$ and $\mathfrak{R}_{\mathrm{SS}}^{\mathcal{R}_{\mathrm{LTI}}}(\mathcal{S})$ have a relatively huge representation error in the magnitude of 35% even for these very smooth and slowly varying p and u, which mainly comes from a scheduling dependent phase and gain lag with respect to y. □

4.2 From State-Space to the Input-Output Domain

Equivalence transformations between SS and IO representations in the LPV behavioral framework are of paramount importance. Such transformations are not only necessary to provide representations of a given system in these domains, but they are also the key elements to compare LPV model structures in terms of representation capabilities, to compare identified models in different representation domains,

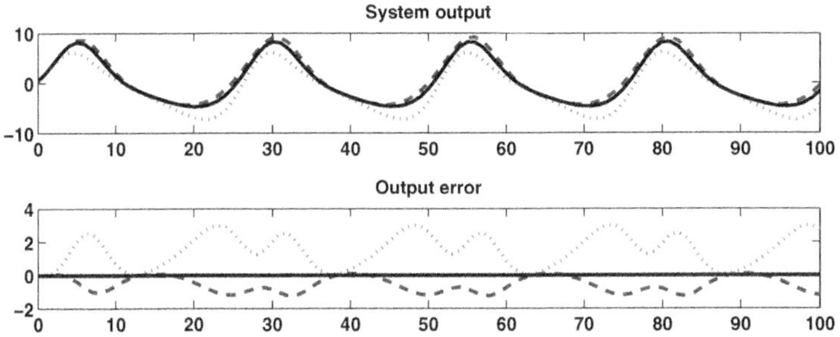

Fig. 4.2 Comparison of global and frozen canonical representations of Example 4.7 in terms of their output response, i.e. their output error with respect to the original representation. $\mathfrak{R}_{\mathrm{SS}}^{\mathcal{O}_{\mathrm{LTI}}}(\mathcal{S})$ (dashed line), $\mathfrak{R}_{\mathrm{SS}}^{\mathcal{R}_{\mathrm{LTI}}}(\mathcal{S})$ (dotted line), $\mathfrak{R}_{\mathrm{SS}}^{\mathcal{O}}(\mathcal{S})$ (solid line), $\mathfrak{R}_{\mathrm{SS}}^{\mathcal{R}}(\mathcal{S})$ (identical to the solid line), original representation (identical to the solid line).

and also to convert results of IO identification approaches to SS models applicable for control.

In the following we develop these transformations based on the behavioral framework. We introduce algorithms to obtain an IO realization of a given LPV-SS representation and vice-versa, solving the core problem of the existing LPV system theory (see also [202]). We show that the results generalize the theory presented for LTI systems in case the coefficient dependence on the scheduling vector is meromorphic and dynamic. It is also shown that the canonical forms, developed in the previous part, give special cases of the transformation problem, thus they serve as a simple gateway between the SS and the IO representation domains. Again, it is proved that the common practice of the current LPV literature, namely to apply LTI theory to convert IO models to SS models, provides results which have an unequal manifest behavior.

At first the equivalence transformation from the SS to the IO representation domain is considered. The equivalence transformation in this context means the search for an equivalence class of IO realizations with the same manifest behavior. Theorem 3.3 shows that the state in an LPV-SS representation can always be eliminated as a latent variable without changing the manifest behavior. As a consequence of this elimination property on $\mathcal{R}[\xi]^{\cdot\times\cdot}$ the following corollary holds:

Corollary 4.3 (Latent variable elimination). *For any latent representation* (3.39) *with manifest behavior* $\mathfrak{B}$ *and polynomial matrices* $R_{\mathrm{W}} \in \mathcal{R}[\xi]^{n_{\mathrm{r}}\times n_{\mathbb{W}}}$, $R_{\mathrm{L}} \in \mathcal{R}[\xi]^{n_{\mathrm{r}}\times n_{\mathbb{X}}}$, *there exists a unimodular matrix* $M \in \mathcal{R}[\xi]^{n_{\mathrm{r}}\times n_{\mathrm{r}}}$ *such that*

$$M(\xi)R_{\mathrm{L}}(\xi) = \begin{bmatrix} R'_{\mathrm{L}}(\xi) \\ 0 \end{bmatrix}, \quad M(\xi)R_{\mathrm{W}}(\xi) = \begin{bmatrix} R'_{\mathrm{W}}(\xi) \\ R''_{\mathrm{W}}(\xi) \end{bmatrix}, \tag{4.19}$$

with R'_{L} *of full row rank. The manifest behavior defined by* $(R''_{\mathrm{W}}(\xi)\diamond p)w = 0$ *is equal (almost everywhere) with* $\mathfrak{B}$.

Due to the latent nature of the variable w_{L}, such a transformation in terms of Th. 3.3 is always possible and does not change the manifest behavior, hence it is called an *equivalence transformation*.

The next step to establish an IO realization is to formulate the unimodular transformation of Th. 4.3 and the resulting R'' in the form of an output side polynomial R_{y} and an input side polynomial R_{u}. Write the LPV state-space representation $\mathfrak{R}_{\mathrm{SS}}(\mathcal{S})$ with matrix functions (A,B,C,D) into the latent form (3.39):

$$\underbrace{\begin{bmatrix} I\xi - A \\ -C \end{bmatrix}}_{R_{\mathrm{L}}(\xi)} x = \underbrace{\begin{bmatrix} 0 & B \\ -I & D \end{bmatrix}}_{R_{\mathrm{W}}(\xi)} \begin{bmatrix} y \\ u \end{bmatrix}. \tag{4.20}$$

The resulting polynomials $R \in \mathcal{R}[\xi]^{(n_{\mathbb{X}}+n_{\mathbb{Y}})\times(n_{\mathbb{Y}}+n_{\mathbb{U}})}$ and $R_{\mathrm{L}} \in \mathcal{R}[\xi]^{(n_{\mathbb{X}}+n_{\mathbb{Y}})\times n_{\mathbb{X}}}$ give an equivalent representation of the full behavior of $\mathfrak{R}_{\mathrm{SS}}(\mathcal{S})$. According to Cor. 4.3, there exists a unimodular matrix

$$M(\xi) = \begin{bmatrix} M_{11}(\xi) & M_{12}(\xi) \\ M_{21}(\xi) & M_{22}(\xi) \end{bmatrix} \in \mathcal{R}[\xi]^{(n_{\mathbb{X}}+n_{\mathbb{Y}})\times n_{\mathbb{X}}+n_{\mathbb{Y}}} \tag{4.21}$$

which in terms of $M(\xi)R_{\mathrm{L}}(\xi) = \begin{bmatrix} * & 0 \end{bmatrix}^{\top}$ in (4.19) satisfies

$$M_{21}(\xi)(I\xi - A) - M_{22}(\xi)C = 0.$$

This yields that

$$\underbrace{\begin{bmatrix} * & * \\ -M_{21}(\xi) & M_{21}(\xi)B + M_{22}(\xi)D \end{bmatrix}}_{M(\xi)R_{\mathrm{W}}(\xi)} \begin{bmatrix} y \\ u \end{bmatrix} = \underbrace{\begin{bmatrix} * \\ 0 \end{bmatrix}}_{M(\xi)R_{\mathrm{L}}(\xi)} x,$$

and $R''_{\mathrm{W}}(\xi) = \begin{bmatrix} -M_{21}(\xi) & M_{21}(\xi)B + M_{22}(\xi)D \end{bmatrix}$ is in the form of an output side polynomial $R_{\mathrm{y}}(\xi) = M_{21}(\xi)$ and an input side polynomial $R_{\mathrm{u}}(\xi) = M_{21}(\xi)B + M_{22}(\xi)D$. Note that there is a particular freedom in choosing M to satisfy (4.19), however by restricting $M_{22}(\xi)$ to be monic and $\mathrm{Deg}(M_{22}) = n_{\mathbb{X}}$, $M_{21}(\xi)$ with $\mathrm{Deg}(M_{21}) \leq n_{\mathbb{X}} - 1$ and $M_{22}(\xi)$ can be uniquely determined.

Corollary 4.4 (IO Equivalence transformation). *Let $\mathfrak{R}_{\mathrm{SS}}(\mathcal{S})$ be a state-space representation with manifest behavior $\mathfrak{B}$ and system matrices (A,B,C,D) where $A \in \mathcal{R}^{n_{\mathbb{X}}\times n_{\mathbb{X}}}$. Then there exists a unique monic polynomial $\bar{R}_{\mathrm{y}} \in \mathcal{R}[\xi]^{n_{\mathbb{Y}}\times n_{\mathbb{Y}}}$ with $\mathrm{Deg}(\bar{R}_{\mathrm{y}}) = n_{\mathbb{X}}$ and a unique $\bar{R}_{\mathrm{u}} \in \mathcal{R}[\xi]^{n_{\mathbb{Y}}\times n_{\mathbb{X}}}$ with $\mathrm{Deg}(\bar{R}_{\mathrm{u}}) \leq n_{\mathbb{X}} - 1$ such that*

$$\bar{R}_{\mathrm{y}}(\xi)C = \bar{R}_{\mathrm{u}}(\xi)(I\xi - A). \tag{4.22}$$

Let $R_{\mathrm{com}} = \mathrm{Diag}(r_1, \ldots, r_{n_{\mathbb{Y}}})$, $r_i \in \mathcal{R}[\xi]$, be the greatest common divisor of $\bar{R}_{\mathrm{y}}$ and $\bar{R}_{\mathrm{u}}$ such that there exist $R_{\mathrm{y}}, R_{\mathrm{u}} \in \mathcal{R}[\xi]$ satisfying

$$R_{\mathrm{com}}(\xi)R_{\mathrm{y}}(\xi) = \bar{R}_{\mathrm{y}}(\xi) \quad \textit{and} \quad R_{\mathrm{com}}(\xi)R_{\mathrm{u}}(\xi) = \bar{R}_{\mathrm{u}}(\xi)B + \bar{R}_{\mathrm{y}}(\xi)D. \tag{4.23}$$

Then the IO representation of $\mathcal{S}$, denoted by $\mathfrak{R}_{\mathrm{IO}}(\mathcal{S})$, is given by

$$(R_{\mathrm{y}}(\xi)\diamond p)y = (R_{\mathrm{u}}(\xi)\diamond p)u. \tag{4.24}$$

The algorithm defined by (4.22) and (4.23) is structurally similar to the LTI case (see [153, 146]), but it is more complicated as it involves multiplication with the time operators on the coefficients. Thus, this transformation can result in an increased complexity (like dynamic dependence) of the coefficient functions in the equivalent IO representation. The following property also holds:

Corollary 4.5. *Assume that $\mathfrak{R}_{\mathrm{SS}}(\mathcal{S})$ is minimal, i.e. structurally state-observable. Then the polynomials $\bar{R}_{\mathrm{u}}$ and $\bar{R}_{\mathrm{y}}$ satisfying (4.22) are left-coprime (their greatest common divisor R_{com} is 1).*

Corollary 4.5 means that the equivalence transformation between the SS and IO domain results in the elimination of dynamics related to unobservable states. Thus in case of a structurally state-observable SS representation, like the observability canonical forms, the equivalence transformation simplifies. However, dynamics related to unreachable states are preserved. This underlines the validity of the proposed minimality concept of LPV-SS representations, namely that minimality is equivalent with structural state-observability.

Example 4.8 (IO equivalence transformation). Consider the LPV-SS representation derived in Example 3.22:

$$\left[\begin{array}{c|c} A & B \\ \hline C & D \end{array}\right]\diamond p = \left[\begin{array}{cc|c} 1 & -\mathrm{T}_{\mathrm{d}}^2\mathrm{k}_{\mathrm{s}} & \mathrm{T}_{\mathrm{d}}^2 \\ \frac{1}{p} & 1 & 0 \\ \hline 0 & 1 & 0 \end{array}\right].$$

Let r be the identity function so $r\diamond p = p$. In terms of (4.22), we are looking for a $\bar{R}_{\mathrm{u}} \in \mathcal{R}[\xi]^{1\times 2}$ with $\mathrm{Deg}(\bar{R}_{\mathrm{u}}) = 1$ and a monic polynomial $\bar{R}_{\mathrm{y}} \in \mathcal{R}[\xi]$ with $\mathrm{Deg}(\bar{R}_{\mathrm{y}}) = 2$. Parameterize these polynomials as

$$\bar{R}_{\mathrm{y}}(\xi) = \xi^2 + a_1\xi + a_0, \quad \bar{R}_{\mathrm{u}}(\xi) = \begin{bmatrix} b_{11}\xi + b_{12} & b_{21}\xi + b_{22} \end{bmatrix}.$$

Then in terms of (4.22):

$$(\xi^2 + a_1\xi + a_0)\begin{bmatrix} 0 & 1 \end{bmatrix} = \begin{bmatrix} b_{11}\xi + b_{12} & b_{21}\xi + b_{22} \end{bmatrix} \underbrace{\begin{bmatrix} \xi - 1 & \mathrm{T}_{\mathrm{d}}^2\mathrm{k}_{\mathrm{s}} \\ -\frac{1}{r} & \xi - 1 \end{bmatrix}}_{I\xi - A}.$$

Solving this equation system it follows that

$$a_1 = -\tfrac{r}{\overrightarrow{r}} - 1,\ b_{11} = 0,\ b_{12} = \tfrac{1}{\overrightarrow{r}},$$
$$a_0 = \tfrac{\mathrm{T}_{\mathrm{d}}^2\mathrm{k}_{\mathrm{s}} + r}{\overrightarrow{r}}, \quad b_{21} = 1,\ b_{22} = -\tfrac{r}{\overrightarrow{r}}.$$

The resulting polynomials $\bar{R}_{\mathrm{u}}$ and $\bar{R}_{\mathrm{y}}$ are left coprime, hence

$$R_{\mathrm{y}}(\xi) = \bar{R}_{\mathrm{y}}(\xi) = \xi^2 + a_1\xi + a_0, \tag{4.25a}$$
$$R_{\mathrm{u}}(\xi) = \bar{R}_{\mathrm{u}}(\xi)B + \bar{R}_{\mathrm{y}}(\xi)D = \tfrac{\mathrm{T}_{\mathrm{d}}^2}{\overrightarrow{r}}. \tag{4.25b}$$

After left-multiplying these polynomials with $\overrightarrow{r}$, the IO representation in the form of (3.60) with $n_a = 2$ and $n_b = 0$ has the coefficients

$$a_2 \diamond p = \mathrm{q}p \qquad a_1 \diamond p = -\mathrm{q}p - p,$$
$$a_0 \diamond p = \mathrm{T}_\mathrm{d}^2 \mathrm{k}_\mathrm{s} + p, \quad b_0 \diamond p = \mathrm{T}_\mathrm{d}^2.$$

In terms of $w = \mathrm{Col}(y, u)$, the resulting LPV-IO representation is equal to (3.56) which shows its equivalence with the LPV-SS representation in Example 3.22. □

4.3 From the Input-Output to the State-Space Domain

As a next step, the equivalence transformation from the IO to the SS representation domain is considered. To derive such a transformation, a vital ingredient is to construct a state-map for a given IO representation that defines an equivalent SS representation. This construction can be seen as the counterpart of the previous latent variable elimination. The actual aim is to introduce a latent variable into (4.24) such that it satisfies the state-property, ergo it defines a SS representation of the original system via Th. 3.4.

Generation of a state-map is more involved in the LPV case than for time-invariant systems as the scheduling dependence of the coefficients does not commute with time operators as integration, derivation, and time-shift. In order to improve readability of the upcoming rather technical discussion, we first investigate the intuitive idea behind the used state-construction mechanism, the so-called *cut-and-shift* operation. Then this operation is formally defined over the ring $\mathcal{R}[\xi]$ both in CT and in DT. As a next step, state-maps generated by the cut-and-shift procedure are constructed and it is shown that they introduce latent variables satisfying the state property. Minimality of the resulting state kernel forms is also shown in the SISO case. Finally, algorithms are derived that provide realization of a given IO representation in terms of the previously introduced canonical forms.

4.3.1 *The Idea of Recursive State-Construction*

First, we explain the intuitive idea of the cut-and-shift-map based state-construction through a simple example. Assume that, in continuous-time, a kernel representation $\mathfrak{R}_\mathrm{K}(\mathcal{S})$ with $\mathbb{P} = \mathbb{R}$ is given. $\mathfrak{R}_\mathrm{K}(\mathcal{S})$ is described by $R \in \mathcal{R}[\xi]$, providing the differential equation

$$(r_0 \diamond p)w + (r_1 \diamond p)\frac{\mathrm{d}}{\mathrm{d}t}w + (r_2 \diamond p)\frac{\mathrm{d}^2}{\mathrm{d}t^2}w = 0. \tag{4.26}$$

In the following we construct an equivalent SS representation of $\mathfrak{R}_\mathrm{K}(\mathcal{S})$. This requires a state-construction, where our aim is the elimination of the derivatives of w in (4.26) through the introduction of state-variables. Introduce the latent variable $x_1 : \mathbb{R} \to \mathbb{R}$, defined by

$$x_1 = ((r_1 - \dot{r}_2) \diamond p)w + (r_2 \diamond p)\frac{\mathrm{d}}{\mathrm{d}t}w. \tag{4.27}$$

Then equation (4.26) can be rewritten as

$$\frac{\mathrm{d}}{\mathrm{d}t}x_1 + ((r_0 - \dot{r}_1 + \ddot{r}_2) \diamond p)w = 0, \tag{4.28}$$

using the rule of chain-derivation. It is obvious, that the resulting equations (4.27) and (4.28) define the same manifest behavior as (4.26). In this way, we have eliminated the derivatives of w from (4.26), however the resulting extra equation (4.27) still contains a first order derivative of w. Thus, introduce the latent variable $x_2 : \mathbb{R} \to \mathbb{R}$ such that

$$x_2 = (r_2 \diamond p)w. \tag{4.29}$$

Then, equation (4.27) can be rewritten as

$$\frac{\mathrm{d}}{\mathrm{d}t}x_2 - x_1 + ((r_1 - 2\dot{r}_2) \diamond p)w = 0. \tag{4.30}$$

We have arrived at the following equation system

$$-\left(\begin{bmatrix} r_0 - \dot{r}_1 + \ddot{r}_2 \\ r_1 - 2\dot{r}_2 \\ r_2 \end{bmatrix} \diamond p\right) w = \begin{bmatrix} \frac{\mathrm{d}}{\mathrm{d}t} & 0 \\ -1 & \frac{\mathrm{d}}{\mathrm{d}t} \\ 0 & -1 \end{bmatrix} \begin{bmatrix} x_1 \\ x_2 \end{bmatrix}, \tag{4.31}$$

which is equivalent with (4.26). Due to the fact that the left side is a zero-order while the right side is first-order polynomial, (4.31) is a state-kernel form and $x = [x_1 \;\; x_2]^\top$ trivially fulfills the property of state (see Th. 3.4). Additionally, the algebraic equivalence of the introduced state-relations (4.27) and (4.29) implies that there exists a unimodular transformation in terms of Th. 4.3, which can eliminate x from (4.31) such that (4.26) is reobtained. Thus, the manifest behavior of (4.31) is equivalent with the behavior of $\mathfrak{R}_\mathrm{K}(\mathcal{S})$. Note that in this way we have defined an equivalent state-map of the original system. Using this state-map, one choice of an SS realization follows through the use of (4.29) as the output equation and by the substitution of this equation into (4.31) to obtain the state-equation:

$$\frac{\mathrm{d}}{\mathrm{d}t}x = \left(\begin{bmatrix} 0 & -\frac{r_0 - \dot{r}_1 + \ddot{r}_2}{r_2} \\ 1 & -\frac{r_1 - 2\dot{r}_2}{r_2} \end{bmatrix} \diamond p\right) x, \tag{4.32a}$$

$$w = \left(\begin{bmatrix} 0 & \frac{1}{r_2} \end{bmatrix} \diamond p\right) x, \tag{4.32b}$$

which is a companion observability canonical form if $r_2 = 1$, i.e. if R is monic. This SS realization is equivalent with (4.26) only for those scheduling trajectories where $(r_2 \diamond p)(t) \neq 0$ (see (4.29)). Thus, (4.32a–b) is equivalent in an almost everywhere sense with (4.26). If r_2 is nonzero for all scheduling trajectories, then full equivalence holds.

The intuitive idea behind the state construction that we exposed in this example can be formalized as the following recursive scheme. For a given continuous-time SISO kernel representation $\mathfrak{R}_\mathrm{K}(\mathcal{S})$, the behavior is described by

$$(R_0(\frac{\mathrm{d}}{\mathrm{d}t}) \diamond p)w = 0. \tag{4.33}$$

Let $l = 1$ and define a latent variable as

$$x_l := (R_l(\frac{\mathrm{d}}{\mathrm{d}t}) \diamond p)w, \quad \text{s.t.} \quad R_{(l-1)}(\xi) := \bar{R}_l + \xi R_l(\xi), \tag{4.34}$$

where $\bar{R}_l \in \mathcal{R}$, $R_l \in \mathcal{R}[\xi]$, and due to the multiplication rules, $\bar{R}_0$ is chosen as:

$$\bar{R}_0 := r_0 + \sum_{i=1}^{n_\xi} (-1)^i \dot{r}_i^{(i)}, \tag{4.35}$$

where $\dot{r}^{(i)}$ denotes the dot operation applied to $r \in \mathcal{R}$ for i-times (see equations (4.28) and (4.30) as examples). According to this mechanism

$$\begin{cases} \text{for } l = 1, & \frac{\mathrm{d}}{\mathrm{d}t} x_1 = -(\bar{R}_0 \diamond p)w, \\ \text{for } 1 < l \leq n_\xi, & \frac{\mathrm{d}}{\mathrm{d}t} x_l = x_{l-1} - (\bar{R}_{l-1} \diamond p)w. \end{cases} \tag{4.36}$$

holds and (4.34) with (4.36) give a latent variable representation of $\mathcal{S}$. Repeat these steps recursively on (4.34) till $l = n_\xi$ which results in $R_{n_\xi}(\xi) = 0$. Then the obtained polynomials $\{R_l(\xi)\}_{l=1}^{n_\xi}$ define a state map and $\{\bar{R}_l\}_{l=1}^{n_\xi}$ give the coefficients of the associated SS representation, as it holds that

$$\begin{cases} \text{for } 1 \leq l < n_\xi, & \frac{\mathrm{d}}{\mathrm{d}t} x_{(l+1)} = x_l - (\bar{R}_l \diamond p)w, \\ \text{for } l = n_\xi, & (\bar{R}_{n_\xi} \diamond p)w = x_{n_\xi}. \end{cases} \tag{4.37}$$

The procedure, defined by (4.34), gives the algorithm of state-construction which we call the parameter-varying cut-and-shift map $\delta_- : \mathcal{R}[\xi]^{\cdot \times \cdot} \to \mathcal{R}[\xi]^{\cdot \times \cdot}$. In this terminology, $\bar{R}_l$ corresponds to the cut term while $\xi R_l(\xi)$ is the shift term. Due to the different non-commutative multiplication rules of differentiation and time-shift with respect to the scheduling dependent coefficients, the cut-and-shift-map is defined differently for the CT and DT time-axis. In the following, the technical definition of δ_- is given both in the CT and DT cases, using the idea of the recursive scheme introduced before.

4.3.2 *Cut-and-Shift in Continuous-Time*

In the CT case, the indeterminant ξ is associated with $\frac{\mathrm{d}}{\mathrm{d}t}$, implying that multiplication with ξ on $\mathcal{R}[\xi]^{\cdot \times \cdot}$ gives the non-commutative rule of (3.32). The reverse operation, multiplication by ξ^{-1}, results in integration, which yields:

$$\xi^{-1}\left(r\xi^i + (-1)^i \dot{r}^{(i)}\right) = \sum_{j=0}^{i-1} (-1)^j \dot{r}^{(j)} \xi^{i-j-1}. \tag{4.38}$$

This operation is the same as what is used in (4.34). Based on this, the cut-and-shift-map in continuous-time is defined on $\mathcal{R}[\xi]^{\cdot\times\cdot}$ as

$$\delta_{-}\underbrace{(r_0 + r_1\xi + \ldots + r_n\xi^n)}_{R(\xi)} = \underbrace{r_1' + \ldots + r_n'\xi^{n-1}}_{R'(\xi)}, \tag{4.39}$$

where $R, R' \in \mathcal{R}[\xi]^{\cdot\times\cdot}$ and each new meromorphic coefficient function of R' is computed as the result of elementary cut-and-shift operations:

$$\delta_{-}(r\xi^i) = \begin{cases} \sum_{j=0}^{i-1}(-1)^j r^{(j)}\xi^{i-j-1}, & \text{if } i > 1; \\ 0, & \text{if } i = 0. \end{cases} \tag{4.40}$$

This yields that

$$r_i' = \sum_{j=0}^{n-i}(-1)^j r_{i+j}^{(j)}. \tag{4.41}$$

4.3.3 Cut-and-Shift in Discrete-Time

In the DT case, ξ is associated with the forward time-shift operator q, implying that multiplication with ξ on $\mathcal{R}[\xi]^{\cdot\times\cdot}$ gives the non-commutative rule of (3.51). The reverse operation, multiplication by ξ^{-1}, results in backward time-shift, giving:

$$\xi^{-1}(r\xi^i) = \overleftarrow{r}\,\xi^{i-1}, \tag{4.42}$$

where $\overleftarrow{\cdot}$ is the backward shift operation on $\mathcal{R}$. This implies that in discrete-time, the cut-and-shift-map is defined on $\mathcal{R}[\xi]^{\cdot\times\cdot}$ in the form of (4.39), where each new meromorphic coefficient function is computed as the result of elementary cut-and-shift operations:

$$\delta_{-}(r\xi^i) = \begin{cases} \overleftarrow{r}\,\xi^{i-1}, & \text{if } i > 1; \\ 0, & \text{if } i = 0. \end{cases} \tag{4.43}$$

giving that $r_i' = \overleftarrow{r_i}$.

4.3.4 State-Maps and Polynomial Modules

As a next step we formulate the construction of a state-map, i.e. the generation of state variables, for a given kernel representation, as the recursive use of the cut-and-shift operation on the polynomials of $\mathcal{R}[\xi]$. This procedure is the analog of the introduced recursive-scheme of state-construction in the CT case (see (4.36)). The resulting state-map characterizes a state-kernel representation of the system, which is minimal in the SISO case. In order to describe all equivalent (minimal) state-kernel representations of the system, equivalence classes of the state-map are established in terms of polynomial modules over $\mathcal{R}[\xi]$.

Let $R \in \mathcal{R}[\xi]^{n_r \times n_W}$ be the associated polynomial of the kernel presentation $\mathfrak{R}_K(\mathcal{S})$. Assume that R is monic and given as

$$R(\xi) = r_0^{[0]} + r_1^{[0]}\xi + \ldots + r_{n-1}^{[0]}\xi^{n-1} + \xi^n, \tag{4.44}$$

where superscript $\bullet^{[0]}$ denotes an additional index of the coefficients. Repeated use of δ_- on R and stacking the resulting polynomial matrices leads to

$$\Sigma_-(R) = \begin{bmatrix} \delta_-(R) \\ \delta_-^2(R) \\ \vdots \\ \delta_-^{n-2}(R) \\ \delta_-^{n-1}(R) \end{bmatrix} = \begin{bmatrix} r_1^{[1]} + \ldots + r_{n-1}^{[1]}\xi^{n-2} + \xi^{n-1} \\ r_2^{[2]} + \ldots + r_{n-1}^{[2]}\xi^{n-3} + \xi^{n-2} \\ \vdots \\ r_{n-1}^{[n-1]} + \xi \\ 1 \end{bmatrix}. \tag{4.45}$$

where each coefficient function $r_i^{[j]}$ is computed according to the local cut-and-shift rules based on $\{r_i^{[j-1]}\}_{i=n-j+1}^{n}$ recursively. It is obvious, that if $n_\mathrm{r} = 1$ (SISO case if additionally $n_\mathbb{W} = 2$), then the rows of Σ_- are independent. Thus, it can be shown that $X = \Sigma_-(R)$ defines a minimal state-map in the form of

$$x = (X(\xi) \diamond p)w. \tag{4.46}$$

Later it is shown, that such a state-map implies a unique SS representation. Before that, we characterize all possible minimal state-maps that lead to an equivalent SS representation.

Denote the multiplication by ξ as δ_+, which acts in the same way as as defined by (3.32) and (3.51). Consequently

$$\delta_+\left(\begin{bmatrix} \Sigma_-(R) \\ 0 \end{bmatrix}\right) = \begin{bmatrix} R \\ \Sigma_-(R) \end{bmatrix} - \begin{bmatrix} r_0^{[0]} \\ r_1^{[1]} \\ \vdots \\ 1 \end{bmatrix}. \tag{4.47}$$

Note that $\delta_-\delta_+ = I$, while $\delta_+(\delta_-(R)) = R(\xi) - R(0)$.

Denote by $\mathrm{Span}_\mathcal{R}^{\mathrm{row}}(R)$ the subspace spanned by the rows of $R \in \mathcal{R}[\xi]^{\cdot\times\cdot}$, viewed as a linear space of polynomial vector functions with coefficients in $\mathcal{R}^{\cdot\times\cdot}$. Also introduce $\mathrm{Module}_{\mathcal{R}[\xi]}(R)$ as the left-module in $\mathcal{R}[\xi]^{n_\mathrm{r}\times n_\mathbb{W}}$ generated by the rows of $R \in \mathcal{R}[\xi]^{n_\mathrm{r}\times n_\mathbb{W}}$:

$$\mathrm{Module}_{\mathcal{R}[\xi]}(R) = \mathrm{Span}_\mathcal{R}^{\mathrm{row}}\left(\begin{bmatrix} R \\ \delta_+(R) \\ \vdots \end{bmatrix}\right). \tag{4.48}$$

This module represents the set of equivalence classes on $\mathrm{Span}_\mathcal{R}^{\mathrm{row}}(\Sigma_-(R))$. Let $X \in \mathcal{R}[\xi]^{\cdot\times n_\mathbb{W}}$ be a polynomial matrix with independent rows (full row-rank) such that

$$\mathrm{Span}_\mathcal{R}^{\mathrm{row}}(X) \oplus \mathrm{Module}_{\mathcal{R}[\xi]}(R) = \mathrm{Span}_\mathcal{R}^{\mathrm{row}}(\Sigma_-(R)) + \mathrm{Module}_{\mathcal{R}[\xi]}(R), \tag{4.49}$$

where $\oplus$ denotes direct sum. Based on arguments used in the LTI case (see [231, 153]), X is a minimal state-map of the LPV system $\mathcal{S}$ and it defines a state variable by (4.46). This way it is possible to obtain all minimal SS realizations of $\mathcal{S}$ that are equivalent with the kernel representation associated with R.

4.3.5 State-Maps Based on Kernel Representations

In the previous part, we have established state-map constructions for kernel representations based on the cut-and-shift operation and characterized the class of all state-maps that result in an equivalent SS representation. The next step is to characterize these SS representations with respect to an IO partition. We develop an algorithm which, based on a given state-map, provides a SS realization of a kernel representation for a chosen IO partition. We show that, for specific choices of the state-map, the algorithm provides the SS realization in terms of the previously introduce observability and companion observability canonical forms.

For a given kernel representation $\mathfrak{R}_{\mathrm{K}}(\mathcal{S})$ associated with the polynomial $R \in \mathcal{R}[\xi]^{n_{\mathrm{r}} \times n_{\mathrm{W}}}$, the input-output partition of R is characterized by choosing a selector matrix $S_{\mathrm{u}} \in \mathbb{R}^{\cdot \times n_{\mathrm{W}}}$ giving $u = S_{\mathrm{u}} w$ and a complementary matrix $S_{\mathrm{y}} \in \mathbb{R}^{\cdot \times n_{\mathrm{W}}}$ giving $y = S_{\mathrm{y}} w$. In case of an unknown IO partition, the construction of S_{u} follows by computing the subspace of $\mathcal{R}^{1 \times n_{\mathrm{W}}}$ consisting of the $\mathcal{R}$-span of the elements with degree zero in $\mathrm{Span}_{\mathcal{R}}^{\mathrm{row}}(\Sigma_{-}(R)) + \mathrm{Module}_{\mathcal{R}[\xi]}(R)$ and choosing S_{u} such that the rows of S_{u} span a complement of this subspace relative to $\mathcal{R}^{n_{\mathrm{W}}}$. Then, S_{y} is chosen complementary to S_{u}.

Assume that a full row rank $X \in \mathcal{R}[\xi]^{\cdot \times n_{\mathrm{W}}}$ is given which satisfies (4.49). Then the matrix polynomial X and the matrix S_{u} jointly lead to the direct sum decomposition:

$$\begin{aligned}\mathrm{Span}_{\mathcal{R}}^{\mathrm{row}}(I_{n_{\mathrm{W}} \times n_{\mathrm{W}}}) + \mathrm{Span}_{\mathcal{R}}^{\mathrm{row}}(\Sigma_{-}(R)) + \mathrm{Module}_{\mathcal{R}[\xi]}(R) = \\ \mathrm{Span}_{\mathcal{R}}^{\mathrm{row}}(\Sigma_{-}(S_{\mathrm{u}})) \oplus \mathrm{Span}_{\mathcal{R}}^{\mathrm{row}}(\Sigma_{-}(X)) \oplus \mathrm{Module}_{\mathcal{R}[\xi]}(R). \quad (4.50)\end{aligned}$$

From (4.47), it follows that

$$\begin{aligned}\mathrm{Span}_{\mathcal{R}}^{\mathrm{row}}(\delta_{+}(\Sigma_{-}(X))) \subseteq \mathrm{Span}_{\mathcal{R}}^{\mathrm{row}}(I_{n_{\mathrm{W}} \times n_{\mathrm{W}}}) + \\ \mathrm{Span}_{\mathcal{R}}^{\mathrm{row}}(\Sigma_{-}(R)) + \mathrm{Module}_{\mathcal{R}[\xi]}(R), \quad (4.51)\end{aligned}$$

which implies

$$\mathrm{Span}_{\mathcal{R}}^{\mathrm{row}}(\delta_{+}(X)) \subseteq \mathrm{Span}_{\mathcal{R}}^{\mathrm{row}}(X) \oplus \mathrm{Span}_{\mathcal{R}}^{\mathrm{row}}(S_{\mathrm{u}}) \oplus \mathrm{Module}_{\mathcal{R}[\xi]}(R). \quad (4.52)$$

On the other hand, S_{y} gives

$$\mathrm{Span}_{\mathcal{R}}^{\mathrm{row}}(S_{\mathrm{y}}) \subseteq \mathrm{Span}_{\mathcal{R}}^{\mathrm{row}}(X) \oplus \mathrm{Span}_{\mathcal{R}}^{\mathrm{row}}(S_{\mathrm{u}}) \oplus \mathrm{Module}_{\mathcal{R}[\xi]}(R). \quad (4.53)$$

These inclusions imply, that there exist unique matrix functions (A,B,C,D) in $\mathcal{R}^{\cdot\times\cdot}$ and polynomial matrix functions $X_{\mathrm{u}},X_{\mathrm{y}}\in\mathcal{R}[\xi]^{\cdot\times\cdot}$ with appropriate dimensions such that

$$\xi X(\xi)=AX(\xi)+BS_{\mathrm{u}}+X_{\mathrm{u}}(\xi)R(\xi), \tag{4.54a}$$
$$S_{\mathrm{y}}(\xi)=CX(\xi)+DS_{\mathrm{u}}+X_{\mathrm{y}}(\xi)R(\xi). \tag{4.54b}$$

Then

$$\left[\begin{array}{c|c} A & B \\ \hline C & D \end{array}\right]\in\left[\begin{array}{c|c} \mathcal{R}^{n_{\mathbb{X}}\times n_{\mathbb{X}}} & \mathcal{R}^{n_{\mathbb{X}}\times n_{\mathbb{U}}} \\ \hline \mathcal{R}^{n_{\mathbb{Y}}\times n_{\mathbb{X}}} & \mathcal{R}^{n_{\mathbb{Y}}\times n_{\mathbb{U}}} \end{array}\right], \tag{4.55}$$

is a minimal state-representation of the LPV system $\mathcal{S}$. This algorithm provides an SS realization of both LPV-IO and LPV-KR representations.

As a next step, we show that specific choices of X lead to the construction of the observability and the reachability canonical forms via algorithm (4.54a–b). Consider the SISO case. Assume that $\mathfrak{R}_{\mathrm{IO}}(\mathcal{S})$ is given with polynomial matrices $R_{\mathrm{y}},R_{\mathrm{u}}\in\mathcal{R}$:

$$R_{\mathrm{y}}(\xi)=a_0^{[0]}+a_1^{[0]}\xi+\ldots+a_{n_{\mathrm{a}}}^{[0]}\xi^{n_{\mathrm{a}}}, \tag{4.56}$$
$$R_{\mathrm{u}}(\xi)=b_0^{[0]}+b_1^{[0]}\xi+\ldots+b_{n_{\mathrm{b}}}^{[0]}\xi^{n_{\mathrm{b}}}. \tag{4.57}$$

where $n_{\mathrm{a}}=n_{\mathrm{b}}$. Additionally, let R be monic, i.e. $a_{n_{\mathrm{a}}}^{[0]}=1$, otherwise redefine the polynomials by dividing the coefficients with $a_{n_{\mathrm{a}}}^{[0]}$. Then $\Sigma_{-}([R_{\mathrm{y}}\quad -R_{\mathrm{u}}])$ gives:

$$\begin{bmatrix} a_1^{[1]}+\ldots+a_{n_{\mathrm{a}}-1}^{[1]}\xi^{n_{\mathrm{a}}-2}+\xi^{n_{\mathrm{a}}-1} & -b_1^{[1]}-\ldots-b_{n_{\mathrm{a}}-1}^{[1]}\xi^{n_{\mathrm{a}}-1} \\ a_2^{[2]}+\ldots+a_{n_{\mathrm{a}}-1}^{[2]}\xi^{n_{\mathrm{a}}-3}+\xi^{n_{\mathrm{a}}-2} & -b_2^{[2]}-\ldots-a_{n_{\mathrm{a}}-1}^{[2]}\xi^{n_{\mathrm{a}}-2} \\ \vdots & \vdots \\ a_{n_{\mathrm{a}}-1}^{[n_{\mathrm{a}}-1]}+\xi & -b_{n_{\mathrm{a}}-1}^{[n_{\mathrm{a}}-1]}-b_{n_{\mathrm{a}}}^{[n_{\mathrm{a}}-1]}\xi \\ 1 & -b_{n_{\mathrm{a}}}^{[n_{\mathrm{a}}]} \end{bmatrix}. \tag{4.58}$$

Obviously, this $n_{\mathrm{a}}\times 2$ matrix has independent rows and the span of these rows is linearly independent from $\mathrm{Module}_{\mathcal{R}[\xi]}([R_{\mathrm{y}}\quad -R_{\mathrm{u}}])$. Thus, the construction of the state-map in terms of (4.49) requires to choose $X\in\mathcal{R}[\xi]^{n_{\mathrm{a}}\times 2}$ such that $\mathrm{Span}_{\mathcal{R}}^{\mathrm{row}}(X)$ equals the rowspan of (4.58). As all rows of (4.58) are independent, therefore X can be easily constructed. The selector matrices are also evident: $S_{\mathrm{u}}=[0\quad 1]$ and $S_{\mathrm{y}}=[1\quad 0]$.

A convenient choice for X is to take the rows of (4.58) in the given order (top-to-bottom). Application of the algorithm defined by (4.54a–b) with such a X leads to the companion-observability canonical form $\mathfrak{R}_{\mathrm{SS}}^{\mathcal{O}_{\mathrm{c}}}(\mathcal{S})$. This can be shown by solving the corresponding equation system of (4.54a–b).

To derive a realization in terms of the observability canonical form, define $\beta_0,\ldots,\beta_{n_{\mathrm{a}}}\in\mathcal{R}$ such that

$$R_{\mathrm{u}}(\xi)=\beta_{n_{\mathrm{a}}}^{[0]}R_{\mathrm{y}}(\xi)+\beta_{n_{\mathrm{a}}-1}^{[0]}\xi^{-1}R_{\mathrm{y}}(\xi)+\ldots+\beta_0^{[0]}\xi^{-n_{\mathrm{a}}}R_{\mathrm{y}}(\xi)+\ldots \tag{4.59}$$

These functions are the resulting expansion coefficients (left-fractions) of R_u in terms of R_y. Then, by choosing X as

$$X(\xi) = \begin{bmatrix} 1 & -\beta_{n_\mathrm{a}}^{[n_\mathrm{a}]} \\ \xi & -\beta_{n_\mathrm{a}-1}^{[n_\mathrm{a}-1]} - \beta_{n_\mathrm{a}}^{[n_\mathrm{a}-1]}\xi \\ \vdots & \vdots \\ \xi^{n_\mathrm{a}-2} & -\beta_2^{[2]} - \ldots - \beta_{n_\mathrm{a}}^{[2]}\xi^{n_\mathrm{a}-2} \\ \xi^{n_\mathrm{a}-1} & -\beta_1^{[1]} - \ldots - \beta_{n_\mathrm{a}}^{[1]}\xi^{n_\mathrm{a}-1} \end{bmatrix}, \tag{4.60}$$

results in $\mathfrak{R}_{\mathrm{SS}}^{\mathrm{O}}(\mathcal{S})$ via the algorithm defined by (4.54a–b). The resulting coefficients in this case are

$$\alpha_i^{\mathrm{o}} = a_i^{[0]}, \quad \beta_j^{\mathrm{o}} = \beta_j^{[j]}, \tag{4.61}$$

where $i, j \in \mathbb{I}_1^{n_\mathbb{X}}$, $n_\mathbb{X} = n_\mathrm{a}$ and $\beta_{n_\mathrm{a}}^{\mathrm{o}} = b_{n_\mathrm{a}}^{[n_\mathrm{a}]}$. The following claim follows from the structural properties of the canonical forms:

Claim 4.1 *The $\mathfrak{R}_{\mathrm{SS}}^{\mathrm{O}}(\mathcal{S})$ and $\mathfrak{R}_{\mathrm{SS}}^{\mathrm{O_c}}(\mathcal{S})$ SS realizations of a SISO $\mathfrak{R}_{\mathrm{IO}}(\mathcal{S})$ via (4.54a–b) are completely state-observable and state-trim hence they are minimal. They are also structurally state-reachable iff R_y and R_u are left-coprime on $\mathcal{R}[\xi]$.*

4.3.6 State-Maps Based on Image-Representations

The previously developed algorithm provides SS realizations based on kernel representations. However it is also possible to derive another algorithm that is based on state-maps generated from the so-called image representations. In this part, we develop this algorithm and we show that for specific choices of the state-map it provides the SS realization in terms of the previously introduced reachability and companion reachability canonical forms.

To deduce reachability canonical forms, investigate $\mathfrak{R}_{\mathrm{IO}}(\mathcal{S})$ in the following, so-called *image representation*:

$$\begin{bmatrix} u \\ y \end{bmatrix} = (\underbrace{\begin{bmatrix} R'_\mathrm{y}(\xi) \\ R'_\mathrm{u}(\xi) \end{bmatrix}}_{\breve{X}(\xi)} \diamond p) w_\mathrm{L}, \tag{4.62}$$

with ξ either equal to $\frac{\mathrm{d}}{dt}$ or q and $R'_\mathrm{u}, R'_\mathrm{y} \in \mathcal{R}[\xi]$ with R'_y monic. Note that any LPV system has an image representation in the form of (4.62) with equal manifest behavior (see [239] for a proof). Applying the cut-and-shift based state-construction mechanism on (4.62) with system variables (w_L, u, y) leads to

$$\Sigma_-([\breve{X}\ I_{2\times 2}]) = [\Sigma_-(\breve{X})\ 0\ 0],$$

where

$$\Sigma_-(\breve{X}) = \begin{bmatrix} a_1^{[1]} + a_2^{[1]}\xi + \ldots + \xi^{n_a-1} \\ b_1^{[1]} + b_2^{[1]}\xi + \ldots + b_{n_b}^{[1]}\xi^{n_b-1} \\ a_2^{[2]} + \ldots + \xi^{n_a-2} \\ b_2^{[2]} + \ldots + b_{n_b}^{[2]}\xi^{n_b-2} \\ \vdots \\ 1 \\ b_{n_a}^{[n_a]} \end{bmatrix}.$$

A minimal state for (4.62) is therefore given by

$$x = (X(\xi) \diamond p)w_L, \tag{4.63}$$

where $X \in \mathcal{R}[\xi]^{n_a \times 1}$ has independent rows and satisfies

$$\mathrm{Span}_{\mathcal{R}}^{\mathrm{row}}(X) = \mathrm{Span}_{\mathcal{R}}^{\mathrm{row}}(\Sigma_-(\breve{X})). \tag{4.64}$$

The input is given as $u = S_u(\breve{X}(\xi) \diamond p)w_L$ with S_u a selector matrix such that $\mathrm{Span}_{\mathcal{R}}^{\mathrm{row}}(X)$ and $\mathrm{Span}_{\mathcal{R}}^{\mathrm{row}}(S_u\breve{X})$ are direct summands. This implies that $S_u = [1 \quad 0]$ and $S_y = [0 \quad 1]$. Then again, it can be seen that

$$\mathrm{Span}_{\mathcal{R}}^{\mathrm{row}}(\delta_+(X)) \subseteq \mathrm{Span}_{\mathcal{R}}^{\mathrm{row}}(X) \oplus \mathrm{Span}_{\mathcal{R}}^{\mathrm{row}}(S_u\breve{X}), \tag{4.65a}$$
$$\mathrm{Span}_{\mathcal{R}}^{\mathrm{row}}(S_y\breve{X}) \subseteq \mathrm{Span}_{\mathcal{R}}^{\mathrm{row}}(X) \oplus \mathrm{Span}_{\mathcal{R}}^{\mathrm{row}}(S_u\breve{X}). \tag{4.65b}$$

These inclusions imply the existence of unique matrices (A, B, C, D) in $\mathcal{R}^{\cdot\times\cdot}$ and a polynomial matrix $X \in \mathcal{R}[\xi]^{\cdot\times\cdot}$ with appropriate dimensions such that

$$\xi X(\xi) = AX(\xi) + BS_u\breve{X}(\xi), \tag{4.66a}$$
$$S_y\breve{X}(\xi) = CX(\xi) + DS_u\breve{X}(\xi), \tag{4.66b}$$

giving a state-representation of the LPV system $\mathcal{S}$.

Consider again the SISO case. By choosing X as

$$X(\xi) = \begin{bmatrix} 1 & \xi & \ldots & \xi^{n_a-2} & \xi^{n_a-1} \end{bmatrix}^\top, \tag{4.67}$$

algorithm (4.66a–b) results in the companion-reachability canonical form $\mathfrak{R}_{\mathrm{SS}}^{\mathcal{R}_c}(\mathcal{S})$, while $\mathfrak{R}_{\mathrm{SS}}^{\mathcal{R}}(\mathcal{S})$ is obtained via

$$X(\xi) = \Sigma_-(R_y') = \begin{bmatrix} a_1^{[1]} + \ldots + \xi^{n_a-1} \\ a_2^{[2]} + \ldots + \xi^{n_a-2} \\ \vdots \\ a_{n_a-1}^{[n_a-1]} + \xi \\ 1 \end{bmatrix}. \tag{4.68}$$

Claim 4.2 *The $\mathfrak{R}_{\mathrm{SS}}^{\mathcal{R}}(\mathcal{S})$ and $\mathfrak{R}_{\mathrm{SS}}^{\mathcal{R}_c}(\mathcal{S})$ SS realizations of a SISO $\mathfrak{R}_{\mathrm{IO}}(\mathcal{S})$ via (4.66a–b) are completely state-reachable. They are also structurally state-observable iff R'_{y} and R'_{u} are coprime.*

Example 4.9 (SS equivalence transformation). Consider the LPV-IO representation derived in Example 4.8:

$$R_{\mathrm{y}}(\xi) = \xi^2 - \left(1 + \tfrac{r}{\overrightarrow{r}}\right)\xi + \tfrac{\mathrm{T}_{\mathrm{d}}^2 \mathrm{k}_{\mathrm{s}} + r}{\overrightarrow{r}}, \quad R_{\mathrm{u}}(\xi) = \tfrac{\mathrm{T}_{\mathrm{d}}^2}{\overrightarrow{r}}. \tag{4.69}$$

In the following we derive a SS realization of this representation in a companion-observability canonical form. Denote $R(\xi) = \begin{bmatrix} R_{\mathrm{y}}(\xi) & -R_{\mathrm{u}}(\xi) \end{bmatrix}$, and generate the state-map

$$X(\xi) = \Sigma_{-}(R(\xi)) = \begin{bmatrix} \xi - (1 + \frac{\overleftarrow{r}}{r}) & 0 \\ 1 & 0 \end{bmatrix}.$$

Now with $S_{\mathrm{y}} = \begin{bmatrix} 1 & 0 \end{bmatrix}$ and $S_{\mathrm{u}} = \begin{bmatrix} 0 & 1 \end{bmatrix}$, equations (4.54a-b) read as

$$\underbrace{\begin{bmatrix} \xi^2 - \left(1 + \frac{r}{\overrightarrow{r}}\right)\xi & 0 \\ \xi & 0 \end{bmatrix}}_{\xi X(\xi)} = \underbrace{\begin{bmatrix} \alpha_{11} & \alpha_{12} \\ \alpha_{21} & \alpha_{22} \end{bmatrix}}_{A} \cdot \underbrace{\begin{bmatrix} \xi - (1 + \frac{\overleftarrow{r}}{r}) & 0 \\ 1 & 0 \end{bmatrix}}_{X(\xi)} + \underbrace{\begin{bmatrix} 0 & \beta_1 \\ 0 & \beta_2 \end{bmatrix}}_{BS_{\mathrm{u}}} + \begin{bmatrix} X_{\mathrm{u}1}(\xi) \\ X_{\mathrm{u}2}(\xi) \end{bmatrix} R(\xi),$$

$$\underbrace{\begin{bmatrix} 1 & 0 \end{bmatrix}}_{S_{\mathrm{y}}} = \underbrace{\begin{bmatrix} c_1 & c_2 \end{bmatrix}}_{C} \cdot \underbrace{\begin{bmatrix} \xi - (1 + \frac{\overleftarrow{r}}{r}) & 0 \\ 1 & 0 \end{bmatrix}}_{X(\xi)} + \underbrace{\begin{bmatrix} 0 & d_1 \end{bmatrix}}_{DS_{\mathrm{u}}} + X_{\mathrm{y}}(\xi) R(\xi).$$

By solving these equations, it follows that

$$\begin{aligned}
\alpha_{11} &= 0 & \alpha_{12} &= -\tfrac{\mathrm{T}_{\mathrm{d}}^2 \mathrm{k}_{\mathrm{s}} + r}{\overrightarrow{r}} & \beta_1 &= \tfrac{\mathrm{T}_{\mathrm{d}}^2}{\overrightarrow{r}} \\
\alpha_{21} &= 1 & \alpha_{22} &= 1 + \tfrac{\overleftarrow{r}}{r} & \beta_2 &= 0 \\
c_1 &= 0 & d_1 &= 0 & X_{\mathrm{u}1}(\xi) &= 1 \\
c_2 &= 1 & X_{\mathrm{y}}(\xi) &= 0 & X_{\mathrm{u}2}(\xi) &= 0
\end{aligned}$$

Then, the companion-observability canonical form results as

$$\mathfrak{R}_{\mathrm{SS}}^{\mathcal{O}_c}(\mathcal{S}) = \left[\begin{array}{cc|c} 0 & -\frac{\mathrm{T}_{\mathrm{d}}^2 \mathrm{k}_{\mathrm{s}} + p}{\mathrm{q}p} & \frac{\mathrm{T}_{\mathrm{d}}^2}{\mathrm{q}p} \\ 1 & 1 + \frac{\mathrm{q}^{-1} p}{p} & 0 \\ \hline 0 & 1 & 0 \end{array}\right].$$

By applying the state transformation

$$T \diamond p = \begin{bmatrix} p & \mathrm{q}^{-1} p \\ 0 & 1 \end{bmatrix},$$

on $\mathfrak{R}_{\mathrm{SS}}^{\mathrm{O_c}}(\mathcal{S})$ its equivalence with the LPV-SS representation of Example 3.22 follows. The latter proves that the IO representation given by R_{y} and R_{u} has the same manifest behavior as $\mathfrak{R}_{\mathrm{SS}}(\mathcal{S})$.

Next we develop a SS realization of (4.69) in a reachability canonical form. For this we first transform the IO representation (4.69) into an image representation form (4.62). Introduce the latent variable

$$w_{\mathrm{L}} = \frac{\overleftarrow{r}}{\mathrm{T_d^2}} y \quad \rightarrow \quad y = \frac{\mathrm{T_d^2}}{\overleftarrow{r}} w_{\mathrm{L}}.$$

By substituting w_{L} into $(R_{\mathrm{y}}(\mathrm{q}) \diamond p)y = (R_{\mathrm{u}}(\mathrm{q}) \diamond p)u$, it follows that

$$u = \left(\frac{\mathrm{T_d^2 k_s} + r}{\overleftarrow{r}} \diamond p\right) w_{\mathrm{L}} - \left(\frac{\overrightarrow{r} + r}{r} \diamond p\right) \mathrm{q} w_{\mathrm{L}} + \mathrm{q}^2 w_{\mathrm{L}}.$$

This concludes that

$$\begin{bmatrix} u \\ y \end{bmatrix} = \begin{bmatrix} R_{\mathrm{y}}'(\mathrm{q}) \diamond p \\ R_{\mathrm{u}}'(\mathrm{q}) \diamond p \end{bmatrix} w_{\mathrm{L}}, \quad \text{where} \quad \begin{bmatrix} R_{\mathrm{y}}'(\xi) \\ R_{\mathrm{u}}'(\xi) \end{bmatrix} = \underbrace{\begin{bmatrix} \frac{\mathrm{T_d^2 k_s}+r}{\overleftarrow{r}} - \frac{\overrightarrow{r}+r}{r}\xi + \xi^2 \\ \frac{\mathrm{T_d^2}}{\overleftarrow{r}} \end{bmatrix}}_{\breve{X}(\xi)}, \tag{4.70}$$

is an image representation of $\mathcal{S}$. To introduce a state-map which leads to the reachability form, in terms of (4.68) let

$$X(\xi) = \Sigma_{-}(R_{\mathrm{y}}') = \begin{bmatrix} -\frac{r+\overleftarrow{r}}{\overleftarrow{r}} + \xi \\ 1 \end{bmatrix}.$$

Now with $S_{\mathrm{y}} = [0 \quad 1]$ and $S_{\mathrm{u}} = [1 \quad 0]$, equations (4.66a-b) read as

$$\underbrace{\begin{bmatrix} -\frac{\overrightarrow{r}+r}{r}\xi + \xi^2 \\ \xi \end{bmatrix}}_{\xi X(\xi)} = \underbrace{\begin{bmatrix} \alpha_{11} & \alpha_{12} \\ \alpha_{21} & \alpha_{22} \end{bmatrix}}_{A} \cdot \underbrace{\begin{bmatrix} -\frac{r+\overleftarrow{r}}{\overleftarrow{r}} + \xi \\ 1 \end{bmatrix}}_{X(\xi)} + \underbrace{\begin{bmatrix} \beta_1 R_{\mathrm{y}}'(\xi) \\ \beta_2 R_{\mathrm{y}}'(\xi) \end{bmatrix}}_{BS_{\mathrm{u}}\breve{X}(\xi)},$$

$$\underbrace{R_{\mathrm{u}}'(\xi)}_{S_{\mathrm{y}}\breve{X}(\xi)} = \underbrace{[c_1 \quad c_2]}_{C} \cdot \underbrace{\begin{bmatrix} -\frac{r+\overleftarrow{r}}{\overleftarrow{r}} + \xi \\ 1 \end{bmatrix}}_{X(\xi)} + \underbrace{d_1 R_{\mathrm{y}}'(\xi)}_{DS_{\mathrm{y}}\breve{X}(\xi)}.$$

By solving these equations, it follows that

$$\begin{aligned} \alpha_{11} &= 0 & \alpha_{12} &= -\frac{\mathrm{T_d^2 k_s}+r}{\overleftarrow{r}} & \beta_1 &= 1 \\ \alpha_{21} &= 1 & \alpha_{22} &= 1 + \frac{r}{\overleftarrow{r}} & \beta_2 &= 0 \\ c_1 &= 0 & c_2 &= \frac{\mathrm{T_d^2}}{\overleftarrow{r}} & d_1 &= 0 \end{aligned}$$

The corresponding reachability form is

$$\mathfrak{R}_{\text{SS}}^{\mathcal{R}}(\mathcal{S}) = \left[\begin{array}{cc|c} 0 & -\frac{\mathrm{T}_{\mathrm{d}}^2 \mathrm{k}_{\mathrm{s}}+p}{\mathrm{q}^{-1}p} & 1 \\ 1 & 1+\frac{p}{\mathrm{q}^{-1}p} & 0 \\ \hline 0 & \frac{\mathrm{T}_{\mathrm{d}}^2}{\mathrm{q}^{-1}p} & 0 \end{array}\right].$$

By applying the constant state transformation

$$T = \begin{bmatrix} -\mathrm{T}_{\mathrm{d}}^2 & 0 \\ 0 & -\frac{1}{\mathrm{T}_{\mathrm{d}}^2} \end{bmatrix},$$

on $\mathfrak{R}_{\text{SS}}^{\mathcal{R}}(\mathcal{S})$ its equivalence with the LPV-SS representation of Example 3.17 follows. This proves equivalence with the LPV-SS representation of Example 3.22 and hence also with the previously derived $\mathfrak{R}_{\text{SS}}^{\mathcal{O}_{\text{c}}}(\mathcal{S})$. □

4.3.7 *State-Construction in the MIMO Case*

In the MIMO case, algorithms (4.54a–b) and (4.66a–b) also provide SS realizations of IO representations, however with different selector matrices (due to the multi-dimension) and with a more complicated path to select independent rows from the shift-map for X. It is only guaranteed that at least n_{a} number of rows of the shift-map are independent, thus such a selection is not evident. Similar to the selection schemes generating the SS MIMO canonical representations, only certain selection strategies for X lead to the MIMO observability and reachability canonical forms. Thus minimality of the obtained SS realizations via algorithm (4.54a–b) and (4.66a–b) is not guaranteed in the general sense.

4.4 Conclusions

In this chapter, we have established equivalence transformations between the state-space and the input-output representation domains. These transformations and the corresponding algorithms have been introduced to enable comparison of LPV model structures and identified models later on and also to provide essential tools to develop LPV identification via OBF model structures. Additionally, we have defined observability and reachability canonical SS representations of LPV systems and we have shown that they provide a simple gateway for the conversion between the representation domains. Furthermore, it has been shown that the transpose of a LPV-SS representation does not have the same manifest behavior, which is a notable difference with respect to the LTI theory. This property also proves that coefficients of canonical forms that structurally seem to be the transpose of each other (like the reachability and observability forms), are not equal. The connection of the introduced canonical forms with the applied theories of the current LPV literature has also been investigated. This lead to the conclusion that the common practice to use LTI theory to compute canonical forms or provide a SS realization for LPV systems yields SS representations that do not have an equal manifest behavior. We next explore if a series-expansion representation of LPV systems is available and can be used to establish truncated series-expansion models for the identification of LPV systems just as in the LTI case.

Chapter 5
LPV Series-Expansion Representations

Abstract. In this chapter, series-expansion representations of LPV systems for a given IO partition are developed using the framework of the behavioral approach. In fact, expansion of DT asymptotically stable LPV systems is considered in terms of OBFs and the connection between this type of expansion and the gain-scheduling principle is explored. It is shown that this series-expansion representation is unique and always exists for the considered system class and in some cases only a finite number of the expansion coefficients are nonzero. This implies that finite truncation of a OBFs-based series-expansion can be used as a model structure for the identification of asymptotically stable DT-LPV systems similar to the LTI case.

5.1 Relevance of Series-Expansion Representations

In the LTI framework, series-expansion representations have proved their usefulness in a number of contexts. They not only characterize a unique representation of the *input-output* (IO) system dynamics, like impulse-response representations, but they also provide model structures, like *orthonormal basis functions* (OBFs)-based models, that provide an efficient alternative for LTI system identification. Using such model structures for the identification of LPV systems has a number of attractive proprieties and it would also allow the extension of the OBFs based identification approaches to the LPV case (see the argument of Chap. 1). Based on this, we develop in this chapter the concept of series-expansion of LPV systems in terms of OBFs, providing a unique perspective on the representation of the system dynamics with respect to an IO partition. This enables the formulation of the one-step ahead predictor in the LPV prediction-error identification framework and also the introduction of model structures as the finite truncation of a OBF series-expansion with respect to a LPV system. These model structures are vital ingredients of the LPV identification approach developed later.

To simplify the discussion and to avoid cases that are complicated but unimportant for the intended identification approach, we restrict the discussion to *discrete-time* (DT) asymptotically-stable LPV systems.

R. Tóth: Modeling and Identification of Linear Parameter-Varying Systems, LNCIS 403, pp. 131–141.
springerlink.com

5.2 Impulse Response Representation of LPV Systems

As a first step, we develop the series-expansion representation of DT asymptotically stable LPV systems based on a pulse basis (see Sect. 2.1 for the definition of the pulse basis). LPV systems can not be handled in the frequency domain, thus we develop the concept of series-expansion in terms of the time operator form of the pulse basis. As we will see this results in a convolution based description of the system dynamics. To develop such a description, we first introduce the filter form of an LPV-IO representation. Then we generate the expansion in terms of the pulse basis by recursive substitution of the filter form. We show the uniqueness and the convergence of the resulting expansion coefficients. Additionally we briefly cover how this series-expansion can be generalized to unstable and *continuous-time* (CT) systems.

5.2.1 *Filter Form of LPV-IO Representations*

Based on the previously given line of discussion, first the filter form of LPV-IO representations is introduced. Let a *discrete-time* (DT) LPV system $\mathcal{S} = (\mathbb{Z}, \mathbb{P}, \mathbb{W}, \mathfrak{B})$ be given with an IO partition $w = \mathrm{Col}(u, y)$ and scheduling variable p. Assume that $\mathcal{S}$ is asymptotically dynamically stable and that a minimal IO representation $\mathfrak{R}_{\mathrm{IO}}(\mathcal{S})$ of $\mathcal{S}$ is given, characterized by $R_{\mathrm{u}} \in \mathcal{R}[\xi]^{n_{\mathbb{Y}} \times n_{\mathbb{U}}}$ and a full rank $R_{\mathrm{y}} \in \mathcal{R}[\xi]^{n_{\mathbb{Y}} \times n_{\mathbb{Y}}}$ with $\mathrm{Deg}(R_{\mathrm{y}}) = n_{\mathrm{a}} \geq n_{\mathrm{b}} = \mathrm{Deg}(R_{\mathrm{u}})$. In this way, the behavior $\mathfrak{B}$ is described by the relation

$$\sum_{i=0}^{n_{\mathrm{a}}} (a_i \diamond p) \mathrm{q}^i y = \sum_{j=0}^{n_{\mathrm{b}}} (b_j \diamond p) \mathrm{q}^j u, \tag{5.1}$$

for every $(u, y, p) \in \mathfrak{B}$ with left compact support. Without loss of generality, assume that R_{y} is monic and multiply (5.1) by $\mathrm{q}^{-n_{\mathrm{a}}}$ according to the non-commutative multiplication rules in discrete-time (see Def. 3.16). The resulting expression is

$$y = -\sum_{i=0}^{n_{\mathrm{a}}-1} \left(\overleftarrow{a_i}^{(n_{\mathrm{a}})} \diamond p \right) \mathrm{q}^{i-n_{\mathrm{a}}} y + \sum_{j=0}^{n_{\mathrm{b}}} \left(\overleftarrow{b_j}^{(n_{\mathrm{a}})} \diamond p \right) \mathrm{q}^{j-n_{\mathrm{a}}} u, \tag{5.2}$$

where $\overleftarrow{\cdot}^{(n_{\mathrm{a}})}$ denotes the backward shift operator applied on the coefficient function for n_{a} times. As only the backward time-shifted versions of y appear on the right side of (5.2), the relation (5.2) is called the filter form of (5.1).

5.2.2 *Series Expansion in the Pulse Basis*

Assume that $n_{\mathrm{a}} = n_{\mathrm{b}}$, which can be realized by including extra coefficients $\{b_j\}_{j=n_{\mathrm{b}}+1}^{n_{\mathrm{a}}}$ that are zero functions. By substituting the relation (5.2) recursively into itself to eliminate the shifted versions of y we obtain:

$$y = (\mathrm{g}_0 \diamond p) u + (\mathrm{g}_1 \diamond p) \mathrm{q}^{-1} u + (\mathrm{g}_2 \diamond p) \mathrm{q}^{-2} u + \ldots \tag{5.3}$$

where

$$\begin{aligned}
g_0 &= \overleftarrow{b_{n_a}}^{(n_a)}, \\
g_1 &= \overleftarrow{b_{n_a-1}}^{(n_a)} - \overleftarrow{a_{n_a-1}}^{(n_a)}\overleftarrow{b_{n_a}}^{(n_a+1)}, \\
g_2 &= \overleftarrow{b_{n_a-2}}^{(n_a)} - \overleftarrow{a_{n_a-1}}^{(n_a)}\overleftarrow{b_{n_a-1}}^{(n_a+1)} - \overleftarrow{a_{n_a-2}}^{(n_a)}\overleftarrow{b_{n_a}}^{(n_a+2)} + \\
&\quad + \overleftarrow{a_{n_a-1}}^{(n_a)}\overleftarrow{a_{n_a-1}}^{(n_a+1)}\overleftarrow{b_{n_a}}^{(n_a+2)}.
\end{aligned}$$

It is obvious that $\{g_0, g_1, g_2, \ldots\}$ are meromorphic coefficient functions. Furthermore, they are backward-shifted combinations of the coefficients of $\mathfrak{R}_{\mathrm{IO}}(\mathcal{S})$ and due to the minimality of $\mathfrak{R}_{\mathrm{IO}}(\mathcal{S})$ they are unique with respect to the considered IO partition of $\mathcal{S}$. In addition, the signal trajectories $(u, y, p) \in \mathfrak{B}$ described by (5.1) have left-compact support. This means, that there exists a $n \in \mathbb{N}$, such that after n substitutions of the relation (5.2) recursively into itself, y vanishes from the expression (5.3). This shows, that by this recursive substitution we have obtained an infinite expansion of (5.2) in terms of the LTI pulse basis $\{1, q^{-1}, q^{-2}, \ldots\}$ with coefficients $g_i \in \mathcal{R}^{n_Y \times n_U}$ for $i = 1, 2, \ldots$. If the coefficients of $\mathfrak{R}_{\mathrm{IO}}(\mathcal{S})$ have static dependence or they are all dependent only on the backward shifted versions of p, then each g_i depends only on the past values of p.

Furthermore, in case of a given scheduling trajectory $p \in \mathfrak{B}_{\mathbb{P}}$, and a pulse input at $k = 0$, the output trajectory of $\mathcal{S}$ satisfies

$$y(0) = (g_0 \diamond p)(0), \quad y(1) = (g_1 \diamond p)(1), \quad y(2) = (g_2 \diamond p)(2), \quad \ldots$$

thus $\{g_i\}_{i=0}^{\infty}$ can be considered as the *impulse response coefficients* of $\mathcal{S}$ for the considered IO partition. Additionally, the asymptotic stability of $\mathcal{S}$ implies that

$$y(k) \to 0, \quad \text{as } k \to \infty, \tag{5.4}$$

which means that the sequence of coefficients $\{g_i\}_{i=0}^{\infty}$ converges to zero along every scheduling trajectory $p \in \mathfrak{B}_{\mathbb{P}}$. Thus

$$\lim_{i \to \infty} (g_i \diamond p)(i) = 0, \quad \forall p \in \mathfrak{B}_{\mathbb{P}}. \tag{5.5}$$

This implies that due to the shift-invariant property of $\mathfrak{B}$:

$$\lim_{i \to \infty} (g_i \diamond p) = 0, \quad \forall p \in \mathfrak{B}_{\mathbb{P}}, \tag{5.6}$$

holds, which means that the sequence of coefficient functions $\{g_0, g_1, g_2, \ldots\}$ converges to the zero function with respect to $\mathfrak{B}_{\mathbb{P}}$. It is also important that asymptotic stability of $\mathcal{S}$ implies BIBO stability in the ℓ_∞ norm:

$$\sup_{k \geq 0} \|u(k)\| < \infty \;\Rightarrow\; \sup_{k \geq 0} \|y(k)\| < \infty.$$

As (5.3) holds for any $(u,y,p) \in \mathfrak{B}$ with left compact support,

$$\left(\sup_{k\geq 0} \|u(k)\| < \infty \quad \text{and} \quad \sup_{k\geq 0} \|y(k)\| < \infty \right) \Rightarrow \sup_{k\geq 0} \sum_{i=0}^{\infty} \|(\mathrm{g}_i \diamond p)(k)\| < \infty. \quad (5.7)$$

These properties yield the following theorem:

Theorem 5.1 (Existence of series-expansion representation, pulse basis). *Any asymptotically stable, discrete-time LPV system $\mathcal{S} = (\mathbb{Z}, \mathbb{P}, \mathbb{W}, \mathfrak{B})$ with an IO partition (u,y) has a unique, convergent series-expansion in terms of the pulse-basis $\{q^{-i}\}_{i=0}^{\infty}$ and coefficients $\mathrm{g}_i \in \mathcal{R}^{n_\mathbb{Y} \times n_\mathbb{U}}$, such that*

$$y = \sum_{i=0}^{\infty} (\mathrm{g}_i \diamond p) q^{-i} u, \quad (5.8)$$

is satisfied for all $(u,y,p) \in \mathfrak{B}_\mathbb{P}$ with left compact support.

For a proof see Appendix A.2. The LPV series-expansion in terms of the pulse basis is similar to the series-expansion in the LTI case (see (2.4)). This means that the LPV system has a convergent series-expansion in terms of an LTI basis, which has a strong connection to the gain-scheduling concept of LPV systems (see Sect. 5.5.1 and [194]). Furthermore it is a general property of the expansion coefficients $\{\mathrm{g}_i\}_{i=0}^{\infty}$ that they have dynamic dependence even if the original IO representation, used for their computation, has coefficients with static dependence.

Note that in case of a uniformly frozen unstable LPV system ($\mathcal{S}$ is unstable for every constant scheduling trajectory in $\mathfrak{B}_\mathbb{P}$), a series-expansion representation can also be derived in terms of the pulse basis $\{q^1, q^2, q^3, \ldots\}$. If the system is only non-uniformly frozen stable, then the series-expansion follows by taking the two sided pulse basis $\{\ldots, q^{-1}, 1, q^1, \ldots\}$. These cases are not treated here as they would require additional technicalities. Additionally, we will restrict the scope to asymptotically-stable LPV systems for the identification approaches described in Chaps. 8–9. Furthermore, the continuous-time case of series expansions is also not covered. One of the reasons is that in the introduced behavioral framework it would be cumbersome to handle ξ^{-1}, which corresponds to an integral operator in CT. An additional problem is that even if CT systems admit an impulse response form, the unit-step function, which is associated with $\xi^{-1} = s^{-1}$ in the LTI case, is not an $\mathcal{H}_2$ function, thus it does not correspond to a basis function sequence. Remember, that a CT basis is related to its DT counterpart by the bilinear transformation (2.24). Based on these considerations, we only treat the asymptotically stable DT case.

5.2.3 *The Impulse Response Representation*

Based on Th. 5.1, it is possible to define the series-expansion representation of $\mathcal{S}$ in terms of the LTI pulse basis $\{1, q^{-1}, q^{-2}, \ldots\}$ as follows:

Definition 5.1 (LPV series-expansion representation, pulse basis). The pulse basis series-expansion representation of a discrete-time asymptotically stable $\mathcal{S} = (\mathbb{Z}, \mathbb{P} \subseteq \mathbb{R}^{n_\mathrm{P}}, \mathbb{R}^{n_\mathrm{U}+n_\mathrm{Y}}, \mathfrak{B})$ with scheduling signal p and IO partition (u,y) is denoted by $\mathfrak{R}_\mathrm{IM}(\mathcal{S})$ and defined as:

$$y = \sum_{i=0}^{\infty} (\mathrm{g}_i \diamond p)\mathrm{q}^{-i}u \tag{5.9}$$

where $\mathrm{g}_i \in \mathcal{R}^{n_\mathrm{Y} \times n_\mathrm{U}}$, $i \in \mathbb{I}_0^\infty$ are the meromorphic expansion coefficients. □

In the following, we call this representation the *impulse response representation* (IRR) of the LPV system. Note that $\mathfrak{R}_\mathrm{IM}(\mathcal{S})$, similar to the IO representations, describes the behavior of $\mathcal{S}$ restricted to signal trajectories with left-compact support. However, in contrast with other LPV representations, it is unique. It is also important that an equivalence transformation exists from IO representations to the IRR domain, i.e. $\mathfrak{R}_\mathrm{IO}(\mathcal{S})$ can be always transformed to $\mathfrak{R}_\mathrm{IM}(\mathcal{S})$ if $\mathcal{S}$ is asymptotically stable. See Example 5.1 for the construction of an IRR. On the other hand, realization of an LPV-IO representation from the IRR in the general case is unsolved yet. Alternatively, the use of the LPV extension of the Ho-Kalman algorithm provides a LPV-SS realization of a given sequence of PV impulse response coefficients for a priori chosen class of functional dependencies (see [1] for further details).

Example 5.1 (Pulse basis series-expansion representation of an DT-LPV system). Consider the DT-LPV-IO representation $\mathfrak{R}_\mathrm{IO}(\mathcal{S})$ given in the following filter form:

$$y = -0.2\mathrm{q}^{-1}y - 0.1p\mathrm{q}^{-2}y + \sin(p)\mathrm{q}^{-1}u,$$

with $\mathbb{P} = [0,1]$. By recursive substitution of this equation for $\mathrm{q}^{-1}y$, $\mathrm{q}^{-2}y$, ..., the following series-expansion in terms of the pulse basis functions $\{\mathrm{q}^{-1}, \mathrm{q}^{-2}, \dots\}$ results:

$$y = \underbrace{\sin(p)}_{\mathrm{g}_1 \diamond p}\mathrm{q}^{-1}u - \underbrace{10^{-1} \cdot 2\sin(\mathrm{q}^{-1})}_{-\mathrm{g}_2 \diamond p}\mathrm{q}^{-2}u + \underbrace{10^{-1} \cdot (0.4 - 1p)\sin(\mathrm{q}^{-2}p)}_{\mathrm{g}_3 \diamond p}\mathrm{q}^{-3}u + \dots$$

The resulting expansion coefficients are uniquely defined by the above expression with $\mathrm{g}_0 = 0$. As $\mathfrak{R}_\mathrm{IO}(\mathcal{S})$ corresponds to an asymptotically stable behavior, the above series expansion is convergent. This can be seen from the decreasing magnitude of the Markov parameters. □

5.3 LPV Series Expansion by OBFs

As a next step, we generalize the LPV series-expansion concept to general OBFs in $\mathcal{RH}_{2-}(\mathbb{E})$. To do so, we first show that each element of the pulse basis sequence can be written in a series-expansion form of an orthonormal basis Φ_∞ of $\mathcal{RH}_{2-}(\mathbb{E})$. By substituting each pulse function by its expansion in terms of a Φ_∞, the series-expansion of a $\mathfrak{R}_\mathrm{IO}(\mathcal{S})$ can be obtained for general orthonormal basis functions in a similar way as before.

Consider an OBF set $\Phi_\infty \subset \mathcal{RH}_{2-}(\mathbb{E})$. Then, based on the LTI transfer function theory, a pulse basis function q^{-i}, $i > 0$ has a unique series-expansion in terms of $\Phi_\infty = \{\phi_i\}_{i=1}^\infty$:

$$\mathrm{q}^{-i} = \sum_{j=1}^{\infty} \mathrm{w}_{ij}\phi_j(\mathrm{q}), \tag{5.10}$$

where $\mathrm{w}_{ij} \in \mathbb{R}$, $i, j \in \mathbb{I}_1^\infty$, are the expansion coefficients. This series-expansion is convergent and it is a well-known property that for all $i \in \mathbb{I}_1^\infty$, the sequence $\{\mathrm{w}_{ij}\}_{j=1}^\infty$ is an ℓ_2 sequence. Additionally, the same property holds for each sequence $\{\mathrm{w}_{ij}\}_{i=1}^\infty$.

As a next step, by using the relation derived in the previous part, we develop the series-expansion of a minimal $\mathfrak{R}_{\mathrm{IO}}(\mathcal{S})$ in terms of an OBF set $\Phi_\infty = \{\phi_j\}_{j=1}^\infty$ in $\mathcal{RH}_{2-}(\mathbb{E})$. Let $\mathcal{S}$ be asymptotically stable and assume that the pulse basis expansion of $\mathfrak{R}_{\mathrm{IO}}(\mathcal{S})$ has been derived in the form of (5.9). Then by substituting (5.10) into the expansion (5.9), we obtain

$$y = (\mathrm{g}_0 \diamond p)u + (\mathrm{g}_1 \diamond p)\sum_{j=1}^{\infty} \mathrm{w}_{1j}\phi_j(\mathrm{q})u + (\mathrm{g}_2 \diamond p)\sum_{j=1}^{\infty} \mathrm{w}_{2j}\phi_j(\mathrm{q})u + \ldots$$

By rearranging this expression we arrive at

$$y = (\mathrm{g}_0 \diamond p)u + \underbrace{\left(\sum_{i=1}^{\infty} \mathrm{w}_{i1}\mathrm{g}_i \diamond p\right)}_{w_1 \diamond p}\phi_1(\mathrm{q})u + \underbrace{\left(\sum_{i=1}^{\infty} \mathrm{w}_{i2}\mathrm{g}_i \diamond p\right)}_{w_2 \diamond p}\phi_2(\mathrm{q})u + \ldots$$

where $\{w_1, w_2, \ldots\}$ are the coefficient functions of the new series-expansion in terms of the basis Φ_∞. Note that for each $j \in \mathbb{I}_1^\infty$, $\{\mathrm{w}_{ij}\}_{i=1}^\infty$ is an ℓ_2 sequence. Furthermore the expansion coefficients satisfy (5.6) and (5.7) for every $p \in \mathfrak{B}_{\mathbb{P}}$. Thus, each w_i exists and it is an element of $\mathcal{R}^{n_{\mathrm{Y}} \times n_{\mathrm{U}}}$. Furthermore, zero convergence of all $\{\mathrm{w}_{ij}\}_{j=1}^\infty$ and (5.7) implies that the sequence $\{w_1, w_2, \ldots\}$ converges to zero for every $p \in \mathfrak{B}_{\mathbb{P}}$ similarly as in the pulse-basis case. Based on this, the following theorem holds:

Theorem 5.2 (Existence of series-expansion representation, OBFs). *Let $\Phi_\infty = \{\phi_i\}_{i=1}^\infty$ be a collection of orthonormal basis functions in $\mathcal{RH}_{2-}(\mathbb{E})$ with poles $\{\lambda_1, \lambda_2, \ldots\}$ satisfying the completeness condition $\sum_{i=1}^\infty (1 - |\lambda_i|) = \infty$. Then any asymptotically stable, discrete-time LPV system $\mathcal{S} = (\mathbb{Z}, \mathbb{P}, \mathbb{W}, \mathfrak{B})$ with an IO partition (u, y) has a unique series-expansion in terms of $\Phi_\infty = \{\phi_j\}_{j=1}^\infty$ with coefficients $w_i \in \mathcal{R}^{n_{\mathrm{Y}} \times n_{\mathrm{U}}}$, such that*

$$y = (w_0 \diamond p)u + \sum_{i=1}^{\infty} (w_i \diamond p)\phi_i(\mathrm{q})u, \tag{5.11}$$

is satisfied for all $(u, y, p) \in \mathfrak{B}_{\mathbb{P}}$ with left compact support.

For a proof see Appendix A.2. Note that the rate of convergence of the series-expansion directly depends on the basis sequence Φ_∞. Moreover, it is a general property of the expansion coefficients $\{w_i\}_{i=0}^\infty$ that they have dynamic dependence.

In contrast with the LTI case, commonly the analytical computation of these coefficients is only available in an approximative manner by truncating their infinite sum relation with respect to the impulse response coefficients.

5.4 The OBF Expansion Representation

Based on the series-expansion concept, developed in the previous section, we can extend Def. 5.1. We call (5.11) the discrete-time OBF *expansion representation* of the LPV system $\mathcal{S} = (\mathbb{Z}, \mathbb{P} \subseteq \mathbb{R}^{n_\mathbb{P}}, \mathbb{R}^{n_\mathbb{U}+n_\mathbb{Y}}, \mathfrak{B})$ in terms of the orthonormal basis functions $\Phi_\infty = \{\phi_i\}_{i=1}^{\infty}$, and we denote this representation as $\mathfrak{R}_{\mathrm{OBF}}(\mathcal{S}, \Phi_\infty)$. Similar to the IRR, $\mathfrak{R}_{\mathrm{OBF}}(\mathcal{S}, \Phi_\infty)$ is unique and describes the behavior of $\mathcal{S}$ restricted to signal trajectories with left-compact support. An additional similarity is that the equivalence transformation from IO representations to the expansion representation domain exists, however realization in the other direction is unsolved. See Example 5.1 for the construction of a LPV-OBF expansion representation.

Example 5.2 (OBF series-expansion representation of an LPV system). Continue Example 5.1 to develop the series-expansion of the considered $\mathfrak{R}_{\mathrm{IO}}(\mathcal{S})$ in terms of the Laguerre basis $\Phi_1^\infty = \{\phi_i\}_{i=1}^{\infty}$ with poles $\Lambda_1 = 0.5$. Based on the LTI transfer function theory, the pulse basis $\{\mathrm{q}^{-1}, \mathrm{q}^{-2}, \dots\}$ has the following series-expansion in terms of Φ_1^∞:

$$
\begin{aligned}
\mathrm{q}^{-1} &= 0.866\phi_1(\mathrm{q}) - 0.433\phi_2(\mathrm{q}) + 0.217\phi_3(\mathrm{q}) + \dots \\
\mathrm{q}^{-2} &= 0.433\phi_1(\mathrm{q}) + 0.433\phi_2(\mathrm{q}) - 0.541\phi_3(\mathrm{q}) + \dots \\
\mathrm{q}^{-3} &= 0.217\phi_1(\mathrm{q}) + 0.541\phi_2(\mathrm{q}) - 0.108\phi_3(\mathrm{q}) + \dots
\end{aligned}
$$

Then by substituting these series expansions into the pulse basis expansion of Example 5.1, it follows that

$$
y = (w_1 \diamond p)\,\phi_1(\mathrm{q})u + (w_2 \diamond p)\,\phi_2(\mathrm{q})u + \dots \tag{5.12}
$$

where

$$
\begin{aligned}
w_1 \diamond p &= 0.866\sin(p) - 0.086\sin(\mathrm{q}^{-1}p) + 0.022\,(0.4 - 1p)\sin(\mathrm{q}^{-2}p) + \dots \\
w_2 \diamond p &= -0.433\sin(p) - 0.086\sin(\mathrm{q}^{-1}p) + 0.054\,(0.4 - 1p)\sin(\mathrm{q}^{-2}p) + \dots
\end{aligned}
$$

Note that the infinite sum expression of the coefficients $\{w_1, w_2, \dots\}$ converges as the expansion coefficients of each Laguerre basis $\phi_i(\mathrm{q})$ with respect to the pulse basis corresponds to an ℓ_2 sequence. In this way, the resulting expression is the series-expansion of $\mathfrak{R}_{\mathrm{IO}}(\mathcal{S})$ in terms of Φ_1^∞. Note that the coefficients $\{w_1, w_2, \dots\}$ of this new series-expansion are linear combinations of the coefficients of the IRR. The weights of these linear combinations are uniquely determined by the series-expansion of the pulse basis functions in terms of Φ_1^∞. □

Similar to the LTI case, expansion representations have the property that the relative contribution of the basis, i.e. the w_i functions, converge to the zero function on $\mathfrak{B}_\mathbb{P}$

as $i \to \infty$. In this way, for an asymptotically stable LPV system, it is always possible to find a finite $\Phi_n \subset \Phi_\infty$, i.e. the truncation of (5.11), with a relatively small number of functions, such that the representation error for all $(u, y, p) \in \mathfrak{B}$ is negligible. This provides an efficient approximation of the system. Based on this, finite truncation of a OBF based expansion representation can be used as a model structure for the identification of asymptotically stable DT-LPV systems. Similar to the LTI case, identification based on this type of model structure reduces to the estimation of the meromorphic coefficient functions $\{w_i\}_{i=0}^{n}$ which appear linearly in the dynamic relation.

5.5 Series Expansions and Gain-Scheduling

There is an interesting relation of the OBF expansion representation and the gain-scheduling principle. This relation helps to understand whether an orthonormal basis is adequate for the series-expansion of the LPV system, i.e. how the rate of convergence of the expansion can be characterized. In the following, we explore this relation by first showing that for constant scheduling trajectories the OBF expansion representation is equivalent with the series-expansion representations of the frozen system set. As a next step we show, that this equivalence implies that for some LPV systems, KnW optimality of the basis with respect to the frozen system set implies an optimal convergence rate of the series-expansion in the LPV case.

5.5.1 *The Role of Gain-Scheduling*

Let $\mathfrak{R}_{\mathrm{OBF}}(\mathcal{S}, \Phi_\infty)$ be the expansion representation of a discrete-time asymptotically stable LPV system $\mathcal{S}$ in terms of an orthonormal basis $\Phi_\infty \subset \mathcal{RH}_{2-}(\mathbb{E})$. Denote by $\mathscr{F}_\mathcal{S}$ the frozen system set of $\mathcal{S}$, and introduce $\mathfrak{F}$ as the set of transfer functions associated with each $\mathcal{F}_\mathrm{p} \in \mathscr{F}_\mathcal{S}$ for the IO partition (u, y). As $\mathcal{S}$ is asymptotically stable, each $F_\mathrm{p} \in \mathfrak{F}$ is in $\mathcal{RH}_2(\mathbb{E})$ and the strictly proper part of F_p has a convergent series-expansion in terms of Φ_∞. It is obvious, that in case of a constant scheduling signal $p(k) = \mathrm{p}$, the expansion coefficients $\{w_i\}_{i=0}^{\infty}$ of $\mathfrak{R}_{\mathrm{OBF}}(\mathcal{S}, \Phi_\infty)$ satisfy that

$$w_i \diamond p = w_{i|\mathrm{p}}, \tag{5.13}$$

where $w_{i|\mathrm{p}} \in \mathbb{R}$, $i \in \mathbb{I}_0^\infty$ are the expansion coefficients of the transfer function F_p with respect to Φ_∞. In this way, Φ_∞ with coefficients $\{w_{i|\mathrm{p}}\}_{i=0}^{\infty}$ characterizes the behavior of each $\mathcal{F}_\mathrm{p} \in \mathscr{F}_\mathcal{S}$ in terms of an LTI expansion representation. We have already discussed that by the gain-scheduling principle, an LPV system can be viewed as a collection of LTI behaviors (frozen system set) and a scheduling signal dependent function set (scheduling functions) that selects one of the behaviors to describe the possible continuation of the signal trajectories at every time instant. Form the previous observation it is clear, that the basis Φ_∞ with coefficients $w_{i|\mathrm{p}}$ characterizes all frozen systems $\mathscr{F}_\mathcal{S}$, while the remaining part of the LPV dynamic relation is in the global expansion coefficients $\{w_i\}_{i=0}^{\infty}$. This provides the conclusion that from the gain-scheduling perspective, series-expansion separates the LPV system into a frozen behavior set, described as the linear combination of the basis functions, and a scheduling function set, which is represented by the expansion coefficients.

5.5.2 Optimality of the Basis in the Frozen Sense

Now we can use this insight to consider the question, how the basis should be chosen to achieve a fast convergence rate of the LPV series-expansion. This problem has a key importance in using truncated series expansions as model structures for LPV system identification as it formulates the optimality of a model structure with respect to a given system. To simplify the following discussion, we only investigate the SISO case.

Consider the optimality concept of OBFs in terms of the Kolmogorov n-width theory discussed in Sect. 2.5. By this concept, for a given transfer function set $\mathfrak{F} \subset \mathcal{RH}_{2-}(\mathbb{E})$ with pole locations

$$\Omega = \{\lambda \in \mathbb{C} \mid \lambda \text{ is a pole of } F \in \mathfrak{F}\}, \tag{5.14}$$

the finite set of OBFs $\Phi_n \subset \mathcal{RH}_{2-}(\mathbb{E})$ is called optimal in the n-width sense, if the subspace

$$\mathcal{M}_n = \text{Span}(\Phi_n), \tag{5.15}$$

has the minimal distance in terms of (2.57) for the worst-case $F \in \mathfrak{F}$. Denote by $\Phi_{n_g}^{\infty}$ the Hambo basis generated by the inner function G_b with n_g number of poles and let $\Phi_{n_g}^{n_e}$ describe the Hambo functions obtained with n_e as the number of extensions of G_b (finite truncation of $\Phi_{n_g}^{\infty}$). It has been shown in Sect. 2.5, that if $\Phi_{n_g}^{n_e}$ with $n_e \geq 0$ is optimal in the $n-(n_e+1)n_g$-width sense with respect to $\mathfrak{F}$, then the rate of convergence of the series-expansion of each $F \in \mathfrak{F}$ in terms of $\Phi_{n_g}^{\infty}$ is optimal and it is bounded by ρ^{n_e+1} where

$$\rho = \sup_{\lambda \in \Omega} |G_b(1/\lambda)|. \tag{5.16}$$

This means that in the series-expansion of any $F \in \mathfrak{F}$ with a $n = (n_e+1)n_g$-width optimal Hambo basis $\Phi_{n_g}^{\infty} = \{\phi_{ij}\}_{j=1,\dots,n_g}^{i=1,\dots,\infty}$, there exists a $\gamma > 0$ such that all expansion coefficients $w_{ij} \in \mathbb{R}$ satisfy:

$$|w_{ij}| \leq \gamma \rho^{(n_e+1)(i+1)j}. \tag{5.17}$$

It is obvious that if $\mathfrak{F}$ corresponds to the transfer function set of $\mathscr{F}_S$ with respect to a given IO partition and the Hambo basis $\Phi_{n_g}^{\infty}$ is $n = (n_e+1)n_g$-width optimal with respect to $\mathfrak{F}$, then there exists a $\gamma > 0$ such that the expansion coefficients of the LPV system in terms of $\Phi_{n_g}^{\infty}$ with respect to the considered IO partition satisfy that

$$|(w_{ij} \diamond p)(k)| = |w_{ij|p}| \leq \gamma \rho^{(n_e+1)(i+1)j} \quad \forall k \in \mathbb{Z}. \tag{5.18}$$

for any constant scheduling trajectory $p(k) = \mathrm{p}$. This means that if the basis is optimal with respect to the frozen transfer function set of the LPV system, then fast convergence rate of the expansion coefficients holds in the frozen sense. However, this does not imply fast convergence for non-frozen scheduling trajectories in the general case. This leads to the conclusion, that to achieve a fast convergence rate for the expansion of general LPV systems, a necessary condition is to have fast convergence with respect to the frozen system set.

5.5.3 *Optimality of the Basis in the Global Sense*

As a next step, we investigate when it is possible to characterize the convergence rate of the global expansion coefficients based on the KnW optimality in the frozen sense. Though, it has been not proven explicitly, empirical results show that optimality with respect to the frozen system set is also a sufficient condition to achieve optimal convergence rate of the expansion if the LPV system has a IO representation with static coefficient dependence. This is formalized in the following conjecture:

Conjecture 5.1 (Optimal basis). Given a discrete-time asymptotically stable SISO system $\mathcal{S} = (\mathbb{Z}, \mathbb{P} \subseteq \mathbb{R}^{n_\mathrm{P}}, \mathbb{R}^2, \mathfrak{B})$ with scheduling vector p and IO partition (u, y). Assume that there exists a minimal $\mathfrak{R}_{\mathrm{IO}}(\mathcal{S})$ such that all coefficient dependencies in $\mathfrak{R}_{\mathrm{IO}}(\mathcal{S})$ are static. Denote by $\mathscr{F}_\mathcal{S}$ the frozen system set of $\mathcal{S}$ and let $\mathfrak{F}$ be its associated transfer function set with respect to the IO partition (u, y). Let $\Phi_{n_\mathrm{g}}^\infty = \{\phi_{ij}\}_{j=1,\dots,n_\mathrm{g}}^{i=1,\dots,\infty}$ be a Hambo basis which is $n = (n_\mathrm{e}+1)n_\mathrm{g}$-width optimal with respect to $\mathfrak{F}$ with convergence rate ρ. Then, there exists a set of meromorphic expansion coefficients $w_{ij} \in \mathcal{R}$ and $\gamma > 0$, such that for all $(u, y, p) \in \mathfrak{B}$ with left compact support:

$$y = (w_{00} \diamond p)u + \sum_{i=0}^{\infty} \sum_{j=1}^{n_\mathrm{g}} (w_{ij} \diamond p)\phi_{ij}(\mathrm{q})u, \tag{5.19}$$

and

$$|(w_{ij} \diamond p)(k)| \leq \gamma \rho^{(i+1)j(n_\mathrm{e}+1)} \quad \forall p \in \mathfrak{B}_\mathbb{P} \text{ and } \forall k \in \mathbb{Z}. \tag{5.20}$$

If additionally,

$$\mathfrak{F} \subseteq \mathrm{Span}\{\Phi_{n_\mathrm{g}}^{n_\mathrm{e}}\}, \tag{5.21}$$

is satisfied, then

$$w_{ij} \diamond p = 0, \tag{5.22}$$

for all $i > n_\mathrm{e}$ and $j \in \mathbb{I}_1^{n_\mathrm{g}}$. □

Conjecture 5.1 is of crucial importance even if its proof is an open problem for the general case. By this concept, under the condition of a minimal IO representation with static dependence, the KnW optimality of the basis with respect to the frozen system set can imply optimality of the basis in the series-expansion of the LPV system. Optimality in the latter case means that the same convergence rate of the expansion coefficients is satisfied both in the global and in the frozen sense. Furthermore, an asymptotically stable LPV system can have a finite series-expansion in terms of a basis that can represent all the frozen transfer functions of the system by linear combinations. For example in case of LPV systems with a minimal SS representation where the A and B matrices are constant and the C and D matrices have static dependence, (5.21) is satisfied for a basis function set with poles equal to the eigenvalues of A. In that case, a series-expansion exists with finite nonzero expansion coefficients. This gives the conclusion, that to achieve a fast convergence rate for the expansion of this subclass of LPV systems, it is both a necessary and a sufficient condition to have fast convergence with respect to the frozen local system set.

5.6 Conclusions

In this chapter, series-expansion representations of DT asymptotically stable LPV systems for a given IO partition have been developed based on the concepts of the behavioral approach. We investigated expansions in terms of orthonormal basis functions with the intention to use finite truncation of these expansion representations as a model structure for the identification of asymptotically stable DT-LPV systems later on. The developed theory enables the use of the LTI OBF-based identification approach in the LPV case, which is of crucial importance in providing an efficient and theoretically well-founded LPV identification method. The connection between series-expansion representation and the gain-scheduling principle has been explored, showing that the expansion coefficients of the basis sequence characterize the scheduling function part of the LPV system, while the basis functions are strongly related to the frozen system set. Based on this connection, it has been proven that KnW optimality of a basis sequence, with respect to the frozen system set of a given LPV system $\mathcal{S}$, implies an optimal convergence rate of the series-expansion in these basis functions in the worst case sense along all possible constant scheduling trajectories. It has been also motivated that for systems with minimal IO representations having static coefficient dependence, optimal convergence rate of the LPV series-expansion with such a basis also holds for every scheduling trajectory. As we will see in Chap. 8, these properties have a paramount importance for OBFs-based model structure selection in LPV identification.

Chapter 6
Discretization of LPV Systems

Abstract. This chapter is devoted to the discretization of LPV systems through the discrete-time projection of their state-space representations. Both exact and approximative approaches are developed in a zero-order hold setting. Primary attention is given to the discretization of state-space representations with static coefficient dependence. For this case, criteria are derived to assist the choice of the sampling-time in terms of preservation of frozen dynamic stability and acceptable upperbound on the discretization error.

6.1 The Importance of Discretization

Transformation between LPV systems represented on the *continuous-time* (CT) and the *discrete-time* (DT) axis, has primary importance both for system identification and control synthesis:

Implementation of LPV control solutions in physical hardware often meets significant difficulties, as CT synthesis approaches (see [134, 160]) are preferred over DT solutions (see [6, 133]). The main motivation for this preference is that stability and performance requirements for LPV systems can be expressed significantly easier in CT, like in a mixed sensitivity setting [240]. As a result, the current design tools focus on continuous-time LPV *state-space* (SS) controller synthesis, often in an LFR form, requiring efficient discretization of such representations for implementation purposes.

In the LPV modeling framework, first principle LPV models of nonlinear systems are also often derived in a CT form. In LPV system identification, such models serve as a primary source of information to assist adequate model structure selection in terms of order, type of coefficient dependence, etc. However, current LPV identification methods are developed exclusively for DT. This means that DT projection of first principle LPV models is required to assist model structure selection for practical identification.

These issues show how crucial is to well explore and analyze general discretization of LPV representations. Surprisingly, the existing literature about LPV discretization is very limited. In the early work of [4], three different approaches have

R. Tóth: Modeling and Identification of Linear Parameter-Varying Systems, LNCIS 403, pp. 143–169.
springerlink.com

been introduced based on an *isolated* (stand-alone discretization of a CT system aiming at only the preservation of the CT input-output behavior) *zero-order-hold* (variation of free CT signals is restricted to be piecewise-constant) setting. These approaches: the complete, the rectangular, and the trapezoidal methods have been developed for the discretization of LPV-SS representations by extending the concepts of the LTI framework (see [119]). However, only a limited discussion on the discretization error of the introduced approaches and on their applicability for specific LPV systems has been provided. In [69], an attempt has been made to characterize the discretization error of the rectangular method by expressing the approximation error of the involved state-space matrices. Other types of discretization techniques or criteria for the selection of sampling-time have not been investigated, leaving the-state-of-the-art of LPV discretization incomplete.

In this chapter, we aim to complete the extension of the isolated discretization approaches of the LTI framework. We also focus on comparing the properties of the resulting methods with questions of sampling-time choice, preservation of stability, and discretization errors. Our main purpose is to give tools for the use of CT first principle knowledge in the model structure selection of DT-LPV identification approaches. In this chapter we restrict attention to SS representations, however the methods discussed can be extended to other representation like LFR's (see [199]).

6.2 Discretization of LPV System Representations

As a first step, the considered discretization setting is established. Similar to the major methods of the LTI case, we consider an isolated approach in an ideal ZOH setting presented in Fig. 6.1 where the following assumption holds:

Assumption 6.1 *Given a CT-LPV system $\mathcal{S}$, with input-output partition (u, y) and scheduling signal p, where u and p are generated by an ideal ZOH and y is sampled in a perfectly synchronized manner with sampling time $\mathrm{T_d} \in \mathbb{R}^+$. The ZOH and the instrument providing the output sampling have infinite resolution (no quantization error) and their processing time is zero.*

The problem we intend to solve in the following part is to find the DT equivalent of a given CT-LPV system according to Assumption 6.1. Introduce subscript "$_\mathrm{d}$" to denote sampled/discretized signals. Then it holds for the signals of Fig. 6.1 that

$$u(t) := u_\mathrm{d}(k), \quad \forall t \in [k\mathrm{T_d}, (k+1)\mathrm{T_d}), \tag{6.1a}$$
$$p(t) := p_\mathrm{d}(k), \quad \forall t \in [k\mathrm{T_d}, (k+1)\mathrm{T_d}), \tag{6.1b}$$
$$y_\mathrm{d}(k) := y(k\mathrm{T_d}), \tag{6.1c}$$

for each $k \in \mathbb{Z}$, meaning that u and p can only change at every sampling-time instant. However, in the LPV framework p is considered to be a measurable external/environmental effect (general-LPV) or some function of the states or outputs of the system $\mathcal{S}$ (quasi-LPV). Therefore, possibly it can not be fully influenced by the digitally controlled actuators of the plant which contain the ZOH. Furthermore in

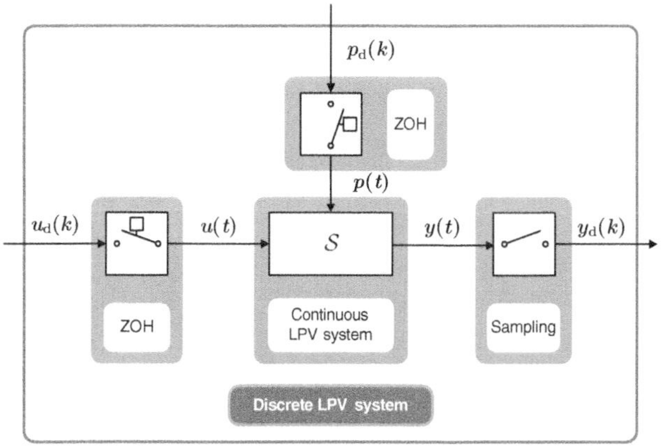

Fig. 6.1 Ideal zero-order hold discretization setting of general LPV systems.

Sect. 3.1.3, equivalence of CT and DT behaviors under a given sampling-time has been established without any restrictions on the variation of u or p.

On the other hand, the variation of u and p inside a sample interval must be restricted to exactly characterize the effect of these signals on the plant. The reason is similar as in the LTI case. In DT, observations of the CT signals u and p are only available at each sampling-time instant. Thus there is no information about the trajectory of these signals during the sample interval. This means that the output signal can not be uniquely determined, unless the variation of the signals u and p is restricted to a certain class of functions. In the ZOH setting, this function class is chosen to be the piecewise constant (zero-order) class. It is also possible to choose this class wider, including linear, 2nd-order polynomial, etc., functions and in this way to define higher-order hold discretization settings of LPV systems. However, in conclusion it holds true that in any LPV discretization setting, the variation of the scheduling signals must be restricted, otherwise the resulting DT description would not be well-posed.

By applying the ZOH setting for the discretization of a given LPV representation, the piecewise-constant variation of p implies that coefficients with dynamic dependence simplify as the derivatives of p are zero inside the sampling interval. This means, that unless the representation has static dependence, the ZOH setting may result in the loss of certain parts of the original behavior. Based on this, we assume that the CT representation to be discretized has static coefficient dependence. Later we explore how this limitation can be overcome, i.e. how the discretization of representations with dynamic dependence can be properly handled.

In conclusion, the introduced discretization setting coincides with the conventional setting of the LTI framework. Moreover, the presented setting is also applicable to closed-loop controllers in the structure given in Fig. 6.2. This closed-loop setting has also been used in [4]. Note that the assumption that the scheduling vector of the continuous LPV controller is affected by the ZOH setting, i.e. it can only vary in piece-wise manner, also holds in this case.

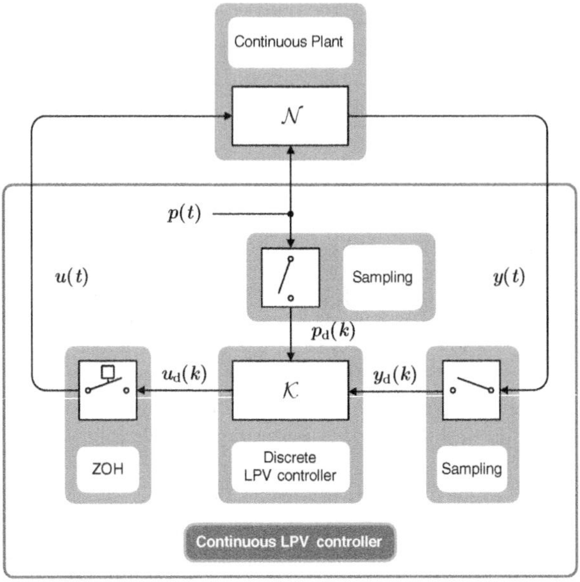

Fig. 6.2 Ideal ZOH discretization setting of closed-loop LPV controllers.

6.3 Discretization of State-Space Representations

Based on the previously introduced discretization setting, the isolated approaches of the LTI framework are extended to the LPV case. The extension only given with respect to LPV-SS representations. In case the system description is available in the form of an other representation, using the equivalence transformations developed in Chap. 4, an equivalent SS realization can always be obtained. Additionally, we assume that the SS representation has only static coefficient dependence. For the case of dynamic dependence see Sec. 6.6. Investigation of the discretization errors and other properties is postponed till Sections 6.4 and 6.5. Before venturing into the derivation of discretization methods, consider the following phenomenon:

For the continuous-time signals u, p defined through (6.1a–b) it holds that

$$u(t) = \sum_{k=-\infty}^{\infty} 1(t - k\mathrm{T_d})\big(u_\mathrm{d}(k) - u_\mathrm{d}(k-1)\big), \tag{6.2a}$$

$$p(t) = \sum_{k=-\infty}^{\infty} 1(t - k\mathrm{T_d})\big(p_\mathrm{d}(k) - p_\mathrm{d}(k-1)\big), \tag{6.2b}$$

where $1(t)$ is the unit-step function (positive zero assumption based Heaviside function) defined as

$$1(t) := \begin{cases} 0, & \text{if } t < 0; \\ 1, & \text{if } t \geq 0. \end{cases} \tag{6.3}$$

The result of $1\,(t-k\mathrm{T_d})$ on $\mathfrak{R}_{\mathrm{SS}}(\mathcal{S})$ in every sampling period is called the switching effect of the ZOH actuation. Based on this, the following important assumption is made:

Assumption 6.2 *The switching behavior of the ZOH actuation has no effect on the CT plant, i.e. the switching of the signals is assumed to take place smoothly.*

The analysis of the consequence of this assumption is postponed till Sect. 6.4.

6.3.1 Complete Method

First the complete signal evolution approach of the LTI framework is extended to the LPV case. Let a continuous-time $\mathfrak{R}_{\mathrm{SS}}(\mathcal{S})$ be given in the ZOH setting. Based on Assumption 6.1, i.e. p and u are constant signals inside each sampling interval, the state-equations (3.44a–b) of $\mathfrak{R}_{\mathrm{SS}}(\mathcal{S})$ can be written as

$$\dot{x}(t) = (A\diamond p)(k\mathrm{T_d})x(t)+(B\diamond p)(k\mathrm{T_d})\,u(k\mathrm{T_d}), \tag{6.4a}$$

$$y(t) = (C\diamond p)(k\mathrm{T_d})x(t)+(D\diamond p)(k\mathrm{T_d})\,u(k\mathrm{T_d}), \tag{6.4b}$$

for $t\in[k\mathrm{T_d},(k+1)\mathrm{T_d})$ with initial condition $x(k\mathrm{T_d})$. Assume that, $\{A,B,C,D\}$ have static dependence on p. Then the DT signal p_d satisfies that[1]

$$\begin{array}{ll}(A\diamond p)(k\mathrm{T_d})=A(p_\mathrm{d}(k)), & (B\diamond p)(k\mathrm{T_d})=B(p_\mathrm{d}(k)),\\ (C\diamond p)(k\mathrm{T_d})=C(p_\mathrm{d}(k)), & (D\diamond p)(k\mathrm{T_d})=D(p_\mathrm{d}(k)).\end{array} \tag{6.5}$$

The state-equation (6.4a), associated with the kth sampling interval, is an *ordinary differential equation* (ODE). To derive a solution of this ODE, introduce $f(x,u,p)$ as the right hand side of (3.44a). Under Assumptions 6.1 and 6.2 it holds that

$$\int_{k\mathrm{T_d}}^{(k+1)\mathrm{T_d}} f(x,u,p)(\tau)\,\mathrm{d}\tau = \int_{k\mathrm{T_d}}^{(k+1)\mathrm{T_d}} A(p_\mathrm{d}(k))x(\tau)+B(p_\mathrm{d}(k))u(k\mathrm{T_d})\,\mathrm{d}\tau, \tag{6.6}$$

which defines the solution of (6.4a) at $t=(k+1)\mathrm{T_d}$ as

$$x((k+1)\mathrm{T_d}) = x(k\mathrm{T_d}) + \int_{k\mathrm{T_d}}^{(k+1)\mathrm{T_d}} f(x,u,p)(\tau)\,\mathrm{d}\tau. \tag{6.7}$$

Assume that $A(\mathrm{p})$ is invertible for $\forall \mathrm{p}\in\mathbb{P}$. By substituting $x_\mathrm{d}(k)=x(k\mathrm{T_d})$ and $u_\mathrm{d}(k)=u(k\mathrm{T_d})$, the previous formula results in

$$\mathrm{q}x_\mathrm{d} = \mathrm{e}^{\mathrm{T_d}A(p_\mathrm{d})}x_\mathrm{d}+A^{-1}(p_\mathrm{d})\left(\mathrm{e}^{\mathrm{T_d}A(p_\mathrm{d})}-I\right)B(p_\mathrm{d})u_\mathrm{d}, \tag{6.8a}$$

$$y_\mathrm{d} = C(p_\mathrm{d})x_\mathrm{d}+D(p_\mathrm{d})u_\mathrm{d}, \tag{6.8b}$$

[1] Due to the assumed static dependence of A, i.e. $A\in\mathcal{R}|_{n_\mathbb{P}}^{n_\mathbb{X}\times n_\mathbb{X}}$, it holds that $(A\diamond p)(t) = A(p(t))$.

where $y_d(k) = y(kT_d)$. Then the complete method gives[2] that the DT equivalent of $\mathfrak{R}_{SS}(\mathcal{S})$ under Assumptions 6.1 and 6.2 is [4]:

$$\mathfrak{R}_{SS}(\mathcal{S}, T_d) := \left[\begin{array}{c|c} e^{T_d A} & A^{-1}\left(e^{T_d A} - I\right) B \\ \hline C & D \end{array}\right]. \tag{6.9}$$

Example 6.1 (Complete discretization). Consider the CT-SS representation

$$\mathfrak{R}_{SS}(\mathcal{S}) = \left[\begin{array}{cc|c} 2p-1 & p & 0 \\ 0 & -1 & 1 \\ \hline 1 & p & 0 \end{array}\right]$$

with $\mathbb{P} = [-1, 1]$. The above representation has static linear dependence and it can be shown that it is uniformly frozen stable. By applying the complete method to $\mathfrak{R}_{SS}(\mathcal{S})$, the DT equivalent of this representation under the sampling-time T_d is

$$\mathfrak{R}_{SS}(\mathcal{S}, T_d) = \left[\begin{array}{cc|c} e^{T_d(2p_d-1)} & \frac{1}{2}e^{T_d(2p_d-1)} - \frac{1}{2}e^{-T_d} & \frac{e^{T_d(2p_d-1)} + (2p_d-1)e^{-T_d} - 2p_d}{4p_d - 2} \\ 0 & e^{-T_d} & 1 - e^{-T_d} \\ \hline 1 & p_d & 0 \end{array}\right].$$

Even for this simple LPV state-space representation, the DT projection results in a complicated rational/exponential dependence on the samples of p. □

6.3.2 Approximative State-Space Discretization Methods

The complete method is commonly not favored in the LPV literature as it introduces heavy nonlinear dependence on p_d, like $e^{T_d A(p_d)}$, which is the main drawback of this approach. Many identification and control synthesis techniques build on the assumption of linear, polynomial or low order rational static dependence on p_d, and hence it is required to develop approximative discretization methods that try to achieve good representation of the original behavior, but with a low complexity of the coefficient dependence. To do that, we systematically extend the approximative discretization methods of the LTI case, by using different approximations of the integral that describes the state-evolution inside the sample-interval.

6.3.2.1 Rectangular (Euler's forward) Method

The simplest way to avoid the appearance of $e^{T_d A}$ is to apply a first-order approximation:

$$e^{T_d A(p_d(k))} \approx I + T_d A(p_d(k)). \tag{6.10}$$

[2] The condition that $A(\mathrm{p})$ is invertible can be relaxed by using the fact that $A^{-1}(e^A - I)$ can be calculated without using the inverse of $A(\mathrm{p})$ (e.g. see [44]).

Consider $f(x,u,p)$ as defined in the previous section. Then an approximation of the solution (6.7) can be considered by the left-hand rectangular evaluation of the integral (6.6), which gives

$$x((k+1)\mathrm{T_d}) \approx x(k\mathrm{T_d}) + \mathrm{T_d}A(p_\mathrm{d}(k))x(k\mathrm{T_d}) + \mathrm{T_d}B(p_\mathrm{d}(k))u(k\mathrm{T_d}), \tag{6.11}$$

coinciding with the suggested matrix exponential approximation of (6.10). Based on this rectangular approach, the DT approximation of $\mathfrak{R}_{\mathrm{SS}}(\mathcal{S})$ is:

$$\mathfrak{R}_{\mathrm{SS}}(\mathcal{S},\mathrm{T_d}) \approx \left[\begin{array}{c|c} I+\mathrm{T_d}A & \mathrm{T_d}B \\ \hline C & D \end{array}\right]. \tag{6.12}$$

Example 6.2 (Rectangular discretization). Continuing Example 6.1, the rectangular discretization method applied to $\mathfrak{R}_{\mathrm{SS}}(\mathcal{S})$ results in

$$\mathfrak{R}_{\mathrm{SS}}(\mathcal{S},\mathrm{T_d}) \approx \left[\begin{array}{cc|c} 2\mathrm{T_d}p_\mathrm{d}+1-\mathrm{T_d} & \mathrm{T_d}p_\mathrm{d} & 0 \\ 0 & 1-\mathrm{T_d} & \mathrm{T_d} \\ \hline 1 & p_\mathrm{d} & 0 \end{array}\right].$$

Comparing this DT approximation with the result of the complete method (Example 6.1) illustrates the difference in complexity of the resulting coefficient dependencies.

□

6.3.2.2 Polynomial (Hanselmann) Method

It is also possible to develop other methods that achieve better approximation of the complete case but with increasing complexity. As suggested in the LTI case by Hanselmann [70], one way leads through the use of higher order Taylor expansion of the matrix exponential term:

$$\mathrm{e}^{\mathrm{T_d}A(p_\mathrm{d}(k))} \approx I + \sum_{l=1}^{n} \frac{\mathrm{T_d}^l}{l!} A^l(p_\mathrm{d}(k)), \tag{6.13}$$

This results in the extension of the LTI polynomial discretization methods. Substituting (6.13) into (6.8a) gives the following SS representation:

$$\mathfrak{R}_{\mathrm{SS}}(\mathcal{S},\mathrm{T_d}) \approx \left[\begin{array}{c|c} I+\sum_{l=1}^{n}\frac{\mathrm{T_d}^l}{l!}A^l & \mathrm{T_d}\left(I+\sum_{l=1}^{n-1}\frac{\mathrm{T_d}^l}{l+1!}A^l\right)B \\ \hline C & D \end{array}\right]. \tag{6.14}$$

Example 6.3 (Polynomial discretization). Approximating the LPV-SS representation of Example 6.1 by the 2nd-order polynomial method results in

$$\mathfrak{R}_{\mathrm{SS}}(\mathcal{S},\mathrm{T_d}) \approx \left[\begin{array}{cc|c} 1+\mathrm{T_d}(2p_\mathrm{d}-1)+\frac{1}{2}\mathrm{T_d^2}(2p_\mathrm{d}-1)^2 & (\mathrm{T_d}-\mathrm{T_d^2})p_\mathrm{d}+\mathrm{T_d^2}p_\mathrm{d}^2 & \frac{1}{2}\mathrm{T_d^2}p_\mathrm{d} \\ 0 & 1-\mathrm{T_d}+\frac{1}{2}\mathrm{T_d^2} & \mathrm{T_d}-\frac{1}{2}\mathrm{T_d^2} \\ \hline 1 & p_\mathrm{d} & 0 \end{array}\right].$$

Due to the polynomial approximation, the originally linear coefficient dependencies are transformed to polynomial functions with order 2. □

6.3.2.3 Trapezoidal (Tustin) Method

An alternative way of approximation leads through the extension of the Tustin method. By using a trapezoidal evaluation of integral (6.6) we obtain:

$$x((k+1)\mathrm{T_d}) \approx x(k\mathrm{T_d}) + \frac{\mathrm{T_d}}{2}\big(f(x,u,p)(k\mathrm{T_d}) + f(x,u,p)((k+1)\mathrm{T_d})\big), \tag{6.15}$$

Using this approximation, the derivation of the LPV Tustin method can be given similarly as in [4]. The key concept is to apply a change of variables:

$$\breve{x}_\mathrm{d}(k) = \frac{1}{\sqrt{\mathrm{T_d}}}\left(I - \frac{\mathrm{T_d}}{2}A(p_\mathrm{d}(k))\right)x(k\mathrm{T_d}) - \frac{\sqrt{\mathrm{T_d}}}{2}B(p_\mathrm{d}(k))u_\mathrm{d}(k). \tag{6.16}$$

If $I - \frac{\mathrm{T_d}}{2}A(\mathrm{p})$ is invertible for all $\mathrm{p} \in \mathbb{P}$, then substitution of (6.16) into (6.15) gives a DT state-equation after some algebraic manipulations. Based on this state-equation, the resulting SS representation reads as

$$\mathfrak{R}_\mathrm{SS}(\mathcal{S},\mathrm{T_d}) \approx \left[\begin{array}{c|c} \left(I+\frac{\mathrm{T_d}}{2}A\right)\left(I-\frac{\mathrm{T_d}}{2}A\right)^{-1} & \sqrt{\mathrm{T_d}}\left(I-\frac{\mathrm{T_d}}{2}A\right)^{-1}B \\ \hline \sqrt{\mathrm{T_d}}C\left(I-\frac{\mathrm{T_d}}{2}A\right)^{-1} & \frac{\mathrm{T_d}}{2}C\left(I-\frac{\mathrm{T_d}}{2}A\right)^{-1}B+D \end{array}\right].$$

It is important to note that, like in the LTI case, the trapezoidal method approximates only the manifest behavior of $\mathfrak{R}_\mathrm{SS}(\mathcal{S},\mathrm{T_d})$, as it gives an approximative DT-SS representation in terms of the new state variable $\breve{x}_\mathrm{d}$.

Example 6.4 (Trapezoidal discretization). Applying the trapezoidal method to the LPV-SS representation of Example 6.1 results in

$$\mathfrak{R}_\mathrm{SS}(\mathcal{S},\mathrm{T_d}) \approx \left[\begin{array}{cc|c} \frac{2+2\mathrm{T_d}p_\mathrm{d}-\mathrm{T_d}}{2-2\mathrm{T_d}p_\mathrm{d}+\mathrm{T_d}} & \frac{4\mathrm{T_d}p_\mathrm{d}}{(2-2\mathrm{T_d}p_\mathrm{d}+\mathrm{T_d})\cdot(2+\mathrm{T_d})} & \frac{2\mathrm{T_d}^{3/2}p_\mathrm{d}}{(2-2\mathrm{T_d}p_\mathrm{d}+\mathrm{T_d})\cdot(2+\mathrm{T_d})} \\ 0 & \frac{2-\mathrm{T_d}}{2+\mathrm{T_d}} & \frac{2\sqrt{\mathrm{T_d}}}{2+\mathrm{T_d}} \\ \hline \frac{2\sqrt{\mathrm{T_d}}}{2-2\mathrm{T_d}p_\mathrm{d}+\mathrm{T_d}} & 4\sqrt{\mathrm{T_d}}p_\mathrm{d}\frac{\mathrm{T_d}+1-\mathrm{T_d}p_\mathrm{d}}{(2-2\mathrm{T_d}p_\mathrm{d}+\mathrm{T_d})\cdot(2+\mathrm{T_d})} & 2\mathrm{T_d}p_\mathrm{d}\frac{\mathrm{T_d}+1-\mathrm{T_d}p_\mathrm{d}}{(2-2\mathrm{T_d}p_\mathrm{d}+\mathrm{T_d})\cdot(2+\mathrm{T_d})} \end{array}\right].$$

Due to the discretization method, the originally linear coefficient dependencies are transformed to rational functions. □

6.3.2.4 Multi-Step Methods

Next, the multistep approximation of the LTI case is extended. Consider the state evolution as the solution of the differential equation defined by (3.44a), where all the coefficients are assumed to have static dependence. This solution can be numerically approximated via multi-step formulas like the Runge-Kutta, Adams-Moulton, or the Adams-Bashforth type of approaches (see [8]). In commercial engineering software packages, like MATLAB *Simulink*, commonly variable step-size implementation of these algorithms assures accurate simulation of continuous-time systems. However in the considered ZOH discretization setting, the step size, i.e. the sampling rate, is fixed and sampled data is only available at past and present sampling instances. This immediately excludes multi-step implicit methods like the Adams-Moulton approaches. Moreover $f(x,u,p)$ can only be evaluated for integer multiples of the sampling-time, as the input only changes at these time instances and the resulting model must be realized as a single rate (not multi-rate) system. Therefore it is complicated to apply methods like the Runge-Kutta approach. The family of Adams-Bashforth methods does fulfill these requirements (see [8]). The 3-step version of this type of numerical approach yields the following approximation:

$$\begin{aligned} x((k+1)\mathrm{T_d}) \approx x_\mathrm{d}(k+1) = x(k\mathrm{T_d}) + \frac{\mathrm{T_d}}{12}\big(5f(x,u,p)((k-2)\mathrm{T_d}) - \\ -16f(x,u,p)((k-1)\mathrm{T_d}) + 23f(x,u,p)(k\mathrm{T_d})\big). \end{aligned} \tag{6.17}$$

Formulating this state-space equation in an augmented SS form with the new state-variable:

$$\breve{x}_\mathrm{d}(k) = \mathrm{Col}\big(x_\mathrm{d}(k), f(x_\mathrm{d},u,p)((k-1)\mathrm{T_d}), f(x_\mathrm{d},u,p)((k-2)\mathrm{T_d})\big), \tag{6.18}$$

leads to

$$\mathfrak{R}_{\mathrm{SS}}(\mathcal{S},\mathrm{T_d}) \approx \left[\begin{array}{ccc|c} I + \frac{23\mathrm{T_d}}{12}A & -\frac{16\mathrm{T_d}}{12}I & \frac{5\mathrm{T_d}}{12}I & \frac{23\mathrm{T_d}}{12}B \\ A & 0 & 0 & B \\ 0 & I & 0 & 0 \\ \hline C & 0 & 0 & D \end{array}\right].$$

The resulting DT-SS representation is an approximation of $\mathfrak{R}_{\mathrm{SS}}(\mathcal{S},\mathrm{T_d})$ in terms of the new state variable $\breve{x}_\mathrm{d}$. Note that multi-step discretization results in linear conversion rules but the state-dimension is increased.

Example 6.5 (3-step Adams-Bashforth discretization). Applying the 3-step Adams-Bashforth method to the LPV-SS representation of Example 6.1 results in

$$\left[\begin{array}{cccccc|c} 1 + \frac{23}{6}\mathrm{T_d}p_\mathrm{d} - \frac{23}{12}\mathrm{T_d} & \frac{23}{12}\mathrm{T_d}p_\mathrm{d} & -\frac{16}{12}\mathrm{T_d} & 0 & \frac{5}{12}\mathrm{T_d}\ 0 & 0 & 0 \\ 0 & 1 - \frac{23}{12}\mathrm{T_d} & 0 & -\frac{16}{12}\mathrm{T_d} & 0 & \frac{5}{12}\mathrm{T_d} & \frac{23}{12}\mathrm{T_d} \\ 2p_\mathrm{d} - 1 & p_\mathrm{d} & 0 & 0 & 0 & 0 & 0 \\ 0 & -1 & 0 & 0 & 0 & 0 & 1 \\ 0 & 0 & 1 & 0 & 0 & 0 & 0 \\ 0 & 0 & 0 & 1 & 0 & 0 & 0 \\ \hline 1 & p_\mathrm{d} & 0 & 0 & 0 & 0 & 0 \end{array}\right].$$

As expected, the obtained DT projection preserves the originally linear coefficient dependence. However, the resulting DT representation has a 6 dimensional state-variable while the original CT representation has only 2 state-variables. □

6.4 Discretization Errors and Performance Criteria

In the following part, the introduced methods are investigated in terms of the generated discretization error, numerical convergence, and numerical stability. These are used to derive upperbounds on the sampling-time $\mathrm{T_d}$, that guarantee a user-defined bounded discretization error and stability preservation with respect to the original CT system. Moreover, the influence of the assumption that no switching effects result from the ZOH actuation is investigated.

6.4.1 Local Discretization Errors

Characterization of the discretization error for each of the introduced approaches is important in order to study how adequate the used approximation is with respect to the original CT behavior. It has been already emphasized that, due to the considered assumptions, the complete method theoretically gives errorless discretization in terms of the ZOH setting. For other approaches we investigate the discretization error in terms of the *local unit truncation* (LUT) error, which is often applied in numerical analysis (see [8]). This concept describes the error that results in each sampling interval due to discretization.

As a first step, the LUT error, denoted by $\varepsilon_k \in \mathbb{R}$, is formally introduced. Let $\mathfrak{R}_{\mathrm{SS}}(\mathcal{S})$ be the SS representation of the CT-LPV system $\mathcal{S}$, with $\mathbb{P} \subseteq \mathbb{R}^{n_\mathbb{P}}$ and with meromorphic matrix functions

$$\left[\begin{array}{c|c} A & B \\ \hline C & D \end{array}\right] \in \left[\begin{array}{c|c} (\mathcal{R}|_{n_\mathbb{P}})^{n_\mathbb{X} \times n_\mathbb{X}} & (\mathcal{R}|_{n_\mathbb{P}})^{n_\mathbb{X} \times n_\mathbb{U}} \\ \hline (\mathcal{R}|_{n_\mathbb{P}})^{n_\mathbb{Y} \times n_\mathbb{X}} & (\mathcal{R}|_{n_\mathbb{P}})^{n_\mathbb{Y} \times n_\mathbb{U}} \end{array}\right],$$

defining static dependence on p. Denote by $(A_\mathrm{d}, B_\mathrm{d}, C_\mathrm{d}, D_\mathrm{d})$ the SS matrices of the DT representation resulting by the discretization methods of Sect. 6.3.2 applied to $\mathfrak{R}_{\mathrm{SS}}(\mathcal{S})$. Due to static dependence of the original representation, these matrix functions also have static dependence on p_d. In the rectangular and the polynomial case, the state-basis of this representation is equal to the state-basis of the original CT representation. However, in the trapezoidal and in the multistep cases, the resulting DT projection also includes a state-transformation. In order to formulate the approximation error of the discretization methods based on the error of the approximation of the state evolution, the results of those discretization methods which involve state transformation must be brought back to the original state-basis. Introduce the matrix polynomials $R_\mathrm{x} \in \mathcal{R}[\xi]^{n_\mathbb{X} \times n_\mathbb{X}}$ and $R_\mathrm{u} \in \mathcal{R}[\xi]^{n_\mathbb{X} \times n_\mathbb{U}}$, that formulate the state update of the DT approximations on the same state-basis as in $\mathfrak{R}_{\mathrm{SS}}(\mathcal{S})$. In the rectangular and the polynomial case, $R_\mathrm{x}(\mathrm{q}) = A_\mathrm{d}$ and $R_\mathrm{u}(\mathrm{q}) = B_\mathrm{d}$, but in the other cases, they also include the appropriate state-transformation. For example in the trapezoidal case, (6.15) implies that:

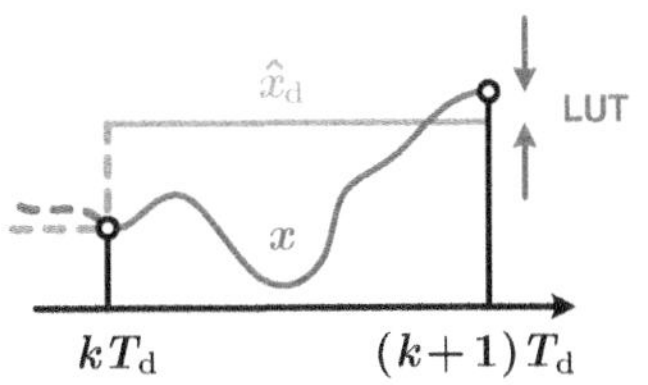

Fig. 6.3 Local unit truncation error of the discretized representation associated state-signal $\hat{x}_d$ with respect to the CT state-signal x at the time-step $(k+1)T_d$.

$$R_x(q) = \left(I + \frac{T_d}{2}A\right)\left(I - \frac{T_d}{2}\overrightarrow{A}\right)^{-1}, \tag{6.19}$$

$$R_u(q) = \frac{T_d}{2}\left(I - \frac{T_d}{2}\overrightarrow{A}\right)^{-1}\left(B + \overrightarrow{B}q\right). \tag{6.20}$$

Then for each sampling interval, ε_k is defined by

$$T_d\varepsilon_{k+1} := \big(qx_d - (R_x(q) \diamond p_d)x_d - (R_u(q) \diamond p_d)u_d\big)(k). \tag{6.21}$$

Note, that LUT represents the relative approximation error of the sampled state signal x_d of the CT representation by the state $\hat{x}_d$ of the DT representation, when past samples of x_d and u_d are used to calculate $\hat{x}_d$ via the DT state-equation (see Fig. 6.3). Hence the name "local". In numerical approximation of differential equations, ε_k is considered as the measure of accuracy. The following definition is important:

Definition 6.1 (N-consistency, based on [8]). Let $\mathfrak{R}_{SS}(\mathcal{S})$ be the SS representation of the LPV system $\mathcal{S}$ with full behavior $\mathfrak{B}_{SS}$. The discrete-time approximation of the state-space equation (3.44a) is called numerically consistent, if for any $(u,x,y,p) \in \mathfrak{B}_{SS}$, it holds that

$$\lim_{T_d \to 0} \sup_{k \in \mathbb{Z}} \|\varepsilon_k\| = 0. \tag{6.22}$$

□

This means that - in case of N-consistency - the local approximation error reduces with decreasing T_d. However, this does not imply that the supremum of the global approximation error,

$$\eta_{k+1} = \big(qx_d - (R_x(q) \diamond p_d)\hat{x}_d - (R_u(q) \diamond p_d)u_d\big)(k), \tag{6.23}$$

where $\hat{x}_d$ is the discrete-time approximation of the state, decreases/converges to zero too. As a next step, the LUT error of each discretization method is investigated together with the N-consistency.

6.4.1.1 LUT Error of the Rectangular Method

For the rectangular method, (6.21) gives

$$x((k+1)T_d) = \big(I + T_dA(p_d(k))\big)x(kT_d) + T_dB(p_d(k))u(kT_d) + T_d\varepsilon_{k+1}. \tag{6.24}$$

Define the first-order Taylor approximation of x around the time instant $k\mathrm{T_d}$ as

$$x(t) = x(k\mathrm{T_d}) + \left(\frac{\mathrm{d}}{\mathrm{d}t}x\right)(k\mathrm{T_d}) \cdot (t - k\mathrm{T_d}) + \frac{1}{2}\left(\frac{\mathrm{d}^2}{\mathrm{d}t^2}x\right)(\tau), \quad \tau \in (k\mathrm{T_d}, t), \tag{6.25}$$

for $t > k\mathrm{T_d}$. Substraction of (6.25) for $t = (k+1)\mathrm{T_d}$ from (6.24) yields that $\mathrm{T_d}\varepsilon_{k+1}$ is equal to the residual term, giving

$$\varepsilon_{k+1} = \frac{\mathrm{T_d}}{2}\left(\frac{\mathrm{d}^2}{\mathrm{d}t^2}x\right)(\tau) \quad \tau \in (k\mathrm{T_d}, (k+1)\mathrm{T_d}). \tag{6.26}$$

This shows that in the ZOH setting, the rectangular method is consistent in first-order (in $\mathrm{T_d}$) if $\left\|\left(\frac{\mathrm{d}^2}{\mathrm{d}t^2}x\right)(\tau)\right\| < \infty$ for all $x \in \mathfrak{B}_{\mathbb{X}}$ and $\tau \in \mathbb{R}$, where $\mathfrak{B}_{\mathbb{X}}$ denotes the projected signal behavior of $\mathfrak{B}_{\mathrm{SS}}$ on the variable x. As the meromorphic coefficients functions in (3.44a) are partially differentiable in p, the state evolution $f(x,u,p)$ is partially differentiable in each variable. Then

$$\frac{\mathrm{d}^2}{\mathrm{d}t^2}x = \frac{\partial f}{\partial x}\underbrace{\frac{\mathrm{d}}{\mathrm{d}t}x}_{f} + \frac{\partial f}{\partial u}\frac{\mathrm{d}}{\mathrm{d}t}u + \frac{\partial f}{\partial p}\frac{\mathrm{d}}{\mathrm{d}t}p. \tag{6.27}$$

Due to Assumptions 6.1 and 6.2, it holds true that $\left(\frac{\mathrm{d}}{\mathrm{d}t}u\right)(t) = 0$ and $\left(\frac{\mathrm{d}}{\mathrm{d}t}p\right)(t) = 0$ in each sampling interval. Thus, (6.27) gives that for $\tau \in (k\mathrm{T_d}, (k+1)\mathrm{T_d})$:

$$\begin{aligned}\left\|\left(\frac{\mathrm{d}^2}{\mathrm{d}t^2}x\right)(\tau)\right\| &= \|\, A(p_\mathrm{d}(\tau)) \cdot f(x,u,p)(\tau)\,\| \\ &\leq \max_{\mathrm{p}\in\mathbb{P},\mathrm{x}\in\mathbb{X},\mathrm{u}\in\mathbb{U}} \left\| A^2(\mathrm{p})\mathrm{x} + A(\mathrm{p})B(\mathrm{p})\mathrm{u} \right\|,\end{aligned} \tag{6.28}$$

where $\|\bullet\|$ is an arbitrary norm. Note that in (6.28), $\mathbb{X}$ and $\mathbb{U}$ must be bounded sets to be able to compute an upperbound. If this is not the case, then commonly $\mathbb{X}$ and $\mathbb{U}$ can be restricted to a bounded subset corresponding to the image of the typical trajectories of the system variables, like it is common in model predictive control (see [20, 21]). Then the previous bound can be formulated for this region of interest. In the sequel, we denote the upperbound (6.28) by $M^{(1)}$ and call it the first-order *numerical sensitivity* (N-sensitivity) constant. Note, that $M^{(1)}$ can be computed via nonlinear optimization or alternatively it can be approximated through gridding. In case of an approximation, gridding of the sets $\mathbb{P}$, $\mathbb{X}$, and $\mathbb{U}$ can be demanding, requiring significant computational power.

6.4.1.2 LUT Error of Other Approximative Methods

Using similar arguments, the LUT error of other discretization methods can be formulated based on [8]. The results are given in the first row of Table 6.1, showing

Table 6.1 Local truncation error ε_k with $\tau \in (k\mathrm{T_d},(k+1)\mathrm{T_d})$ and with $\tau \in ((k-2)\mathrm{T_d},(k+1)\mathrm{T_d})$ in the Adams-Bashforth case, sampling boundary of stability $\check{\mathrm{T}}_\mathrm{d}$, and sampling upper-bound of performance $\hat{\mathrm{T}}_\mathrm{d}$ of LPV-SS ZOH discretization methods.

	Rectangular	n^{th}-polynomial
ε_k	$\frac{\mathrm{T_d}}{2}\left(\frac{\mathrm{d}^2}{\mathrm{d}t^2}x\right)(\tau)$	$\frac{\mathrm{T}_\mathrm{d}^n}{(n+1)!}\left(\frac{\mathrm{d}^{n+1}}{\mathrm{d}t^{n+1}}x\right)(\tau)$
$\check{\mathrm{T}}_\mathrm{d}$	$\min\limits_{\mathrm{p}\in\mathbb{P}}\ \min\limits_{\lambda\in\sigma(A(\mathrm{p}))} -\frac{2\mathrm{Re}(\lambda)}{\lvert\lambda\rvert^2}$	$\arg\min\limits_{\mathrm{T_d}\in\mathbb{R}_0^+}\left\lvert\max\limits_{\mathrm{p}\in\mathbb{P}}\bar{\sigma}\left(\sum_{l=0}^{n}\frac{\mathrm{T}_\mathrm{d}^l}{l!}A^l(\mathrm{p})\right)-1\right\rvert$
$\hat{\mathrm{T}}_\mathrm{d}$	$\sqrt{2\frac{\varepsilon_{\max}M_\mathrm{x}^{\max}}{100M^{(1)}}}$	$\sqrt[n+1]{\frac{\varepsilon_{\max}M_\mathrm{x}^{\max}(n+1)!}{100M^{(n)}}}$

	Trapezoidal	Adams-Bashforth (3-step)
ε_k	$\frac{1}{12}\mathrm{T}_\mathrm{d}^2\left(\frac{\mathrm{d}^3}{\mathrm{d}t^3}x\right)(\tau)$	$\frac{3}{8}\mathrm{T}_\mathrm{d}^3\left(\frac{\mathrm{d}^4}{\mathrm{d}t^4}x\right)(\tau)$
$\check{\mathrm{T}}_\mathrm{d}$	$\max\limits_{\mathrm{p}\in\mathbb{P}}\ \max\limits_{\substack{\lambda\in\sigma(A(\mathrm{p}))\\ \mathrm{Im}(\lambda)=0}}\frac{2}{\mathrm{Re}(\lambda)}$	$\arg\min\limits_{\mathrm{T_d}\in\mathbb{R}_0^+}\left\lvert\max\limits_{\mathrm{p}_0,\dots,\mathrm{p}_{n-1}\in\mathbb{P}}\bar{\lambda}\left(\check{R}_{\mathrm{p}_0,\dots,\mathrm{p}_{n-1}}(\xi,\mathrm{T_d})\right)-1\right\rvert$
$\hat{\mathrm{T}}_\mathrm{d}$	$\sqrt[3]{\frac{12\varepsilon_{\max}M_\mathrm{x}^{\max}}{100M^{(2)}}}$	$\sqrt[4]{\frac{8\varepsilon_{\max}M_\mathrm{x}^{\max}}{300M^{(3)}}}$

that each method is consistent with varying orders [190]. Moreover, using (6.27) and the chain rule of differentiation, higher order N-sensitivity constants can be derived:

$$M^{(n)} = \max_{\mathrm{p}\in\mathbb{P},\mathrm{x}\in\mathbb{X},\mathrm{u}\in\mathbb{U}} \left\| A^{n+1}(\mathrm{p})\mathrm{x} + A^n(\mathrm{p})B(\mathrm{p})\mathrm{u} \right\|. \tag{6.29}$$

6.4.2 *Global Convergence and Preservation of Stability*

So far, only the LUT error of the introduced methods has been investigated, giving basic proofs of consistency. As a next step, we investigate global convergence of approximative methods together with their *numerical stability* (N-stability). The latter concept means that small errors in the initial condition of the discrete-time approximation do not cause the solution to diverge. We show that for the single-step approximative discretization methods, N-stability is identical with the preservation of the frozen stability of the original representation. This means that in case of numerical stability, the discretization method does not changes the frozen stability properties of the discretized model, which is a prime requirement of a successful DT approximation of a CT system. To derive adequate criteria for the largest sampling-time for which this property holds (N-stability radius), each method is analyzed and computable formulas are derived. Define $\mathbb{Z}_0^-$ as the set of all negative integer numbers including 0, and $\mathbb{Z}^+ = \mathbb{Z}\setminus\mathbb{Z}_0^-$. According to the previously explained line of discussion, we introduce the following concepts:

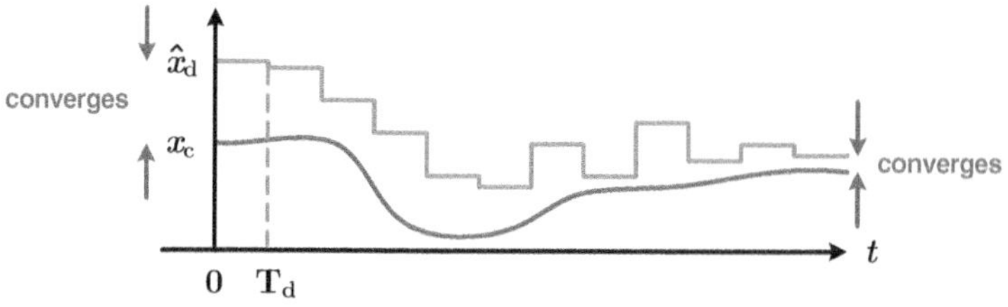

Fig. 6.4 N-convergence of the DT approximation. The DT state-signal $\hat{x}_d$ converges to the CT state-signal x of the approximated representation, if the error on the initial conditions (past) of the approximation converges to zero.

Definition 6.2 (N-convergence, based on [8]). Let $\mathfrak{R}_{SS}(\mathcal{S})$ be the SS representation of the CT-LPV system $\mathcal{S}$ with state-signal x and projected state behavior $\mathfrak{B}_{\mathbb{X}}$, and let $\hat{x}_d$ denote the DT approximation of x with $T_d \in \mathbb{R}^+$. Then a discretization method is called numerically convergent, if for any $x \in \mathfrak{B}_{\mathbb{X}}$

$$\lim_{T_d \to 0} \sup_{k \in \mathbb{Z}_0^-} \| \hat{x}_d(k) - x(kT_d) \| = 0 \quad \Rightarrow \quad \lim_{T_d \to 0} \sup_{k \in \mathbb{Z}^+} \| \hat{x}_d(k) - x(kT_d) \| = 0. \qquad \square$$

Note that in the trapezoidal and multi-step cases, $\hat{x}_d$ is the appropriate state-transform of $\breve{x}_d$ with respect to x. In terms of Def. 6.2, N-convergence means that the discretized solution of the state-equation can get arbitrary close to the original CT behavior by decreasing T_d (see Fig. 6.4).

Definition 6.3 (N-stability, based on [8]). A discretization method is called numerically stable, if for sufficiently small values of T_d and ε, any two state-trajectories $\hat{x}_d$ and $\hat{x}_d'$ of the discretized representation associated with the same input and scheduling trajectory on the half line $\mathbb{Z}^+$ and with $\|\hat{x}_d'(0) - \hat{x}_d(0)\| < \varepsilon$ implies the existence of a $\gamma \geq 0$ such that $\|\hat{x}_d'(k) - \hat{x}_d(k)\| < \gamma\varepsilon$ for $\forall k \in \mathbb{Z}^+$. $\square$

The notion of N-stability means that small errors in the initial condition do not cause divergence as the solution is iterated (see Fig. 6.5). For the approximative methods, N-convergence and N-stability are questions of main importance. To analyze these notions for the introduced discretization approaches, first consider the single-step methods. Introduce the characteristic polynomial $\breve{R}_p \in \mathbb{R}[\xi]$ of the frozen aspects associated with the DT approximation of $\mathfrak{R}_{SS}(\mathcal{S})$ as

$$\breve{R}_p(\xi, T_d) = \mathrm{Det}(\xi I - A_d(p_d)), \quad p_d(k) = \mathrm{p}, \ \forall k \in \mathbb{Z}, \tag{6.30}$$

where the indeterminant ξ is associated with q. Due to the multi-step nature of the Adams-Bashforth method - to avoid conservatism of the upcoming analysis - $\breve{R}$ is defined to reflect the multi-step nature of the state-evolution. In the n-step Adams-Bashforth case, the state evolution with respect to discretized original state x_d is characterized by

$$\xi^n I - T_d \sum_{l=0}^{n-1} \gamma_l \xi^l A, \tag{6.31}$$

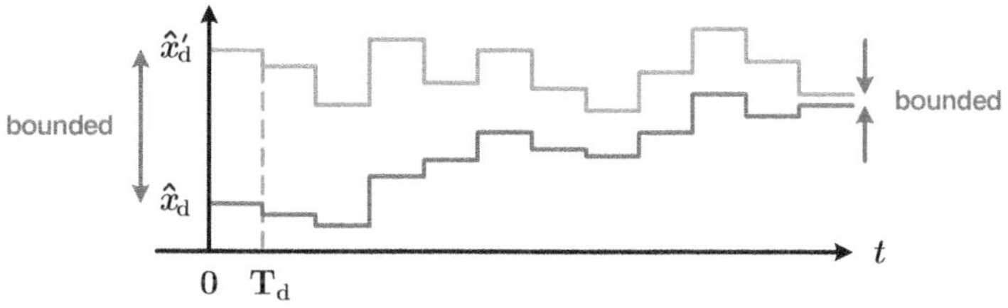

Fig. 6.5 N-stability of the DT approximation. If the difference between the initial conditions of two state trajectories $\hat{x}_{\mathrm{d}}$ and $\hat{x}'_{\mathrm{d}}$, provided by the approximation method for the same input and scheduling on the half line $\mathbb{Z}^+$, is bounded, then the difference of the two trajectories on $\mathbb{Z}^+$ is also bounded.

with $\{\gamma_l\}_{l=0}^{n-1} \subset \mathbb{R}$ the Adams-Bashforth approximation coefficients (values of these coefficients for any $n > 0$ are given in [8]). The form of (6.31) results due to the augmented state vector $\breve{x}_{\mathrm{d}}$. However, even if $\mathfrak{R}_{\mathrm{SS}}(\mathcal{S})$ has static dependence, the resulting polynomial in (6.31) becomes dynamically dependent on p_{d}. To express this, the following local characteristic polynomial is introduced in the "frozen" sense for a scheduling sequence $\mathrm{p}_0, \dots, \mathrm{p}_{n-1} \in \mathbb{P}$:

$$\breve{R}_{\mathrm{p}_0,\dots,\mathrm{p}_{n-1}}(\xi, \mathrm{T_d}) = \mathrm{Det}\left(\xi^n I - \mathrm{T_d}\sum_{l=0}^{n-1}\gamma_l A(\mathrm{p}_l)\xi^l\right), \tag{6.32}$$

Now we can formulate the following theorem to characterize N-stability of the introduced discretization methods:

Theorem 6.1 (Strong root-condition, based on [8]). *Discretization methods are N-convergent and N-stable, if for all $\lambda \in \mathbb{C}$ satisfying*

$$\exists \mathrm{p}_0, \dots, \mathrm{p}_{n-1} \in \mathbb{P} \text{ such that } \breve{R}_{\mathrm{p}_0,\dots,\mathrm{p}_{n-1}}(\lambda, 0) = 0, \tag{6.33}$$

it holds that $|\lambda| \leq 1$ or if $|\lambda| = 1$, then $\frac{\partial}{\partial \xi}\breve{R}_{\mathrm{p}_0,\dots,\mathrm{p}_{n-1}}(\lambda, 0) \neq 0$.

It can be shown, that all of the introduced LPV-SS discretization methods satisfy Th. 6.1, as the proofs given in [8] also hold in this case. This means that all methods discussed in the previous part are N-convergent and N-stable. Now we can extend the root-condition to compute an exact upperbound $\breve{\mathrm{T}}_{\mathrm{d}}$ of the "sufficiently small" $\mathrm{T_d}$ that provides N-stability (see Def. 6.3):

Definition 6.4 (N-stability radius, [8]). The N-stability radius $\breve{\mathrm{T}}_{\mathrm{d}}$ is defined as the largest $\mathrm{T_d} \in \mathbb{R}_0^+$ for which all $\lambda \in \mathbb{C}$ with $\exists \mathrm{p}_0, \dots, \mathrm{p}_{n-1} \in \mathbb{P}$ such that

$$\breve{R}_{\mathrm{p}_0,\dots,\mathrm{p}_{n-1}}(\lambda, \mathrm{T_d}) = 0, \tag{6.34}$$

satisfy that $|\lambda| \leq 1$. □

This definition has an interesting consequence for the discretization of LPV-SS representations. Namely, that through the characteristic polynomial $\breve{R}$, it implies that, if $\mathrm{T_d} \leq \check{\mathrm{T}}_\mathrm{d}$, then in the single-step cases, the resulting DT representation defines a uniformly frozen stable system (see Sect. 3.3.2), as for this $\mathrm{T_d}$ it is satisfied that

$$\max_{\mathrm{p}\in\mathbb{P}} \bar{\sigma}\left(A_\mathrm{d}(\mathrm{p})\right) \leq 1, \tag{6.35}$$

where $\bar{\sigma}(\cdot) = \max|\sigma(\cdot)|$ is the spectral radius and $\sigma(\cdot)$ is the eigenvalue operator. If the original CT system $\mathcal{S}$ is globally stable (dynamic, BIBO, etc.), then commonly it is desirable that its DT approximation is also globally stable. For such a property, it is necessary that uniform frozen stability of $\mathfrak{R}_\mathrm{SS}(\mathcal{S})$:

$$\max_{\mathrm{p}\in\mathbb{P}} \max_{\lambda\in\sigma(A(\mathrm{p}))} \mathrm{Re}(\lambda) \leq 0, \tag{6.36}$$

is preserved, resulting in uniform frozen stability of the DT representation. This gives the important observation that, for the introduced single-step discretization methods, preservation of global stability of the original system and N-stability of the discretization method both require uniform frozen stability of the resulting DT representation [190]. For N-stability it is a sufficient, for preservation of global stability of $\mathcal{S}$ it is a necessary condition. Note, that this condition of N-stability is sufficient only for representations with static dependence. In the following, we analyze the N-stability radius of each discretization method to give computable bounds for the selection of $\mathrm{T_d}$ by which the discretization method preserves frozen stability of the original system, i.e. it has N-stability.

6.4.2.1 Stability Radius of the Rectangular Method

In case of the rectangular method, (6.35) is equivalent with

$$\max_{\mathrm{p}\in\mathbb{P}} \bar{\sigma}\left(I + \mathrm{T_d}A(\mathrm{p})\right) \leq 1. \tag{6.37}$$

Due to the basic properties of eigenvalues, it can be shown that (6.37) holds iff

$$\max_{\mathrm{p}\in\mathbb{P}} \max_{\lambda\in\sigma(A(\mathrm{p}))} \left|\frac{1}{\mathrm{T_d}} + \lambda\right| < \frac{1}{\mathrm{T_d}}. \tag{6.38}$$

From (6.38), the stability radius is

$$\check{\mathrm{T}}_\mathrm{d} = \max\left(0, \min_{\mathrm{p}\in\mathbb{P}} \min_{\lambda\in\sigma(A(\mathrm{p}))} -\frac{2\mathrm{Re}(\lambda)}{|\lambda|^2}\right). \tag{6.39}$$

Note that $\check{\mathrm{T}}_\mathrm{d} = 0$ in case of non-uniformly frozen stable $\mathfrak{R}_\mathrm{SS}(\mathcal{S})$, meaning that the rectangular DT approximation of non-uniformly frozen stable systems is not N-stable. Computation of the bound (6.39) is a nonlinear optimization problem for which an approximative solution may follow by the gridding of $\mathbb{P}$.

6.4.2.2 Stability Radius of the Polynomial Method

In case of the polynomial method, (6.35) translates to

$$\max_{\mathrm{p}\in\mathbb{P}} \bar{\sigma}\left(I+\sum_{l=1}^{n}\frac{\mathrm{T}_\mathrm{d}^l}{l!}A^l(\mathrm{p})\right)\leq 1. \tag{6.40}$$

From (6.40), the stability radius reads as

$$\check{\mathrm{T}}_\mathrm{d} = \arg\min_{\mathrm{T}_\mathrm{d}\in\mathbb{R}_0^+}\left|\max_{\mathrm{p}\in\mathbb{P}} \bar{\sigma}\left(\sum_{l=0}^{n}\frac{\mathrm{T}_\mathrm{d}^l}{l!}A^l(\mathrm{p})\right)-1\right|. \tag{6.41}$$

Again, a possible approximation of $\check{\mathrm{T}}_\mathrm{d}$ can be given by applying bisection based search in T_d on (6.41) over a grid of $\mathbb{P}$. Note, that in case of non-uniform frozen stability, $\check{\mathrm{T}}_\mathrm{d}=0$ with this method as well.

6.4.2.3 Stability Radius of the Trapezoidal Method

For the trapezoidal method, condition (6.35) becomes quite complicated due to the inverse term $(I-\frac{\mathrm{T}_\mathrm{d}}{2}A)^{-1}$ in A_d. First it must be guaranteed that this inverse exists for all scheduling signals, meaning that

$$\mathrm{Det}\left(I-\tfrac{\mathrm{T}_\mathrm{d}}{2}A(\mathrm{p})\right)\neq 0, \quad \forall \mathrm{p}\in\mathbb{P}, \tag{6.42}$$

or equivalently

$$\min_{\mathrm{p}\in\mathbb{P}} \underline{\sigma}\left(I-\tfrac{\mathrm{T}_\mathrm{d}}{2}A(\mathrm{p})\right)>0, \tag{6.43}$$

where $\underline{\sigma}(\centerdot)=\min|\sigma(\centerdot)|$. Again, the eigenvalue properties yield that (6.43) is equivalent with

$$\min_{\mathrm{p}\in\mathbb{P}}\ \min_{\lambda\in\sigma(A(\mathrm{p}))}\left|\frac{2}{\mathrm{T}_\mathrm{d}}-\lambda\right|>0,$$

which is guaranteed for every $0\leq \mathrm{T}_\mathrm{d}<\check{\mathrm{T}}_\mathrm{d}$, where

$$\check{\mathrm{T}}_\mathrm{d}=\max_{\mathrm{p}\in\mathbb{P}}\ \max_{\substack{\lambda\in\sigma(A(\mathrm{p}))\\ \mathrm{Im}(\lambda)=0}}\frac{2}{\mathrm{Re}(\lambda)}. \tag{6.44}$$

Instead of N-stability, here $\check{\mathrm{T}}_\mathrm{d}$ ensures the existence of the DT projection (existence of A_d). It is shown later, that if the DT projection exists, then N-stability and N-convergence hold. Note that, in case of $\mathrm{Im}(\lambda)\neq 0$ for all $\lambda\in\sigma(A(\mathrm{p}))$ and $\mathrm{p}\in\mathbb{P}$, meaning that every frozen representation of the original CT system has only complex poles, condition (6.42) is guaranteed for arbitrary T_d, resulting in $\check{\mathrm{T}}_\mathrm{d}=\infty$. Similarly, uniform frozen stability of $\mathfrak{R}_{\mathrm{SS}}(\mathcal{S})$, meaning that every frozen representation has poles with only negative or zero real part, gives $\check{\mathrm{T}}_\mathrm{d}=\infty$. In [4], the condition

$$\mathrm{T_d} \leq \max_{\mathrm{p}\in\mathbb{P}} \frac{2}{\bar{\sigma}(A(\mathrm{p}))}, \tag{6.45}$$

was proposed to guarantee invertibility, which is a rather conservative upperbound of (6.44). In case $0 \leq \mathrm{T_d} < \check{\mathrm{T}}_\mathrm{d}$ holds and $\mathfrak{R}_{\mathrm{SS}}(\mathcal{S})$ has uniform frozen stability, then (6.35) holds, as in this case

$$\max_{\mathrm{p}\in\mathbb{P}} \bar{\sigma}\left(\left(I + \frac{\mathrm{T_d}}{2}A(\mathrm{p})\right)\left(I - \frac{\mathrm{T_d}}{2}A(\mathrm{p})\right)^{-1}\right) \leq 1. \tag{6.46}$$

See [60] for the proof. Thus, for uniformly frozen stable LPV-SS representations with static dependence, the trapezoidal method always guarantees N-stability and N-convergence if $\mathrm{T_d}$ satisfies condition (6.44).

6.4.2.4 Stability Radius of the Adams-Bashforth Method

In case of the Adams-Bashforth method, the concept of N-stability means that for a given $\mathrm{T_d}$,

$$\max_{\mathrm{p}_0,\ldots,\mathrm{p}_{n-1}\in\mathbb{P}} \bar{\lambda}\left(\check{R}_{\mathrm{p}_0,\ldots,\mathrm{p}_{n-1}}(\xi,\mathrm{T_d})\right) \leq 1, \tag{6.47}$$

where

$$\bar{\lambda}(R(\xi)) = \max_{\substack{\lambda\in\mathbb{C}\\ R(\lambda)=0}} |\lambda|. \tag{6.48}$$

A necessary condition for (6.47) is that the resulting DT representation has uniform frozen stability:

$$\max_{\mathrm{p}\in\mathbb{P}} \bar{\sigma}(A_\mathrm{d}(\mathrm{p})) \leq 1. \tag{6.49}$$

This means, that in the multi-step case, preservation of frozen stability is not sufficient to imply N-stability. From (6.47) it follows that the N-stability radius reads as

$$\check{\mathrm{T}}_\mathrm{d} = \arg\min_{\mathrm{T_d}\in\mathbb{R}_0^+} \left| \max_{\mathrm{p}_0,\ldots,\mathrm{p}_{n-1}\in\mathbb{P}} \bar{\lambda}\left(\check{R}_{\mathrm{p}_0,\ldots,\mathrm{p}_{n-1}}(\xi,\mathrm{T_d})\right) - 1 \right|, \tag{6.50}$$

which is a too complicated expression to be further analyzed. However in practice, it can be solved in an approximative manner based on gridding and bisection based search.

6.4.3 *Guaranteeing a Desired Level of Discretization Error*

In the previous part we have investigated the numerical properties of the introduced discretion methods and derived criteria on $\mathrm{T_d}$ in order to guarantee the preservation of frozen stability of the original CT system. However, the appropriate choice of $\mathrm{T_d}$ to arrive at a specific performance in terms of the discretization error is also important from a practical point of view. By utilizing the LUT error expressions developed in Sect. 6.4.1, in this section upperbounds of $\mathrm{T_d}$ are derived that guarantee a certain bound on the LUT error in terms of a chosen norm $\|\bullet\|$. Then it is investigated, how such expressions can be used to achieve a level of global discretization error.

As a first step, we formulate the concept of desired performance in terms of the LUT error. For a given continuous-time full behavior $\mathfrak{B}_{\mathrm{SS}}$, which is approximated via a discretization method, define

$$\varepsilon_* := \sup_{x\in\mathfrak{B}_{\mathbb{X}}} \sup_{k\in\mathbb{Z}} \|\varepsilon_k\| \tag{6.51}$$

as the maximal LUT error in a given, arbitrary norm $\|\bullet\|$. Note that his quantity describes the maximum of the truncation error with respect to all possible state trajectories of $\mathfrak{B}_{\mathrm{SS}}$. Also introduce

$$M_{\mathrm{x}}^{\max} := \sup_{x\in\mathfrak{B}_{\mathbb{X}}} \max_{t\in\mathbb{R}} \|x(t)\| = \max_{\mathrm{x}\in\mathbb{X}} \|\mathrm{x}\|, \tag{6.52}$$

as the maximum "amplitude" of the state signal for any u and p in $\mathfrak{B}_{\mathrm{SS}}$. Denote $\varepsilon_{\max}$ as the maximal acceptable relative local error of the discretization in terms of percentage. Then a $\mathrm{T_d} \in \mathbb{R}^+$ is searched for, that satisfies

$$\varepsilon_* \leq \frac{\varepsilon_{\max} M_{\mathrm{x}}^{\max}}{100 \cdot \mathrm{T_d}}. \tag{6.53}$$

Here $1/\mathrm{T_d}$ is introduced on the right side of (6.53) as ε_k is scaled by $\mathrm{T_d}$ (see (6.24)). Next, we formulate upperbounds of $\mathrm{T_d}$ with respect to each method, such that (6.53) is satisfied for the desired $\varepsilon_{\max}$ percentage. To derive these criteria, (6.52) must be bounded, i.e. $\mathbb{X}$ must be confined in a ball of $\mathbb{R}^{n_{\mathbb{X}}}$. Such an assumption is not unrealistic in case of global asymptotic stability of S and bounded $\mathbb{P}$ and $\mathbb{U}$.

6.4.3.1 Performance Criterion for the Rectangular Method

Based on (6.26), it holds in the rectangular case that

$$\varepsilon_* = \sup_{x\in\mathfrak{B}_{\mathbb{X}}} \sup_{\tau\in\mathbb{R}} \frac{\mathrm{T_d}}{2} \left\| \left(\frac{\mathrm{d}^2}{\mathrm{d}t^2} x\right)(\tau) \right\|. \tag{6.54}$$

By using the sensitivity constant $M^{(1)}$ (see (6.28)), inequality (6.53) holds for any $0 \leq \mathrm{T_d} \leq \hat{\mathrm{T}}_\mathrm{d}$ where

$$\hat{\mathrm{T}}_\mathrm{d} = \sqrt{2\frac{\varepsilon_{\max} M_{\mathrm{x}}^{\max}}{100 \cdot M^{(1)}}}. \tag{6.55}$$

Criterion (6.55) gives an upperbound estimate of the required $\mathrm{T_d}$, that achieves $\varepsilon_{\max}$ percentage local discretization error of the state variable of the approximated representation in terms of a chosen norm $\|\bullet\|$.

6.4.3.2 Performance Criteria of other Approximative Methods

Similar criteria can be developed for the other methods by using the LUT error expressions of Table 6.1 and the higher-order sensitivity constants $M^{(n)}$. These upper-bounds are presented in Table 6.1. Note that similar expressions can also be worked out with respect to the approximation error of the output trajectories.

6.4.3.3 Guaranteeing Bounds on the Global Error

In practical situations, one may be concerned about the maximum relative global error as a performance measure. Define

$$\eta_* := \sup_{x\in\mathfrak{B}_\mathbb{X}} \sup_{k\in\mathbb{Z}} \|\eta_k\| \tag{6.56}$$

as the maximum global error (see (6.23) for the definition of η_k). Also define $\eta_{\max}$ as the maximal acceptable relative global error of the discretization in terms of percentage. Then one would like to choose $\mathrm{T_d}$, such that the global error η_k satisfies

$$\eta_* \leq \frac{\eta_{\max} M_\mathbb{X}^{\max}}{100}. \tag{6.57}$$

Unfortunately, characterization of η_* for the introduced discretization methods requires the introduction of serious restrictions of the considered CT behaviors. However, in case of $\mathrm{T_d} \leq \check{\mathrm{T}}_\mathrm{d}$ i.e. N-stability, $\varepsilon_{\max}$ can be often used as a good approximation of $\eta_{\max}$. Therefore, the performance bound $\hat{\mathrm{T}}_\mathrm{d}$ can be used to approximate/guarantee a global error bound as well.

6.4.4 *Switching Effects*

In the previous part, the effect of neglecting the switching phenomena of the ZOH actuation has not been considered. Here we investigate the case when the signals u and p described by (6.2a–b) are applied to $\mathfrak{R}_{\mathrm{SS}}(\mathcal{S})$. First we show the effect of these discontinuous signals on the state evolution of $\mathfrak{R}_{\mathrm{SS}}(\mathcal{S})$ inside a sample interval and the error that results by neglecting these terms. Then we motivate why this phenomenon is negligible in practical situations.

Consider the ODE corresponding to (3.44a) in the kth sample interval. By using the bilateral Laplace transform of this differential equation with reference time $t_0 = k\mathrm{T_d}$ and assuming that the dependence on p is commutative under addition[3], it follows that for a fixed k:

$$\begin{aligned} sX(s) = x_\mathrm{d}(k) + \left[\tfrac{(A\diamond p_\mathrm{d})(k)+(s-1)(A\diamond p_\mathrm{d})(k-1)}{s}\right] X(s) + \tfrac{(B\diamond p_\mathrm{d})(k)}{s} u_\mathrm{d}(k) + \\ + \left[\tfrac{(B\diamond p_\mathrm{d})(k)+(s-1)(B\diamond p_\mathrm{d})(k-1)}{s}\right] u_\mathrm{d}(k-1). \end{aligned} \tag{6.58}$$

[3] Without this assumption, the formulation of the Laplace transform becomes complicated, but the core problem that results in the general case is illustrated well by (6.58).

Table 6.2 Properties of the derived discretization methods in terms of: (a) consistency/convergence; (b) preservation of stability/N-stability; (c) preservation of instability; (d) existence; (e) complexity; (f) preservation of linear dependence; (g) computational load; (h) system order.

Prop.	Complete	Rectangular	n^{th}-polynomial	Trapezoidal	Adams-Bashforth
(a)	always	1st-order	n^{th}-order	2nd-order	3rd-order
(b)	always global	frozen with $\breve{\mathrm{T}}_{\mathrm{d}}$	frozen with $\breve{\mathrm{T}}_{\mathrm{d}}$	always frozen	frozen with $\breve{\mathrm{T}}_{\mathrm{d}}$
(c)	+	-	-	+	-
(d)	always	always	always	conditional	always
(e)	exponential	linear	polynomial	rational	linear
(f)	-	+	-	-	+
(g)	high	low	moderate	high	low
(h)	preserved	preserved	preserved	preserved	increased

where $X(s)$ is the Laplace transform of the solution of the ODE (the behavior of the state in the kth sample interval). It is immediate that in the given sample interval, (6.58) does not correspond to (6.8a). (6.58) has a dynamic dependence, and it is not realizable as a LPV-SS representation directly without associating $\mathrm{q}^{-1}u_{\mathrm{d}}$ with a new state-variable. In this way, it becomes clear that neglecting the switching effects introduces discretization errors in the LPV case, which can be even more significant if T_{d} is decreased (more discontinuous switches in the dynamics). On the other hand, it is true that the discontinuous phenomenon which is described by (6.58) never happens in reality. One reason is that usually p is not actuated by ZOH and it changes smoothly/relatively slowly with respect to the actual dynamics of the plant. Additionally, ZOH actuation has a transient in practice as the underlying physical device needs to build up the new signal value, preventing sudden changes of the signals. In conclusion, for the considered class of LPV representations, the introduced discretization methods of this section give no step-invariant discretization in the ZOH setting (meaning equivalence even in case of switching effects), however they are well applicable methods for practical use. It is important to note that derivation of LPV discretization methods with step-invariant property is also possible, however the resulting discretization approaches are technical and their actual performance gain compared to the previously developed approaches is negligible in practice.

6.5 Properties of the Discretization Approaches

Beside stability and discretization error characteristics, there are other properties of the derived discretization methods which could assist or hinder further use of the resulting DT model. With the previously obtained results, these vital properties are summarized in Table 6.2. From this table it is apparent that the complete method gives errorless conversion at the price of heavy nonlinear dependence of the DT model on p_{d}. As in LPV control synthesis low complexity of the p-dependence is assumed (like linear, polynomial, or rational functions, see [160]), both for modeling and controller discretization purposes - beside the preservation of stability - the preservation of linear dependence over the scheduling is preferred. This favors approximative methods that give acceptable performance, but with less complexity of the new coefficient dependence on the scheduling. Complicated coefficient functions, like inversion or matrix exponential, also results in a serious increase of the computation time, which gives a preference towards the linear methods like the rectangular or the Adams-Bashforth approach. In the latter case, the order increase of the DT representation requires extra memory storage or more complicated controller design depending on the intended use. If the quality of the DT model has priority, then the trapezoidal and the polynomial methods are suggested due to their fast convergence and large stability radius. In terms of identification, linear dependence of the suggested model structures is also important as it simplifies parametrization.

6.6 Discretization and Dynamic Dependence

So far, the discretization of LPV-SS representations with static dependence has been considered in a ZOH setting. It has been already discussed that using the ZOH setting for the discretization of representations with dynamic dependence may result in the loss of significant parts of the original behavior. These parts, which are associated with the dynamic nature of the coefficient dependencies, are lost because in each sample interval the derivatives of p are assumed to be zero. In this way, dynamic dependence of the original coefficients simplifies to a static dependence.

To show this phenomenon, consider the case when $A \diamond p = rp\frac{\mathrm{d}}{\mathrm{d}t}p$ with $r \in \mathbb{R}$ and $\mathbb{P} = \mathbb{R}$. Then in the ZOH setting, the following holds in each sample interval:

$$(A \diamond p)(t) = \begin{cases} 0, & \text{if } t \neq k\mathrm{T_d},\ k \in \mathbb{Z}; \\ \pm\infty, & \text{if } t = k\mathrm{T_d},\ k \in \mathbb{Z}; \end{cases} \tag{6.59}$$

If the switching effect is neglected, then A is approximated in DT as an identity matrix by all of the introduced discretization methods. Thus, the original behavior of the CT representation is lost due to the ZOH setting. However in practice, one would try to use the approximation

$$\frac{\mathrm{d}}{\mathrm{d}t}p(t) \approx \frac{p((k+1)\mathrm{T_d}) - p(k\mathrm{T_d})}{\mathrm{T_d}}, \tag{6.60}$$

for each $t \in [k\mathrm{T_d}, (k+1)\mathrm{T_d})$. In fact, (6.60) means that p is assumed to be a linear function in the sample interval. Then, using this assumption, a better DT approximation of the original CT representation can be derived. This shows that in case of dynamic dependence, the ZOH assumption on p is not appropriate and instead of that, a first or higher order hold discretization setting is necessary for p.

Based on the previous example, consider the case when (u,y) are assumed to satisfy the ZOH setting, but p varies linearly in $t \in [k\mathrm{T_d}, (k+1)\mathrm{T_d})$:

$$p(t) = \underbrace{\frac{p_{\rm d}(k+1) - p_{\rm d}(k)}{{\rm T_d}}}_{{\rm p}_{1k}}(t - k{\rm T_d}) + p_{\rm d}(k). \tag{6.61}$$

This assumption on the scheduling is called the *first-order hold* setting. Additionally, define ${\rm p}_{0k} = (k+1)p_{\rm d}(k) - kp_{\rm d}(k+1)$. Note that, $p(t) = {\rm p}_{1k}t + {\rm p}_{0k}$ for $t \in [k{\rm T_d}, (k+1){\rm T_d}))$. Let $\mathfrak{R}_{\rm SS}(\mathcal{S})$ be a continuous-time LPV-SS representation and consider it in the above defined first-order hold setting. In case the matrices of $\mathfrak{R}_{\rm SS}(\mathcal{S})$ are dependent on p and $\frac{\rm d}{{\rm d}t}p$, like $A \diamond p = A(p, \frac{\rm d}{{\rm d}t}p)$ (dynamic dependence), then the state-evolution in the $k^{\rm th}$ sampling interval satisfies:

$$\frac{\rm d}{{\rm d}t}x(t) = A({\rm p}_{1k}t + {\rm p}_{0k}, {\rm p}_{1k})x(t) + B({\rm p}_{1k}t + {\rm p}_{0k}, {\rm p}_{1k})u_{\rm d}(k). \tag{6.62}$$

The solution of this ODE can be obtained in the time interval $t \in [k{\rm T_d}, (k+1){\rm T_d}))$ for particular meromorphic functions A and B. Similar to the complete method of the ZOH setting in Sect. 6.3, this analytic solution results in a complete type of discretization of the continuous-time LPV-SS representation. The resulting DT counterpart has also dynamic dependence on $p_{\rm d}$ and its time-shifted versions, and yields a better approximation of the CT representation than what would result in a pure ZOH setting. This suggests that for the discretization of LPV representations with dynamic dependence, the order of the hold setting with respect to p should be greater or equal than the maximal order of derivatives in the coefficient dependencies. With some trivial modifications, the approximative methods treated in this paper, except the trapezoidal method, can be extended to this hybrid higher-order hold case. Unfortunately, for the extended approaches, the deduced formulas for the approximation error and the step-size bounds do not apply. Solving discretization of LPV representations with dynamic dependence in a general sense and giving compact formulas of discretization remains the objective of further research.

6.7 Numerical Example

In the following a simple example is presented to visualize/compare the properties of the analyzed discretization methods and the performance of the sample-bound criteria. Consider the following state-space representation of a continuous-time SISO LPV system $\mathcal{S}$ with IO partition (u, y):

$$\mathfrak{R}_{\rm SS}(\mathcal{S}) = \left[\begin{array}{c|c} A \diamond p & B \diamond p \\ \hline C \diamond p & D \diamond p \end{array}\right] = \left[\begin{array}{cc|c} 19.98p - 20 & 202 - 182p & 1+p \\ 45p - 50 & 0 & 1+p \\ \hline 1+p & 1+p & \frac{1+p}{10} \end{array}\right]$$

where $\mathbb{P} = [-1, 1]$. The above representation has static linear dependence on the scheduling signal p. Furthermore, for a constant scheduling $p(t) = {\rm p}$ for all $t \in \mathbb{R}$, $\mathfrak{R}_{\rm SS}(\mathcal{S})$ is equivalent with an LTI representation that has poles

Table 6.3 Discretization error of $\mathcal{S}$, given in terms of the achieved average MSE of y_d and $\hat{\eta}_\mathrm{max} = 100 \cdot \eta_* / M_\mathrm{x}^\mathrm{max}$ (relative worst-case η_k) for 100 simulations. $^{(*)}$ indicates unstable projection to the discrete domain.

MSE of y_d

$\mathrm{T_d}$	Complete	Rectangular	2nd-polynom.	Trapezoidal	Adams-Bash.
$2 \cdot 10^{-2}$, (50Hz)	$1.68 \cdot 10^{-10}$	(*)	(*)	$1.97 \cdot 10^{-3}$	(*)
$5 \cdot 10^{-3}$, (0.2kHz)	$1.69 \cdot 10^{-10}$	(*)	$4.70 \cdot 10^{-4}$	$3.81 \cdot 10^{-5}$	$2.14 \cdot 10^{-1}$
10^{-4}, (10kHz)	$1.68 \cdot 10^{-10}$	$2.27 \cdot 10^{-6}$	$1.05 \cdot 10^{-10}$	$1.53 \cdot 10^{-8}$	$1.60 \cdot 10^{-8}$

$\hat{\eta}_\mathrm{max}$ of $\hat{x}_\mathrm{d}$

$\mathrm{T_d}$	Complete	Rectangular	2nd-polynom.	Trapezoidal	Adams-Bash.
$2 \cdot 10^{-2}$, (50Hz)	0.053%	(*)	(*)	106.12%	(*)
$5 \cdot 10^{-3}$, (0.2kHz)	0.060%	(*)	40.31%	8.02%	665.94%
10^{-4}, (10kHz)	0.063%	2.62%	0.06%	0.19%	0.76%

$$\sigma(A(\mathrm{p})) = 9.99\mathrm{p} - 10 \pm \mathrm{i}\sqrt{10^4 - 17990.2\mathrm{p} + 8090.2\mathrm{p}^2}. \tag{6.63}$$

From (6.63), it is obvious that $\mathcal{S}$ is uniformly frozen stable on $\mathbb{P}$.

Assume that $\mathcal{S}$ is in a ZOH setting with sampling rate $\mathrm{T_d} = 0.02$. By applying the discretization methods of Sect. 6.2, approximative discrete-time representations of $\mathcal{S}$ have been calculated. In order to show the performance of the investigated discretization methods, the output of the original and its discrete approximations have been simulated on the $[0,1]$ time interval for zero initial conditions and for 100 different realizations of white u_d and p_d with uniform distribution $\mathcal{U}(-1,1)$. For fair comparison, the achieved MSE of the resulting output signals $\hat{y}_\mathrm{d}$ has been calculated with respect to the output y of $\mathfrak{R}_\mathrm{SS}(\mathcal{S})$ and presented in Table 6.3. Beside the MSE of the output evolution, the relative worst-case maximum global error $\hat{\eta}_\mathrm{max} = 100 \cdot \eta_* / M_\mathrm{x}^\mathrm{max}$ of the DT state-signals $\hat{x}_\mathrm{d}$ associated with the discrete-time SS representations has also been computed with respect to the state signal x of $\mathfrak{R}_\mathrm{SS}(\mathcal{S})$ and presented in Table 6.3. From these error measures it is immediate that, except for the complete and the trapezoidal method, all approximations diverge. As expected, the error of the complete method is extremely small and the trapezoidal method gives a moderate, but acceptable performance. Note that the response of the original CT $\mathfrak{R}_\mathrm{SS}(\mathcal{S})$ has been calculated via a 5th-order Runge-Kutta numerical approximation (see [8]) with step size 10^{-8}. This implies that the switching effect of the ZOH actuation does not show up in the calculated response.

As a second step, we calculate sampling bounds $\check{\mathrm{T}}_\mathrm{d}$ and $\hat{\mathrm{T}}_\mathrm{d}$ by choosing the Euclidian norm as an error measure and $\varepsilon_\mathrm{max} = 1\%$, with the intention to achieve $\eta_\mathrm{max} = 1\%$. The calculated sampling bounds are presented in Table 6.4. During the calculation of $\hat{\mathrm{T}}_\mathrm{d}$ it has been assumed that $\mathbb{X} = [-0.1, 0.1]^2$, which has been

Table 6.4 Stability ($\check{T}_d$) and performance ($\hat{T}_d$) bounds provided by the criterion functions of Table 6.1. The results here are presented in terms of the Euclidian norm and $\varepsilon_{max} = 1\%$.

	Rectangular	2nd-polynom.	Trapezoidal	Adams-Bash.
$\check{T}_d$	$2 \cdot 10^{-4}$, (5kHz)	$5.60 \cdot 10^{-3}$, (0.2kHz)	∞	$1.77 \cdot 10^{-3}$, (0.6kHz)
$\hat{T}_d$	$6.87 \cdot 10^{-5}$, (15kHz)	$1.73 \cdot 10^{-3}$, (0.6kHz)	$1.28 \cdot 10^{-3}$, (0.8kHz)	$1.21 \cdot 10^{-3}$, (0.8kHz)

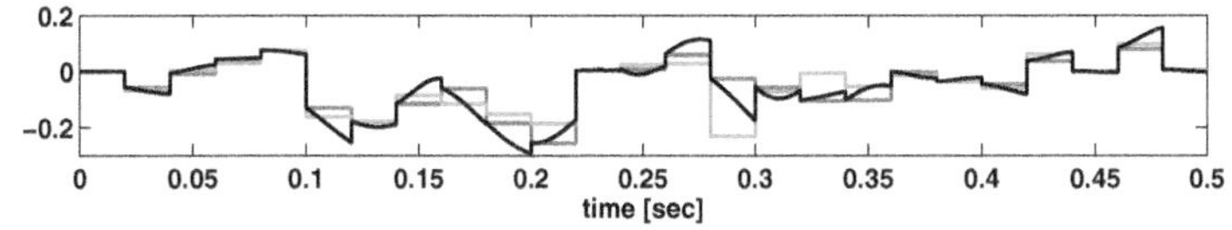

(a) $T_d = 0.02$, complete (grey), trapezoidal (light grey)

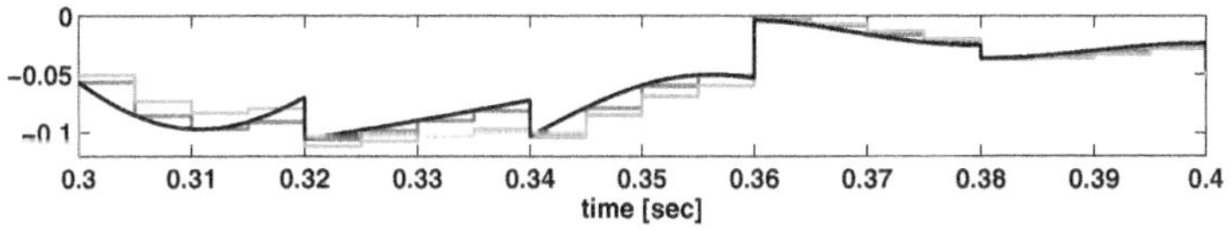

(b) $T_d = 0.005$, complete (grey), trapezoidal (light grey)

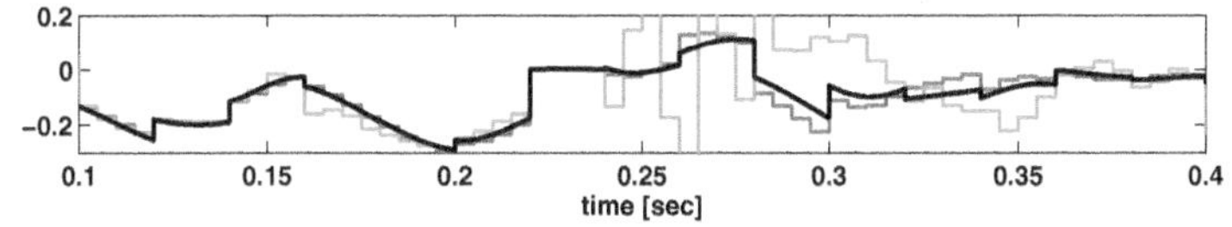

(c) $T_d = 0.005$, 2nd order polynomial (grey), Adams-Bashforth (light grey)

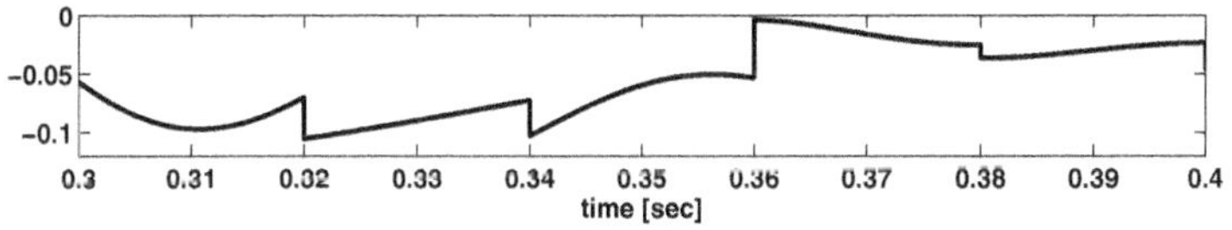

(d) $T_d = 10^{-4}$, complete (grey), trapezoidal (light grey)

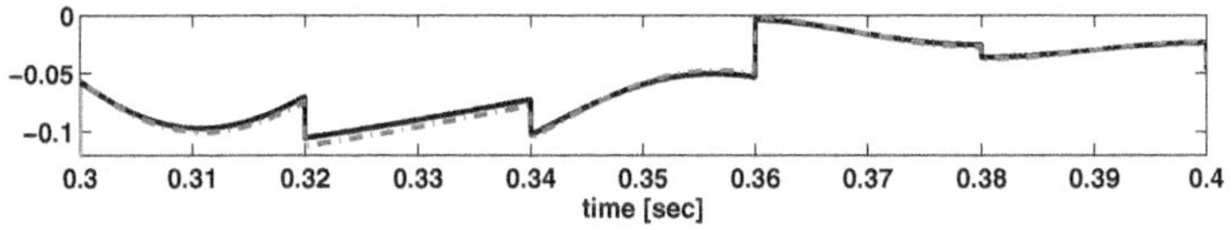

(e) $T_d = 10^{-4}$, 2nd order polynomial (grey), Adams-Bashforth (light grey), rectangular (dash-dotted grey)

Fig. 6.6 Output signal y of $\mathfrak{R}_{SS}(\mathcal{S})$ (black) in a ZOH setting with $T_d = 0.02$ and its discrete-time approximations with different sampling-times.

verified by several simulations of $\mathfrak{R}_{SS}(\mathcal{S})$ based on $u_d, p_d \in \mathcal{U}(-1,1)$. By these results, the rectangular method needs a fast sampling rate to achieve a stable projection and even a faster sampling to obtain the required performance. The 2nd-order polynomial projection has significantly better bounds due to the 2nd-order accuracy of this method. For the trapezoidal case, the existence of the transformation is always guaranteed because $\mathfrak{R}_{SS}(\mathcal{S})$ is uniformly frozen stable. For comparison, the bound (6.45) proposed in [4], would have resulted in $\check{T}_d = 0.2$.

Now we use the derived bounds to choose a new T_d for the calculation of the discrete projections. As the $\check{T}_d$ bounds of Table 6.4 represent the boundary of stability, therefore $T_d < \check{T}_d$ is used as a new sampling-time in each case. Discretization of $\mathfrak{R}_{SS}(\mathcal{S})$ with $T_d = 0.005$, almost the stability bound of the polynomial method, provides the simulation results given in the second row of Table 6.3. The rectangular method again results in an unstable projection, while the Adams-Bashforth method is on the brink of instability due to frozen instability of A_d for some $p \in \mathbb{P}$. The polynomial method gives a stable, convergent approximation, in accordance with its $\check{T}_d$ bound. The trapezoidal method also improves significantly in performance. The achieved $\hat{\eta}_{max}$ of each approximative method is above the aimed 1% which is in accordance with their $\hat{T}_d$.

As a next step, discretizations of $\mathfrak{R}_{SS}(\mathcal{S})$ with $T_d = 10^{-4}$, the half of the $\check{T}_d$ bound of the rectangular method, are calculated. The results are given in the third row of Table 6.3. Finally, the rectangular method converges and also the approximation capabilities of the other methods improve. By looking at the achieved $\hat{\eta}_{max}$, all the methods, except the rectangular, obtain the aimed 1% error performance which is in accordance with their $\hat{T}_d$ bound, while in the rectangular case the achieved $\hat{\eta}_{max}$ is larger than 1% as 10^{-4} is larger than its $\hat{T}_d$ bound. An interesting phenomenon is that the approximation error of the complete method is non-zero and it is slightly increasing by lowering the sampling-time. This increasing approximation error is due to numerical errors of the digital computation. However, the resulting approximation error is significantly less than the step size of the numerical approximation used for the simulation of $\mathfrak{R}_{SS}(\mathcal{S})$, thus it can be considered zero.

6.8 Conclusions

In this chapter, discretization of LPV state-space representations has been investigated in a zero-order hold setting, where the continuous-time input and scheduling trajectories are restricted to be piecewise constant. It has been shown that the ZOH setting provides an adequate discretization concept for SS representations with static dependence. Extending the approaches of the LTI discretization theory, both exact and approximative methods have been developed in the ZOH setting for the discretization of LPV-SS representations. These approaches have been developed to provide tools that assist DT model structure selection based on first-principle CT models. Using the results of the numerical analysis field, the introduced methods have been investigated in terms of numerical consistency, convergence, and stability. Criteria have been developed for the choice of sampling-time that guarantees a user defined maximum of the truncation error and also the preservation of the frozen stability characteristics of the original system. As a next step we will explore how can we arrive at a CT-LPV form of general nonlinear systems, which through the discretization approaches developed in this chapter can be used to assist model structure selection.

Chapter 7
LPV Modeling of Physical Systems

Abstract. With the motivation to provide tools for model-structure selection in LPV identification, in this section modeling of physical systems described by nonlinear differential equations is studied in the LPV framework. It is investigated how such a nonlinear system can be realized or approximated by a LPV system, giving models of the original behavior in terms of LPV representations. First an overview is presented of the available LPV modeling methods. Then, an algorithmic approach is introduced that ensures errorless conversion of nonlinear differential equations into LPV kernel representations. The approach explores adequate choices of the scheduling variable. Based on equivalent sate-space, input-output or orthonormal basis functions based representation of the resulting kernel form, an adequate choice of a model structure can be obtained for the LPV identification of the physical system.

7.1 Towards Model Structure Selection

A crucial ingredient of any system identification procedure is the choice of model structure, that describes the model set in which the optimal candidate is to be found. If this structure is well-founded with respect to the system to be identified, then all other ingredients, like experiment design, criterion selection, estimation, etc. can contribute successfully to the validity of the end result. Commonly, a poor choice of the model set directly results in a poor estimate of the system.

LPV models have considerably more freedom in parametrization than the class of LTI models. This is due to the presence of functional and often dynamic dependence of the model parameters on the scheduling. Thus LPV model-structure selection is even a more sensitive question than in the LTI case. Furthermore, due to the lack of a general LPV validation theory, mis-modeled dynamics only show up in the performance loss of the designed controllers.

In LPV identification, initiatives has only recently appeared to assist model-structure selection based on measured data [75, 200]. Commonly, the sole source of information available for this decision is in the form of first principle laws or

R. Tóth: Modeling and Identification of Linear Parameter-Varying Systems, LNCIS 403, pp. 171–195.
springerlink.com

expert's knowledge. Such knowledge is mostly presented in terms of nonlinear differential equations, which must be transformed to or approximated by an LPV representation to facilitate model structure selection in this setting. In this chapter, the concept of LPV modeling of nonlinear dynamical systems is investigated. The intention is to give a practical way of using first principle laws and knowledge in the decision process of LPV model structures, appropriate for the identification of the underlying system. Such a process is inevitable for successful identification in the general LPV framework, as it interprets key information about the order and type of functional dynamical dependencies on the scheduling signal and even about which signal(s) can be used for scheduling purposes.

7.2 General Questions of LPV Modeling

Before the first papers on LPV identification methods appeared, modeling of physical systems in a LPV form was dominated by the gain-scheduling principle (see Chap. 1). Motivated by the early approaches of LPV control, the available *nonlinear* (NL) description of the system was linearized in several operating/equilibrium points resulting in a collection of local LTI descriptions of the plant. These "local " descriptions were interpolated to obtain a global approximation of the physical system on the entire operation regime. However, unknown coefficients/relations of the NL model still had to be estimated beforehand. This concept has resulted in many linearization-based LPV modeling approaches trying to approximate the NL system in a LPV *state-space* (SS) form.

For direct approximation of the NL system with an LPV model, LPV identification methods has appeared soon (see Chap. 1). However, these approaches only aim at the identification of an underlying "true LPV system" with completely known structural information. It has been seemingly forgotten that the true aim has always been the approximation of the NL physical system. Additionally, many of these approaches has followed the gain-scheduling principle by identifying local LTI models and interpolating them.

Due to higher performance demands and the still existing gap between LPV identification methods and practical applications, a new generation of LPV modeling approaches have appeared, formulating LPV modeling from a different perspective than gain-scheduling. These approaches aim at the transformation of the original NL representation into a particular LPV form by using substitution or other mathematical manipulations. However, the resulting methods are only able to handle certain sub-classes of NL systems. So the natural question, what kind of systems can the LPV framework describe accurately, has remained unanswered.

Till today, both the control and identification literature on LPV systems typically takes the existence of the plant in a LPV form as a starting point. It is commonly not pointed out how the underlying nonlinear system is transformed to this LPV form. On the other hand, the available LPV modeling approaches are focusing only on particular subclasses of LPV representations. The question whether a significant loss of generality is introduced by the used assumptions usually remains uninvestigated. This shows that LPV modeling of NL systems deserves much more attention and research in order to understand what can be represented by LPV systems and how the best description for a given NL system can be found in the LPV system class. This is why we give in this section an overview on the available approaches, comparing and evaluating their modeling concepts.

To set the stage for the upcoming discussion we state the following questions that are intrinsical for the analysis or development of an LPV description with respect to a physical system:

- The scheduling variable, that governs the dynamics of LPV models, has a crucial role in the validity or in the approximation quality of the LPV description. Thus, in the process of formulating an LPV model, one of the most important questions is which variables of the original system can be selected as the scheduling variable in order to obtain an equivalent LPV description.
- Another question is whether LPV modeling needs to be formulated in terms of systems or in terms of particular representations. The latter concept is commonly used to consider LPV modeling as an approximation of a given state-space equation, which is a subjective description of the NL system. On the other hand, the former concept focuses on the approximation/description of the original behavior. Thus formulating modeling on the level of systems provides more freedom and focuses on the "natural aim" of a modeling problem.

These general considerations are the guidelines by which we explore in the following how a a reliable model-transformation approach can be formulated to support both identification and control design in the LPV framework.

7.3 Modeling of Nonlinear Systems in the LPV Framework

In the following, an overview is presented about the state-of-the-art methods of approximation/equivalent realization of NL systems by an LPV form. Our intention is to give a general picture about the difficulties and the available solutions of this task. To do so, first we define the class of NL systems we consider. These nonlinear systems characterize the first-principle laws of the behavior that we want to capture in the LPV framework. Then we define a particular class of LPV systems, the so-called quasi-LPV systems, that commonly result in the presented modeling approaches. As a next step, we give a SS representation of the introduced NL systems. This representation is often the starting point of the existing LPV modeling methods. We also motivate that using such a representation as a starting point implies the assumption of prior chosen state variables and IO partition, which might restrict the generality of the resulting LPV description. After this, we systematically present the available approaches, sorted into categories.

7.3.1 First Principle Models

As first principle laws are commonly (exclusively) available in continuous-time, we restrict our attention in the sequel to this domain. We consider real, finite dimensional, continuous-time NL systems in the following form:

Definition 7.1 (Nonlinear dynamic systems). A dynamical system $\mathcal{G} = (\mathbb{R}, \mathbb{W}, \mathfrak{B})$ is called a nonlinear, continuous-time, dynamical system denoted as $\mathcal{G}_{\mathrm{NL}}$, if $\mathbb{W} = \mathbb{R}^{n_{\mathrm{W}}}$, $\mathfrak{B} \subseteq \mathbb{W}^{\mathbb{T}}$, and there exists a meromorphic (nonlinear) function $f : \mathbb{R}^{(n+1)n_{\mathrm{W}}} \to \mathbb{R}^{n_{\mathrm{r}}}$, such that

$$\mathfrak{B} = \left\{ w \in \mathcal{L}_1^{\mathrm{loc}}(\mathbb{R}, \mathbb{W}) \;\middle|\; f\left(w, \frac{\mathrm{d}}{\mathrm{d}t} w, \frac{\mathrm{d}^2}{\mathrm{d}t^2} w, \ldots, \frac{\mathrm{d}^n}{\mathrm{d}t^n} w \right) = 0 \text{ holds weakly} \right\}.$$

□

Weak solutions of

$$f\left(w, \frac{\mathrm{d}}{\mathrm{d}t} w, \frac{\mathrm{d}^2}{\mathrm{d}t^2} w, \ldots, \frac{\mathrm{d}^n}{\mathrm{d}t^n} w \right) = 0, \tag{7.1}$$

are defined in terms of distributions. Definition 7.1 is wide enough to encompass of many physical systems and represent their dynamic behavior, however it does not describe every system which is non-LTI. This class of systems is considered with the main purpose to illustrate the problem of transformation of first-principle laws to LPV representations. In the sequel we consider the problem of finding an equivalent or a well approximating LPV system with respect to a given $\mathcal{G}_{\mathrm{NL}}$. Without going into details, for the considered class of NL systems we can also define the set of free variables and IO partitions, similarly as in the LTI case. During transformation of these systems to an LPV form, such free signals are the prime candidates for scheduling or input signals depending on their role in the NL relationship (7.1).

In the LPV framework, models commonly originate from nonlinear dynamic system representations via the gain-scheduling approach. However, linearization of $\mathcal{G}_{\mathrm{NL}}$ at different operating points in $\mathbb{W}$ and then interpolating the resulting LTI models by an operating point dependent function implies that the scheduling signal p of the obtained LPV description is dependent on w. In this way, the fundamental assumption of the LPV framework, namely the property of freedom for p does not hold in this case. Therefore, these descriptions are often referred to as *quasi-LPV* systems (see [102]). In the developed LPV framework, we define these systems as follows:

Definition 7.2 (Quasi-LPV system). A PV dynamical system $\mathcal{S} = (\mathbb{T}, \mathbb{P}, \mathbb{W}, \mathfrak{B})$ with signals w and scheduling variable p is called quasi-LPV, if it satisfies Definition 3.3 without p being a free variable, i.e. $\mathfrak{B}_p$ is not a linear subspace of $\mathbb{W}^{\mathbb{T}}$. □

Note that if some components of p are free signals, then the system is still considered to be quasi-LPV. During the dawn of the gain-scheduling era, LPV systems have been defined with the concept of exogenous/external and thus free scheduling signal (see [165]), opening the possibility of the later development of popular and theoretically well-founded optimal control solutions for such systems. However in the practical application, this assumption has been commonly neglected, treating LPV models with non-free scheduling signal as if their scheduling signal would be a free variable of the system [17]. Even if such an assumption introduces conservatism into the model, and thus the control design applied to it, this masking of the dynamical relation made reliable control of many heavily nonlinear processes possible. In the upcoming analysis, we also intend to follow this tradition, by seeking out ways of transformation of NL systems into a quasi-LPV or a true LPV form.

The class of NL systems of Definition 7.1 is often found to be too broad in the literature. Commonly a subclass of state-space realizations is investigated with a prior selection of the state variable x and the IO partition (u, y):

$$\frac{\mathrm{d}}{\mathrm{d}t}x = f(x, u), \tag{7.2a}$$

$$y = g(x, u), \tag{7.2b}$$

where f and g are partially differentiable (smooth) functions and $w = \mathrm{Col}(u, y)$. In general, not every (7.1) is realizable in the form of (7.2a–b) due to the loss of possible smoothness constraints [146]. Beside the realization problem, it is even more important that with respect to obtaining an equivalent or a well approximating LPV form of a given $\mathcal{G}_{\mathrm{NL}}$, prior selection of a state variable may severely restrict the search space, i.e. the transformation properties. Choosing a state variable or an IO partition in a priori sense is also not motivated from the viewpoint of first principle laws as in the laws of physics or chemistry there are no dedicated state variables, nor predefined inputs or outputs. There are only system variables that are connected by algebraic and differential equations. Therefore, it must be pointed out that any latent-variable-based representation is just a particular and subjective description of the system. As we will see later, such a state-variable based description is not necessary to arrive at an equivalent LPV description of the considered NL system. Additionally, an important concept for the linearization based methods is the equilibrium point of (7.2a), which is defined as $(\mathtt{x}, \mathtt{u}) \in (\mathbb{X} \times \mathbb{U})$ satisfying:

$$0 = f(\mathtt{x}, \mathtt{u}). \tag{7.3}$$

Note that equilibrium points can be stable or unstable depending on the partial derivatives of f in their neighborhood. Using the previously introduced concepts and notions, the existing LPV modeling approaches fall into the following categories:

7.3.2 Linearization Based Approximation Methods

The family of these methods applies linearization theory on a given SS representation (7.2a–b) of the NL system to obtain local LTI models in a state-space form and then interpolates these models to derive an LPV approximation. Thus, the scheduling of the resulting LPV description is equal to those components of x and u that the linearization is based on. If we introduce selector matrices $S_{\mathrm{x}} \in \mathbb{R}^{n_1 \times n_{\mathbb{X}}}$ and $S_{\mathrm{u}} \in \mathbb{R}^{n_2 \times n_{\mathbb{U}}}$ (see Sect. 4.3.5), which select these components, then we can write that $p = \mathrm{Col}(S_{\mathrm{x}}x, S_{\mathrm{u}}u)$ with $n_{\mathbb{P}} = n_1 + n_2$. The following subcategories of these methods are distinguished:

7.3.2.1 Linearization around a Set of Equilibrium Points

Application of classical linearization theory requires that each linearization corresponds to an equilibrium of (7.2a). Thus it is a common approach in the LPV modeling process to use first-order (Jacobian) linearization of (7.2a–b) at a set of equilibrium points. This approach is followed in many works like [7, 77, 164, 157, 95, 102]. More details about first-order linearization and its delicacies, e.g. when it provides appropriate results can be found in [146]. Based on the linearization concept, an LPV model of (7.2a–b) is formulated as follows:

First, the equilibrium points $\mathrm{p}_i = \mathrm{Col}(\mathrm{x}_i, \mathrm{u}_i) \in (\mathbb{X} \times \mathbb{U})$, $i \in \mathbb{I}_1^n$ of (7.2a) are computed, then

$$A_i = \frac{\partial f}{\partial x}(\mathrm{p}_i), \quad B_i = \frac{\partial f}{\partial u}(\mathrm{p}_i), \quad C_i = \frac{\partial g}{\partial x}(\mathrm{p}_i), \quad D_i = \frac{\partial g}{\partial u}(\mathrm{p}_i), \tag{7.4}$$

are obtained. Around each equilibrium point p_i, the state and output evolution are approximated by applying a first-order Taylor expansion of f and g:

$$\frac{\mathrm{d}}{\mathrm{d}t}x \approx A_i(x - \mathrm{x}_i) + B_i(u - \mathrm{u}_i), \tag{7.5a}$$

$$y \approx C_i(x - \mathrm{x}_i) + D_i(u - \mathrm{u}_i) + g(\mathrm{p}_i). \tag{7.5b}$$

Then (7.5a–b) can be seen as a local LTI model of the system. Define $p = \mathrm{Col}(x, u)$ and $\mathbb{P} = \mathbb{X} \times \mathbb{U}$ with a set of normalized interpolation (scheduling) functions $\mathfrak{g}_i : \mathbb{P} \to [0, 1]$, like radial basis, triangular functions, etc. with $\sum_{i=1}^n \mathfrak{g}_i(\mathrm{p}) = 1$ for all $\mathrm{p} \in \mathbb{P}$ and $\mathfrak{g}_i(\mathrm{p}_i) = 1$ for each $i \in \mathbb{I}_1^n$. Then, the PV model is formulated as

$$\frac{\mathrm{d}}{\mathrm{d}t}\breve{x} = \sum_{i=1}^n \mathfrak{g}_i(p) A_i \breve{x} + \sum_{i=1}^n \mathfrak{g}_i(p) B_i u - \gamma_{\mathrm{x}}(p), \tag{7.6a}$$

$$\breve{y} = \sum_{i=1}^n \mathfrak{g}_i(p) C_i \breve{x} + \sum_{i=1}^n \mathfrak{g}_i(p) D_i u - \gamma_{\mathrm{y}}(p), \tag{7.6b}$$

where $\breve{x}$ and $\breve{y}$ are the approximations of the original x and y and the remainder terms are given as

$$\gamma_{\mathrm{x}}(p) = \sum_{i=1}^n \mathfrak{g}_i(p)(A_i \mathrm{x}_i + B_i \mathrm{u}_i), \tag{7.7a}$$

$$\gamma_{\mathrm{u}}(p) = \sum_{i=1}^n \mathfrak{g}_i(p)(C_i \mathrm{x}_i + D_i \mathrm{u}_i - g(\mathrm{p}_i)). \tag{7.7b}$$

It is obvious that for $p(t) = \mathrm{p}_i$, (7.6a–b) is equivalent with (7.5a–b), i.e. for each equilibrium point the global PV model is equal to the local LTI description. Note that in some cases all the partial derivatives in (7.4) are constants with respect to some elements of x and u. This means that these partial derivatives have the same value for all equilibrium points. This observation implies that the associated elements of x and u can be left out from p, i.e. from the interpolation space to formulate (7.6a–b).

Additionally, it is an important observation that the PV differential equation (7.6a–b) is not an LPV-SS representation as it contains the remainder terms γ_{x} and γ_{u}. These remainder terms can not be eliminated in general due to their time dependent nature. In many cases, the remainder terms are either considered as an additional input/disturbance in (7.6a–b) or eliminated locally from (7.5a–b) by subtracting them from the signal variables (u,y,x), then the locally altered LTI models are interpolated to obtain a global model. While the previous approach leads to a conservative model, the latter method results in an alteration of the state-space as, due to the local transformations, the local states do not have the same meaning any more. This may lead to a complete misfit of the approximation. An exceptional case is when $n=1$, so the system is linearized in only one point. In such case no interpolation is needed, i.e. $\mathfrak{g}_1 = 1$. Then by redefining the state as $\breve{x} = x - \mathrm{x}_1$, the input as $\breve{u} = u - \mathrm{u}_1$, and the output as $\breve{y} = y - g(\mathrm{p}_1)$ an LTI-SS approximation of $\mathcal{G}_{\mathrm{NL}}$ results via (7.6a–b). Proper elimination of the remainder terms is only available in this case.

In the general sense, linearization in the equilibrium points is a serious restriction that may lead to poor transient performance and inability to preserve stability characteristics of (7.2a–b). Therefore, an adequate approximation capability requires the assumption of slowly varying scheduling. However, such an assumption is often unrealistic as x and u are not slowly varying signals. To improve the approximation capabilities of this approach many alternative linearization methods have been considered like higher order series-expansion based linearization [14] or the reformulation of the mean value theorem [31].

Furthermore, interpolation has its own pitfalls as well. If the local LTI models resulting from the linearization are transformed to a canonical form to accomplish interpolation, then the effect of local transformations can completely alter the behavior (see [130] for an example). In this way, the resulting LPV representation may not be able to reproduce the dynamical aspects of the original NL form (see Example 7.1 and [189]). Similar errors result if the interpolation is applied through the transfer functions of the local LTI models to get a mixed scheduling and frequency dependent description and then applying LTI realization theory to obtain the LPV form (see [125] for an example). In [142], the interpolation problem of the local models is formulated in an optimization sense to reduce the complexity of the resulting dependence of the SS matrices on p. As the procedure avoids any local manipulations of the SS matrices, it provides a reliable way for efficient generation of (7.6a–b).

Example 7.1 (Pitfalls of Interpolation). In this example, one of the merits of local transformations is illustrated. Assume that the linearization of the NL system has resulted in two local LTI-SS representations

$$\left[\begin{array}{c|c}\alpha_1 & 1\\ \hline \beta_1 & 0\end{array}\right] \text{ if } \mathrm{p}=0, \qquad \left[\begin{array}{c|c}\alpha_2 & 1\\ \hline \beta_2 & 0\end{array}\right] \text{ if } \mathrm{p}=1,$$

with $\mathbb{P} = [0,1]$. If $\mathcal{S}_1$ represents an LPV approximation of the considered NL system, then a $\mathfrak{R}_{\mathrm{SS}}(\mathcal{S}_1)$ can be formulated as

$$\left[\begin{array}{c|c} A \diamond p & B \diamond p \\ \hline C \diamond p & D \diamond p \end{array}\right] = \sum_{i=1}^{2} \left[\begin{array}{c|c} \mathfrak{g}_i(p)\alpha_i & 0.5 \\ \hline \mathfrak{g}_i(p)\beta_i & 0 \end{array}\right], \tag{7.8}$$

where $\mathfrak{g}_1(p) = 1 - p$ and $\mathfrak{g}_2(p) = p$ are the linear interpolation functions. The two local models can also be interpolated based on their IO representation. By using the same interpolation functions, the resulting $\mathfrak{R}_{\mathrm{IO}}(\mathcal{S}_2)$ is

$$\frac{\mathrm{d}}{\mathrm{d}t} y - (a_0 \diamond p) y = (b_0 \diamond p) u, \tag{7.9}$$

with $a_0 = \sum_{i=1}^{2} \mathfrak{g}_i \alpha_i$ and $b_0 = \sum_{i=1}^{2} \mathfrak{g}_i \beta_i$. However, the IO representation of $\mathfrak{R}_{\mathrm{SS}}(\mathcal{S}_1)$ reads as:

$$\frac{\mathrm{d}}{\mathrm{d}t} y - \left(\left(a_0 + \frac{\dot{b}_0}{b_0} \right) \diamond p \right) y = (b_0 \diamond p) u, \tag{7.10}$$

where $\dot{b}_0 \diamond p = (\beta_2 - \beta_1) \frac{\mathrm{d}}{\mathrm{d}t} p$. One can conclude that $\mathfrak{R}_{\mathrm{SS}}(\mathcal{S}_1)$ and $\mathfrak{R}_{\mathrm{IO}}(\mathcal{S}_2)$ are the representations of two different LPV systems $\mathcal{S}_1$ and $\mathcal{S}_2$ unless $\beta_1 = \beta_2$. This phenomenon clearly emphasizes that interpolation must be carried out in the representation where the linearization was performed, otherwise unexpected alteration of the behavior can occur. □

7.3.2.2 Multiple Linearizations around a Single Equilibrium Point

This approach originates from the Fuzzy control framework, where it is used to obtain linear *Takagi-Sugeno* (TS) dynamic fuzzy models, which under certain restrictions can be viewed as quasi-LPV systems [86]. The basic idea is to linearize the NL-SS representation at multiple points of $\mathbb{X} \times \mathbb{U}$ around a single equilibrium point. Then the resulting local models are interpolated in a similar fashion as in the previous part. The method leads to a LPV model that performs well during transient operation, because the framework allows some of the local LTI models to be associated with transient operating regimes. If the system stays close to the used equilibrium point, no restrictions concerning slowly varying trajectories is needed. However, a principal disadvantage is that this approach is not suited for NL models with multiple equilibria.

7.3.2.3 Linearization along a Nominal Trajectory

This approach was introduced in the early 1990s when LPV controllers were typical scheduled in a open-loop sense based on a chosen reference trajectory (intended operation trajectory of the plant) or fixed auxiliary input variables (typical operation trajectories). These user-defined signals were used to describe a nominal trajectory of system operation. Linearizing the NL model along this nominal signal trajectory gives an LPV model (see [164, 102] for examples of this approach). It is an advantage of this approach that the resulting LPV description can cover transient operation of the plant along the used nominal trajectory, however such a description also suffers from the drawback that the performance may be poor when the system is operating far away from it. As the time-variation of the system is considered along a pre-chosen trajectory of p, the resulting models resemble a LTV system rather than a LPV system.

7.3.2.4 Off-Equilibrium Linearization around a Set of Operating Points

Linearization of the NL system at points in the state space that may not be equilibria has been considered in numerous approaches (see [76, 102, 121, 122]) The benefit of this concept is that the transient (off-equilibrium) dynamics of the LPV approximation may be significantly improved. The LPV model is obtained by selecting a set of linearization points $\mathrm{p}_i = \mathrm{Col}(\mathrm{x}_i, \mathrm{u}_i)$ with $(\mathrm{x}_i, \mathrm{u}_i) \in \mathbb{X} \times \mathbb{U}$, $i \in \mathbb{I}_1^N$ and using Jacobian linearization of the nonlinear functions f and g around these points in the sense of (7.5a–b). The only difference is that in (7.5a) an extra term $f(\mathrm{p}_i)$ appears if p_i is not an equilibrium point. Then the local models are interpolated as described by (7.6a–b) and the remainder terms are considered as disturbances, extra input channels or they are locally eliminated to form a global LPV model. This approach has the same features as the equilibrium-points-based linearization with all the pitfalls of interpolation and local state transformations. An additional problem rises however from the selection of linearization points $\{\mathrm{p}_i\}_{i=1}^n$. Equidistant selection on the space $\mathbb{X} \times \mathbb{U}$ may seem tempting, but it might happen that due to the dynamical changes of $\mathcal{G}_{\mathrm{NL}}$, a non-equidistant selection with dense samples in specific regions of $\mathbb{X} \times \mathbb{U}$ can lead to far better approximations (see [123] for details).

7.3.3 *Multiple Model Design Procedures*

Multiple model design techniques investigate the LPV approximation of the NL system given by the state-space representation

$$\frac{\mathrm{d}}{\mathrm{d}t}x = A(x,u)x + B(x,u)u, \tag{7.11a}$$

$$y = C(x,u)x + D(x,u)u. \tag{7.11b}$$

The LPV modeling of (7.11a–b) is accomplished by selecting a set of interpolation functions $\{\mathfrak{g}_i\}_{i=1}^n$. Then the model approximation problem becomes a search for a set of constant matrices $\{(A_i, B_i, C_i, D_i)\}_{i=1}^n$ such that the nonlinear functions (A, B, C, D) are approximated optimally by the weighted sum of the constant matrices, like

$$A(x,u) \approx \sum_{i=1}^{n} \mathfrak{g}_i(p) A_i, \quad \text{where } p = \mathrm{Col}(x,u). \tag{7.12}$$

However, these kind of techniques are most often considered to be model reduction tools rather than LPV model transformation methods as with the scheduling signal $p = \mathrm{Col}(x,u)$ the SS equations (7.11a–b) already define a quasi-LPV model. Notable approaches that fall into this category are the orthogonal decomposition based methods like [29, 163, 236], the convex polytope methods like [82, 86], and the radial basis functions based optimization techniques of [85, 213].

A unique exception among these methods is [141], where it is only assumed that some samples $\{(\hat{A}_j, \hat{B}_j, \hat{C}_j, \hat{D}_j)\}_{j=1}^{n_s}$ of (A, B, C, D) are available for certain scheduling points $\{\mathsf{p}_j\}_{j=1}^{n_s}$ of $\mathbb{X} \times \mathbb{U}$. Then, such an LPV model with affine or polynomial dependence is searched whose frozen transfer functions at each p_j have the minimal error in the $\mathcal{H}_2$ sense with respect to the transfer functions associated with $\{(\hat{A}_j, \hat{B}_j, \hat{C}_j, \hat{D}_j)\}_{j=1}^{n_s}$. In this way the resulting procedure can even be used to accomplish the interpolation step of various LPV local identification approaches avoiding the local alteration of the state basis common in many SS methods (see Sec. 1.3.4.3).

7.3.4 *Substitution Based Transformation Methods*

Methods of this category use substitution techniques on an available SS representation of the NL system to generate a quasi-LPV model without the need of approximation. The PV coefficients appear as substituted functions of (x, u) and the resulting scheduling is $p = \mathrm{Col}(S_\mathrm{x} x, S_\mathrm{u} u)$, where S_x and S_u are the selector matrices of the components of (x, u) used for the substitution. The subcategories of these approaches are the following:

7.3.4.1 The "State-Transformation" Method

The "state-transformation" method was first introduced in [167] followed by many successful applications in aerospace engineering (see [138, 168, 137]). A class of nonlinear systems that qualifies for this method is described by the following SS equation:

$$\frac{\mathrm{d}}{\mathrm{d}t}\begin{bmatrix} x_1 \\ x_2 \end{bmatrix} = \begin{bmatrix} f_1(x_1) \\ f_2(x_1) \end{bmatrix} + \begin{bmatrix} A_{11}(x_1) & A_{21}(x_1) \\ A_{21}(x_1) & A_{22}(x_1) \end{bmatrix} \begin{bmatrix} x_1 \\ x_2 \end{bmatrix} + \begin{bmatrix} B_1(x_1) \\ B_2(x_1) \end{bmatrix} u, \tag{7.13}$$

where $A_{11}, \ldots, A_{22}$ with B_1, B_2 are (nonlinear) matrix functions. Additionally, f_1, f_2 represent matrix function terms which cannot be written as $f_i(x_1) = \tilde{f}_i(x_1)x_1$ with $\tilde{f}_i$ bounded in the origin. Assume that there exist differentiable functions γ_x and γ_u, such that:

$$\begin{bmatrix} 0 \\ 0 \end{bmatrix} = \begin{bmatrix} f_1(x_1) \\ f_2(x_1) \end{bmatrix} + \begin{bmatrix} A_{11}(x_1) & A_{12}(x_1) \\ A_{21}(x_1) & A_{22}(x_1) \end{bmatrix} \begin{bmatrix} x_1 \\ \gamma_\mathrm{x}(x_1) \end{bmatrix} + \begin{bmatrix} B_1(x_1) \\ B_2(x_1) \end{bmatrix} \gamma_\mathrm{u}(x_1), \tag{7.14}$$

holds for all $x_1 \in \mathcal{L}_1^{\mathrm{loc}}(\mathbb{R}, \mathbb{R}^{n_1})$, which are the solutions of (7.13), i.e. for which there exist signals $(x_2, u) \in \mathcal{L}_1^{\mathrm{loc}}(\mathbb{R}, \mathbb{R}^{n_2} \times \mathbb{U})$ such that (7.13) is satisfied. Subtracting (7.14) from (7.13) with some rearrangement of the signals yields:

$$\frac{\mathrm{d}}{\mathrm{d}t}\begin{bmatrix} x_1 \\ \breve{x}_2 \end{bmatrix} = \begin{bmatrix} 0 & A_{12}(x_1) \\ 0 & A_{22}(x_1) - \frac{\partial \gamma_\mathrm{x}(x_1)}{\partial x_1} A_{12}(x_1) \end{bmatrix} \begin{bmatrix} x_1 \\ \breve{x}_2 \end{bmatrix} + \begin{bmatrix} B_1(x_1) \\ B_2(x_1) - \frac{\partial \gamma_\mathrm{x}(x_1)}{\partial x_1} B_1(x_1) \end{bmatrix} \breve{u}, \tag{7.15}$$

where $\breve{x}_2 = x_2 - \gamma_x(x_1)$ and $\breve{u} = u - \gamma_u(x_1)$. In this way the state-transformation method has transformed the state-equation (7.13) into the quasi-LPV form (7.15) with scheduling signal $p = x_1$. Note that there are no approximations involved in this procedure, but it is only applicable to a limited class of NL-SS representations and often the resulting LPV representation is non-minimal. Furthermore, no constructive procedure to find γ_x and γ_u is available.

7.3.4.2 Substitution by Virtual Scheduling

Based on specific mathematical manipulations, some nonlinear equations in the form of (7.2a–b) can be rewritten as (7.11a–b). Then, assigning virtual scheduling signals for each nonlinear function element of the resulting matrix functions (A, B, C, D), a quasi-LPV SS representation of the system results with static linear dependence. Due to the several possibilities of assignment, the result of the transformation is non-unique. Commonly, the methods that fall into this category can only be applied for specific NL systems without any generality. In most cases, the number of associated scheduling signals increases rapidly with the system order and it remains to the skill and insight of the modeler to find an economical representation. Despite its ad-hoc nature, this method is preferred for mildly-nonlinear systems as it involves no approximation of the system dynamics, delivering efficient modeling solutions in many applications (see [55, 191, 156] for examples). At the same time, many pitfalls are present for unexperienced users. To illustrate this, consider

$$\frac{\mathrm{d}}{\mathrm{d}t}x = x^2 - 1, \tag{7.16}$$

and represent this differential equation in a LPV form

$$\frac{\mathrm{d}}{\mathrm{d}t}x = px, \quad \text{where } p = x - \frac{1}{x}. \tag{7.17}$$

Seemingly, the two differential equations are equivalent. However, the resulting behaviors are different for $x(t) = 0$. Thus using the LPV model (7.17), to design a stabilizing LPV controller for the original NL system is dangerous as it is unpredictable how the closed loop system will behave when x approaches 0.

7.3.4.3 Velocity-Based Scheduling Technique

The velocity-based method of [100, 101] associates a linear system with every operating point of a NL system, rather than just the equilibrium points or pre-specified reference points. This is called *local linear equivalence* in [99]. Assume that a representation of the NL system is given in the form

$$\frac{\mathrm{d}}{\mathrm{d}t}x = Ax + Bu + \breve{f}(\gamma(x,u)), \tag{7.18a}$$

$$y = Cx + Du + \breve{g}(\gamma(x,u)), \tag{7.18b}$$

where (A,B,C,D) are constant matrices, $\breve{f}:\mathbb{R}^n \to \mathbb{R}^{n_\mathrm{X}}$, $\breve{g}:\mathbb{R}^n \to \mathbb{R}^{n_\mathrm{U}}$ are partially differentiable nonlinear functions, and the function γ is given by

$$\gamma(x,u) = E_\mathrm{x}x + E_\mathrm{u}u, \tag{7.19}$$

where $E_\mathrm{x} \in \mathbb{R}^{n\times n_\mathrm{X}}$ and $E_\mathrm{u} \in \mathbb{R}^{n\times n_\mathrm{U}}$. This reformulation of (7.2a–b) can always be achieved. Introduce $p = \mathrm{Col}(x,u)$ as the scheduling signal. Differentiating equations (7.18a–b) gives the following alternative reformulation:

$$\frac{\mathrm{d}^2}{\mathrm{d}t^2}x = \left(A + \frac{\partial}{\partial\gamma}\breve{f}(\gamma(p))E_\mathrm{x}\right)\frac{\mathrm{d}}{\mathrm{d}t}x + \left(B + \frac{\partial}{\partial\gamma}\breve{f}(\gamma(p))E_\mathrm{u}\right)\frac{\mathrm{d}}{\mathrm{d}t}u, \tag{7.20a}$$

$$\frac{\mathrm{d}}{\mathrm{d}t}y = \left(C + \frac{\partial}{\partial\gamma}\breve{g}(\gamma(p))E_\mathrm{x}\right)\frac{\mathrm{d}}{\mathrm{d}t}x + \left(D + \frac{\partial}{\partial\gamma}\breve{g}(\gamma(p))E_\mathrm{u}\right)\frac{\mathrm{d}}{\mathrm{d}t}u. \tag{7.20b}$$

By restricting the behavior of (7.2a–b) to signals that are differentiable, the set of solutions satisfying (7.20a–b) is equivalent with the solution set of (7.2a–b) for appropriate initial conditions. Substitution by $\breve{x} = \frac{\mathrm{d}}{\mathrm{d}t}x$, $\breve{u} = \frac{\mathrm{d}}{\mathrm{d}t}u$, and $\breve{y} = \frac{\mathrm{d}}{\mathrm{d}t}y$ delivers a quasi-LPV form of (7.20a–b), suggesting the conclusion that every nonlinear system (7.18a–b) can be reformulated in this way as a quasi-LPV SS representation. However, by masking differentiation of the system signals into new variables, the behavior of the resulting quasi-LPV system is different. Additionally, if instead of $\frac{\mathrm{d}}{\mathrm{d}t}u$ and $\frac{\mathrm{d}}{\mathrm{d}t}y$, only the measurements of u and y are available in the physical system, then in the practical use of the suggested LPV description the amplification of noise is inevitable by the differentiation of u and y. Such a phenomenon can have serious impact on identification or control of the underlying system. Moreover, there is often little hope of controlling the original NL system only via its differentiated description.

7.3.4.4 Function Substitution

Another way of quasi-LPV model generation leads through the idea of approximating the nonlinear functions f and g in (7.2a–b) by a linear combination of scheduling dependent functions multiplied by x and u. Consider the NL system described by the state-equation

$$\frac{\mathrm{d}}{\mathrm{d}t}x = A(x_1)x + B(x_1)u + f(x_1), \tag{7.21}$$

where $x = \mathrm{Col}(x_1,x_2)$. To perform the substitution method, choose an equilibrium point $(\mathrm{x}_1,\mathrm{x}_2,\mathrm{u})$ and transform the variables as $\breve{x}_1 = x_1 - \mathrm{x}_1$, $\breve{x}_2 = x_2 - \mathrm{x}_2$, $\breve{u} = u - \mathrm{u}$. Using these new variables, (7.21) can be rewritten as

$$\frac{\mathrm{d}}{\mathrm{d}t}\begin{bmatrix}\breve{x}_1\\ \breve{x}_2\end{bmatrix} = A(x_1)\begin{bmatrix}\breve{x}_1\\ \breve{x}_2\end{bmatrix} + B(x_1)u + \breve{f}(x_1), \tag{7.22}$$

where

$$\breve{f}(x_1) = A(x_1)\begin{bmatrix}\mathrm{x}_1\\ \mathrm{x}_2\end{bmatrix} + B(x_1)\mathrm{u} + f(x_1). \tag{7.23}$$

The next step is to reformulate $\breve{f}$ as

$$\breve{f}(x_1) \approx \Gamma(p)\breve{x}_1, \tag{7.24}$$

where Γ is an unknown matrix function and $p = x_1$. Now, the goal of the modeling approach is to determine Γ such that the approximation (7.24) is adequate for every trajectory of x_1. It is obvious that solutions of (7.24) are not unique since this is an under-determined problem. In many applications of this idea like [185, 186, 170, 112], Γ is calculated based on a linear parametrization with prior chosen functions of p, to minimize the approximation error of (7.24) on the entire operating envelope $\mathbb{X}_1$. The solution of this minimization problem is obtained by linear programming. Then, the final quasi-LPV approximation of (7.21) is given as

$$\frac{\mathrm{d}}{\mathrm{d}t}\breve{x} = \left(A(p) + \begin{bmatrix} \Gamma(p) & 0 \end{bmatrix}\right)\breve{x} + B(p)\breve{u} \tag{7.25}$$

with $p = x_1$. The behavior of (7.25) can approximate the behavior of the original NL representation if (7.24) is satisfied adequately. A disadvantage of this method is the strong dependence on the equilibrium point (with different reference points different representations can be obtained) and that the model may not capture the local stability of the original NL model at other equilibrium points. In [169], an improved version of the method has been developed to preserve local stability over the entire operation envelope. In this modified approach, the search for Γ is formulated as a bilinear-matrix-inequality based optimization problem including stability constraints.

7.3.5 Automated Model Transformation

Automated model transformation is based on the exploration of all possible ways of reformulating the NL system as a quasi-LPV model with the smallest possible conservatism. Such a technique can also be seen as a substitution method. Recently the approach, introduced in [90], has appeared in this context, formulating the conversion concept in terms of an algorithm.

The proposed procedure starts with the description (7.2a) and by relying on symbolic manipulations it assumes the separation of f to irreducible additive terms, i.e. the ith row of f is decomposed as:

$$[f(x,u)]_i = \sum_{j=1}^{n_i} f_{ij}(x,u). \tag{7.26}$$

Then, each summand f_{ij} is written in a rational form and the numerator is factorized as the product of powers of state and input elements and a remainder term that is a non-factorizable function of x and u. Summands are assigned to state-space matrices A and B based on which elements of x or u appear in the product part of their factorized form. For example, if x_k is a factor of f_{ij}, then it is assigned to the kth state as a coefficient $\alpha_{ijk}(x,u) = f_{ij}(x,u)/x_k$ in the ith row and k^{th} column of A. As in the factorized form of the summands multiple elements of x and u can be presented in the product terms, thus such an assignment is non-unique. If the summand is non-factorizable, then it is divided by x_k or u_l to be able to write it for example as $f_{ij}(x,u) = \frac{f_{ij}(x,u)}{x_k} x_k$, and assign the resulting term to A or B. Then, each of the assigned coefficient functions, i.e. $\{\alpha_{ijk}\}$ and $\{\beta_{ijk}\}$, are associated with a virtual scheduling signal p_l. Finally, the SS matrices are formed as the linear combination of these virtual scheduling signals.

By using all assignment possibilities, a set of LPV-SS representations is generated, each corresponding to (7.2a). These representations are tested for complexity and the most adequate LPV description is selected by the user. The exact way of this test and the selection of the most suitable assignment is not formulated in the approach. Thus, these tests remain to the intuition of the user. The same procedure can be executed for the output equation (7.2b) as well. The whole approach depends on how well the simplification of the nonlinear terms can be achieved. As simplification of symbolic terms is not unique in general, the complexity of the resulting model can vary with different symbolic solvers (see [71] for an overview of the required symbolic manipulation techniques). A more serious problem, which has been already mentioned in the linearization part, arises when non-factorizable terms are divided by signal components. Such operations can result in the alteration of the behavior around the origin. It can also happen that no element of u can be lifted out from any summands and thus the generation procedure results in an autonomous quasi-LPV description.

Another technique that falls into this category develops an automated transformation of a nonlinear model to an *linear fractional representation* (LFR) by using symbolic manipulation techniques (see [71, 211]). A disadvantage of this method, implemented in the *LFR toolbox* of MATLAB, is that it assumes the nonlinear model to be in the form (7.11a–b) with also a set of algebraic constraints. In this way it avoids the crucial part of the modeling, namely the generation of the quasi-LPV form (7.11a–b). This technique shows resemblance with other multiple modeling methods and implements a particular way of model reduction with respect to LFRs.

Using the tools offered by the LFR toolbox, the previous approach has been recently improved in [46] and [37] by establishing a transformation method between quasi-LPV forms and nonlinear models formulated in the MODELICA environment as first order *differential algebraic equations* (DAEs) with uncertain parameters. The transformation is based on triangulization of the DAEs and solving them recursively to obtain a set of system equations. Then uncertain parameters and remaining nonlinear terms are extracted to the Δ block to form the scheduling dependency of the resulting description. The latter implies that the original latent variables of the DAEs are forced to be the state variables of the resulting LFR and that there is no control over what is chosen to be the scheduling variable. Despite these problems and some restrictions in the allowed nonlinear relationships between the signals themselves and the uncertain parameters, this approach offers an attractive way of automated model conversion.

7.3.6 Summary of Existing Techniques

In conclusion, the existing techniques for transformation of NL systems to a quasi-LPV form are either based on linearization-based approximation or substitution techniques. Linearization techniques are commonly easily applicable for this purpose, but they suffer from serious disadvantages in terms of non-eliminatable affine reminder terms, pitfalls of interpolation, selection of adequate linearization points, and the loss of general representation of the nonlinear dynamics. On the other hand, substitution techniques are based on mathematical manipulations that are only applicable for some class of NL systems. Commonly they preserve the original dynamic behavior, however for the general class of NL systems they may result in loss of validity or even stability (e.g. division by signal elements).

A common feature of all approaches is that they use a SS representation of the system as a starting point, thus they try to achieve good approximation with respect to a prior chosen state variable. As a consequence the scheduling variable of the resulting LPV description is composed from the priori chosen state and input variables. This restriction can severely reduce the search-space where an adequate LPV representation of the original NL behavior can be found as rewriting (7.2a–b) to another state-basis may result in a simplified/better transformation to a LPV form.

All of the approaches do not pay attention how the scheduling variable is chosen and what kind of effects a particular choice of p has on the obtained LPV behavior. Seemingly it does not matter that components of x which are inner variables or components of u which are free variables are used for p.

As a general conclusion, existing transformation possibilities to a LPV form are conservative, non-unique, and the validity of the resulting model is based on the skill of the user. On the other hand, to support LPV identification and control of physical systems, the LPV modeling phase must be accomplished carefully, exploring the best possibility of transformation of the first principle laws into an LPV form, without ad hoc selection of state signals and scheduling variables. Based on this conclusion, in the next section the possibilities to accomplish this task are investigated and a model transformation procedure is proposed, that can tackle this problem using the powerful theoretical framework of the developed LPV behavioral approach.

7.4 Translation of First Principle Models to LPV Systems

As a next step, a transformation method is investigated that converts first principle laws represented by (7.1) into a LPV *kernel* (KR) representation. The procedure gives the freedom to consider all possibilities of transformation, not restricted by preselected state or IO partition or particular formulation of the nonlinear dynamical relationship like a SS representation. In fact, the method explores all possible transformations that are applicable for general NL dynamical systems, to convert the specific NL behavior into an LPV behavior. Then the obtained LPV-KR representations are categorized by complexity and transformed to an LPV-SS or IO realization based on the equivalence transformation theory of Chap. 3. To assist LPV model structure selection, the discretization theory developed in Chap. 6 is applied to obtain DT descriptions of the original behavior.

The developed transformation mechanism is based on similar concepts like the approach of [90], however the proposed method is constructed in a more structured way, where the validity of the transformation is guaranteed in the applied formulas. In the following, a brief outline of the procedure is presented to give insight into the theoretical concept instead of technicalities. First we define the exact problem setting we consider.

7.4.1 Problem Statement

Consider a NL dynamical system $\mathcal{G}_{\mathrm{NL}} = (\mathbb{R}, \mathbb{W}, \mathfrak{B})$ with signal space $\mathbb{W} = \mathbb{R}^{n_{\mathrm{W}}}$ and behavior $\mathfrak{B} \subseteq \mathbb{W}^{\mathbb{T}}$, where $\mathfrak{B}$ is represented by (7.1). Assume that f is a meromorphic function: $f \in \mathcal{R}^{n_{\mathrm{r}} \times 1}$. Then as a short hand notation, introduce

$$f \diamond w = f\left(w, \frac{\mathrm{d}}{\mathrm{d}t}w, \frac{\mathrm{d}^2}{\mathrm{d}t^2}w, \ldots, \frac{\mathrm{d}^n}{\mathrm{d}t^n}w\right), \tag{7.27}$$

as the evaluation of f along the signal trajectory $w \in \mathfrak{B}$. In this way, we associate variables of f with specific signal elements of w and their derivatives, similar to the mechanism of Chap. 3. Furthermore, assume that each element of the variable w is of prime interest to the user (they are non-latent variables) and the functional relation described by f can not be simplified without changing the behavior $\mathfrak{B}$. The latter assumption means that (7.1) is minimal in the sense that no equation can be eliminated from (7.1) by simple row operations like addition or multiplication by functional terms, similar to left-side unimodular transformations in the LPV case. Additionally, if for any $i \in \mathbb{I}_1^{n_{\mathrm{W}}}$ there exists a $\hat{f} \in \mathcal{R}^{n_{\mathrm{r}} \times 1}$, a partition $w = \mathrm{Col}(w_1, w_2)$ with $\mathrm{Dim}(w_2) = n_2$ and a invertible holomorphic function $g : \mathbb{R}^{n_{\mathrm{W}}} \to \mathbb{R}^{n_2}$ such that

$$f \diamond w = \hat{f} \diamond \mathrm{Col}(w_1, g(w)), \tag{7.28}$$

for all $w \in \mathfrak{B}$ and $\hat{f}$ is a less complicated function than f, then redefine w_2 as $g(w)$ to achieve simplification of f. This operation is similar to right-side unimodular transformations in the LPV case. By applying such a simplification, the resulting behavior is isomorphic with $\mathfrak{B}$. Now we define our idealistic transformation problem as follows:

Problem 7.1 (Translation of dynamic NL systems to LPV systems). For a given NL dynamical system $\mathcal{G}_{\mathrm{NL}} = (\mathbb{R}, \mathbb{R}^{n_{\mathrm{W}}}, \mathfrak{B})$ with signal variable w, find an LPV system $\mathcal{S}' = (\mathbb{R}, \mathbb{P}', \mathbb{W}', \mathfrak{B}')$ with signal variable w' and scheduling variable p' such that there exist selector matrices $S_{\mathrm{p}}, S_{\mathrm{w}} \in \mathbb{R}^{\cdot \times n_{\mathrm{W}}}$ satisfying $w' = S_{\mathrm{w}} w$ and $p' = S_{\mathrm{p}} w$ and it holds that

$$w \in \mathfrak{B} \cap \mathcal{C}^\infty(\mathbb{R}, \mathbb{W}) \quad \Leftrightarrow \quad (w', p') \in \mathfrak{B}' \cap \mathcal{C}^\infty(\mathbb{R}, \mathbb{W}' \times \mathbb{P}'). \qquad \square$$

Based on this problem setting, we are looking for such an LPV system that has a behavior equal to the behavior of the original NL system. We will see that in most cases the if and only if condition of Problem 7.1 has to be relaxed such that $w \in \mathfrak{B} \cap \mathcal{C}^\infty(\mathbb{R}, \mathbb{W}) \Rightarrow (w', p') \in \mathfrak{B}' \cap \mathcal{C}^\infty(\mathbb{R}, \mathbb{W}' \times \mathbb{P}')$.

7.4.2 The Transformation Algorithm

In the following, an algorithm is applied on the simplified differential equation to explore all possibilities of its transformation to an LPV form and in this way to solve Problem 7.1. We follow a similar strategy as the algorithm of [90] but in a different setting. First we separate the rows of f in (7.1) into summands. Then we factorize the nominator of these summands in a specific way to lift out signal variables in a product form. Next we collect all factorization possibilities and their associated coefficients in terms of factors into a decision tree. By selecting a route in the resulting tree we assign elements of w to signals or scheduling variables and use their associated coefficients to form a LPV-KR representation. In the process we assume that all summands are factorizable. This assumption is relaxed later. The proposed algorithm reads as follows:

Algorithm 7.1 (Translation to LPV-KR representations)

Step 1. Write f in (7.1) as a summation of additive functional terms (summands) for each row separately. The ith row is written as

$$[f \diamond w]_i = \sum_{j=1}^{n_i} f_{ij} \diamond w, \tag{7.29}$$

where each $f_{ij} \in \mathcal{R}$ is not separable to further summands. This assumes that ideal symbolic recognition of additive terms is available. The results of this operation are unique up to multiplication by a constant. Store the summands in a graph structure, as shown in Fig. 7.1 where each node represents a specific summand or a row of f.

Step 2. Each summand is written in the form of

$$f_{ij} = \frac{\hat{g}_{ij}}{\breve{g}_{ij}}, \tag{7.30}$$

where $\hat{g}_{ij}$, $\breve{g}_{ij} : \mathbb{R}^{(n+1)n_\mathrm{W}} \to \mathbb{R}$ are coprime holomorphic functions with n denoting the highest derivative order in (7.27). Such a formulation is again unique up to multiplication by a constant.

Step 3. For each $i \in \mathbb{I}_1^{n_\mathrm{r}}$, $j \in \mathbb{I}_1^{n_i}$, $k \in \mathbb{I}_1^{n_\mathrm{W}}$, and $l \in \mathbb{I}_0^{n}$, it is investigated if the summand f_{ij} is factorizable to the form

$$f_{ij} \diamond w = \left(\frac{\breve{g}_{ijlk}}{\breve{g}_{ij}} \diamond w \right) \frac{\mathrm{d}^l}{\mathrm{d}t^l} w_k, \tag{7.31}$$

where $\breve{g}_{ijkl} : \mathbb{R}^{nn_\mathbb{W}} \to \mathbb{R}$ is holomorphic. Contrary to the algorithm described in [90], factorization in this case only involves first-order product terms, as any higher order relation is not interesting for the further procedure. Denote

$$\breve{f}_{ijkl} = \frac{\breve{g}_{ijlk}}{\breve{g}_{ij}}. \tag{7.32}$$

Note that $\breve{f}_{ijkl} \in \mathcal{R}$. Assume for the moment that each f_{ij} can be factorized by at least one $\frac{\mathrm{d}^l}{\mathrm{d}t^l} w_k$. The results of the factorization are stored in the graph structure of Fig. 7.1. In this graph each node representing a summand f_{ij} gets a leaf for each $\frac{\mathrm{d}^l}{\mathrm{d}t^l} w_k$ it can be factorized with. The edges, that connect the leafs to their associated summand, receive a label, which is the set of the specific variables $\{w_1, \ldots, w_n\}$ that are involved in the remaining $\breve{f}_{ijkl}$ expression. All other edges of the graph get a label of an empty set. Note that leafs with an edge having an empty set label are the linear terms of the nonlinear equation. The resulting graph describes a decision tree.

Step 4. As a next step, possible LPV-KR representations are generated based on the previously developed decision tree. All possible routes in this graph are considered, which involve all nodes and a single leaf for each node if it has one. Routes are only considered to be different if they consist of different leafs. This yields all realization possibilities of (7.1) as a PV differential equation in the following way. For a specific route (see Fig. 7.1), compute the union of the sets of variables associated with labels along the edges of the route. This gives a subset of all variables $\{w_1, \ldots, w_{n_\mathbb{W}}\}$. These variables are recognized as scheduling signals and denoted as $\{p_1, \ldots, p_{n_\mathbb{P}}\}$ where $n_\mathbb{P} \leq n_\mathbb{W}$. Define $p = [\, p_1 \ \ldots \ p_{n_\mathbb{P}} \,]$ as the scheduling variable for the specific route. Additionally, let $\mathbb{P}$ be the projected subspace of $\mathbb{W}$ with respect to p. Consider all leaves in the route. Define index sets $\mathcal{I}_{ikl}$, containing all indexes $j \in \mathbb{I}_1^{n_i}$ for which the node, associated with the summand f_{ij}, has a leaf of $\frac{\mathrm{d}^l}{\mathrm{d}t^l} w_k$ in the considered route. For each leaf in the route, collect the remainder terms $\{\breve{f}_{ijlk}\}$ of the factorization into meromorphic coefficient functions $r_l \in \mathcal{R}^{n_\mathrm{r} \times n_\mathbb{W}}$, $l \in \mathbb{I}_0^n$ where

$$[r_l]_{ik} = \sum_{j \in \mathcal{I}_{ikl}} \breve{f}_{ijlk}, \quad \forall (i,k) \in \mathbb{I}_1^{n_\mathrm{r}} \times \mathbb{I}_1^{n_\mathbb{W}}. \tag{7.33}$$

With the resulting coefficients $\{r_i\}_{i=1}^n$ the NL differential equation (7.1) is formulated as an LPV-KR representation:

$$(R(\frac{\mathrm{d}}{\mathrm{d}t}) \diamond p) w = \sum_{i=0}^{n} (r_i \diamond p) \frac{\mathrm{d}^i}{\mathrm{d}t^i} w = 0, \tag{7.34}$$

where $R \in \mathcal{R}[\xi]^{n_\mathrm{r} \times n_\mathbb{W}}$. Now define $\breve{w}$ as the vector containing the subset of the variables $\{w_1, \ldots, w_{n_\mathbb{W}}\}$, such that for each variable w_i in $\breve{w}$, there is a a leaf in the route where the label of the leaf contains w_i or its derivative.

Those signals that do not satisfy this property are simply presented with 0 weights in (7.34), thus they do not participate in the signal relation as variables. To eliminate such superfluous terms, (7.34) is rewritten in terms of $\breve{w}$ by deleting from R the zero columns associated with the additional variables. Furthermore the signal space associated with $\breve{w}$ is defined as the projected subspace of $\mathbb{W}$ with respect to the variables presented in $\breve{w}$.

Step 5. As a result of the previous step, numerous PV differential equation forms of (7.1) are formulated based on all possible routes in the graph structure. However, it is not guaranteed that all of them preserve the dynamical aspects of the original nonlinear behavior $\mathfrak{B}$. To ensure validity of the transformation, the freedom of the remaining signal variables $\breve{w}$ for each PV description has to be checked. If for every IO partition $\breve{w} = (\breve{u}, \breve{y})$ of the resulting LPV-KR representation $\mathfrak{R}_{\mathrm{K}}(\mathcal{S})$, there exists an IO partition $w = (u, y)$ for the original NL system such that $\breve{y} = y$ and the elements of $\breve{u}$ are a subset of u and the maximal order of derivatives of y in each row of $\mathfrak{R}_{\mathrm{K}}(\mathcal{S})$ are the same as in f, then the LPV representation can be considered as a valid transform of the original system. Otherwise alteration of the dynamical behavior occurred during the process by masking essential dynamics into coefficients. If it is also true for every IO partition of $\mathfrak{R}_{\mathrm{K}}(\mathcal{S})$, that $u = \mathrm{Col}(p, \breve{u})$, meaning that all elements of p are free variables of the original system and they are independent from $\breve{u}$, then the model corresponds to a true LPV system $\mathcal{S}$, not just a quasi-LPV, and it is a prime candidate for representing the original system behavior.

Step 6. The resulting LPV representations can be ranked based on the corresponding routes in the graph. Representations which involve the smallest cost in terms of the number of variables associated with scheduling signals give the simplest models of the NL system. From these candidates, representations with free scheduling signals have priority. Further distinction can be the maximal order of derivatives of the leafs. Based on these, the resulting valid representations can be ordered in terms of complexity, to assist selection by the user.

For each LPV-KR representation, that has been found to be a valid representation of $\mathcal{G}_{\mathrm{NL}}$, the PV behavior can be considered to be equal with the original NL behavior $\mathfrak{B}$. Then the most attractive representation can be selected by the user based on the complexity ordering derived in Step 6. In this way, a solution for our transformation problem, i.e. for Problem 7.1 is obtained. Next, the behavioral approach is applied on the chosen representation to obtain a full row rank KR representation together with an SS or IO realization of the resulting LPV system. With the introduced discretization theory, this methodology serves as a model selection tool for DT-LPV identification routines.

Example 7.2 (Transformation of an NL model to LPV). Consider the nonlinear system $\mathcal{G}_{\mathrm{NL}} = (\mathbb{R}, \mathbb{R}^4, \mathfrak{B})$ with $w = \begin{bmatrix} w_1 & w_2 & w_3 & w_4 \end{bmatrix}^\top$ in the form:

$$f\left(w, \frac{\mathrm{d}}{\mathrm{d}t}w, \frac{\mathrm{d}^2}{\mathrm{d}t^2}w\right) = 0,$$

where $f \diamond w$ is equal to

$$\begin{bmatrix} \overbrace{w_1 \cos^2(w_3)}^{f_{11}} + \overbrace{\sin^2(w_3)\frac{\mathrm{d}}{\mathrm{d}t}w_1}^{f_{12}} + \overbrace{2w_3w_4}^{f_{13}} + \overbrace{2\frac{\mathrm{d}}{\mathrm{d}t}w_2}^{f_{14}} + \overbrace{\left(\frac{\mathrm{d}}{\mathrm{d}t}w_2\right)\frac{\mathrm{d}}{\mathrm{d}t}w_3}^{f_{15}} + \overbrace{w_3\frac{\mathrm{d}^2}{\mathrm{d}t^2}w_1}^{f_{16}} \\ \underbrace{w_3\frac{\mathrm{d}}{\mathrm{d}t}w_1}_{f_{21}} + \underbrace{w_2}_{f_{22}} \end{bmatrix}.$$

Note that for this nonlinear system the only available IO partition is $y = [\,w_1 \;\; w_2\,]^\top$ and $u = [\,w_3 \;\; w_4\,]^\top$. This is easy to show by rewriting the second row as $w_2 = -w_3\frac{\mathrm{d}}{\mathrm{d}t}w_1$, and substituting it into the first row of f. Then a differential equation results where only the derivatives of w_1 appear. Thus w_1 is an obvious output of the system and therefore w_2 is also an output as its trajectory is described by the used substitution rule. The remaining variables w_3 and w_4 are free in the resulting description and thus they are the obvious inputs of the system. However, the second equation written as $w_3 = -w_2/\frac{\mathrm{d}^2}{\mathrm{d}t^2}w_1$, cannot be used for substitution as it would exclude trajectories of w_1 with $\frac{\mathrm{d}}{\mathrm{d}t}w_1 = 0$. Hence w_3 cannot be the output of the system instead of w_2.

Additionally, the nonlinear equations of the above given representation have already been separated to minimal summand terms and it can be shown that no further symbolic simplification of the equations is possible. A further property is that all summand terms are factorizable. By applying the decision tree generation procedure described in the previous part, the resulting tree is presented in Fig. 7.1. This completes Step 1 to Step 3 of the proposed algorithm.

In terms of Step 4, now we generate all possible routes that contain all nodes and one leaf for each node. There are $2^5 = 32$ possibilities. One of these routes is given by bold lines in Fig. 7.1. By using this specific route, a LPV-KR representation is generated in terms of Step 4. The resulting scheduling variable is the collection of variables in the labels along the route: $p = w_3$ and the new signal variable is the collection of variables of the leafs along the route: $\breve{w} = [\,w_1 \;\; w_2 \;\; w_4\,]^\top$. Then by these choices, the LPV-KR representation has the following form

$$\begin{bmatrix} \cos^2(p_1) & 0 & 2p_1 \\ 0 & 1 & 0 \end{bmatrix}\breve{w} + \begin{bmatrix} \sin^2(p_1) & 2+\frac{\mathrm{d}}{\mathrm{d}t}p_1 & 0 \\ p_1 & 0 & 0 \end{bmatrix}\frac{\mathrm{d}}{\mathrm{d}t}\breve{w} + \begin{bmatrix} p_1 & 0 & 0 \\ 0 & 0 & 0 \end{bmatrix}\frac{\mathrm{d}^2}{\mathrm{d}t^2}\breve{w} = 0.$$

Note that the IO partition of the obtained LPV-KR representation is $y = [\,w_1 \;\; w_2\,]^\top$ and $u = w_4$, and the maximal orders of the derivatives of w_1 and w_2 in the representation are the same as in f, thus it corresponds a valid LPV model of the original nonlinear system. As the chosen scheduling variable is a free variable in the original system and is independent from $\breve{w}$, the obtained representation corresponds to a true (non-quasi) LPV system. This completes Step 5 of the proposed algorithm.

Another choice of route is similar to the previous one except taking the right branch at f_{16}. The resulting scheduling variable is $p = \begin{bmatrix} w_3 & w_1 \end{bmatrix}^\top$ and the signal variable is $\breve{w} = \begin{bmatrix} w_1 & w_2 & w_3 & w_4 \end{bmatrix}^\top$. Then by these choices, the following LPV-KR representation results:

$$\begin{bmatrix} \cos^2(p_1) & 0 & \frac{d^2}{dt^2}p_2 & 2p_1 \\ 0 & 1 & 0 & 0 \end{bmatrix} \breve{w} + \begin{bmatrix} \sin^2(p_1) & 2+\frac{d}{dt}p_1 & 0 & 0 \\ p_1 & 0 & 0 & 0 \end{bmatrix} \frac{d}{dt}\breve{w} = 0.$$

A valid IO partition of the obtained LPV-KR representation is $y = \begin{bmatrix} w_1 & w_2 \end{bmatrix}^\top$ and $u = \begin{bmatrix} w_3 & w_4 \end{bmatrix}^\top$, the IO partition of the original nonlinear system. However, the maximal order of derivatives with respect to w_1 is 1. This means that the dynamics of f has been simplified, i.e. masked into coefficient dependence, during the transformation process. Thus the resulting LPV representation is not a valid representation of the nonlinear system.

A third choice is to use the route given with bold lines but to take the left branch at f_{13}. The resulting scheduling variable is $p = \begin{bmatrix} w_3 & w_4 \end{bmatrix}^\top$ and the signal variable is $\breve{w} = \begin{bmatrix} w_1 & w_2 & w_3 \end{bmatrix}^\top$. Then by these choices, the following LPV-KR representation results:

$$\begin{bmatrix} \cos^2(p_1) & 0 & 2p_2 \\ 0 & 1 & 0 \end{bmatrix} \breve{w} + \begin{bmatrix} \sin^2(p_1) & 2+\frac{d}{dt}p_1 & 0 \\ p_1 & 0 & 0 \end{bmatrix} \frac{d}{dt}\breve{w} + \begin{bmatrix} p_1 & 0 & 0 \\ 0 & 0 & 0 \end{bmatrix} \frac{d^2}{dt^2}\breve{w} = 0.$$

Note that the resulting LPV-KR representation with IO partition $y = \begin{bmatrix} w_1 & w_2 \end{bmatrix}^\top$ and $u = w_3$ is a valid representation of the nonlinear system similarly as the representation associated with the bold lines, but with an increased scheduling dimension and a free but not independent scheduling variable p.

The remaining choices of available routes are the combinations of the previous ones with no interesting further property. Thus in conclusion, the best choice of LPV-KR representation of the system follows through the decision route indicated by bold lines in Fig. 7.1. This concludes the final step of the algorithmic scheme. □

7.4.3 Handling Non-Factorizable Terms

Now we investigate the case, when not all summand terms $\{f_{ij}\}$ are factorizable. In that case, there is little chance for the elimination of these terms and to enable the use of the previously introduced mechanism without any approximation. Variable substitution to eliminate these terms (see Example 7.3) generally does not work, as the substitution must satisfy the equation for all derivative relations of the substituted variables. This may result in ad hoc operations, cancelations of terms and alteration of the behavior. This unfortunate phenomenon even holds for constant terms in general (see Example 7.3).

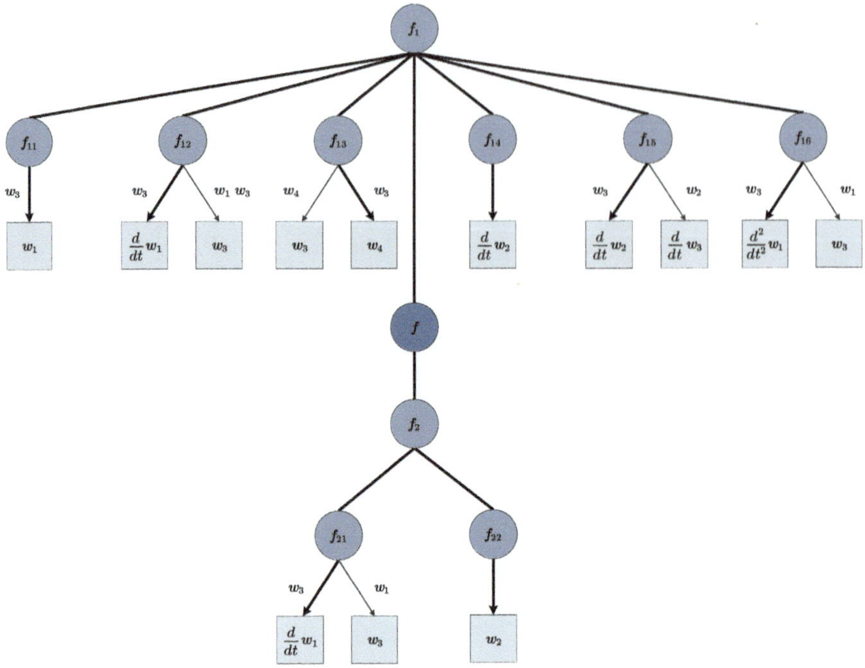

Fig. 7.1 Decision tree of NL model transformation to LPV-KR representations

Example 7.3 (Elimination of non-factorizable terms). Consider the case when $f_{ij} = 1$. Then this constant term is a non-factorizable expression in the nonlinear differential equation described by f. If f reads as

$$f \diamond w = w_2 \frac{\mathrm{d}}{\mathrm{d}t} w_1 + w_1 + 1,$$

then the non factorizable term 1 can be eliminated by substituting w_1 with $\breve{w}_1 = w_1 + 1$, which gives

$$w_2 \frac{\mathrm{d}}{\mathrm{d}t} \breve{w}_1 + \breve{w}_1 = 0.$$

Such an expression contains only factorizable summands thus the previously described procedure can be applied on this new differential equation to obtain an LPV representation in terms of $(\breve{w}_1, w_2)$. However, if f is

$$f \diamond w = \begin{bmatrix} w_2 \frac{\mathrm{d}}{\mathrm{d}t} w_1 + w_1 + 1 \\ w_1 + \frac{\mathrm{d}}{\mathrm{d}t} w_2 \end{bmatrix},$$

then this elimination cannot be applied on the upper equation as it would introduce a nonfactorizable term -1 in the lower one. In case of other non-factorizable terms, it is generally true that elimination through variable substitution can be applied if it satisfies all the equations without introducing new non-factorizable terms. □

A sound possibility is however to approximate the non-factorizable terms in the following way:

1. Let f_{ij} be a non-factorizable summand. Let $\mathcal{I}_{ij}$ be an index set containing all indexes $(k,l) \in \mathbb{I}_1^{n_\mathrm{W}} \times \mathbb{I}_0^n$ such that f_{ij} is dependent on $\frac{\mathrm{d}^l}{\mathrm{d}t^l} w_k$.
2. For each $(k,l) \in \mathcal{I}_{ij}$, try to approximate the function f_{ij} as

$$(f_{ij} \diamond w)(t) \approx \left((\breve{f}_{ijkl} \diamond w) \frac{\mathrm{d}^l}{\mathrm{d}t^l} w_k \right)(t) \quad \forall w \in \mathfrak{B} \ \forall t \in \mathbb{R}, \tag{7.35}$$

where $\breve{f}_{ijkl} \in \mathcal{R}$.
3. If such approximation exists for some $(k,l) \in \mathcal{I}_{ij}$, then treat them as valid factorizations of f_{ij} and proceed with the original algorithm, else transformation to a PV form does not exist with the given precision.

The resulting LPV representation can be considered as the approximation of the original NL system if the conditions described in Step 5 are satisfied. Note that what is considered to be an appropriate approximation is highly dependent on the intended accuracy. For some specific non-factorizable functions like $\sin(\centerdot)$, such an approximation can be carried out even in an exact sense (see Example 7.4). If the non-factorizable term is a constant $f_{ij} = \gamma \in \mathbb{R} \setminus \{0\}$, then no sound approximation exists and one may risk to either use an approximation $f_{ij} = \frac{\gamma}{w_k} w_k$ or multiply the ith row of f with w_k and restart the transformation procedure from the first step. Both approaches may result in alteration of the original NL behavior.

Note that in the LPV behavioral approach, the terminology of almost everywhere equivalence has been introduced to handle singularities of the coefficient functions that result or change due to transformations in $\mathcal{R}[\xi]$. One can sense that in case of non-factorizable terms, we face the same situation as by dividing with a w_i any non-factorizable term can be turned to a factorized relation. However, the price to be paid is an almost everywhere equivalence of the resulting representation if w_i is a free signal. However in case w_i is non-free, division by w_i can destabilize the origin of the signal-space, critically altering the dynamical behavior of the system.

Example 7.4 (Approximation of non-factorizable terms). Consider the case, when the non-factorizable term is $\sin(w)$. Then by writing this term as

$$\sin(w) = \mathrm{sinc}(w) w,$$

where $\mathrm{sinc}(w) = \frac{\sin(w)}{w}$ and $\mathrm{sinc}(0) = 1$, we get an exact factorization. However in case of $\cos(w)$,

$$\cos(w) = \frac{\cos(w)}{w} w,$$

where $\lim_{\mathrm{x} \to 0} \frac{\cos(\mathrm{x})}{\mathrm{x}} = \infty$. Thus this form can only be considered a factorization if w is a non-zero signal. As an alternative, a finite Taylor expansion of $\cos(\centerdot)$ is suggested

$$\cos(w) = 1 + \frac{1}{2}w^2 + \frac{1}{24}w^4 + \ldots$$

where if the non-factorizable term 1 can be eliminated, then the remaining tail contains only factorizable parts. This holds for all common functions like trigonometric, exponential, logarithmic, etc. It holds in general that approximation problems of non-factorizable terms relate to the question, how constant terms can be eliminated or approximated with a factorized expression. □

7.4.4 Properties of the Transformation Procedure

It can be concluded that the proposed method can transform a wide class of NL systems satisfying Definition 7.1. An advantage of the method is that it provides a systematic way of conversion, examining all possibilities of an equivalent LPV realization. By checking validity of the derived LPV representation with respect to the behavior of the original NL system, it provides a successful tool to solve Problem 7.1. Additionally, the algorithm gives a structured selection of the scheduling variable p highlighting when not only a quasi-LPV but a true LPV formulation is possible.

However, the approach has disadvantages as well. One of them is the heavy dependence on symbolic recognition of summands and possible ways of factorization. Thus the performance of the algorithm is clearly limited by the available symbolic computational tools. Another disadvantage is that investigating all possible LPV-KR representations that can be obtained from the generated decision tree can be quite demanding in case of large signal dimensions or a complicated f. Moreover, in case of large scale systems, computing and comparing the possible IO partitions of the resulting LPV descriptions is hopeless.

It may also happen for some cases that no valid transformation of the NL dynamic system is available via the proposed algorithm. Commonly, there is little chance of any ad-hoc transformation to succeed if the proposed method does not work as the original system description is minimal and all possibilities are investigated. That means that some sort of approximation technique must be applied in advance. In principle, it must be accepted that not every NL system can be appropriately transformed to an LPV form, which is especially true for systems not satisfying Definition 7.1. Examples for such dynamical relations are systems with delays, hysteresis, or non-functional signal relations like if-then rules. In such cases, the LPV framework may be inappropriate for dealing with the system dynamics without considerable approximation.

Comparing the proposed approach of model transformation to the available approaches presented in Sect. 7.3, it can be concluded that this method gives adequate transformation for a much wider class of NL systems. The selection of the scheduling variable follows a more structured procedure, without the ad hoc selection of input, output, or latent variables. This gives the flexibility to find the most efficient form of transformation. In case of non-factorizable terms, approximation is inherently involved in the transformation, similar to the methods of Sect. 7.3, which result in an approximation of the original behavior up to a specific precision.

7.5 Conclusions

In this chapter, modeling of physical systems described by nonlinear differential equations has been studied in the LPV framework. Our motivation has been to derive tools that can assist model structure selection in LPV identification.

First we investigated the general questions of LPV modeling and the available approaches. One of the major conclusions has been that instead of the transformation of a given mathematical description of the NL system to a LPV description, the modeling problem needs to be approached from the perspective of a search for a LPV system with equal behavior. This also implies that there is a prime emphasis on the selection of the scheduling vector, i.e. which variables of the original system are used as scheduling variables in order to find an equivalent LPV description. The latter gives that the common starting point of the available approaches: the use of a priori chosen state-space representation of the NL system, has the danger that the prespecified state variable can seriously lower the realization possibilities in the LPV system class. Furthermore, it has been shown that for linearization based methods any local transformation of the obtained linearized models can seriously alter the result of their interpolation, i.e. the behavior of the obtained LPV description may not even resemble the original NL system.

By showing the open problems of LPV modeling, in Sect. 7.4 a transformation algorithm has been proposed that aims at the exploration of all possible LPV systems that are equivalent with respect to a given dynamical NL system and selecting that specific description which preserves the original behavior the best. The approach is formulated in terms of conversion of a NL differential equation to a PV form, avoiding the problems related to other conversion tools. The developed technique provides a useful tool to assist the identification methods of the LPV field with structural information about the plant and ease the choice of parametrization of the coefficient dependencies.

Chapter 8
Optimal Selection of OBFs

Abstract. In this chapter, selection of the optimal basis, i.e. the basis with the fastest convergence rate, for the series-expansion of LPV systems is investigated. In fact, we consider the situation when information about the system is only available in terms of measured data records of the frozen signal behavior. Solution of this problem is crucial to provide a model structure selection tool for LPV identification based on truncated series-expansion models. In case of an optimal basis, a fast convergence rate of the expansion representation implies that only the estimation of a few expansion coefficients is necessary for a good approximation of the system. By using the concept of Kolmogorov n-width optimality of the basis with respect to the frozen behaviors, we derive a practically applicable algorithm, that provides optimal basis selection based on fuzzy clustering of estimated "frozen" poles. The pole estimates are the results of LTI identification of the system with constant scheduling trajectories. To consider the effect of noise on the estimation of the frozen poles, a robust version of the algorithm is also developed using the strong relation between the Kolmogorov n-width theory and hyperbolic geometry.

8.1 Perspectives of OBFs Selection

The concept of modeling discrete-time asymptotically stable LPV systems with *orthonormal basis functions* (OBFs) based truncated series expansions has been introduced in Chap. 5 to develop an effective model structure for LPV identification. However, practical application of this concept requires the selection of basis functions that guarantees a fast convergence rate of the LPV expansion representation of a system. The reason is that using a truncated expansion with a fast convergence rate, i.e. only a finite number of OBFs from the basis sequence, a model results that approximates the system well with only a few expansion coefficients.

In Sect. 5.5, it has been motivated that to achieve a fast convergence rate of the LPV expansion representation, a necessary condition is to use a basis which has fast convergence rate with respect to $\mathfrak{F}_S$, the set of transfer functions of the frozen system set for the considered *input-output* (IO) partition. In order to characterize optimality of the convergence rate with respect to a given transfer function set, like $\mathfrak{F}_S$, we

R. Tóth: Modeling and Identification of Linear Parameter-Varying Systems, LNCIS 403, pp. 197–231.
springerlink.com

have introduced the worst-case concept of the *Kolmogorov n-width* (KnW) theory in Sect. 2.5. However, to use this concept to choose an n-width optimal basis, it is required to know the pole locations of $\mathfrak{F}_S$, i.e. the region $\Omega_\mathbb{P}$, which contains the points where not all transfer functions in $\mathfrak{F}_S$ are analytic. If reliable first-principle knowledge is available to calculate $\Omega_\mathbb{P}$, then by solving the min-max problem of (2.62), optimal basis selection in the KnW sense can be achieved for the system. However, in an identification scenario, such knowledge might not be available. Therefore, if no reliable first-principle information is available about $\Omega_\mathbb{P}$, selection of the basis has to be based on measured data records of the system. In this chapter, we consider the situation when data records of some frozen behaviors of the LPV system are available. Estimating LTI models based on these data records, result in pole samples of $\Omega_\mathbb{P}$. Based on these sample poles, we aim at the derivation of a basis selection mechanism, that is capable to accomplish the following objectives:

- Reconstruction of $\Omega_\mathbb{P}$ based on the sample pole locations.
- Determination of the set of KnW optimal OBFs with respect to $\Omega_\mathbb{P}$.

The method we propose to solve these objectives simultaneously is based on the fusion of the KnW theory and the *fuzzy c-means* (FcM) clustering approach (see [22, 79]). These theories are applied together to derive KnW optimal basis functions by the clustering of the sample pole locations. The resulting mechanism guarantees optimality in the KnW sense for the obtained basis functions with poles at the cluster centers in case the fuzzyness parameter approaches infinity. In this way it provides a trade-off between the optimality of the chosen basis and the complexity of the optimization.

8.2 Kolmogorov n-Width Optimality in the Frozen Sense

As a first step, we investigate the KnW optimality concept of orthonormal basis functions with respect to a LPV system in a frozen sense. To do so we revisit the KnW theory presented in Sect. 2.5 and we highlight properties that are important for the discussion of the basis selection mechanism. As the presented KnW theory is formulated in the SISO case, we restrict the discussion to SISO asymptotically stable LPV systems in the following. The basis selection problem for MIMO LPV systems is postponed till Chap. 9.

Let $\mathfrak{F}_S$ denote the set of transfer functions corresponding to the frozen system set $\mathscr{F}_S$ for a given IO partition of the LPV system $\mathcal{S}$. In Sect. 2.5 it has been already discussed that the KnW concept provides the selection of n_g poles of an inner function $G_\mathrm{b} \in \mathcal{H}_{2-}(\mathbb{E})$, such that the Hambo basis sequence $\Phi_{n_\mathrm{g}}^{\infty}$ generated by G_b (see Sect. 2.3) is optimal in the $n = n_\mathrm{g}(n_\mathrm{e}+1)$-width sense with respect to a given transfer function set. This optimality means that among all Hambo bases, the linear combination of the set of $n_\mathrm{g}(n_\mathrm{e}+1)$ functions $\Phi_{n_\mathrm{g}}^{n_\mathrm{e}}$ has the smallest worst-case representation error (in the $\mathcal{H}_2$ norm). In this sense, optimality means also the fastest worst-case convergence rate ρ of the expansion of these transfer functions with the basis sequence $\Phi_{n_\mathrm{g}}^{\infty}$. In other words, a K$n$W optimal basis for $\mathfrak{F}_S$ provides series-expansion representations of all frozen systems $\mathscr{F}_S$, such that the convergence rate of the coefficients is optimal for the considered width. Note that n, i.e. the width in which the optimality of the basis is considered represents a particular freedom. In fact, it is a trade of between ρ and the number of poles required for $G_\mathfrak{b}$ to achieve it. By using a KnW optimal basis where both the optimal convergence rate ρ and the width n is small, it is guaranteed that truncated expansion representations of all $\mathscr{F}_\mathbb{S}$ need only a few expansion coefficients to approximate each frozen behavior adequately. This means that beside finding n-width optimal OBFs for a fixed n, it is also important to search for an adequate n. The latter problem indirectly refers to the question how many basis functions are required for the truncated expansion representation to achieve a good approximation.

Introduce the pole manifest set

$$\Omega_\mathbb{P} = \left\{\lambda \in \mathbb{C} \mid \exists \mathrm{p} \in \mathbb{P}, \text{ such that } \lambda \text{ is a pole of } F_\mathrm{p} \in \mathfrak{F}_\mathbb{S}\right\}, \tag{8.1}$$

the collection of pole locations belonging to $\mathfrak{F}_\mathbb{S}$. For a given $\Omega_\mathbb{P}$ and a fixed $n = n_\mathrm{g}$, the KnW basis selection problem with respect to $\mathfrak{F}_\mathbb{S}$ comes down to the inverse Kolmogorov problem (see Sect. 2.5): finding the best fitting hull of $\Omega_\mathbb{P}$ in the form

$$\Omega(\Lambda_{n_\mathrm{g}}, \rho) := \{z \in \mathbb{C} \mid |G_\mathfrak{b}(z^{-1})| \leq \rho\}, \tag{8.2}$$

where $G_\mathfrak{b}$ is defined by the poles $\Lambda_{n_\mathrm{g}} = \left[\lambda_1 \ \dots \ \lambda_{n_\mathrm{g}}\right]$ and $\rho > 0$ is as small as possible. Then, in terms of Proposition 2.1, the inner function $G_\mathfrak{b}$, associated with the best fitting $\Omega(\Lambda_{n_\mathrm{g}}, \rho)$, generates the n-width optimal basis functions with respect to $\mathfrak{F}_\mathbb{S}$. For a given inner function $G_\mathfrak{b}$ with poles Λ_{n_g}, define

$$\kappa_{n_\mathrm{g}}(z, \Lambda_{n_\mathrm{g}}) := |G_\mathfrak{b}(z^{-1})| = \prod_{j=1}^{n_\mathrm{g}} \left| \frac{z - \lambda_j}{1 - z\lambda_j^*} \right|, \tag{8.3}$$

the so-called *Kolmogorov cost*. Then for a given $n_\mathrm{g} > 0$, the solution of the inverse Kolmogorov problem is equivalent with the min-max problem (2.62), formulated in this case as an optimization:

$$\begin{aligned} &\text{minimize} && \rho = \max\nolimits_{z \in \Omega_\mathbb{P}} \ \kappa_{n_\mathrm{g}}(z, \Lambda_{n_\mathrm{g}}), \\ &\text{such that} && \Lambda_{n_\mathrm{g}} = \left[\lambda_1 \ \dots \ \lambda_{n_\mathrm{g}}\right] \in \mathbb{D}^{n_\mathrm{g}}. \end{aligned} \tag{8.4}$$

The minimizer of (8.4), is an array of pole locations Λ_{n_g} which defines the best inner function, i.e. Hambo basis in the KnW sense. If the resulting $\Omega(\Lambda_{n_\mathrm{g}}, \rho)$ is equal to $\Omega_\mathbb{P}$, then in terms of Proposition 2.1, the generated basis is also optimal in the $k \cdot n_\mathrm{g}$-width sense for any $k \in \mathbb{N}$. Otherwise, higher width optimality of the basis does not hold in the general case (see Example 8.1). This underlines that to find an optimal OBF set for $\Omega_\mathbb{P}$, an optimal choice of n_g is also needed. We will see that by using powerful results in hyperbolic geometry, in some cases, an optimal choice of the width is available.

Example 8.1 (Optimal n-widths). In this example, we show that a basis sequence which is optimal in the KnW sense is not necessary optimal in the $2n$-width

Fig. 8.1 Kolmogorov 2-width optimal basis functions Φ_2^0 with poles Λ_2 (denoted by □) and Kolmogorov 4-width optimal basis functions Φ_4^0 with poles Λ_4 (denoted by ×) with respect to the pole manifest region $\Omega_{\mathbb{P}}$ (shaded grey area). The perimeter of the associated minimal regions $\Omega(\Lambda_2, 0.5333)$ (thick black line) and $\Omega(\Lambda_4, 0.1)$ (identical to $\Omega_{\mathbb{P}}$) is indicated with contour lines.

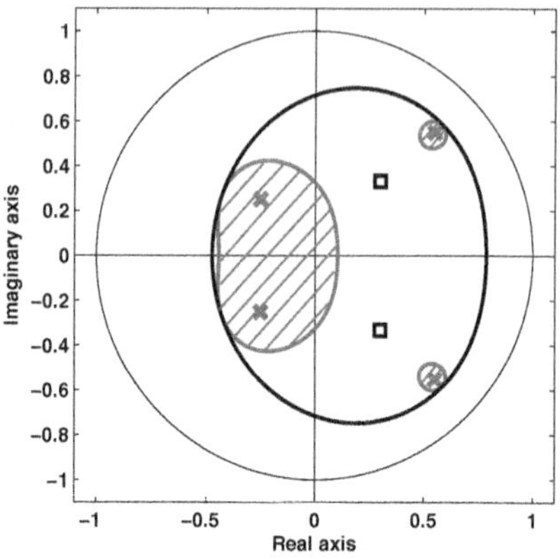

or other higher width cases. Let an inner function G_b be given with poles $\Lambda_4 = [0.55 \pm 0.55i \;\; -0.25 \pm 0.25i]$. For $\rho = 0.1$, G_b defines the complex region $\Omega(\Lambda_4, \rho)$ whose perimeter is given with a grey line in Fig. 8.1. Any strictly proper transfer function (and associated LTI system) with all poles in $\Omega(\Lambda_4, \rho)$ has a series-expansion in terms of the inner function G_b generated Hambo basis Φ_4^∞ with a worst-case convergence rate 0.1. Denote by $\Omega_{\mathbb{P}} = \Omega(\Lambda_4, \rho)$, the region for which we would like to find optimal OBFs in the KnW sense. It is obvious that in the 4-width sense, the Hambo functions Φ_4^0 are optimal with respect to $\Omega_{\mathbb{P}}$ and the convergence rate is 0.1.

Now we would like to derive a 2-width optimal basis with respect to $\Omega_{\mathbb{P}}$. By applying nonlinear optimization in terms of (8.4), the resulting optimal Hambo functions Φ_2^0 are associated with the pole locations $\Lambda_2 = [0.2995 \pm 0.3318i]$. In this case, the convergence rate, i.e. the minimal $\rho > 0$ such that $\Omega_{\mathbb{P}} \subseteq \Omega(\Lambda_2, \rho)$ is 0.5333. With this ρ, the perimeter bound of $\Omega(\Lambda_2, \rho)$ is given with a thick black line in Fig. 8.1.

Now we can see that series-expansion of any strictly proper transfer function with all poles in $\Omega_{\mathbb{P}}$ has a convergence rate 0.1 in case of the 4-width optimal basis Φ_4^∞ and 0.5333 in case of the 2-width optimal basis Φ_2^∞. This clearly shows that the worst-case decay rate of a Hambo basis $\Phi_4^{\prime\infty}$ generated by a inner function with poles $\Lambda_4^\prime = [\Lambda_2 \; \Lambda_2]$ (repetition of the 2-width optimal poles) has a convergence rate $(0.5333)^2 = 0.2844$. Comparing this to the convergence rate 0.1 of Φ_4^0 it is clear that the Hambo functions $\Phi_4^{\prime 0}$ are not optimal in the 4-width sense. In general it is true that higher width optimality is only guaranteed for basis functions with $\Omega(\Lambda, \rho) = \Omega_{\mathbb{P}}$. From Fig. 8.1 it follows that this is not the case for Λ_2. However in case of Λ_4, this equivalence is satisfied, thus with respect to $\Omega_{\mathbb{P}}$, Φ_4^1 is optimal in the 8-width sense, Φ_4^2 is optimal in the 12-width sense, etc. □

If $\Omega_{\mathbb{P}}$ is known, then a solution of (8.4) for a fixed n_g can be obtained via the gradient-search based method of [73]. However, in case of an identification scenario, when $\Omega_{\mathbb{P}}$ is not available, the gradient approach is not applicable unless $\Omega_{\mathbb{P}}$

is reconstructed from some estimated samples. This raises the need for an approach that solves both objectives (reconstruction and KnW optimization) and gives a practical solution for the basis selection step.

8.3 The Fuzzy-Kolmogorov c-Max Clustering Approach

In the following we introduce a particular data clustering algorithm, proposed in [196], which is the extension of the *fuzzy c-means* (FcM) clustering approach used in a wide collection of applications like pattern recognition, data analysis, image processing and fuzzy modeling (see [22, 79]). As we will show, the proposed approach is able to effectively handle the reconstruction problem of $\Omega_{\mathbb{P}}$ jointly with the solution of (8.4). First, in section 8.3.1, we formulate the exact problem setting for the clustering approach to interpret our basis selection goal. Then we characterize the optimal solution of the clustering problem and introduce the extension of the FcM algorithm. As a next step, in Sect. 8.3.2, we show how the derived optimal solution provides an answer for the original basis selection problem of Sect. 8.2. Numerical properties of the derived algorithm are investigated together with practical aspects, like the use of *adaptive cluster merging* (ACM) in this setting. At last, a simulation example is presented to show the effectiveness of the introduced basis selection mechanism.

8.3.1 The Pole Clustering Algorithm

Objective-function-based fuzzy clustering algorithms, such as FcM, partitions the data into overlapping groups, so-called clusters, where each data element is associated with a set of membership levels with respect to these clusters. These levels indicate the strength of the association between that data element and a particular cluster. In this way, fuzzy clustering is a process of assigning these membership levels such that the resulting clusters describe the underlying structure within the data [79]. This enables the determination of the region $\Omega_{\mathbb{P}}$ on the basis of the observed poles by exploring the underlying data coherency.

Let $n_c > 1$ be the number of clusters (pole regions) to be used to reconstruct $\Omega_{\mathbb{P}}$. Note that due to the uniform frozen asymptotic stability of $\mathscr{F}_S$, it is guaranteed that $\Omega_{\mathbb{P}} \subseteq \mathbb{D}$, i.e. all sample poles z_k are in the unit disc $\mathbb{D}$. Thus, let $\mathbb{D}$ be the clustering space and let $Z = \{\mathrm{z}_k\}_{k=1}^{N_\mathrm{z}} \subset \mathbb{D}$, be the set of observed poles for clustering. In fuzzy clustering, a cluster is defined by two ingredients: a center (or *prototype*) $\upsilon_i \in \mathbb{D}$, $i \in \mathbb{I}_1^{n_c}$ and a membership function $\mu_i : \mathbb{D} \to [0,1]$. While the former defines the central point, the latter describes the "degree of membership" to the cluster for all $\mathrm{z} \in \mathbb{D}$. Note that the shapes of fuzzy clusters are not described by hard borders but by the commonly bell-like shape of the membership functions on $\mathbb{D}$. Introduce also the so-called dissimilarity measure $d := \mathbb{D} \times \mathbb{D} \to \mathbb{R}_0^+$. This dissimilarity measure is used to define the memberships of the clusters, i.e. how the function set $\{\mu_i\}_{i=1}^{n_c}$ is distributed on $\mathbb{D}$. By using a *threshold value* ε, we can obtain a set

$$\Omega_\varepsilon = \{z \in \mathbb{D} \mid \exists i \in \mathbb{I}_1^{n_c}, \mu_i(z) \geq \varepsilon\}. \tag{8.5}$$

This set characterizes the region approximated by the clusters for the minimal membership level ε. Now we formulate the clustering problem that is considered:

Problem 8.1 (Pole clustering problem). For a set of sampled pole locations Z and for a given number of clusters n_c, find a set of cluster centers $\{\upsilon_i\}_{i=1}^{n_c}$, a set of membership functions $\{\mu_i\}_{i=1}^{n_c}$, and the maximum of ε, such that

- Ω_ε contains Z and it describes the underlying distribution of Z in terms of a chosen dissimilarity measure d.
- With respect to Ω_ε, the OBFs, with poles Λ_{n_c} in the cluster centers $\{\upsilon_i\}_{i=1}^{n_c}$, are optimal in the K$n$W sense, where $n = n_c$. □

The solution is based on finding clusters in accordance with the KnW concept and subsequently finding a maximal value for ε, such that all sampled poles are inside Ω_ε. The latter is equivalent to minimizing ρ in the optimization problem of (8.4). Note that optimality of the OBFs is considered with $n_e = 0$. According to the principle of the KnW theory, this might result in repetitive optimal poles and therefore similar clusters. In the following we focus on finding n-width-based clusters. Additionally, in case $n_c \geq N_z$, the solution of Problem 8.1 is trivial: the cluster centers are associated with the sample poles. Thus only the case when $n_c < N_z$ is considered in the sequel.

Denote by $V = [\upsilon_i]_{i=1}^{n_c}$ the vector of cluster centers and introduce the membership matrix $U = [\mu_{ik}]_{n_c \times N_z}$, where μ_{ik} is the degree of membership of z_k to cluster i, i.e. $\mu_{ik} := \mu_i(\mathrm{z}_k)$. In order to achieve the clustering goal, the dissimilarity measure d is formulated in terms of the 1-width version of the n-width Kolmogorov cost (2.61):

Definition 8.1 (Kolmogorov measure)

$$\kappa_1(\mathrm{x},\mathrm{y}) := \left|\frac{\mathrm{x}-\mathrm{y}}{1-\mathrm{x}\mathrm{y}^*}\right| : \mathbb{D} \times \mathbb{D} \to \mathbb{R}_0^+, \tag{8.6}$$

is called the *Kolmogorov measure* (KM) on $\mathbb{D}$. □

Later we show that κ_1 is a metric in $\mathbb{D}$. As notation,

$$d_{ik} = \kappa_1(\upsilon_i, \mathrm{z}_k), \tag{8.7}$$

is introduced. In the following discussion it is shown how the KM relates the FcM clustering asymptotically to the KnW theory, and in this way to the solution of Problem 8.1. In order to uniquely associate each d_{ik} with a membership level μ_{ik}, the set of membership functions we consider must be restricted. A particular way is to restrict them to $\sum_{i=1}^{n_c} \mu_i(z) = 1$ for all $z \in \mathbb{D}$, i.e. requiring that $U \in \mathcal{U}_{n_c}^{N_z}$, where $\mathcal{U}_{n_c}^{N_z}$, defined as

$$\left\{ U \in [0,1]^{n_c \times N_z} \mid \sum_{i=1}^{n_c} \mu_{ik} = 1,\ \forall k \in \mathbb{I}_1^{N_z} \text{ and } 0 < \sum_{k=1}^{N_z} \mu_{ik},\ \forall i \in \mathbb{I}_1^{n_c} \right\}, \tag{8.8}$$

characterizes the *fuzzy constraints*. The solution of Problem 8.1 can be viewed as a minimization of the fuzzy-functional $J_m(U,V) : \mathcal{U}_{n_c}^{N_z} \times \mathbb{D}^{n_c} \to \mathbb{R}_0^+$ formulated as

$$J_m(U,V) := \max_{k\in\mathbb{I}_1^{N_z}} \sum_{i=1}^{n_c} \mu_{ik}^m d_{ik}. \tag{8.9}$$

This functional defines the cost function, i.e. the criterion of the expected solution for Problem 8.1. It can be observed that (8.9) corresponds to a *worst-case (max) sum-of-error* criterion, contrary to the *mean-squared-error* criterion of the original FcM, see [22]. Hence, we call the algorithm that minimizes the fuzzy-functional (8.9) *fuzzy-Kolmogorov c-max* (FKcM) clustering. The exact relation of (8.9) with the KnW optimality of a partition (U,V) is explained later (see Th. 8.2). The design parameter $m \in (1,\infty)$, which is called the *fuzzyness*, determines the sharpness of the cluster separation in the global minima of (8.9). This means that for low values of m, the clusters in the optimal partition (U,V) are separated, i.e. even for a low value of ε they contribute disjoint regions to Ω_ε. For large values of m, the contribution of the regions are indistinguishable in almost every point of $\mathbb{D}$. This gives the intuition that for low m, we try to achieve the reconstruction of $\Omega_\mathbb{P}$ with the clusters in an "individual" sense, while for large m in a "cooperative" sense. Based on this, the following theorem yields the ingredients to solve Problem 8.1:

Theorem 8.1 (Optimal partition). *Let $m > 1$, a data set $Z \subset \mathbb{D}$ with N_z elements, and a fuzzy partition $(U,V) \in \mathcal{U}_{n_c}^{N_z} \times \mathbb{D}^{n_c}$ be given. Denote $[V]_i = \upsilon_i$ and $[U]_{ij} = \mu_{ij}$. Define $\gamma_i(v,U)$ as the minimal value of $\tau \in [0,1]$ fulfilling the quadratic constraints:*

$$\begin{bmatrix} |1 - \mathrm{z}_k^* \mathrm{v}|^2 & \mu_{ik}^m \cdot (\mathrm{z}_k - \mathrm{v}) \\ \mu_{ik}^m \cdot (\mathrm{z}_k - \mathrm{v})^* & \tau^2 \end{bmatrix} \succeq 0, \quad \forall \mathrm{z}_k \in Z, \tag{8.10}$$

where $\mathrm{v} \in \mathbb{D}$. Additionally, let $d_{ik} = \kappa_1(\upsilon_i, \mathrm{z}_k)$ be the dissimilarity measure of z_k with respect to V and $\mathbb{S}(V,\mathrm{z}_k) = \{i \in \mathbb{I}_1^{n_c} \mid d_{ik} = 0\}$ be the singularity set of z_k with $\mathrm{Card}(\mathbb{S}(V,\mathrm{z}_k)) = n_k$ *(number of elements). If (U,V) is a local minimum of J_m, then for any $(i,k) \in \mathbb{I}_1^{n_c} \times \mathbb{I}_1^{N_z}$:*

$$\mu_{ik} = \begin{cases} \left(\sum\limits_{j=1}^{n_c} \left(\frac{d_{ik}}{d_{jk}} \right)^{\frac{1}{m-1}} \right)^{-1} & \textit{if } \mathbb{S}(V,\mathrm{z}_k) = \emptyset; \\ \frac{1}{n_k} & \textit{if } i \in \mathbb{S}(V,\mathrm{z}_k); \\ 0 & \textit{if } i \notin \mathbb{S}(V,\mathrm{z}_k) \neq \emptyset; \end{cases} \tag{8.11a}$$

$$\upsilon_i = \arg\min_{\mathrm{v}\in\mathbb{D}} \ \gamma_i(\mathrm{v},U). \tag{8.11b}$$

The proof is given in Appendix A.3. Using the approach of the FcM case, minimization of (8.9) subject to (8.8) is tackled by alternating optimization (Picard iteration), steering the solution towards a settling partition in the sense of Th. 8.1. For the FKcM, this yields Algorithm 8.1.

Algorithm 8.1 (FKcM clustering)

Step 1. Fix n_c and m; and initialize $V_0 \in \mathbb{D}^{n_c}$, $l = 0$.

Step 2. With (8.11a), solve $U_{l+1} = \arg \min_{U \in \mathcal{U}_{n_c}^{N_z}} J_m(U, V_l)$.

Step 3. With (8.11b), solve $V_{l+1} = \arg \min_{V \in \mathbb{D}^{n_c}} J_m(U_{l+1}, V)$.

Step 4. If $J_m(U_{l+1}, V_{l+1})$ has converged, then stop, else $l = l + 1$ and goto Step 2.

8.3.2 *Properties of the FKcM*

Next,we investigate the properties of the introduced algorithm, showing that KnW optimality of the resulting cluster centers (if the solution is the global minima of (8.9)) holds in an asymptotic sense ($m \longrightarrow \infty$). It is also discussed how Algorithm 8.1 can be implemented in practice, how convergence of the solution can be detected, and how numerical problems can be avoided.

8.3.2.1 Asymptotic Property

In order to explain the specific choices for the fuzzy functional (8.9) and the dissimilarity measure d_{ik}, we use the following theorem.

Theorem 8.2 (Asymptotic property of J_m). *Given a data set $Z \subset \mathbb{D}$ with N_z elements, and a vector of cluster centers $V \in \mathbb{D}^{n_c}$, such that $d_{ik} = \kappa_1(\upsilon_i, z_k) \neq 0$ for all $(i,k) \in \mathbb{I}_1^{N_z} \times \mathbb{I}_1^{n_c}$ (no singularity). Define U_m as a membership matrix of V satisfying (8.11a) for $m > 1$. Then*

a. $\lim_{m \to 1} J_m(U_m, V) = \max_{k \in \mathbb{I}_1^{N_z}} \min_{i \in \mathbb{I}_1^{n_c}} \{d_{ik}\}$, *which corresponds to the hard partitioning of Z, i.e. $\mu_{ik} \in \{0, 1\}$, $\forall (i,k) \in \mathbb{I}_1^{n_c} \times \mathbb{I}_1^{N_z}$.*

b. $J_2(U_2, V) = \max_{k \in \mathbb{I}_1^{N_z}} \left(\sum_{i=1}^{n_c} d_{ik}\right)^{-1}$, *which is the maximum of the harmonic-means-based distance of each z_k with respect to the clusters.*

c. $J_m(U_m, V) = n_c^{1-m} \max_{k \in \mathbb{I}_1^{N_z}} \left(\prod_{i=1}^{n_c} d_{ik}\right)^{1/n_c} + \mathcal{O}(e^{-m})$. *Furthermore, $J_m(U_m, V)$ decreases monotonically with m, and* $\lim_{m \to \infty} J_m(U_m, V) = 0$.

The proof is presented in Appendix A.3. Th. 8.2 shows that the value of m has great impact on what the minimization of the fuzzy-functional (8.9) represents. If $m = 1$, each sample pole is assigned exactly to one cluster. Thus, minimizing the KM distance of the cluster center with respect to only the assigned poles yields that the resulting cluster center is the pole of the 1-width optimal basis function with respect to the assigned sample poles. In this way, the optimal partition corresponds to a collection of 1-width optimal basis functions with respect to each separated groups of the sample poles. In case $m > 1$, each of the sample poles belongs to all clusters with different membership levels. Thus minimizing the KM distance of each cluster center with respect to the sample poles with these membership weights (see (8.9)), yields a set of pole locations that approximates the poles of the KnW optimal solution. If $m \to \infty$, then these weights/memberships become equal, and all cluster centers try to decrease the KM distance for all sampled poles in a cooperative manner, which is equivalent with the KnW optimization problem (8.4). In this way, the minimization of J_m corresponds to a close approximation of (8.4) for large m. This property enables the FKcM to solve Problem 8.1 directly and explains all the particular choices[1] (dissimilarity measure, modified fuzzy-functional) we made during its introduction. In this way, as a clustering mechanism, the algorithm solves the reconstruction of the possible $\Omega_{\mathbb{P}}$ and at the same time it solves the optimization problem (8.4) in an approximative manner.

It should be noted that, in case $m \to \infty$, $\mu_{ik} \to 1/n_c$ for all $(i,k) \in \mathbb{I}_1^{n_c} \times \mathbb{I}_1^{N_z}$ in the optimal partition, which can cause numerical problems in the minimization of (8.11b). Therefore m acts as a trade-off parameter: to obtain a well approximating solution of Problem 8.1, an appropriately large value of $m \in (1,\infty)$ should be used, but at the same time m must be as low as possible to reduce the complexity of the optimization. Based on experience in the application of the algorithm, $m \in [5,10]$ usually yields satisfactory results (see [196]).

For $m > 1$, the FKcM-functional (8.9) is a bounded $(0 \leq J_m \leq 1)$ monotonically decreasing function both in $\{d_{ik}\}$ and U, which allows Algorithm 8.1 to converge in practice[2]. The convergence point, which directly depends on the initial V_0, can either be a local minimum or a saddle point of J_m, fulfilling Th. 8.1. Therefore, just like for FcM clustering, it is advisable to repeat the algorithm multiple times with different initial choices for V_0 and then select the best resulting set of OBFs. The performance comparison of the resulting clusters is available by computing the Kolmogorov cost (8.3), i.e. associated decay rate, $\breve{\rho}$ of the cluster centers with respect to the sample poles:

$$\breve{\rho} := \max_{z\in\Omega} \kappa_{n_g}(z,\Lambda_{n_g}) = \max_{z\in Z} \prod_{i=1}^{n_c} \left| \frac{z - \upsilon_i}{1 - z\upsilon_i^*} \right|. \tag{8.12}$$

An other performance indicator is how tight the boundary region $\Omega(\breve{\rho},\Lambda_{n_c} = V)$ (see (8.1)) is with respect to Z. In practice, uniformly random choices for V_0 are suggested. Initial partitions based on the distribution of Z can also be used (V_0 chosen as random elements of Z, V_0 is given as the points of a circle around the mean of Z, etc.) however they limit the possibilities to explore all local minima, while random initialization based on a uniform distribution gives equal probability.

In the rare case of singularity of the resulting partition (some $d_{ik} = 0$), Th. 8.2 does not hold. Such a phenomenon can only happen in extreme situations when for example $n_c \approx N_z$. In that case, an optimal partition can contain some clusters whose cluster center is equal to sample pole locations. Singularity of the partition can also result if the samples of $\Omega_{\mathbb{P}}$ do not describe any data coherency, suggesting that $\Omega_{\mathbb{P}}$

[1] See [192] for a comparison between different choices of the dissimilarity measure and the fuzzy-functional to solve Problem 8.1.

[2] For the standard FcM, convergence to a local minimum can be shown [22], but the underlying reasoning does not hold for the FKcM case as J_m is discontinuous on $\mathcal{U}_{n_c}^{N_z}$.

is not a region but a finite set of isolated points. In such cases, the best solution in the KnW sense is to assign a dedicated OBF with respect to some sampled poles. However, when $\Omega_{\mathbb{P}}$ is a region, this solution should be avoided, as the reconstruction of $\Omega_{\mathbb{P}}$ based on the pole samples is required before choosing the basis functions. Thus, to assist the validity of the reconstruction, n_c must be chosen a priori such that it correctly describes the separated pole regions of $\Omega_{\mathbb{P}}$. This can be achieved by visual inspection of the sampled poles, by trial-and-error or by using automated selection (see Sect. 8.3.2.4).

8.3.2.2 Optimization and Numerical Conditioning

While Step 3 in Algorithm 8.1, i.e. the membership update, can be analytically computed through (8.11a), Step 4, i.e. the cluster center update, requires the solution of (8.11b). The optimization defined by (8.11b) is a minimization problem with *quadratic constraints* (QC)s, where γ is the optimization variable and ν is the decision variable. Based on [162], it is possible to derive *sum-of-squares* (SoS) relaxations of such constraints, through which (8.10) is turned into a *linear matrix inequality* (LMI). The resulting convex minimization of γ, based on these constrains, is a *linear semi definite programming* (LSDP) problem that can be efficiently solved by a variety of (interior-point-based) solvers like SeDuMi [183] or CSDP etc. Alternatively, bisection-based recursive search (see [8]) can also be used to obtain the minimization of $\{\gamma_i\}$ with respect to (8.10). In each step of this bisection-based minimization, every QCs with a fixed τ is rewritten as a LMI constraint. Checking feasibility of the constraints indicates how to proceed with the minimization of γ_i (see [193] for the details).

For high values of m, the QCs (8.10) become numerically ill-conditioned, which can be avoided by the normalization of $\{\mu_{ik}^m\}_{k=1}^{N_z}$:

$$\bar{\mu}_{ik} = \frac{\mu_{ik}^m}{\breve{\mu}_i}, \quad \text{with} \quad \breve{\mu}_i = \sum_{k=1}^{N_z} \mu_{ik}^m. \tag{8.13}$$

8.3.2.3 Termination Criterion

In Algorithm 8.1, the cost function J_m "flattens" when m increases. This yields that for high values of m, J_m is almost constant for all points of $\mathcal{U}_{n_c}^{N_z} \times \mathbb{D}^{n_c}$ except in the close neighborhood of its local minima where its value decreases quickly. To avoid unnecessary termination of the algorithm on the flat surface of the cost function, the relative evolution of J_m, in each iteration step l, has to be checked in a windowed sense:

$$1 - \frac{\max_k \left(J_m(U_k, V_k) - J_m(U_{k-1}, V_{k-1})\right)}{\max_k J_m(U_k, V_k)} < \varepsilon_t \tag{8.14}$$

where $k \in \mathbb{I}_{l-n}^{l}$, $n > 0$ is the length of the window, and $0 \ll \varepsilon_t < 1$ is a user defined termination constant. If in step l (8.14) is satisfied, then the relative evolution of J_m has been small in the considered window, so the optimization is terminated. Experience has shown, that for $m \in [5, 10]$, the threshold $\varepsilon_t = 0.99$ with $n = 3$ usually works well.

8.3.2.4 Cluster Merging

The determination of the number of "natural" pole groups in Z, i.e. the best suitable n_c for clustering, is important for the successful application of the FK*c*M method. Similarity-based *adaptive cluster merging* (ACM) can be effectively used for this purpose [196]. The basic idea of ACM is the following: a measure of similarity is introduced with respect to cluster pairs. Then, in each iteration step of Algorithm 8.1, a cluster pair is merged when its similarity between iterations changes negligibly (characterized with a threshold $\varepsilon_s \in [0, 1]$) and if also this pair is the most similar of all cluster pairs. However, merging is only applied if the similarity measure exceeds a certain threshold value, $\varepsilon_a \in [0, 1]$. In F*c*M clustering, commonly the so-called *inclusion similarity measure* is applied (see [83]) and ε_a is taken as an adaptive threshold $\varepsilon_a^{(l)} = (n_c^{(l)} - 1)^{-1}$, where $\varepsilon_a^{(l)}$ is the value of the threshold and $n_c^{(l)}$ the remaining number of clusters in iteration l. In [83] this adaptive threshold has been observed empirically to work well if the initial number of clusters $n_c^{(0)}$ satisfies: $n_c^{(0)} < \frac{1}{2} N_z$.

In [196] it has been shown that starting from a large n_c, the use of the ACM strategy in the FK*c*M algorithm gives the possibility to "adequately"choose the number of clusters to describe the region $\Omega_{\mathbb{P}}$, associated with the samples Z. In the considered problem setting, adequateness means that for the reconstructed region Ω_ε, in which width sense we search for K*n*W optimal OBFs, where the basis functions are associated with the cluster centers. If the available information, the sampling of $\Omega_{\mathbb{P}}$, is dense, then in the optimal solution of Problem 8.1: $\Omega_\varepsilon \approx \Omega_{\mathbb{P}}$. This implies that the ACM provides an effective choice of the width, i.e. the value of n in which the K*n*W basis is searched for $\Omega_{\mathbb{P}}$ with the proposed algorithm. However in terms of Proposition 2.1, the setting of Problem (8.1) implies that repetitive basis poles can also be part of the optimal solution. With ACM, these solutions are not accessible as repetitive poles result in perfectly similar clusters which are immediately joined. As a result, the ACM only yields convergence to partitions with distinct cluster centers. This means that the choice of the adequate width in the K*n*W sense is restricted such that the optimal basis poles for the adequate width must be distinct.

8.3.2.5 Validity Measures of Fuzzy Partitions

To check the quality of the resulting (U, V) partition in terms of the clustering goal, several measures can be introduced that quantify the compactness, separation, and validity of (U, V) (see [9] and [43]). A measure that can jointly express these concepts and give a common ground of comparison between different FcM partitions is the *Xie-Beni validity index* [234]:

$$\chi = \frac{1}{N_\mathrm{z}} \frac{\sum_{k=1}^{N_\mathrm{z}} \sum_{i=1}^{n_\mathrm{c}} \mu_{ik}^2 d^2(\upsilon_i, \mathrm{z}_k)}{\min_{i,j \in \mathbb{I}_1^{n_\mathrm{c}}} d^2(\upsilon_i, \upsilon_j)}. \tag{8.15}$$

It can be proved that the smaller χ is, the better the corresponding fit of (U,V) with respect to Z is.

Based on the initialization of the FcM algorithm, optimal partitions with different n_c can be attractive solutions of Algorithm 8.1 with ACM. To decide which of the solutions represents the underlying data-structure best, the separation of the clusters can give an indication. To quantify the quality of separation, commonly the *normalized entropy* is used [22]:

$$S_\mathrm{e} = -\frac{\sum_{i=1}^{n_\mathrm{c}} \sum_{k=1}^{N_\mathrm{z}} \mu_{ik} \log(\frac{\mu_{ik}}{N_\mathrm{z}})}{N_\mathrm{z} - n_\mathrm{c}}. \tag{8.16}$$

The smaller S_e is, the more valid the hypothesis is that the clusters match with the natural data groups (if they exist).

8.3.3 Simulation Example

To allow insight into the basis selection method an extensive example is studied.

8.3.3.1 The Data Generating System

Consider an asymptotically stable discrete-time SISO LPV system $\mathcal{S}$ with IO partition (u,y) and scheduling signal p. Let a minimal IO representation of $\mathcal{S}$, $\mathfrak{R}_\mathrm{IO}(\mathcal{S})$ be given as:

$$\sum_{i=0}^{5} (a_i \diamond p)\mathrm{q}^i y = (b_4 \diamond p)\mathrm{q}^4 u, \tag{8.17}$$

with $\mathbb{P} = [0.6, 0.8]$ and coefficients

$$\begin{aligned}
a_0 \diamond p &= -0.003, & a_3 \diamond p &= \tfrac{61}{110} - 0.2\sin(\mathrm{q}^5 p),\\
a_1 \diamond p &= \tfrac{12}{125} - 0.1\sin(\mathrm{q}^5 p), & a_4 \diamond p &= -\tfrac{511+192\mathrm{q}^5 p^2 - 258\left(\cos(\mathrm{q}^5 p) - \sin(\mathrm{q}^5 p)\right)}{860},\\
a_2 \diamond p &= -\tfrac{23}{85} + 0.2\sin(\mathrm{q}^5 p), & a_5 \diamond p &= 0.58 - 0.1\mathrm{q}^5 p,\\
b_4 \diamond p &= \cos(\mathrm{q}^5 p).
\end{aligned}$$

In Fig. 8.2a, the pole manifest set $\Omega_\mathbb{P}$ of $\mathfrak{R}_\mathrm{IO}(\mathcal{S})$ is presented with a solid red line, while in Fig. 8.2b the first 15 Markov parameters of the frozen impulse responses of $\mathfrak{R}_\mathrm{IO}(\mathcal{S})$ are given for all constant scheduling trajectories $p(k) = \mathrm{p}$. These frozen

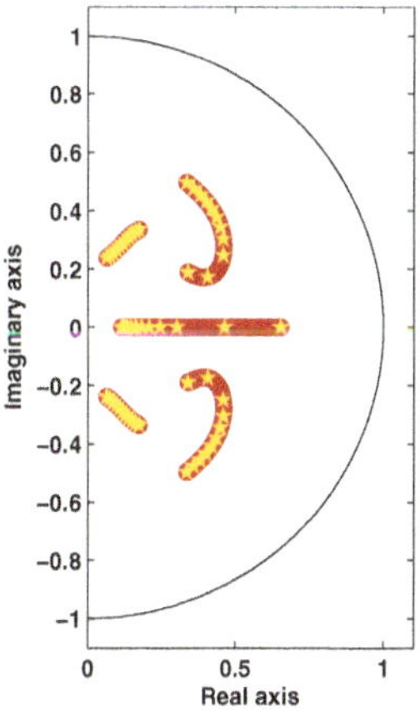

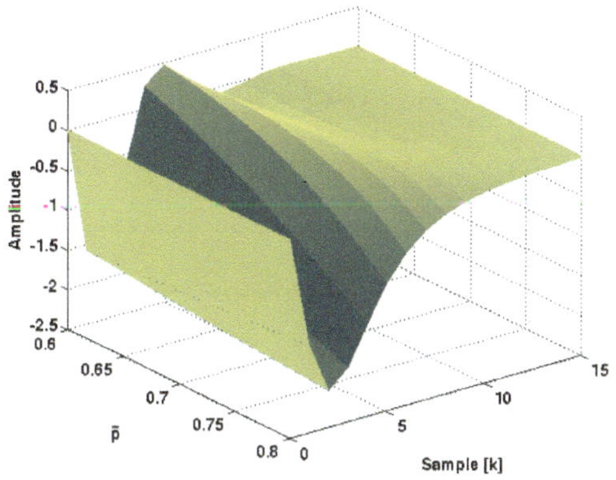

Fig. 8.2 (a) The pole manifest set $\Omega_{\mathbb{P}}$ (solid red line) of the LPV-IO representation $\mathfrak{R}_{\mathrm{IO}}(\mathcal{S})$. Sampled pole locations are denoted by $\star$. (b) First 15 Markov parameters of the impulse responses of $\mathfrak{R}_{\mathrm{IO}}(\mathcal{S})$ with respect to all constant scheduling trajectories $p(k) = \bar{p}$.

impulse responses are associated with the behaviors of the frozen system set $\mathscr{F}_{\mathcal{S}}$ of $\mathcal{S}$ for the considered IO partition. These pictures show that the dynamic changes of $\mathcal{S}$ are quite heavy for different constant scheduling trajectories.

By using constant scheduling signals with values $\{0.6; 0.6+\tau; \ldots; 0.8\}$, where $\tau = 0.02$, 11 frozen LTI representations of $\mathcal{S}$ are obtained, whose pole locations are samples of $\Omega_{\mathbb{P}}$. These samples are given with yellow $\star$ in Fig. 8.2a. In our basis selection approach, these LTI representation are considered to be the results of identification of $\mathcal{S}$ with constant p.

8.3.3.2 FK*c*M Clustering of the Sample Poles

By using the obtained $N_{\mathrm{z}} = 11 \cdot 5$ sample pole locations as a data set Z, the FK*c*M algorithm has been executed with different values of m and both with a fixed number of clusters as with the application of ACM starting from $n_{\mathrm{c}}^{(0)} = 27 \approx N_{\mathrm{z}}/2$. In the fixed case, $n_{\mathrm{c}} = 8$ is used and the obtained solution is denoted as *m2n*$_{\mathrm{c}}$*8* for a fuzzyness $m = 2$. In case of ACM, if the algorithm has resulted in $n_{\mathrm{c}} = 11$ clusters for a fixed m, like $m = 8$, then the solution is denoted by *m8ad11*. Note that in the fixed case we use the particular choice of 8 clusters as this number of clusters agrees with the number of sets by visual inspection (two times 3 sets for the complex and 2 sets for the real poles). It is shown in the sequel that this number of clusters is also selected by the ACM.

The results of the algorithm are presented in Table 8.1 and in Fig. 8.3. The comparison in Table 8.1 is given in terms of N_{av}, the average number of iterations based on 10 runs of the algorithm starting from random V_0; n_{c}, the number of obtained clusters; S_{e}, the Normalized Entropy (see (8.16)); χ, the Xie-Beni validity index

Table 8.1 Comparison of algorithmic results in terms of N_{av}, the average number of iterations based on 10 runs of the algorithm starting from random V_0; n_c, the number of obtained clusters; S_e, the Normalized Entropy; χ, the Xie-Beni validity index; $\breve{\rho}$, the achieved decay rate; and $\varepsilon_{max}^{n_e}$, the worst-case absolute error of the impulse responses of the truncated series-expansion representation of each $\mathcal{F}_p \in \mathscr{F}_S$ in terms of the cluster centers generated OBFs with n_e repetition.

Test case	N_{av}	n_c	χ (dB)	$\breve{\rho}$ (dB)	S_e	$\varepsilon_{max}^{n_e=1}$ (dB)	$\varepsilon_{max}^{n_e=3}$ (dB)
m2n$_c$*8*	21	8	−17.49	−55.86	1.79	−43.73	−146.61
m8ad8	37	8	−12.42	−58.38	2.41	−46.90	−171.41
m8ad11	65	11	−8.44	−83.11	2.94	−77.33	−249.63
m25n$_c$*8*	56	8	−13.20	−61.36	2.43	−45.34	−168.83

(see (8.15)); $\breve{\rho}$, the achieved decay rate (see (8.12)); and $\varepsilon_{max}^{n_e}$, the worst-case absolute error of the impulse responses of the truncated series-expansion representation of each $\mathcal{F}_p \in \mathscr{F}_S$ in terms of the resulting OBFs with n_e repetitions. In Fig. 8.3, the resulting basis poles are given by blue $\times$ for each solutions together with the sampled poles (red $\circ$). By using the cluster centers as basis poles, $\Lambda_{n_c} = V$, the resulting boundary of $\Omega\left(\Lambda_{n_c}, \breve{\rho}\right)$ is also given in Fig. 8.3. Based on these, the following observations can be made:

8.3.3.3 Analysis of the Results

- The values of N_{av}, which are based on the results of 10 runs starting from random V_0, are relatively low, but they are growing with m. Explanation lies in Th. 8.2, by which $J_m \longrightarrow 0$ as $m \longrightarrow \infty$. This property introduces both increased computational error and flat shapes of membership surfaces for large m (compare Fig. 8.4a to 8.4b). Flat surfaces give smaller improvements towards the minimum of J_m in each iteration of Algorithm 8.1, than the smooth slope of the $m = 2$ case.
- The FK*c*M with ACM ($\varepsilon_s = -15$dB), starting from $n_c^{(0)} = N_z/2$, converges to a 8-cluster-based partition for low m, but in case of higher values of m, the optimization has different attractive solutions, like the *m8ad8* and *m8ad11* cases. Here both the 8 and the 11 cluster-based partitions are attractive, depending on the initial position of the cluster centers. However, *m8ad8* achieves a lower entropy S_e than *m8ad11*, suggesting that *m8ad8* corresponds better to the natural data structure. As different initial conditions can drive the FK*c*M with ACM to converge to partitions with different n_c, it is suggested to the user to choose the one with the lowest S_e, as it most likely yields the "best" partition.
- χ is small in all cases, showing that each partition represents the underlying structure well. However, χ is not comparable for different m. χ has a decreasing tendency with growing n_c and an increasing tendency for growing m, therefore the fact that $\chi_{m25n_c8} < \chi_{m8ad8}$ supports that *m25n*$_c$*8* corresponds better to the underlying data structure in the K*n*W sense than *m8ad8*. However, such a comparison can not be made with respect to the *m8ad11*-case.

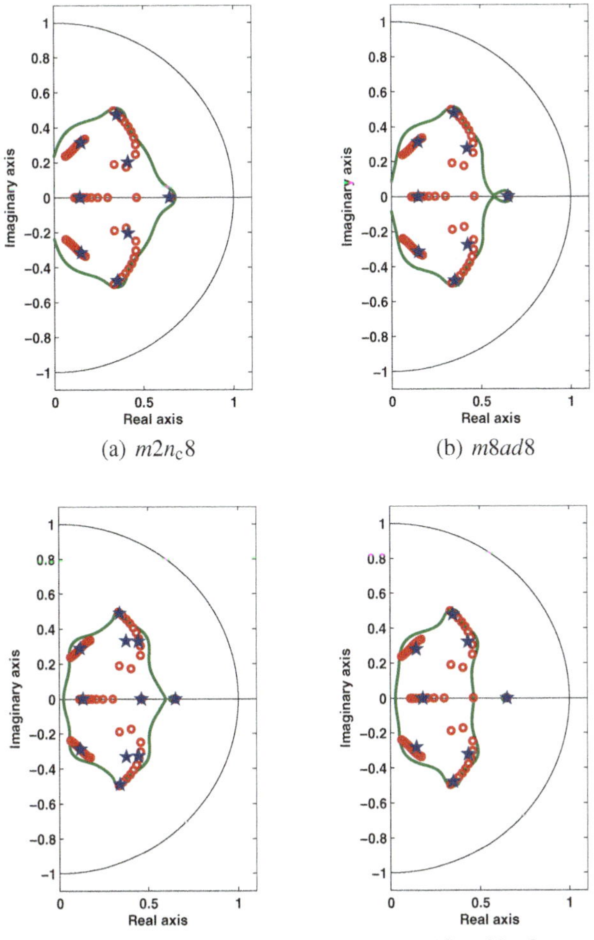

(a) $m2n_c8$ (b) $m8ad8$

(c) $m8ad11$ (d) $m25n_c8$

Fig. 8.3 Results of FKcM clustering in the considered cases: sampled poles (red o), resulting cluster centers (blue $\star$), and boundaries of $\Omega\,(\Lambda_{n_c}, \breve{\rho})$ (green bold lines).

- The region $\Omega\,(\Lambda_{n_c}, \breve{\rho})$ describes the pole locations of all transfer functions that have a series-expansion with a worst-case convergence rate of $\breve{\rho}$ in terms of the Λ_{n_c} associated OBFs. For the resulting cluster centers, the boundary of $\Omega\,(\Lambda_{n_c}, \breve{\rho})$ is relatively tight in all cases except for *m2n*$_c$*8* and it also includes $\Omega_{\mathbb{P}}$ (see Fig. 8.2). This means that the convergence rate of the basis has been focused/optimized for LTI systems with transfer functions that have poles close to the frozen poles of $\mathfrak{R}_{\mathrm{IO}}(\mathcal{S})$. $\breve{\rho}$ is also acceptable, which means small modeling error, i.e. fast convergence rate if the corresponding poles generated orthonormal basis are used for the series-expansion representation of the frozen behaviors of $\mathcal{S}$. In terms of Sect. 5.5, this implies fast convergence rate of the series-expansion of $\mathcal{S}$ with respect to the derived basis. In the *m8ad11*-case, $\breve{\rho}$ is the best, which

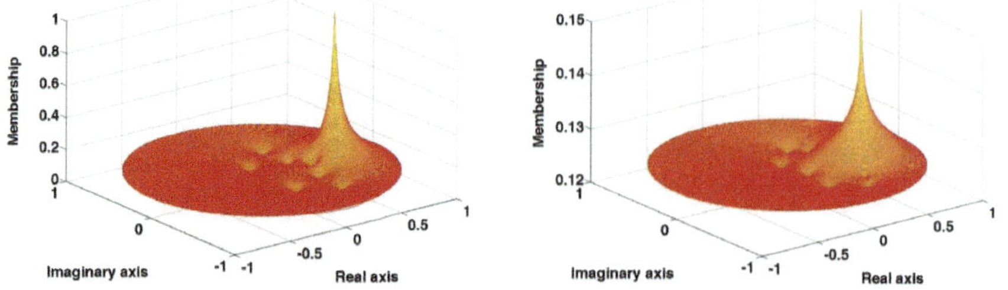

(a) $m = 2$, $n_c = 8$ (b) $m = 25$, $n_c = 8$

Fig. 8.4 Membership functions of the 8th cluster for different m. In the second case the point corresponding to υ_8 is not shown due to the peaking nature of the function to 1 (only a straight line to 1 from a flat surface would be presented).

is the consequence of the larger ($n_c = 11$) number of OBFs only. By repeating the obtained pole sets in the Hambo generation of the basis functions, i.e. using $n_e > 0$, such that the number of generated basis functions are equal, comparison of the KnW performance of these cases becomes available. Based on such a comparison, it follows that $m25n_c8$ is better in the KnW sense, which is in agreement with Th. 8.2. The partition $m2n_c8$ is the worst among these results, which suggests that only larger values of m can ensure the quality of the obtained solution.

- Fig. 8.5 and Table 8.1 show the representation errors of the frozen impulse responses of $\mathfrak{R}_{\mathrm{IO}}(\mathcal{S})$ by the impulse responses of the truncated series-expansion representation of each $\mathcal{F}_p \in \mathscr{F}_{\mathcal{S}}$ in terms of the OBFs generated by the cluster centers with n_e repetitions. From these results it follows that the obtained set of OBFs has negligible representation error with respect to $\mathfrak{F}_{\mathcal{S}}$, which has been our main objective (see Sect. 8.3.1). Among the solutions with 8 basis functions, surprisingly $m8n_c8$ has the lowest representation error of the frozen impulse responses instead of $m25n_c8$. Based on the previous results, one would expect that the representation error of the frozen impulse responses is less for OBFs generated with higher m, however this is not the case here, due to the fact that $\Omega_{\mathbb{P}}$ is sampled. Even if $m25n_c8$ delivers a better choice with respect to the sampled pole locations, it is not guaranteed that the reconstruction of $\Omega_{\mathbb{P}}$, based on the sample poles, resulted in a better estimate than in the other cases. By comparing the results in terms of S_e, such a phenomenon is clearly indicated. The quality of the information, i.e. how well the pole samples describe $\Omega_{\mathbb{P}}$ is highly significant in establishing optimality between the sampled-poles-based OBFs and the original system.

In conclusion, the FKcM solutions for the considered example are converging relatively fast to optimal partitions in terms of Th. 8.1. In accordance with Th. 8.2, as m increases, these partitions give better solutions of Problem 8.1. ACM also ensures proper selection of an efficient number of OBFs in the KnW sense, if the different

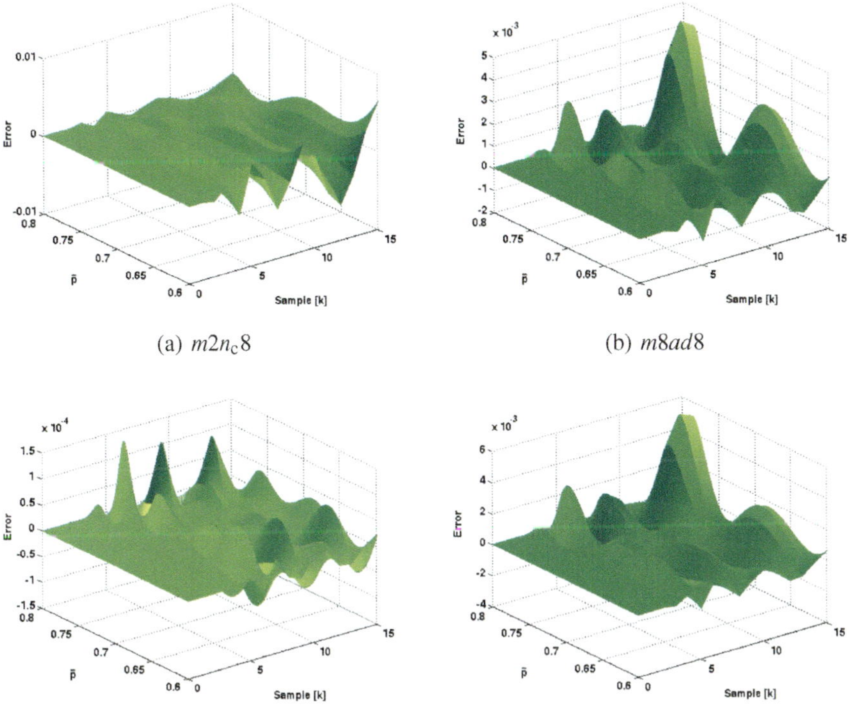

(a) $m2n_c8$ (b) $m8ad8$

(c) $m8ad11$ (d) $m25n_c8$

Fig. 8.5 Approximation error of the frozen impulse responses of $\mathfrak{R}_{IO}(\mathcal{S})$ by the impulse responses of the truncated expansion representations of each $\mathcal{F}_p \in \mathscr{F}_\mathcal{S}$ in terms of the FKcM clustering obtained OBFs.

settling partitions are compared in terms of S_e. Furthermore, validity of the derived partitions is supported by low χ in all cases.

Comparison of results to solutions obtained by the gradient search method [73, Ch. 11], is only possible if the number of available samples of $\Omega_\mathbb{P}$ is so high that there is no need for the reconstruction of $\Omega_\mathbb{P}$. Thus an advantage of the FKcM approach is that it gives a solution for the practical case when only few samples of $\Omega_\mathbb{P}$ are available. In the unrealistic case, when $\Omega_\mathbb{P}$ is known, both algorithms converge to similar solutions, but with a lower computational time in the FKcM case. The two algorithms also have similar properties in the sense that they only yield convergence to local minima. As online selection of the efficient number of OBFs is difficult to implement into the gradient search method, the FKcM approach, with strategies like the ACM, has a second advantage over gradient approaches.

8.4 Robust Extension of the FKcM Approach

During the development of the FKcM approach, we have assumed that some samples of $\Omega_{\mathbb{P}}$ are given. These samples have been used as a data set, i.e. a source of information on which the developed basis selection tool is based on. We have motivated that such samples are available in practice either based on first principle information or as the poles of model estimates of the frozen system set $\mathscr{F}_{\mathbb{S}}$. These model estimates are considered to be the result of LTI identification of the LPV system with constant scheduling signals. However, system identification is in practice always affected by noise, thus the pole locations of the estimated models are the realizations of an underlying probability function of the frozen poles. This means that selecting basis functions, based on estimated poles, carries the risk that the result of the basis selection by the FKcM method is significantly effected by the noise. In this section, we aim to provide a robust solution of the previously considered basis selection problem by reformulating the developed FKcM approach, such that it takes into account the uncertainty of the pole estimates.

8.4.1 Questions of Robustness

In prediction-error identification, each estimated pole can be associated with an uncertainty region in the complex plain for a certain level of confidence. In the classical literature, ellipsoidal regions are quantified that are the results of the linearization of the map from parameters θ to pole locations $\{\lambda\}$. In Sect. 2.4.6, an alternative approach of [223] has been introduced that characterizes pole uncertainty regions of the model estimates without linearization. The approach leads to a (possibly disconnected) uncertainty region $\mathcal{P}(\mathcal{Q}_\theta,\alpha)\subset\mathbb{C}$ for which it holds that

$$\hat{\lambda}_1,\ldots,\hat{\lambda}_n\in\mathcal{P}(\mathcal{Q}_\theta,\alpha),\quad\text{with probability}\geq\alpha \tag{8.18}$$

where $\{\hat{\lambda}_1,\ldots,\hat{\lambda}_n\}$ are the poles of the model estimate. Note that this uncertainty concept is still necessarily conservative as it disregards covariance of the pole estimates. However, later it will be shown that the locations of the possible pole estimates, i.e. possible poles of a transfer function set, are formulated in the same worst-case sense as the KnW theory. Based on this, it is clear that by using the uncertainty regions provided by this approach as data objects in both the reconstruction problem of $\Omega_{\mathbb{P}}$ and the KnW optimization problem, the basis selection task can be solved in a robust sense.

In the original problem of FKcM clustering, the samples of $\Omega_{\mathbb{P}}$ form a finite set of points Z in $\mathbb{D}$. Because of the fact that Z is finite, both the reconstruction problem and the minimization of (8.4) can be analytically computed and solved via the proposed algorithm. However, if Z is not a finite set but a collection of complex regions, i.e. uncertainty sets, it is not trivial how to calculate ρ in (8.4) with respect to these regions or how to obtain the worst-case KM distance used in the FKcM algorithm as the dissimilarity measure d_{ik}. Subsequently the problem rises how to solve the selection problem of the OBFs poles by using the FKcM mechanism.

To provide answers for these questions, in Sect. 8.4.2 we first show that by using results of hyperbolic geometry, (8.3) and its 1-width version, the so-called Kolmogorov measure, can be analytically computed if the regions in Z are hyperbolic circles or hyperbolic segments. However, shapes of uncertainty regions can vary arbitrary, thus to use the developed hyperbolic results, the regions must be covered by a collection of hyperbolic circles or hyperbolic segments. In Sect. 8.4.2 a practically applicable multistep procedure is developed that covers complex regions with the union of hyperbolic circles. In this way the KnW optimization problem (8.4) and also the reconstruction problem of $\Omega_{\mathbb{P}}$ by the clustering of pole uncertainty regions can be solved along a similar line of reasoning as in Sect. 8.3. This robust extension of the FKcM clustering is developed in Sect. 8.4.4 and its properties are investigated again in terms of optimality of the solution, numerical convergence, etc. It is shown that the resulting algorithm provides the selection of asymptotically optimal OBFs in the KnW sense for the local behaviors of $\mathcal{S}$ even in case of significant measurement noise. This property is also illustrated through an example.

8.4.2 Basic Concepts of Hyperbolic Geometry

In order to develop the required geometrical tools, the basic aspects of the 2-dimensional *Poincaré disc model* of hyperbolic geometry are introduced in the sequel. It is also shown how hyperbolic geometry can be used to generalize the results of the KnW theory.

The 2-dimensional Poincaré model provides a conformal disc model, where points of the geometry are in a complex disc. The lines are segments of circles that lie inside the disc, where the circles themselves are orthogonal to the boundary of the disc, or the segments are part of a diameter of the disc (see Fig. 8.6). Before defining these objects, it is motivated why this geometric model has important relations to the KnW theory.

We have already discussed that the Kolmogorov measure (see Definition 8.1) is equal to the cost function of the 1-width Kolmogorov problem ($n_{\mathrm{g}} = 1$) (see (8.3)). Based on this property, the KM had an important role as a dissimilarity measure in the FKcM algorithm to formulate KnW optimality of the obtained solution in an asymptotic sense. Additionally, this observation has an important consequence for geometrical objects in $\mathbb{D}$ which are convex in terms of the KM. Namely, that in the $n_{\mathrm{g}} = 1$ case, the solution of (8.4) with respect to regions equivalent with these objects can be found through LMI's based optimization. We will see that this property provides an efficient way to solve the heavy nonlinear optimization problem that (8.4) represents and is also the key ingredient to handle the robust basis selection problem. In the following, we first introduce some simple convex objects in the KM sense and additional geometrical tools that are required in the second part to develop three key results for the above mentioned property.

The most simplest of the convex objects in the KM sense are the segments of hyperbolic lines which are introduced through the concept of i-lines (see Fig. 8.6.a) and their orthogonality in $\mathbb{C}$:

Definition 8.2 (i-line, [32]). In $\mathbb{C}$, an i-line L is either an Euclidian line E:

$$L = E(\mathtt{e},\mathtt{r}) := \left\{ z \in \mathbb{C} \;\middle|\; \begin{array}{ll} \text{if } \mathtt{r} < \infty, & \mathrm{Im}(z) = \mathtt{r}\mathrm{Re}(z) + \mathtt{e} \\ \text{else,} & \mathrm{Re}(z) = \mathtt{e} \end{array} \right\}, \tag{8.19}$$

with y-intercept $\mathtt{e} \in \mathbb{C}$ and slope $\mathtt{r} \in \mathbb{R}_0^+ \cup \{\infty\}$, or Euclidian circle K

$$L = K(\mathtt{e},\mathtt{r}) := \{z \in \mathbb{C} \mid |z - \mathtt{e}| = \mathtt{r}\}, \tag{8.20}$$

with center $\mathtt{e} \in \mathbb{C}$ and radius $\mathtt{r} \in \mathbb{R}^+$, such that $L \cap \mathbb{D} \neq \emptyset$. □

Definition 8.3 (Orthogonality, [32]). Two i-lines L_1 and L_2 are orthogonal, iff

L_1	L_2	$L_1 \perp L_2$
line	line	Euclidian orthogonality
circle	line	$L_1 = K(\mathtt{e}_1, \mathtt{r}_1)$, then $\mathtt{e}_1 \in L_2$
circle	circle	$\forall z \in L_1 \cap L_2 \neq \emptyset$, the radii of L_1 and L_2 through z are orthogonal

□

Now it is possible to define hyperbolic lines and segments (see Fig. 8.6):

Definition 8.4 (h-line & h-segment, [32]). A hyperbolic line (h-line) is defined as $H = L \cap \mathbb{D}$, where L is an i-line and $L \perp \mathbb{J}$ (orthogonal to the unit circle). The section of H between $\mathrm{x}, \mathrm{y} \in H$ is denoted as H_{xy} and called a hyperbolic segment (h-segment). □

Lemma 8.1 (Uniqueness of h-lines, [32]). *If* $\mathrm{x} \neq \mathrm{y}$ *and* $\mathrm{x}, \mathrm{y} \in \mathbb{D}$*, then there is a unique h-line* H *such that* $\mathrm{x}, \mathrm{y} \in H$.

Hence h-lines are part of Euclidean circles orthogonal to the unit circle or part of Euclidian lines through the origin, where the part is strictly inside $\mathbb{D}$. The concept of h-bisectors of h-segments (see Fig. 8.6.b) is also important to develop connections of Euclidian and hyperbolic geometry.

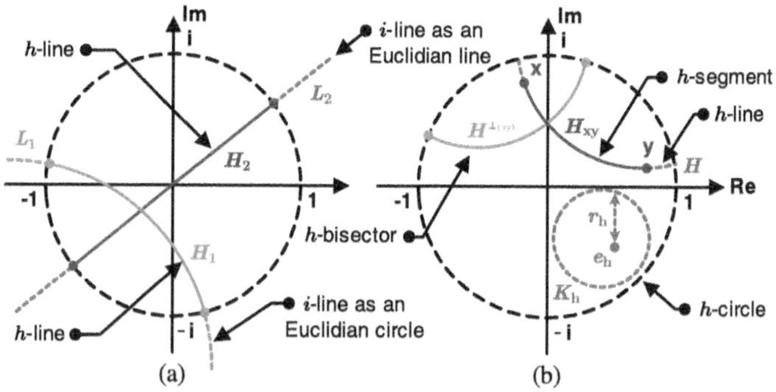

Fig. 8.6 Hyperbolic objects of the Poincaré disc model.

Definition 8.5 (h-bisector, [32]). The h-bisector of the h-segment H_{xy} (segment of h-line H), is an h-line $H^{\perp(\mathrm{xy})}$ containing the midpoint (in the KM sense) of H_{xy} and $H^{\perp(\mathrm{xy})} \perp H$, meaning that their corresponding i-lines are orthogonal. □

Circles are convex geometrical objects in Euclidian geometry. Their counterpart in hyperbolic geometry (see Fig. 8.6.b) is defined as follows:

Definition 8.6 (h-circle & h-disc, [32]). A hyperbolic circle (h-circle) $K_{\mathrm{h}}(\mathrm{e}_{\mathrm{h}}, \mathrm{r}_{\mathrm{h}})$ and a hyperbolic disc (h-disc) $D_{\mathrm{h}}(\mathrm{e}_{\mathrm{h}}, \mathrm{r}_{\mathrm{h}})$ with h-center $\mathrm{e}_{\mathrm{h}} \in \mathbb{D}$ and h-radius $\mathrm{r}_{\mathrm{h}} \in (0,1)$ are defined as

$$K_{\mathrm{h}}(\mathrm{e}_{\mathrm{h}}, \mathrm{r}_{\mathrm{h}}) := \{z \in \mathbb{D} \mid \kappa_{\mathrm{l}}(z, \mathrm{e}_{\mathrm{h}}) = \mathrm{r}_{\mathrm{h}}\}, \tag{8.21a}$$

$$D_{\mathrm{h}}(\mathrm{e}_{\mathrm{h}}, \mathrm{r}_{\mathrm{h}}) := \{z \in \mathbb{D} \mid \kappa_{\mathrm{l}}(z, \mathrm{e}_{\mathrm{h}}) \leq \mathrm{r}_{\mathrm{h}}\}. \tag{8.21b}$$

□

To establish connection with the Euclidian geometry, the following lemmas are important:

Lemma 8.2 (h-circle equivalence, [32]). *For any Euclidian circle $K(\mathrm{e}, \mathrm{r}) \subset \mathbb{D}$ with $\mathrm{r} > 0$, there exists a unique h-circle $K_{\mathrm{h}}(\mathrm{e}_{\mathrm{h}}, \mathrm{r}_{\mathrm{h}})$, such that $K_{\mathrm{h}}(\mathrm{e}_{\mathrm{h}}, \mathrm{r}_{\mathrm{h}}) = K(\mathrm{e}, \mathrm{r})$ and e_{h} is strictly inside $K(\mathrm{e}, \mathrm{r})$, i.e. $\mid \mathrm{e}_{\mathrm{h}} - \mathrm{e} \mid < \mathrm{r}$.*

Lemma 8.3 (h-center relation). *For any h-circle $K_{\mathrm{h}}(\mathrm{e}_{\mathrm{h}}, \mathrm{r}_{\mathrm{h}})$ and its Euclidian equivalent $K(\mathrm{e}, \mathrm{r})$, there exists a $\varphi_{\mathrm{h}} \in \mathbb{R}$, such that $\mathrm{e} = \varphi_{\mathrm{h}} \mathrm{e}_{\mathrm{h}}$.*

The proof of Lemma 8.3 is given in Appendix A.3. From Lemma 8.2, it follows that the same equivalence holds between discs and h-discs. Furthermore, Lemma 8.3 states that the h-center and the Euclidian center of a circle or a disc lie on the same Euclidian line connecting them to the origin. Note that hyperbolic circles are defined through the KM measure. Therefore, based on (2.62) and Proposition 2.1, for any circular pole region $\Omega = D_{\mathrm{h}}(\mathrm{e}_{\mathrm{h}}, \mathrm{r}_{\mathrm{h}})$, the optimal Λ_n in the Kolmogorov n-width sense is $\Lambda_n = [\,\mathrm{e}_{\mathrm{h}} \ \dots \ \mathrm{e}_{\mathrm{h}}\,]_{1 \times n}$ with $\rho = \mathrm{r}_{\mathrm{h}}^n$. This important consequence generalizes the result of [143] for the pulse basis (see Sect. 2.5):

Theorem 8.3 (KnW optimal OBFs for circular regions of non-analyticity). *For the class of stable transfer functions analytical outside the disc $D_{\mathrm{h}}(\mathrm{e}_{\mathrm{h}}, \mathrm{r}_{\mathrm{h}})$ with $\mathrm{e}_{\mathrm{h}} \in \mathbb{D}$ and $\mathrm{r}_{\mathrm{h}} \in (0,1)$, the set of (complex) OBFs*

$$\left\{ \frac{\sqrt{1 - \mid \mathrm{e}_{\mathrm{h}} \mid^2}}{z - \mathrm{e}_{\mathrm{h}}} G_{\mathrm{b}}^i(z) \right\}_{i=0}^{n-1}, \quad \text{with} \quad G_{\mathrm{b}}(z) = \frac{1 - z\mathrm{e}_{\mathrm{h}}^*}{z - \mathrm{e}_{\mathrm{h}}}, \tag{8.22}$$

are optimal in the KnW sense.

The proof of Th. 8.3 is trivial from the previously described motivation. To utilize this property of h-circles in the sequel, the following concepts are crucial:

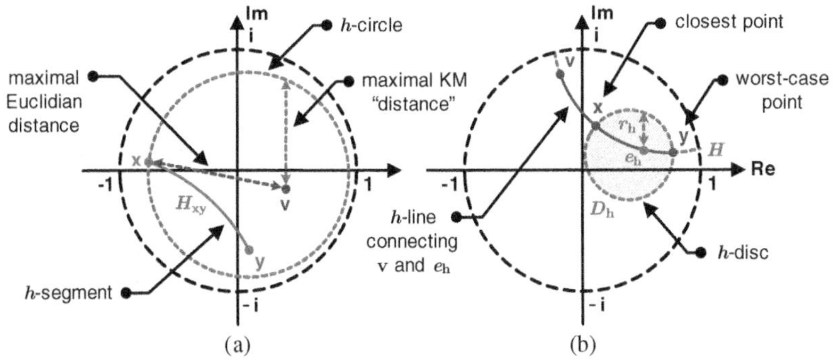

Fig. 8.7 Worst-case distance of a point $v \in \mathbb{D}$ in the KM sense with respect to (a) h-segment H_{xy} or (b) h-disc $D_h(e_h, r_h)$.

Definition 8.7 (Angle of h-lines, [32]). Let H_1 and H_2 be two h-lines intersecting in point x and let H_3 be an h-line intersecting H_1 in y and H_2 in z. Then the angle γ_h between H_1 and H_2 is defined by the cosine rule:

$$\sin(\kappa_1(x,y))\sin(\kappa_1(x,z))\cos(\gamma_h) = \cos(\kappa_1(x,y))\cos(\kappa_1(x,z)) - \cos(\kappa_1(y,z)),$$

By convention, γ_h is the acute angle between H_1 and H_2. □

Definition 8.8 (h-inversion, [32]). $\mathfrak{h}_H : \mathbb{D} \to \mathbb{D}$ is called an inversion with respect to the h-line H, iff $\mathfrak{h}_H$ maps h-lines to h-lines, preserves angles between h-lines, and $\mathfrak{h}_H(z) = z$ for $\forall z \in H$. □

The following properties are immediate:

Lemma 8.4 (h-inversion uniqueness, [32]). *For $\forall x \in \mathbb{D}$, there exists an h-inversion $\mathfrak{h}_H$, such that $\mathfrak{h}_H(x) = 0$ and $\mathfrak{h}_H(0) = x$. If $x \neq 0$, then the h-line H associated with $\mathfrak{h}_H$ is unique and $H = H^{\perp(x,0)}$, otherwise H can be any h-line through the origin.*

In the sequel, the notation $\mathfrak{h}_x$ is used to denote the h-inversion associated with x.

Lemma 8.5 (Hyperbolic group, [32]). *The h-inversions generate a hyperbolic group $\mathcal{D}$ whose elements $\mathfrak{h} : \mathbb{D} \to \mathbb{D}$ are called hyperbolic transformations. Each $\mathfrak{h} \in \mathcal{D}$ is a conformal mapping of $\mathbb{D}$.*

These give the following crucial observations:

Lemma 8.6 (h-circle transformation, [32]). *For any $\mathfrak{h} \in \mathcal{D}$ and h-circle $K_h(e_h, r_h)$,*

$$\mathfrak{h}(K_h(e_h, r_h)) = K_h(\mathfrak{h}(e_h), r_h). \tag{8.23}$$

Corollary 8.1 (κ_1-invariance, [32]). *$\kappa_1(x,y)$ is invariant under $\mathcal{D}$, meaning that $\kappa_1(x,y) = \kappa_1(\mathfrak{h}(x), \mathfrak{h}(y))$, for all $\mathfrak{h} \in \mathcal{D}$ and $x, y \in \mathbb{D}$.*

Now it is possible to develop three key results that are used for the robust solution of the basis selection problem.

Theorem 8.4 (κ_1-metric). *The Kolmogorov measure κ_1 is a metric on $\mathbb{D}$.*

The proof is given in Appendix A.3. This shows that KM is a natural "distance[3]" on $\mathbb{D}$.

Theorem 8.5 (h-segment worst-case distance). *Given $\mathrm{x}, \mathrm{y} \in \mathbb{D}$, $\mathrm{x} \neq \mathrm{y}$, defining the h-segment H_{xy}, then for any $\mathrm{v} \in \mathbb{D}$*

$$\max_{z \in H_{\mathrm{xy}}} \kappa_1(\mathrm{v}, z) = \max_{z \in \{\mathrm{x}, \mathrm{y}\}} \kappa_1(\mathrm{v}, z). \tag{8.24}$$

The proof is given in Appendix A.3. Theorem 8.5 shows (see Fig. 8.7a), that the worst-case KM of any point in $\mathbb{D}$ with respect to an h-segment can be calculated as the maximum of the KMs with respect to the endpoints of the segment. In Sect. 8.4.4, this result is used in the solution of the KnW optimization problem for calculating the worst-case cost of basis pole candidates with respect to uncertainty regions associated with real pole estimates.

Theorem 8.6 (h-disc worst-case distance). *Let $D_{\mathrm{h}}(\mathrm{e}_{\mathrm{h}}, \mathrm{r}_{\mathrm{h}})$ be an h-disc and $\mathrm{v} \in \mathbb{D}$. Denote by $K_{\mathrm{h}}(\mathrm{e}_{\mathrm{h}}, \mathrm{r}_{\mathrm{h}})$ the perimeter circle of $D_{\mathrm{h}}(\mathrm{e}_{\mathrm{h}}, \mathrm{r}_{\mathrm{h}})$ and by H the unique h-line through v and e_{h} if $\mathrm{v} \neq \mathrm{e}_{\mathrm{h}}$. Then,*

$$\max_{z \in D_{\mathrm{h}}} \kappa_1(\mathrm{v}, z) = \begin{cases} \max\limits_{z \in \{\mathrm{x}, \mathrm{y}\}} \kappa_1(\mathrm{v}, z), & \text{if } \mathrm{v} \neq \mathrm{e}_{\mathrm{h}}; \\ \mathrm{r}_{\mathrm{h}}, & \text{if } \mathrm{v} = \mathrm{e}_{\mathrm{h}}; \end{cases} \tag{8.25}$$

where $\{\mathrm{x}, \mathrm{y}\} = K_{\mathrm{h}}(\mathrm{e}_{\mathrm{h}}, \mathrm{r}_{\mathrm{h}}) \cap H$ (see Fig. 8.7b).

The proof is given in Appendix A.3. Again this key result is used in the solution of the KnW optimization problem for calculating the worst-case cost of basis pole candidates with respect to uncertainty regions associated with complex pole estimates. Furthermore:

Theorem 8.7 (Convexity). *h-segments and h-discs are convex in $\mathbb{D}$ in terms of the metric κ_1.*

The proof is given in Appendix A.3. In conclusion, the importance of hyperbolic geometry with respect to the KnW theory is twofold. First of all, it can be shown that the metric associated with this geometry is equal to the cost function of the Kolmogorov $n_{\mathrm{g}} = 1$-width optimization problem described by (8.3). This equality is a key property used during the derivation of the FcM based basis selection algorithm (see Sect. 8.3). The second importance is that by using convex objects of the hyperbolic geometry as h-lines and h-circles, the solution of (8.4) over these regions can be turned into a convex optimization problem with LMI constraints. The latter property is essential to the robust basis selection algorithm developed later.

---

[3] Note that KM is not a distance on $\mathbb{D}$ in the geometrical sense; only $2\operatorname{arctanh}(\kappa_1(\mathrm{x}, \mathrm{y}))$ bears this property and is called the *Poincaré distance* [32].

8.4.3 Pole Uncertainty Regions as Hyperbolic Objects

In the robust basis selection problem the results of hyperbolic geometry will be applied on pole uncertainty regions. However, the pole uncertainty regions of Sect. 2.4.6 do not necessary coincide with hyperbolic objects. Thus, it is required to approximate these regions by a coverage of hyperbolic objects in order to apply the developed results. In the following, a simple pragmatic procedure for this approximation is briefly explained.

In principle it is true that, if an identified transfer function $G(\mathrm{q},\hat{\theta}_{N_\mathrm{d}})$ has a real valued pole, then the pole-uncertainty region $\mathcal{P}(\mathcal{Q}_\theta,\alpha)$ of $G(\mathrm{q},\hat{\theta}_{N_\mathrm{d}})$ (for a given confidence level α) contains segments of the real axis. Therefore, these segment parts of $\mathcal{P}(\mathcal{Q}_\theta,\alpha)$ can be associated with h-segments without the need of any approximation. In case $G(\mathrm{q},\hat{\theta}_{N_\mathrm{d}})$ has a complex pole pair, then $\mathcal{P}(\mathcal{Q}_\theta,\alpha)$ contains complex regions which come in complex conjugate pairs. Depending on the uncertainty of the model estimate, these complex regions can merge into one another, and can also encircle parts of $\mathbb{D}$ which seemingly do not contain any of the estimated or original pole locations (see Sect. 2.4.6). Thus complex pole uncertainty regions can occur in complicated shapes. However, with a simple methodology, quite effective coverage of such shapes can be achieved by h-discs.

Let Ω be a separate, complex region in $\mathcal{P}(\mathcal{Q}_\theta,\alpha)$. Due to the method of [198], the separated regions are distinguished during the calculation of $\mathcal{P}(\mathcal{Q}_\theta,\alpha)$. Assume that an equidistant gridding of $\mathbb{D}$ is given with step size $\tau\in\mathbb{R}^+$ and the set of grid points is denoted by Q. Let $\mathrm{Q}_+=\Omega\cap\mathrm{Q}$ be the grid points in Ω and let $\mathrm{Q}_-=\mathrm{Q}\setminus\mathrm{Q}_+$. Next, define $\mathscr{D}_0$ as a collection of equidistant Euclidian discs with centers at each points of Q_+ and with radii $\mathtt{r}=\frac{\sqrt{2}}{2}\tau$. In this way, a coverage of Ω is obtained by an unnecessary large number of Euclidian discs determining uniquely their h-disc counterparts. This is used as the initialization of the iterative optimization procedure of Algorithm 8.2 to find an efficient disc-coverage based on a fixed number of n discs with unequal radii.

In Step 2 of this algorithm, a disc is selected from the coverage which is the furthest from the points of Q_- and its radii can be increased by $\frac{\sqrt{2}}{2}\tau$, without causing the discs to contain any point of Q_-. In case of a tie, an arbitrary disc is selected from the possible choices. The radius of the selected disc is increased with $\frac{\sqrt{2}}{2}\tau$ in Step 3. If a disc contains other discs after its radius was increased, then those discs are removed from the coverage in Step 4. The procedure is repeated till there is no disc whose radius can be increased by $\frac{\sqrt{2}}{2}\tau$ without containing a point of Q_-. The optimization results in a h-disc coverage that gives an "optimal" coverage for the grid-points Q_+ with a minimal number of circles, however this number can be much larger than desired. Therefore a second optimization is initiated in Step 6, which gives a suboptimal approximation of this coverage by a predefined number of discs n. In Step 7, two discs that have the smallest dissimilarity in terms of $\varepsilon_{ij}=\left|\mathsf{e}_j-\mathsf{e}_i\right|+\left|\mathtt{r}_j-\mathtt{r}_i\right|$, where $\mathsf{e}_j,\mathsf{e}_i$ are the Euclidean centers and $\mathtt{r}_j,\mathtt{r}_i$ the Euclidian radiuses of the discs, are selected. These discs are merged in Step 8 by increasing the radii of the larger disc with ε_{ij}. If a disc contains other discs after merging, then the contained discs are removed from the coverage. Merging is repeated till the desired number of circles is achieved. Example 8.2 shows how the method works in practice.

Algorithm 8.2 (Disc coverage of complex regions)

Step 1. Let $\mathscr{D}_0 := \{D(\mathrm{e}, \frac{\sqrt{2}}{2}\tau)\}_{\mathrm{e}\in\mathrm{Q}_+}$ be a disc coverage of Ω based on the step size $\tau > 0$. Let $n > 0$ and $l := 0$.

Step 2. Find a $D(\mathrm{e},\mathrm{r}) \in \mathscr{D}_l$ such that $\min_{\mathrm{v}\in\mathrm{Q}_-}\min_{\mathrm{z}\in D(\mathrm{e},\mathrm{r})} |\mathrm{v}-\mathrm{z}|$ is maximal on $\mathscr{D}_l$ and $D(\mathrm{e},\mathrm{r}+\frac{\sqrt{2}}{2}\tau)\cap\mathrm{Q}_- = \emptyset$. Let $\hat{D}_l = D(\mathrm{e},\mathrm{r}+\frac{\sqrt{2}}{2}\tau)$ otherwise $\hat{D}_l = \emptyset$.

Step 3. If such D exists, then $\mathscr{D}_{l+1} = (\mathscr{D}_l \setminus D) \cup \hat{D}_l$.

Step 4. Remove all D from $\mathscr{D}_{l+1}$ which satisfy $D \subseteq \hat{D}_l$.

Step 5. If $\hat{D}_l \neq \emptyset$, then set $l := l+1$ and goto Step 2.

Step 6. If $\mathrm{Card}(\mathscr{D}_{l+1}) \leq n$, then stop, else set $l := l+1$ and continue with Step 7.

Step 7. Find $D_i(\mathrm{e}_i,\mathrm{r}_i), D_j(\mathrm{e}_j,\mathrm{r}_j) \in \mathscr{D}_l$ such that $D_i \neq D_j$, $\mathrm{r}_i \geq \mathrm{r}_j$, and $\varepsilon_{ij} = |\mathrm{e}_j - \mathrm{e}_i| + |\mathrm{r}_j - \mathrm{r}_i|$ is minimal.

Step 8. $\hat{D}_l = D(\mathrm{e}_i, \mathrm{r}_i + \varepsilon_{ij})$ and $\mathscr{D}_{l+1} = (\mathscr{D}_l \setminus \{D_i, D_j\}) \cup \hat{D}_l$. Perform Step 4 and goto Step 6.

Example 8.2 (Hyperbolic coverage of pole-uncertainty regions). Continue Example 2.1 by computing the hyperbolic coverage of the pole uncertainty regions with Algorithm 8.2 using $\tau = 10^{-3}$. For regions associated with different confidence levels, the results are given in Fig. 8.8. As can be seen, even the complicated butterfly region is rather well approximated with a small number of h-discs. □

8.4.4 The Robust Pole Clustering Algorithm

In the following, the robust extension of the original FKcM approach is discussed. As the mechanism of the clustering remains the same in the robust extension, we will follow the same line of reasoning as in Sect. 8.3. However, we investigate and derive these results in a different problem setting.

Let $\{\hat{G}_{\mathrm{p}_i}\}_{i=1}^{n_{\mathrm{loc}}}$ be a set of estimated frozen transfer functions of the LPV system $\mathcal{S}$ identified for constant scheduling signals: $\mathrm{p}_i \in \mathbb{P}$ and IO partition (u,y). Each $\hat{G}_{\mathrm{p}_i}$ is associated with a pole uncertainty region $\mathcal{P}_{\mathrm{p}_i}$ containing a number of regions in $\mathbb{D}$. Let $\{\Omega_k\}_{k=1}^{N_\mathrm{z}}$ denote the collection of these regions and introduce $\mathcal{Z}_k$ as the set of hyperbolic objects describing/approximating each Ω_k, as has been discussed in Sect. 8.4.3. If Ω_k is a segment on the real axis (real pole), then associate $\mathcal{Z}_k$ with an h-segment being equal to Ω_k, else associate it with a union of h-discs covering Ω_k (complex case). Denote $Z = \{\mathcal{Z}_k\}_{k=1}^{N_\mathrm{z}}$ and call it the data set. Similar to the original FKcM algorithm, introduce $1 < n_\mathrm{c} < N_\mathrm{z}$ as the number of clusters or data groups, $\upsilon_i \in \mathbb{D}$ as the cluster centers, and $\mu_i : \mathbb{D} \to [0,1]$ as the membership functions of the clusters for all $z \in \mathbb{D}$. By using the threshold value ε, we again obtain the set

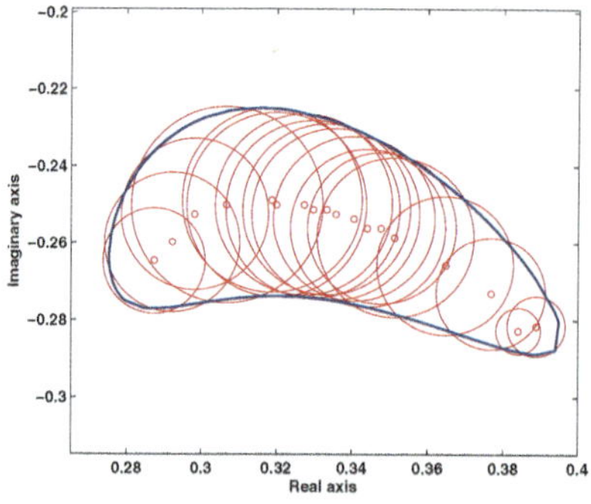

(a)

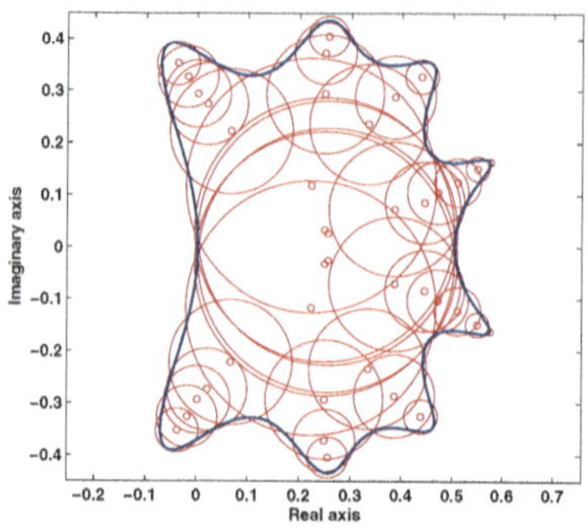

(b)

Fig. 8.8 (a) 20 h-discs based coverage of a non-connected uncertainty region of Fig. 2.2 with confidence level 1%. The overfit in area is 4.55%. (b) 40 h-discs based coverage of a connected uncertainty region of Fig. 2.2 with confidence level 99%. The overfit in area is 4.35%.

$$\Omega_\varepsilon = \{z \in \mathbb{D} \mid \exists i \in \mathbb{I}_1^{n_c},\ \mu_i(z) \geq \varepsilon\}, \tag{8.26}$$

which is used to formulate the robust variant of Problem 8.1:

Problem 8.2 (Robust pole clustering problem). For a set of pole uncertainty regions $\{\mathcal{P}_{p_i}\}_{i=1}^{n_{loc}}$, described by a set of hyperbolic objects Z, and for a given number of clusters n_c, find a set of cluster centers $\{\upsilon_i\}_{i=1}^{n_c}$, a set of membership functions $\{\mu_i\}_{i=1}^{n_c}$, and the maximum of ε, such that

- Ω_ε contains Z.
- With respect to Ω_ε, the OBFs, with poles Λ_{n_c} in the cluster centers $\{\upsilon_i\}_{i=1}^{n_c}$, are optimal in the K$n$W sense, where $n = n_c$. □

Again, the solution is based on finding clusters in accordance with the KnW concept and subsequently finding a maximal value for ε, such that all uncertainty regions are inside Ω_ε. It is obvious that the only difference between the robust and non-robust clustering problem is in terms of the data objects, which are points (poles) of the complex plane in the non-robust case and regions of the complex plane in the robust case.

Denote $V = [\upsilon_i]_{i=1}^{n_c}$ and $U = [\mu_{ik}]_{n_c \times N_z}$, where $\mu_{ik} = \max_{z \in \mathcal{Z}_k} \mu_i(z)$ is the degree of membership of $\mathcal{Z}_k$ to cluster i. Furthermore, "distances" d_{ik} between υ_i and $\mathcal{Z}_k$ are also introduced to measure dissimilarity of Z with respect to each cluster. To derive an algorithmic solution of Problem 8.2, this dissimilarity measure is defined as the Kolmogorov measure, the cost function of the 1-width version of (8.4), but instead of pointwise, it is defined with respect to each $\mathcal{Z}_k$ in a worst-case sense:

$$d_{ik} = \max_{z \in \mathcal{Z}_k} \kappa_1(\upsilon_i, z)\,. \tag{8.27}$$

Note that by applying Theorems 8.5 and 8.6, d_{ik} can be computed analytically with respect to each $\mathcal{Z}_k \in Z$. If $\mathcal{Z}_k$ is a h-segment between points $\mathrm{z}_{k1}, \mathrm{z}_{k2} \in (-1, 1)$, then

d_{ik} is the maximum of the Euclidian distance between υ_i and these end-points. In case $\mathcal{Z}_k$ is a union of h-discs, then first the intersection points are computed for each disc between the perimeter of the disc and the h-line connecting its h-center with υ_i. Then d_{ik} results by calculating the maximum of the KM between these points and υ_i. In all cases $d_{ik} > 0$, as the worst-case "distance" over regions can never drop to zero. In the sequel, it is shown that similar to the original FKcM approach, this specific choice of dissimilarity measure relates the FcM asymptotically to the KnW theory, and in this way to the solution of Problem 8.2. In order to uniquely associate each d_{ik} with a membership level μ_{ik} the set of membership function is restricted to satisfy $\sum_{i=1}^{n_c} \mu_i(z) = 1$, which requires again that $U \in \mathcal{U}_{n_c}^{N_z}$ (see (8.8)).

In the robust FKcM case, the fuzzy-functional $J_m(U,V)$ is formulated as

$$J_m(U,V) := \max_{k \in \mathbb{I}_1^{N_z}} \sum_{i=1}^{n_c} \mu_{ik}^m d_{ik} = \max_{k \in \mathbb{I}_1^{N_z}} \max_{z \in \mathcal{Z}_k} \sum_{i=1}^{n_c} \mu_{ik}^m \kappa_1(\upsilon_i, z), \tag{8.28}$$

where the design parameter $m \in (1, \infty)$ determines the fuzziness of the resulting partition. (8.28) defines the criterion of the expected solution for Problem 8.2, corresponding to a *worst-case (max) sum of error* criterion. Its relation with the KnW optimality of (U,V) is the same as in the FKcM algorithm except that here the KnW problem is considered for regions and not for points of $\mathbb{D}$. Regarding the fuzzyness m, its role in (8.28) is the same as discussed in the non-robust case. Based on this, the following theorem yields the solution of Problem 8.2:

Theorem 8.8 (Optimal robust partition). *Let $m > 1$, a fuzzy partition $(U,V) \in \mathcal{U}_{n_c}^{N_z} \times \mathbb{D}^{n_c}$, and a data set $Z = \{\mathcal{Z}_k\}_{k=1}^{N_z}$ be given, where $\mathcal{Z}_k$ is either an h-segment between points $\mathrm{z}_{k1}, \mathrm{z}_{k2} \in (-1,1)$ or a union of n_k h-discs with h-centers $\{\mathrm{e}_{kl}\}_{l=1}^{n_k}$ and h-radii $\{\mathrm{r}_{kl}\}_{l=1}^{n_k}$. Denote $[V]_i = \upsilon_i$ and $[U]_{ij} = \mu_{ij}$. Define $\gamma_i(\mathrm{v},U)$ as the minimal value of $\tau \in [0,1]$ fulfilling the quadratic constraints:*

$$\begin{bmatrix} |1 - z^* \mathrm{v}|^2 & \mu_{ik}^m \cdot (z - \mathrm{v}) \\ \mu_{ik}^m \cdot (z - \mathrm{v})^* & \tau^2 \end{bmatrix} \succeq 0, \quad \forall z \in \mathcal{Z}_k, \tag{8.29}$$

for all $k \in \mathbb{I}_1^{N_z}$, where $\mathrm{v} \in \mathbb{D}$. Additionally, let $d_{ik} = \max_{z \in \mathcal{Z}_k} \kappa_1(\upsilon_i, z)$ be the dissimilarity measure of $\mathcal{Z}_k$ with respect to V. If (U,V) is a local minimum of J_m, then for any $(i,k) \in \mathbb{I}_1^{n_c} \times \mathbb{I}_1^{N_z}$:

$$\mu_{ik} = \left(\sum_{j=1}^{n_c} \left(\frac{d_{ik}}{d_{jk}} \right)^{\frac{1}{m-1}} \right)^{-1}, \tag{8.30a}$$

$$\upsilon_i = \arg\min_{\mathrm{v} \in \mathbb{D}} \gamma_i(\mathrm{v}, U). \tag{8.30b}$$

The proof is given in Appendix A.3. Similar to the FcM case, minimization of (8.28), subject to (8.8), is tackled by alternating optimization which yields the same algorithm as Algorithm 8.1, except in Step 3 and Step 4 the solutions are obtained via (8.30a) and (8.30b).

8.4.5 Properties of the Robust FK*c*M

As a next step we investigate the properties of the robust extension of the FK*c*M algorithm. It is shown that K*n*W optimality of the resulting cluster centers (if the solution is the global minima of (8.28)) holds again in an asymptotic sense ($m \to \infty$). Practical implementation of the algorithm is also discussed together with how the conservatism of the used pole uncertainty concept influences the procedure.

8.4.5.1 Asymptotic Property

We use the asymptotic properties of J_m to explain the specific choices for the fuzzy functional (8.28) and the dissimilarity measure (8.27).

Theorem 8.9 (Asymptotic property of J_m in the robust FK*c*M). *Given a data set $Z = \{\mathcal{Z}_k\}_{k=1}^{N_z}$, and a set of cluster centers $V \in \mathbb{D}^{n_c}$. Define U_m as a membership matrix of V satisfying (8.30a) for $m > 1$ and let d_{ik} be defined as 8.27. Then*

$$J_m(U_m, V) = n_c^{1-m} \max_{k \in \mathbb{I}_1^{N_z}} \left(\prod_{i=1}^{n_c} d_{ik} \right)^{1/n_c} + \mathcal{O}(e^{-m}) \tag{8.31}$$

Furthermore, $J_m(U_m, V)$ decreases monotonically with m, and $\lim_{m \to \infty} J_m(U_m, V) = 0$.

As the proof is not affected by the specific choice of d_{ik}, the same proof can be exploited as in the original FK*c*M case (see Th. 8.2). Based on Th. 8.9, for large m and for a U satisfying (8.30a), J_m corresponds to

$$J_m(U, V) \approx n_c^{1-m} \max_{k \in \mathbb{I}_1^{N_z}} \max_{z \in \mathcal{Z}_k} \left(\prod_{i=1}^{n_c} \kappa_1(\upsilon_i, z) \right)^{1/n_c}, \tag{8.32}$$

thus its minimization gives a close approximation of (8.4), enabling the FK*c*M to solve Problem 8.2 directly. However, if $m \to \infty$, then again numerical problems can occur in the minimization of (8.30b). Therefore, to obtain a well approximating solution of Problem 8.2, an appropriately large value of $m \in (1, \infty)$ should be used. Just like for the original FK*c*M, $m \in [5, 10]$ usually yields satisfactory results.

Similar to the previous case, for $m > 1$ the FK*c*M-functional (8.28) is a bounded ($0 \leq J_m \leq 1$) monotonically decreasing function both in $\{d_{ik}\}$ and U, which allows Algorithm 8.1 to converge in practice in the same sense as has been discussed in the non-robust case. By using different initial choices of V_0, all local minima of (8.28) can be explored. From these multiple runs, the best set of the obtained OBFs can be selected by comparison of the achieved decay rate, which is formulated in the robust case as:

$$\breve{\rho} = \max_{k \in \mathbb{I}_1^{N_z}} \max_{z \in \mathcal{Z}_k} \prod_{i=1}^{n_c} \left| \frac{z - \upsilon_i}{1 - z\upsilon_i^*} \right|. \tag{8.33}$$

Comparison can also be made by visual inspection of the boundary region of $\Omega(\Lambda_{n_c}, \breve{\rho})$. In practice, uniformly random choices for V_0 are suggested.

8.4.5.2 Optimization

In order to derive an algorithmic solution of Problem 8.2 in terms of Th. 8.8, it is important to define the regions $\mathcal{Z}_k$ as inequality constrains. Such form would enable to use (8.29) as a quadratic constraint and apply the same LMIs based optimization as in the FKcM case. Using the hyperbolic geometry, each uncertainty region can be represented in the following way:

- If $\mathcal{Z}_k$ is an h-segment then for $z \in \mathcal{Z}_k$ it holds that

$$\mathrm{z}_{k1} \preceq z \preceq \mathrm{z}_{k2}. \tag{8.34}$$

- If $\mathcal{Z}_k$ is the union of h-disks then for a $z \in \mathcal{Z}_k$ there exists a $l \in \mathbb{I}_1^{n_k}$ such that

$$\begin{bmatrix} |\,1 - z^* \mathrm{e}_{kl}\,|^2 & z - \mathrm{e}_{kl} \\ z^* - \mathrm{e}_{kl}^* & \mathrm{r}_{kl}^2 \end{bmatrix} \succeq 0. \tag{8.35}$$

As the resulting descriptions define an inequality (see 8.34) or a set of quadratic constraints (see 8.35) thus (8.30b) is equivalent with a minimization problem with QCs where γ is the optimization variable and v is the decision variable. As the structure of these constraints is similar as in the non-robust case, it is possible to derive SoS relaxations through which (8.29) and (8.34) turn into LMIs. The resulting LSDP can be efficiently solved by LMI solvers. Alternatively, bisection-based recursive search can be used to obtain the minimization of γ_i in (8.30b). Also numerical conditioning of U_l can relax the computational need of the LMI's based optimization and the termination criterion of the overall algorithm can be established similar to the original FKcM method. For details on these items see Sect. 8.3.2. Similarity-based ACM (see Sect. 8.3.2.4) can also be used for the determination of the number of "natural" groups in Z, i.e. the best suitable n_c for clustering, which is also important for the successful application of the robust FKcM method.

8.4.5.3 Conservatism of the Pole Uncertainty Concept

The approach of calculating pole uncertainty regions, presented in Sect. 2.4.6, projects ellipsoidal uncertainty regions of parameter estimates to regions in the complex plane. It has been already explained in Sect. 2.4.6 that these regions represent the set of possible pole locations of the model with respect to the parameter uncertainties with the given confidence level. However, the covariance of the poles, i.e. which poles occur together in the model estimate with the given confidence level, is disregarded in this representation. Thus, the projection is inherently conservative as for a pole uncertainty region $\mathcal{P}$ of a 5th order LTI model it is not guaranteed that for any 5 arbitrary chosen points $\{\lambda_1, \ldots, \lambda_5\}$ in $\mathcal{P}$ there exists a parameter vector θ in the ellipsoidal parameter uncertainty region such that the model associated with θ has poles $\{\lambda_1, \ldots, \lambda_5\}$. In other words, neglecting the covariance between the pole estimates introduces conservatism. However, it is true that for any point in $\mathcal{P}$ it is guaranteed that there exist 4 other points in $\mathcal{P}$, such that these 5 points correspond to the pole locations of a model whose parameters are in the ellipsoidal parameter uncertainty region. This means that the pole uncertainty regions obtained by this method contain all possible, including the worst-case pole locations that can occur due to the uncertainty of the obtained model. However, such worst-case conservatism completely matches the worst-case concept of the KnW theory (see Sect. 2.5). Thus the conservatism of the projection does not affect the basis selection mechanism. This is an important observation which underlines the validity of the presented approach.

8.4.6 Simulation Example

In the following an example is given to visualize the applicability of the robust basis selection mechanism and to enable comparison with the original FKcM method. In this example it is shown that disregarding pole uncertainties during the basis selection process can result in a lower worst-case convergence rate of the obtained basis functions.

8.4.6.1 Data Generation

Consider again the asymptotically stable SISO LPV system $\mathcal{S}$ given by the LPV-IO representation (8.17). By using constant scheduling signals with values $\{0.6; 0.6+\tau; \ldots; 0.8\}$, where $\tau = 0.04$, 6 local LTI-IO representations of $\mathcal{S}$ are obtained, whose pole locations are samples of $\Omega_{\mathbb{P}}$ (see Fig. 8.2a). We use fewer constant scheduling points in this case than before, to make the results and the figures more transparent. The LTI models, associated with constant scheduling signals, in this example are estimated. The identification of each frozen model has been based on a OE parametrization and 250 samples long measured IO signals of the system. The measurements contained additive white output noise ε with normal distribution $\mathcal{N}(0, 0.1)$ and a white input signal u with uniform distribution $\mathcal{U}(-1, 1)$. Using the pole uncertainty concept of Sect. 2.4.6, the pole uncertainty regions $\{\mathcal{P}_i\}_{i=1}^{6}$ of the estimated models have been calculated with confidence level $\alpha = 99\%$. The resulting uncertainty regions consist of $N_z = 6 \cdot 5$ complex regions $\{\Omega_k\}_{k=1}^{N_z}$ which are presented in Figures 8.9a–b. In Fig. 8.9a, the estimated frozen poles are denoted by $\circ$ with a color indicating the constant scheduling trajectory they are associated with (deep red $\mathrm{p}_1 = 0.6 \leftrightarrow$ yellow $\mathrm{p}_6 = 0.8$). The perimeter lines of the pole uncertainty regions are given in Fig. 8.9b with a color indicating which poles they are associated with. Based on these figures, the resulting uncertainty regions are relatively large, prognosticating that using only the estimated poles by the non-robust algorithm can result in a serious performance degradation of the obtained solution. In order to derive an OBF set for $\mathcal{S}$ based on the robust FKcM mechanism, a hyperbolic coverage $\{\mathcal{Z}_k\}_{k=1}^{N_z}$ has been generated with respect to each region Ω_k, using gridding step size 0.01 and 20 h-discs per complex region. The average overfit in area has been 4.5% of the resulting coverage (see Sect. 8.4.3 for details on this algorithm).

8.4.6.2 Robust and Non-robust Pole Clustering

For the identified pole locations, the non-robust FK*c*M algorithm has been applied with $m = 8$ and $n_c = 8$ (denoted by $m8n_c8$), while for the obtained hyperbolic coverage, the robust FK*c*M with the same fuzzyness m and number of clusters n_c has been executed (denoted by *rob*-$m8n_c8$). Note that $m = 8$ has been used to guarantee close approximation of the asymptotic case similar to the choices of the example presented in Sect. 8.3.3. Moreover, $n_c = 8$ has been used in both algorithms to enable comparison to the previous example. Note that the estimated poles form more or less 8 groups in this case as well (see Fig. 8.9a), however this is not true for their associated uncertainty regions where only 5 groups can be detected (see Fig. 8.9b). The results of the clustering are presented in Figures 8.10a–b and also in Table 8.2. In these figures, the resulting cluster centers are given by blue $\times$. To visualize the performance of the associated OBFs, the perimeter of their $\Omega(\Lambda_{n_c}, \breve{\rho})$ region with respect to the uncertainty regions/pole estimates is given with a green line. Additionally, to compare the performance of the resulting OBFs, the achieved $\Omega(\Lambda_{n_c}, \breve{\rho}_{id})$ regions (green line) of the basis functions with worst-case convergence rate $\breve{\rho}_{id}$ are given in Figures 8.11a–b with respect to the estimated pole locations. The performance is also compared in terms of their achieved $\Omega(\Lambda_{n_c}, \breve{\rho}_{unc})$ regions given with respect to the uncertainty regions in Figures 8.12a–b, and also in terms of their achieved $\Omega(\Lambda_{n_c}, \breve{\rho}_{true})$ regions given with respect to the true frozen poles in Figures 8.13a–b. These figures have been generated for comparison purposes to show why the use of the robust FK*c*M clustering delivers a better basis for the frozen behaviors than the non-robust solution. In Table 8.2, the comparison of the results is presented in terms of the previously used indicators like χ, the Xie-Beni validity index and S_e, the normalized entropy together with the achieved decay rate with respect to the estimated pole locations: $\breve{\rho}_{id}$, to the uncertainty regions: $\breve{\rho}_{unc}$, and to the true pole location of the local systems: $\breve{\rho}_{true}$. Additionally, $\varepsilon_{max}^{n_e}$, the worst-case absolute error of the impulse responses of the truncated series-expansion representation of each $\mathcal{F}_p \in \mathscr{F}_S$ in terms of the generated OBFs with n_e repetitions, is also given. Based on these and the previous results of Sect. 8.3.3, the following observations can be made:

8.4.6.3 Analyzing the Results

The resulting partitions of the non-robust ($m8n_c8$) and the robust FK*c*M clustering (*rob*-$m8n_c8$) solve the basis selection problem for the identified pole locations / pole uncertainty regions. This follows from the tight fit of the resulting boundary regions (see Figures 8.10a–b) with respect to the data sets and the achieved small value of the worst-case convergence rate with respect to the used estimated poles/uncertainty regions (see $\breve{\rho}_{unc}$ for *rob*-$m8n_c8$ and $\breve{\rho}_{id}$ for $m8n_c8$ in Table 8.2). However, the OBFs represented by the *rob*-$m8n_c8$ partition are better basis functions for $\mathcal{S}$ than the $m8n_c8$ solution, as the worst-case convergence rate $\breve{\rho}_{true}$ of the basis poles with respect to the true frozen pole locations is smaller in the former case. This is also shown in Fig. 8.13a, where the *rob*-$m8n_c8$ solution achieves a

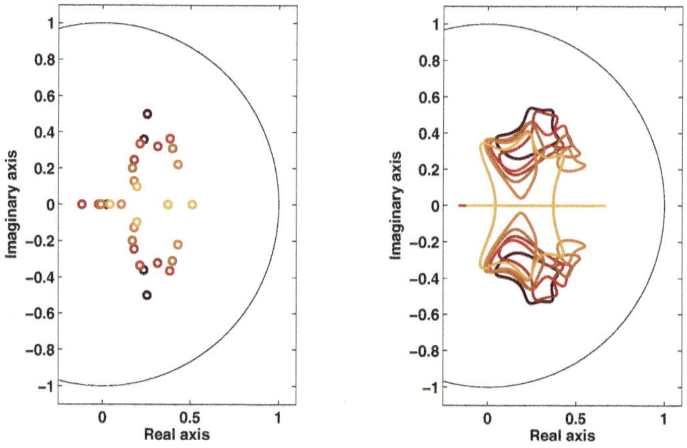

(a) estimated poles (b) uncertainty regions

Fig. 8.9 Estimated frozen poles of $\mathfrak{R}_{\mathrm{IO}}(\mathcal{S})$ and their associated uncertainty regions. The poles are denoted by $\circ$ in subfigure (a) with a color indicating the constant scheduling trajectory they are associated with (deep red $\mathrm{p}_1 = 0.6 \leftrightarrow$ yellow $\mathrm{p}_6 = 0.8$). The perimeter lines of the pole uncertainty regions are given in subfigure (b) with a color indicating that which estimated poles they belong to.

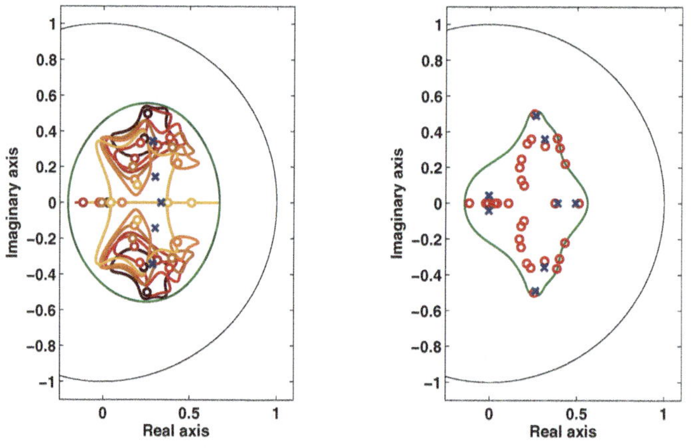

(a) robust algorithm (b) non-robust algorithm

Fig. 8.10 Resulting cluster centers (blue $\times$) of robust FKcM clustering of the pole uncertainty regions (deep red for $\mathrm{p}_1 = 0.6 \leftrightarrow$ light orange for $\mathrm{p}_6 = 0.8$) and non-robust FKcM clustering of the estimated poles (red $\circ$). To visualize the performance of the cluster centers associated OBFs, the perimeter of their $\Omega(\Lambda_{n_c}, \breve{\rho})$ region with respect to the uncertainty regions/pole estimates is given with a green line.

Table 8.2 Comparison of algorithmic results in terms of χ, the Xie-Beni validity index; S_e, the Normalized Entropy; the achieved decay rate with respect to the estimated pole locations: $\breve{\rho}_{id}$, to the uncertainty regions: $\breve{\rho}_{unc}$, and to the true pole location of the local systems: $\breve{\rho}_{true}$; and $\varepsilon_{max}^{n_e}$, the worst-case absolute error of the impulse responses of the truncated series-expansion representation of each $\mathcal{F}_p \in \mathscr{F}_S$ in terms of the cluster centers generated OBFs with n_e repetition. All results are given in dB.

Test case	χ	$\breve{\rho}_{id}$	$\breve{\rho}_{unc}$	$\breve{\rho}_{true}$	S_e	$\varepsilon_{max}^{n_e=1}$	$\varepsilon_{max}^{n_e=3}$
rob-m8n$_c$8	38.958	−52.444	−43.695	−47.012	2.830	−37.326	−140.919
m8n$_c$8	−5.003	−62.809	−32.165	−44.052	2.828	−32.165	−131.207

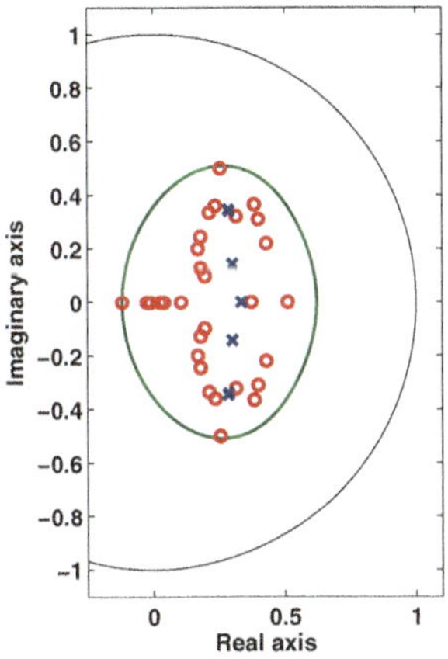

(a) robust algorithm

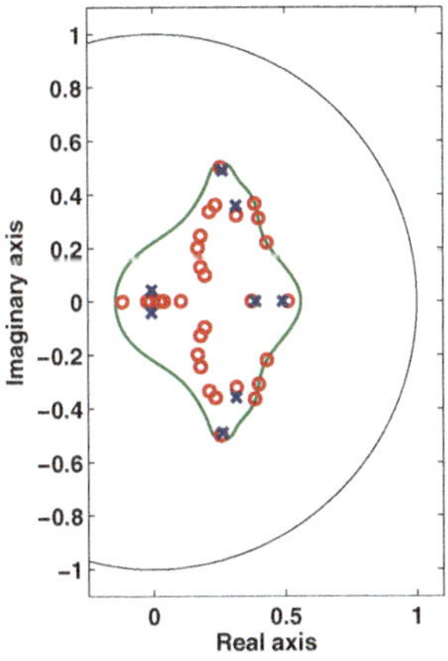

(b) non-robust algorithm

Fig. 8.11 Performance comparison of the resulting cluster centers (blue $\times$) associated OBFs in terms of the perimeter of their $\Omega(\Lambda_{n_c}, \breve{\rho}_{id})$ region (green line) with respect to the estimated frozen poles (red $\circ$) of $\mathfrak{R}_{IO}(\mathcal{S})$.

relatively tight $\Omega(\Lambda_{n_c}, \breve{\rho}_{true})$ on the true pole locations, while the *m8n$_c$8* case has quite high performance degradation resulting in a loose bound (see in Fig. 8.13b). This yields a smaller representation error if the *rob-m8n$_c$8* OBFs are used for a truncated series-expansion representation of $\mathcal{S}$, which is proved by comparing the worst-case impulse response representation errors $\varepsilon_{max}^{n_e}$ in Table 8.2. The reason why *rob-m8n$_c$8* outperforms the *m8n$_c$8* solution, comes from the fact that the *m8n$_c$8* solution is based on only one realization of the *probability density function* (pdf) associated with the pole estimation, while *rob-m8n$_c$8* is based on pole regions associated with a level set of the pdf through the pole uncertainty concept of Sect. 2.4.6. Therefore, if the realization, i.e. the estimated poles, is far from the true pole locations, then there is no guarantee about the true performance of the non-robust OBF selection ($\breve{\rho}_{true} > \breve{\rho}_{id}$). Contrary, the robust solution obtains guaranteed performance for any realization inside the used uncertainty regions, giving a high probability in

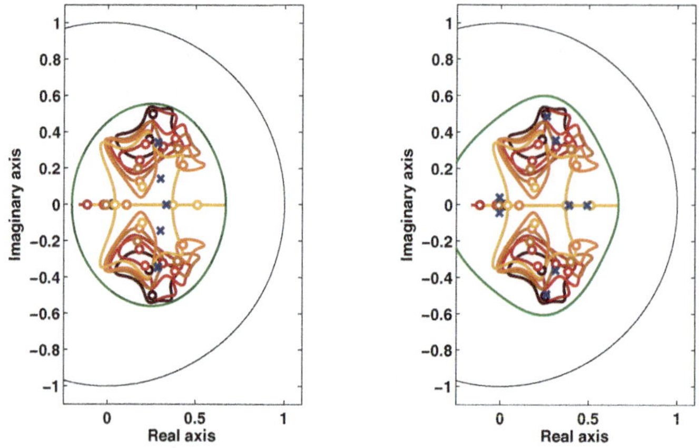

(a) robust algorithm (b) non-robust algorithm

Fig. 8.12 Performance comparison of the resulting cluster centers (blue $\times$) associated OBFs in terms of the perimeter of their $\Omega(\Lambda_{n_c}, \breve{\rho}_{\mathrm{unc}})$ region (green line) with respect to the uncertainty regions of the estimated frozen poles.

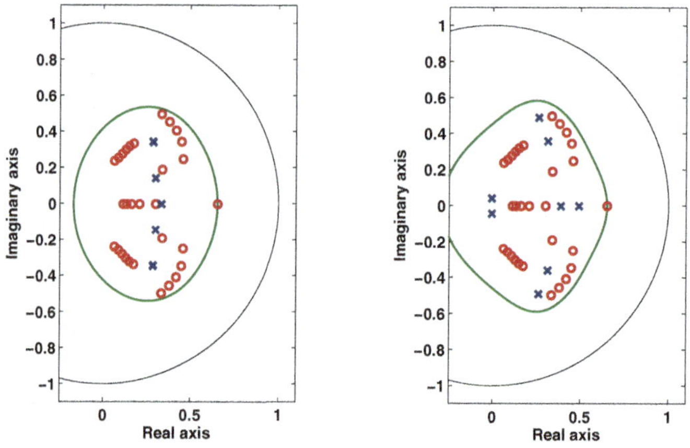

(a) robust algorithm (b) non-robust algorithm

Fig. 8.13 Performance comparison of the resulting cluster centers (blue $\times$) associated OBFs in terms of the perimeter of their $\Omega(\Lambda_{n_c}, \breve{\rho}_{\mathrm{true}})$ region (green line) with respect to the true frozen poles (red $\circ$) of $\mathfrak{R}_{\mathrm{IO}}(\mathcal{S})$.

terms of α, that this guaranteed performance is the upper bound of the achieved performance with respect to the true pole locations. This is clearly shown by comparing $\breve{\rho}_{\mathrm{true}}$ to $\breve{\rho}_{\mathrm{unc}}$ in the robust case. Obviously, the optimal performance of the non-robust solution with respect to the true pole locations, $\breve{\rho}_{\mathrm{true}} = -58.38$ dB (see Table 8.1) can be achieved in the noiseless case only.

χ is quite high in the *rob-m8n*$_c$*8* case and also moderately high in the *m8n*$_c$*8* case, showing that each partition represents the underlying structure inefficiently. Moreover, the relatively high value of S_e in both cases suggests also that a smaller number of clusters is more suitable to describe the underlying data structure, which agrees with the visual inspection of the almost colliding cluster centers (see Fig. 8.10). This phenomenon is due to the noise uncertainty: the originally required 8 clusters (see Sect. 8.3.3) to describe the samples of $\Omega_{\mathbb{P}}$ are more than the number of clusters suggested by the identified poles, i.e. their associated uncertainty regions. Using the ACM, the solution converges to a partition that consist of 7 clusters in the non-robust case and 5 clusters in the robust case which agrees with visual inspection. By comparing properties of clustering with varying m, the same conclusions can be drawn as in the non-robust case in Sect. 8.3.3, except that the computation time increases more rapidly with increasing m.

8.5 Conclusions

In this chapter, optimal basis selection for series expansion of LPV systems has been investigated in the case when only measured data records of the frozen signal behavior are available. The solution of this problem is crucial to provide a practical model structure selection tool for LPV identification based on truncated series-expansion models. In case of an optimal basis, a fast convergence rate of the expansion representation implies that only the estimation of a few coefficients is necessary for a good approximation of the system.

A crucial conclusion that we could draw is that a practically applicable and effective basis selection can be formulated based on estimated pole locations of the system with respect to constant scheduling trajectories. By reconstruction of the pole manifest set of the system based on these sample pole locations and finding the K*n*W optimal basis with respect to the reconstructed regions adequate selection of the basis follows. We could see that both of these tasks can be efficiently solved in one step by using a modified fuzzy clustering approach. This approach is also applicable in situations where the measurement noise has a significant effect on the sample pole estimates, as by the use of hyperbolic geometry the basis selection algorithm can be robustified with respect to pole uncertainty regions. In conclusion the algorithms, developed in this chapter, provide effective tools for model structure selection in terms of LPV truncated series-expansion models. In the next chapter we will see how all tools developed so far, enable the identification of LPV systems through a simple and effective approach.

Chapter 9
LPV Identification via OBFs

Abstract. All the theory that has been introduced so far has served the sole purpose of providing tools to formulate identification of general LPV systems in a well-established manner. Building on the developed tools, a widely applicable identification approach is proposed in this chapter by using model structures that originate from truncated OBF expansion representations of LPV systems. First, under the assumption of static dependence of the expansion coefficients, two identification methods, a local and global one, are developed for the introduced model structures. While the local approach uses the gain-scheduling principle: identification with constant scheduling signals and interpolation of the resulting LTI models, the global approach provides a direct LPV model estimate via linear regression based on data records with varying scheduling trajectories. The approaches are analyzed in terms of variance, bias, consistency, and applicability together with the validation of the model estimates. Finally, to enable the estimation of modes with dynamic coefficient dependencies, a modified feedback-based OBF model structure is proposed and estimation in this framework is formulated through a separable least-squares strategy.

9.1 Aim and Motivation of an Alternative Approach

In the previous chapters we have built up an extensive LPV system theoretical framework to provide understanding of model structures and to develop tools for their analysis. Based on this framework it has been shown that a series-expansion representation of discrete-time asymptotically stable LPV systems is available in terms of *orthonormal basis functions* (OBFs). Based on the motivation given in Chap. 1, we aim to use finite truncations of such representations as models for the identification of LPV systems in the classical *least-squared* (LS) setting. We will show that these structures:

- provide an easily scalable trade-off between model complexity and accuracy of the estimate,
- simplify identification and control design,

R. Tóth: Modeling and Identification of Linear Parameter-Varying Systems, LNCIS 403, pp. 233–284.
springerlink.com

- do not suffer from locally changing system order,
- they extend the results of LTI system identification.

We also formulate the prediction-error setting for LPV identification, in order to analyze the stochastical properties of estimation in these model structures and to be able to investigate possible noise model concepts. This provides the final tool in order to compare identification algorithms of the LPV field.

The aim of this chapter is to give a set of identification approaches that are capable to deliver theoretically well-founded model estimates in the LPV framework, accomplishing our primary objective. Here we do not pursue the proper exploration of the other two steps of the identification cycle: experiment design and model validation, though some basic results about these issues are briefly covered in the analysis of the identification approaches. The proper treatment of these steps is reserved for further studies. Our intention is to open a new and sound alternative for the existing identification literature and to stimulate the development of a future generation of LPV identification approaches.

9.2 OBFs Based LPV Model Structures

LPV model structures based on truncated OBF expansion representations are introduced in this section. The structures are defined by using the concepts of the classical prediction-error setting. Thus first, characterization of this setting is developed in the LPV case. Due to the absence of a transfer function type of description of LPV systems, the process and noise models are formulated based on their impulse-response representation. This gives the possibility to develop one-step-ahead predictors in this framework. As a next step, the proposed model structures, as truncated OBF expansion representations are formulated with an *output-error* (OE) type of noise models. Based on the fact that the coefficients can appear on the left or on the right-side of the basis functions in the expansion, different model structures can be introduced. Finally, important properties of the proposed model structures are investigated and compared to other model structures of the LPV identification field.

9.2.1 The LPV Prediction-Error Framework

As a starting point, we assume that a non-autonomous, SISO, asymptotically stable LPV system $\mathcal{S} = (\mathbb{Z}, \mathbb{P}, \mathbb{W}, \mathfrak{B})$ is given in *discrete-time* (DT) with scheduling signal p. We aim at the identification of this system based on a predefined *input-output* (IO) partition (u, y). We suppose that $\mathcal{S}$ is equivalent with the deterministic part of the data generating physical system (i.e. the deterministic part of the physical system corresponds to an LPV system with the considered scheduling signal). Based on first principles information, it is possible in practice to select scheduling variables for a plant that yield an LPV equivalent (see the procedure developed in Sect. 7.4).

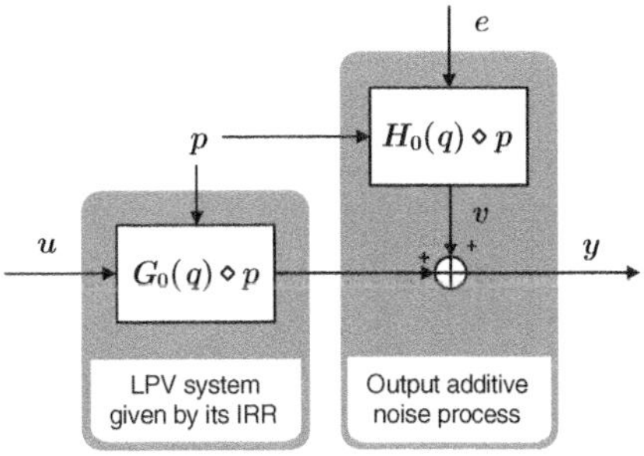

Fig. 9.1 Data generating system in the LPV prediction-error framework.

9.2.1.1 Prediction Error Setting

First we clarify the identification setting in which we position our models. Using the concept of the classical prediction-error identification (see Sect. 2.4), it is assumed that the data generating system, illustrated in Fig. 9.1, is given as

$$y = (G_0(\mathrm{q}) \diamond p)u + v, \tag{9.1}$$

where the process part, i.e. the LPV system $\mathcal{S}$, is represented in a impulse response form $\mathfrak{R}_{\mathrm{IM}}(\mathcal{S})$:

$$G_0(\mathrm{q}) \diamond p = \sum_{i=0}^{\infty} (\mathsf{g}_i \diamond p)\mathrm{q}^{-i}, \tag{9.2}$$

with $\mathsf{g}_i \in \mathcal{R}$. Note that due to the asymptotic stability assumption, every considered discrete-time LPV system has a *impulse response representation* (IRR) (see Sect. 5.3). The reason why we use the IRR to define the data generating equation (9.1) is due to the fact that a transfer function form of the dynamic relation, like in the LTI counterpart (2.25), is not available for LPV systems. Additionally, the noise part of (9.1) is given as v, satisfying

$$(Q_{A_0}(\mathrm{q}) \diamond p)v = (Q_{B_0}(\mathrm{q}) \diamond p)e, \tag{9.3}$$

where e is a zero-mean white noise process with variance σ_{e}^2 and $Q_{A_0}, Q_{B_0} \in \mathcal{R}[\xi]$ are polynomial functions such that they define an LPV-IO representation with IO partition (v, e), implying that $\mathrm{Deg}(Q_{A_0}) \geq \mathrm{Deg}(Q_{B_0})$. Similar to the LTI case, it is assumed that the IO representation (9.3) defines an asymptotically stable system in the deterministic sense, otherwise the identification of G_0 in (9.1) is not meaningful. Under this assumption, the representation of the noise structure (9.3) has a pulse basis series-expansion, which is denoted by $H_0(\mathrm{q})$ and satisfies

$$v = (H_0(\mathrm{q}) \diamond p)e \quad \text{with} \quad H_0(\mathrm{q}) = \sum_{i=0}^{\infty} (\mathsf{h}_i \diamond p)\mathrm{q}^{-i}, \tag{9.4}$$

where $\mathsf{h}_i \in \mathcal{R}$.

9.2.1.2 One-Step-Ahead Prediction of v

Similar to the LTI case, to formulate a one-step-ahead predictor with respect to y, it is required to clarify how we can predict $v(k)$ at a given time step k, if we have observed $v(\tau)$ for $\tau \leq k-1$. A crucial property of (9.4) that we will impose to enable an answer to this question is that it should be invertible, i.e. there exists a stable inverse $H_0^{\dagger}(\mathrm{q},p)$ of $H_0(\mathrm{q},p)$, where $H_0^{\dagger}(\mathrm{q},p)$ is a convergent LPV series-expansion and

$$e = (H_0^{\dagger}(\mathrm{q}) \diamond p)v. \tag{9.5}$$

Note that by taking $Q_{B_0}, Q_{A_0} \in \mathbb{R}[\xi]$, where $\mathbb{R}[\xi]$ (the ring of polynomials with real constant coefficients) is a subspace of $\mathcal{R}[\xi]$, H_0 is equivalent with a transfer function and $H_0^{\dagger}$ is its stable inverse. This results in the LTI case discussed in Sect. 2.4.

As a next step, write (9.4) as

$$v(k) = (\mathsf{h}_0 \diamond p)(k)e(k) + \sum_{i=1}^{\infty}(\mathsf{h}_i \diamond p)(k)e(k-i). \tag{9.6}$$

Now the knowledge of $\{v(\tau)\}_{\tau\leq k-1}$ and a given trajectory of p implies the knowledge of $\{e(\tau)\}_{\tau\leq k-1}$ in the view of (9.6). Based on this relation, there are many ways to define the prediction of $v(k)$, like the maximum a posteriori prediction or the mean value of the distribution in question, etc. The classical approach we use in the following is to view the prediction of $v(k)$ as the conditional expectation of $v(k)$ based on $\{e(\tau)\}_{\tau\leq k-1}$ and a fixed trajectory of $p \in \mathfrak{B}_{\mathbb{P}}$:

$$\hat{v}(k|k-1) := \mathcal{E}\left\{v(k) \mid \{e(\tau)\}_{\tau\leq k-1}, \{p(\tau)\}_{\tau\in\mathbb{Z}}\right\}, \tag{9.7}$$

where $\mathcal{E}$ is the expectation operator. Assume that p is deterministic and $\mathsf{h}_0 = 1$, which also implies that the feedthrough term in $H_0^{\dagger}$ is 1. These imply, that the conditional expectation of $v(k)$ is given as

$$\hat{v}(k|k-1) = \sum_{i=1}^{\infty}(\mathsf{h}_i \diamond p)(k)e(k-i). \tag{9.8}$$

It is also easy to establish that the conditional expectation minimizes the mean-squared error of the prediction [105]:

$$\hat{v}(k|k-1) = \arg\min_{\hat{v}(k)} \mathcal{E}\left\{v(k) - \hat{v}(k)\right\}^2, \tag{9.9}$$

where the minimization is carried out over all functions $\hat{v} \in \mathbb{R}^{\mathbb{Z}}$. Additionally, using (9.5) we can write

$$\hat{v} = \big((H_0(\mathrm{q}) \diamond p) - 1\big)e = \big(1 - (H_0^{\dagger}(\mathrm{q}) \diamond p)\big)v, \tag{9.10}$$

which gives the classical one-step-ahead predictor result of v. In case p is a stochastic process, it is a difficult problem to establish conditional expectation of $v(k)$, as each Markov parameter h_i can be a nonlinear function of $p(k-l)$ where $l \in \mathbb{Z}$, i.e. it contains forward and backward samples of p. Due to this fact, in the upcoming analysis p is considered to be a deterministic signal.

9.2.1.3 One-Step-Ahead Prediction of y

As a next objective, we develop the one-step-ahead prediction of $y(k)$ based on $\{y(\tau)\}_{\tau \leq k-1}$, $\{u(\tau)\}_{\tau \leq k}$ and a given scheduling trajectory $p \in \mathfrak{B}_{\mathbb{P}}$. Since

$$v(k) = y(k) - \sum_{i=0}^{\infty} (\mathrm{g}_i \diamond p)(k) u(k-i), \tag{9.11}$$

this means that also $v(\tau)$ is known for $\tau \leq k-1$. We would like to predict the value of $y(k)$ based on this information. Using the reasoning of the previous discussion, the conditional expectation $\hat{y}(k|k-1)$ of $y(k)$ is

$$\begin{aligned}
\hat{y} &= (G_0(\mathrm{q}) \diamond p)u + \hat{v} \\
&= (G_0(\mathrm{q}) \diamond p)u + \left(1 - (H_0^{\dagger}(\mathrm{q}) \diamond p)\right)v \\
&= (G_0(\mathrm{q}) \diamond p)u + \left(1 - (H_0^{\dagger}(\mathrm{q}) \diamond p)\right)\left(y - (G_0(\mathrm{q}) \diamond p)u\right) \\
&= \left((H_0^{\dagger}(\mathrm{q})G_0(\mathrm{q})) \diamond p\right)u + \left(1 - (H_0^{\dagger}(\mathrm{q}) \diamond p)\right)y.
\end{aligned} \tag{9.12}$$

This gives that in the view of the developed IRR representation of LPV systems, the classical result of the one-step-ahead predictor also holds in the LPV case, giving a powerful tool to develop and analyze identification methods.

9.2.1.4 Prediction Error Models

Following a similar reasoning as in the LTI case, we introduce the parameterized model

$$(G(\mathrm{q}, \theta), H(\mathrm{q}, \theta)), \tag{9.13}$$

where $\theta \in \mathcal{R}^n$ represents the "parameter vector", the collection of meromorphic coefficients associated with G and H (in case H is not dependent on p, then θ contains the real constant coefficients of H). Note that these coefficients are not necessarily associated with Markov parameters. So θ can correspond to the coefficients of the process and the noise models given in a SS or IO representation. Then, these parameterized structures are represented in a series-expansion form by G and H. The parameterized model of (9.13) leads to the following one-step-ahead *parameter-varying* (PV) predictor based on (9.12):

$$\hat{y}_\theta := \left(1 - H^{\dagger}(\mathrm{q}, \theta) \diamond p\right)y + \left((H^{\dagger}(\mathrm{q}, \theta)G(\mathrm{q}, \theta)) \diamond p\right)u. \tag{9.14}$$

In this case, the allowed parameter space is $\Theta \subseteq \mathcal{R}^n$. Now again, we are looking for an estimate of θ such that $\hat{y}_\theta$ is a good approximation of y, i.e. the *prediction error*:

$$\varepsilon(k,\theta) := y(k) - \hat{y}_\theta(k), \tag{9.15}$$

is minimized. Just like in the LTI case, it is possible to apply the LS criterion (2.30) for this purpose based on an available data record $\mathcal{D}_{N_\mathrm{d}} = \{y(k), u(k), p(k)\}_{k=0}^{N_\mathrm{d}-1}$.

9.2.2 *The Wiener and the Hammerstein OBF Models*

In the introduced prediction-error setting, we aim to develop model structures in which the process model G is a finite truncation of a OBF expansion representation and the noise model H is equal to identity. The motivation is similar as in the LTI case (see Sect. 2.4.5), namely that with this particular choice of the process and noise models the coefficients of G appear linearly in (9.14) and the noise model H is parameterized independently from G.

Assume that we are given a set of Hambo orthonormal basis functions $\Phi_{n_\mathrm{g}}^\infty$, defined as

$$\Phi_{n_\mathrm{g}}^\infty := \left\{\phi_j(z) G_\mathrm{b}^i(z)\right\}_{j=1,\dots,n_\mathrm{g}}^{i=0,\dots,\infty} \tag{9.16}$$

where G_b is an inner function in $\mathcal{H}_2(\mathbb{E})$. Note that $\Phi_{n_\mathrm{g}}^\infty$ can be arbitrary or chosen by a basis selection mechanism, as described in Chap. 8. In terms of these basis functions, a series-expansion representation of $\mathcal{S}$, i.e. the process part of (9.1), is available in the form of (5.19), implying that:

$$y = e + (\mathscr{w}_{00} \diamond p)u + \sum_{i=0}^{\infty} \sum_{j=1}^{n_\mathrm{g}} (\mathscr{w}_{ij} \diamond p) \phi_j(\mathrm{q}) G_\mathrm{b}^i(\mathrm{q}) u, \tag{9.17}$$

where $\mathscr{w}_{ij} \in \mathcal{R}$ and the feedthrough-term $\mathscr{w}_{00} \in \mathcal{R}$ are meromorphic coefficient functions. As a next step, we define model structures as the finite truncation of the series-expansion in (9.17). Here, one should realize that (9.17) can also be formulated in an alternative way, where the expansion coefficients appear *after* the basis functions. Since multiplication by the time operator q is non-commutative with respect to $\mathscr{w}_{ij}$ (see Sect. 3.1.3), a finite truncation of that alternative form would lead to a model with different approximation capabilities. To show that the alternative formulation of (9.17) exists, consider (5.3). It is obvious that this pulse basis expansion can be also written as

$$y = (\mathrm{g}_0 \diamond p)u + \mathrm{q}^{-1}(\overrightarrow{\mathrm{g}}_1 \diamond p)u + \mathrm{q}^{-2}(\overrightarrow{\overrightarrow{\mathrm{g}}}_2 \diamond p)u + \dots \tag{9.18}$$

where $\overrightarrow{\bullet}$ is the forward-shift operator on $\mathcal{R}$ (see Definition 3.16). Using this alternative formulation of (5.3) to derive OBF expansions of asymptotically LPV systems via the substitution rule (5.10), leads to

$$y = e + (\mathscr{w}_{00} \diamond p)u + \sum_{i=0}^{\infty} \sum_{j=1}^{n_\mathrm{g}} \phi_j(\mathrm{q}) G_\mathrm{b}^i(\mathrm{q}) (\mathscr{w}_{ij} \diamond p)u. \tag{9.19}$$

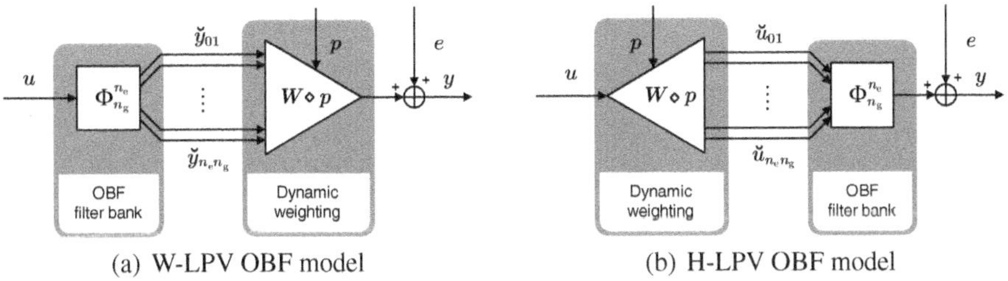

Fig. 9.2 IO signal flow graph of (a) the W-LPV OBF model described by (9.21) and (b) the H-LPV OBF model described by (9.24) with $W = [w_{01} \;\cdots\; w_{n_e n_g}]$ and excluding the feedthrough term ($w_{00} = 0$).

Note that in (9.19), the coefficient functions $\{w_{ij}\}$ are generally not equal to the coefficients of (9.17) due to the non-commutativity of q under multiplication.

Now consider the finite truncation of the Hambo basis $\Phi_{n_g}^{\infty}$:

$$\Phi_{n_g}^{n_e} := \left\{\phi_j(z)G_b^i(z)\right\}_{j=1,\ldots,n_g}^{i=0,\ldots,n_e} \tag{9.20}$$

with $n_e < \infty$. In terms of this truncation, (9.17) or (9.19) provide an approximation of the data generating system (9.1), i.e the LPV system $\mathcal{S}$. The resulting structures, presented in Fig. 9.2a–b, can be viewed as a filter bank of OBFs, which is a LTI system, followed or preceded by a meromorphic weighting function set with dynamic dependence on p. Thus, these structures have some resemblance with *nonlinear Wiener* (NW) and *nonlinear Hammerstein* (NH) models, important model classes for chemical, biological, and sensor/actuator systems [25]. An LTI model with static nonlinearity on its output is called a Wiener model while an LTI model with static nonlinearity on its input is called a Hammerstein model. The consequences of this similarity are investigated later on, but for the time being, to respect this relation, the introduced structures are called a *Wiener LPV OBF model* (W-LPV OBF) and a *Hammerstein LPV OBF model* (W-LPV OBF). These model structures are formally defined as follows:

- Wiener LPV OBF model (W-LPV OBF)

$$G(q,\theta)\diamond p = w_{00}\diamond p + \sum_{i=0}^{n_e}\sum_{j=1}^{n_g}(w_{ij}\diamond p)\phi_j(q)G_b^i(q), \quad H(q,\theta) = 1, \tag{9.21}$$

with $\theta = \begin{bmatrix} w_{00} & w_{01} & \ldots & w_{n_e n_g}\end{bmatrix}^\top \in \mathcal{R}^{1+(n_e+1)n_g}$ and $p \in \mathfrak{B}_{\mathbb{P}}$, where $\mathfrak{B}_{\mathbb{P}}$ is considered to be known. This model, given in Fig. 9.2a, is called the W-LPV OBF model and denoted by $\mathfrak{M}_W(\Phi_{n_g}^{n_e}, \theta, \mathfrak{B}_{\mathbb{P}})$. In terms of (9.14), the predictor form of (9.21) reads as

$$\hat{y}_\theta = (w_{00}\diamond p)\underbrace{u}_{\breve{y}_{00}} + \sum_{i=0}^{n_e}\sum_{j=1}^{n_g}(w_{ij}\diamond p)\underbrace{\phi_j(q)G_b^i(q)u}_{\breve{y}_{ij}}. \tag{9.22}$$

Let (A,B,C,D) be a minimal balanced LTI *state-space* (SS) realization of $\Phi_{n_g}^{n_e}$. Using this realization, the SS equivalent representation of (9.22), is given by

$$q x = Ax + Bu, \tag{9.23a}$$
$$\hat{y}_\theta = (W \diamond p)x + (w_{00} \diamond p)u, \tag{9.23b}$$

where $x = \begin{bmatrix} \breve{y}_{01} & \dots & \breve{y}_{n_e n_g} \end{bmatrix}^\top$ and $W = \begin{bmatrix} w_{01} & \dots & w_{n_e n_g} \end{bmatrix}$.

- Hammerstein LPV OBF model (H-LPV OBF)

$$G(q,\theta) \diamond p = w_{00} \diamond p + \sum_{i=0}^{n_e} \sum_{j=1}^{n_g} \phi_j(q) G_b^i(q)(w_{ij} \diamond p), \quad H(q,\theta) = 1, \tag{9.24}$$

with $\theta = \begin{bmatrix} w_{00} & w_{01} & \dots & w_{n_e n_g} \end{bmatrix}^\top \in \mathcal{R}^{1+(n_e+1)n_g}$ and $p \in \mathfrak{B}_\mathbb{P}$, where $\mathfrak{B}_\mathbb{P}$ is considered to be known. This model, given in Fig. 9.2b, is called the H-LPV OBF model and denoted by $\mathfrak{M}_H(\Phi_{n_g}^{n_e}, \theta, \mathfrak{B}_\mathbb{P})$. In this case, the predictor reads as

$$\hat{y}_\theta = \underbrace{(w_{00} \diamond p)u}_{\breve{u}_{00}} + \sum_{i=0}^{n_e} \sum_{j=1}^{n_g} \phi_j(q) G_b^i(q) \underbrace{(w_{ij} \diamond p)u}_{\breve{u}_{ij}}. \tag{9.25}$$

The SS equivalent representation of (9.25), is given by

$$q x = Ax + (W \diamond p)^\top u, \tag{9.26a}$$
$$\hat{y}_\theta = Cx + (w_{00} \diamond p)u, \tag{9.26b}$$

where $\begin{bmatrix} \breve{u}_{01} & \dots & \breve{u}_{n_e n_g} \end{bmatrix}^\top = (W \diamond p)^\top u$.

These model structures are the PV forms of the (A,B) and the (A,C) invariant Hambo OBFs based model parameterizations in the LTI case (see Sect. 2.1). Those LTI model parameterizations are considered to be equivalent in a SISO setting, as their coefficients are equivalent up to a linear transformation (Sect. 2.4.5). However, this does not hold in the LPV case due to the absence of the transposition property (Sect. 4.1.4), therefore the coefficients $\{w_{ij}\}_{j=1,\dots,n_g}^{i=0,\dots,n_e}$ are generally not equivalent in (9.21) and (9.24). Thus, these model structures are distinguished in the sequel.

9.2.3 Properties of Wiener and Hammerstein OBF Models

As a next step, important properties of the introduced models are investigated from the viewpoint of system identification. First we prove that the W-LPV OBF and H-LPV OBF models are general approximators of LPV systems, so by these models, LPV systems can be approximated with arbitrary precision. Then we explore how a locally changing McMillan degree of the system, which is a crucial problem in the interpolation-based identification methods, affects these model structures. Next we show why these model structures are beneficial in the prediction-error setting and how estimates in these structural forms can be used for control design, i.e. that these models are compatible with the existing LPV control theory.

9.2.3.1 General Approximation Property

It has been show in Chap. 5 that an asymptotically stable discrete-time LPV systems has a series-expansion representation in terms of an arbitrary basis for $\mathcal{RH}_{2-}(\mathbb{E})$. For such an approximation-free representation, infinitely many basis functions, $\Phi_{n_g}^{\infty}$, are required in the general case. This implies that using a finite number of basis functions, $\Phi_{n_g}^{n_e}$, restricts the class of realizable LPV systems. Thus, it is important to investigate the approximation capabilities of these models.

In the work of [30], it has been proved for nonlinear Wiener models that, if the LTI part is an OBF filter bank, then such models are general approximators of nonlinear systems with fading memory (NL dynamic systems with convolution representation). This means that, if the number of the OBFs in the filter bank tends to infinity, then the best achievable approximation error of the output trajectories converges to zero in an arbitrary norm, under the condition that the correct static nonlinearity is used in the weighting block.

Now consider the LPV case. In Chap. 5 it has been shown that the series-expansion of asymptotically stable LPV systems in terms of an orthonormal basis $\Phi_{n_g}^{\infty}$ is convergent. It has also been highlighted, that such series expansions in general only exist, if the expansion coefficients have dynamic dependence. Furthermore, it has been shown that the expansion coefficients yield a sequence, which converges to zero for each scheduling trajectory. For a given basis sequence $\Phi_{n_g}^{\infty}$, the worst-case approximation error of a truncated expansion representation is

$$\sup_{(y,u,p)\in\mathfrak{B}} \left\| \sum_{i=n_e+1}^{\infty} \sum_{j=1}^{n_g} (w_{ij} \diamond p)\phi_j(\mathrm{q})G_\mathrm{b}^i(\mathrm{q})u \right\|. \tag{9.27}$$

Based on the previous, it holds that this worst-case approximation error satisfies that

$$\lim_{n_e\to\infty} \sup_{(y,u,p)\in\mathfrak{B}} \left\| \sum_{i=n_e+1}^{\infty} \sum_{j=1}^{n_g} (w_{ij} \diamond p)\phi_j(\mathrm{q})G_\mathrm{b}^i(\mathrm{q})u \right\| = 0. \tag{9.28}$$

W-LPV and H-LPV OBF models are formulated based on finite truncation of LPV expansion representations, thus (9.28) proves the following property:

Property 9.1 (General LPV approximation) *W-LPV and H-LPV OBF models are general approximators of LPV systems.*

This result means that, by extending the basis function set of these models, approximation of general LPV systems can be achieved with arbitrary precision. An additional property is that, in practice, careful selection of the basis functions can ensure almost error free representation of the frozen transfer function set $\mathfrak{F}_\mathcal{S}$ of $\mathcal{S}$ with a limited number of OBFs (see Chap. 8). Such a basis function set has fast convergence rate in the series-expansion of $\mathcal{S}$. This provides the conclusion that with the general approximator property, the proposed model structures offer an efficient approximation structure for identification.

There is also an important difference with respect to the previously mentioned nonlinear counterpart of this result. In the nonlinear case, a necessary condition of the general approximator property is that the static nonlinearity must contain all possible combinations of the products of the output signals $\breve{y}_{ij}$ of the filter bank. Due to the linear signal relation of the LPV setting, this condition is not required for Property 9.1, i.e. the linear combination of $\breve{y}_{ij}$ with meromorphic PV coefficients is sufficient.

9.2.3.2 McMillan Degree Property

An additional property of the introduced model structures is that they are well structured against changes of the McMillan degree in the frozen system set $\mathscr{F}_\mathcal{S}$ of $\mathcal{S}$. It is generally true that, if for a constant scheduling signal $p(t) = \mathrm{p}$ the associated $\mathcal{F}_\mathrm{p} \in \mathscr{F}_\mathcal{S}$ has a lower McMillan degree than the rest of the systems in $\mathscr{F}_\mathcal{S}$, then this does not imply that any of the coefficient functions $\{w_{ij}\}$ of the OBF series-expansion is zero for p. This shows that these model parameterizations are not affected by problems that are common for LPV-SS or IO representations based model structures (see Chap. 1).

9.2.3.3 Linear in the Coefficients Property

The third, but equally important property of W-LPV OBF and H-LPV OBF models is that they are linear in the coefficients θ, i.e. both in predictor equations (9.22) and (9.25) the coefficients $\{w_{ij}\}$ appear linearly. This means that for a LS identification criterion and with a linear parametrization of $\{w_{ij}\}$, the estimation problem of these coefficient functions has an analytic solution.

9.2.3.4 Models for Control

The proposed models are also efficiently applicable for LPV control design. Through (9.23a–b) and (9.26a–b), a SS realization of the estimated models is available. Such SS forms have a trivial LFR realization if the coefficients $\{w_{ij}\}$ have polynomial or rational dependence on p. Therefore the existing LPV control approaches, which are exclusively formulated for SS representations, can be directly applied. Due to the fact, that both in (9.23a) and (9.26a) the matrix A is constant, optimal control design greatly simplifies with respect to these model estimates. The reason is that global dynamic stability of models with constant A can be always expressed by a Lyapunov equation with non-parameter dependent, i.e. constant P (see Sect. (3.3.2)). The only disadvantage is that many control approaches assumes static dependence of the matrices. This implies, that dependence of $\{w_{ij}\}$ must be restricted to static dependence, i.e. $w_{ij} \in \mathcal{R}|_{n_\mathbb{P}}$, to provide models to which these control solutions can be applied. However, for the general approximator property, dynamic dependence of the coefficients is required, which means that the restriction to static dependence reduces the representation capabilities of the models. This issue is explored further in Sect. 9.3 and 9.4.

9.2.4 OBF Models vs. Other Model Structures

The prediction-error setting with the one-step-ahead predictor (9.14) enables the comparison with other model structures used in the LPV identification literature. Hence in the following, the properties of LPV-IO and SS models, introduced in Chap. 1, are discussed in the prediction-error setting and compared to OBFs models. As we will show, there are hidden assumptions in these LPV model structures in terms of the noise models.

9.2.4.1 Comparison to LPV-IO Models

Consider

$$y = -\sum_{i=1}^{n_a}(a_i \diamond p)\mathrm{q}^{-i}y + \sum_{j=0}^{n_b}(b_j \diamond p)\mathrm{q}^{-i}u + e, \tag{9.29}$$

the LPV-IO filter model of the IO identification approaches (see Chap. 1). The parameter vector θ for this so-called LPV-ARX model consists of the coefficients in (9.29), $\theta = [a_1 \ \dots \ a_{n_a} \ b_0 \ \dots \ b_{n_b}]$, where each coefficient has static dependence. It can be easily shown that the one-step-ahead predictor of y reads as

$$\hat{y}_\theta = -\sum_{i=1}^{n_a}(a_i \diamond p)\mathrm{q}^{-i}y + \sum_{j=0}^{n_b}(b_j \diamond p)\mathrm{q}^{-i}u, \tag{9.30}$$

if the noise model H is chosen in a way that

$$H^{\dagger}(\mathrm{q},\theta)\diamond p = 1 + \sum_{i=1}^{n_a}(a_i \diamond p)\mathrm{q}^{-i}. \tag{9.31}$$

Note that, similar to the OBF models, this model structure is also linear in the coefficients, but its noise model is not independently parameterized from the processes part, as it is well-known for the LTI case. However, the suggested noise model

$$e = v + \sum_{i=1}^{n_a}(a_i \diamond p)\mathrm{q}^{-i}v, \tag{9.32}$$

reveals that

$$H(\mathrm{q},\theta)\diamond p = 1 - \sum_{i=1}^{n_a}(a_i \diamond p)\mathrm{q}^{-i} + \left(\sum_{i=1}^{n_a}(a_i \diamond p)\mathrm{q}^{-i}\right)\left(\sum_{i=1}^{n_a}(a_i \diamond p)\mathrm{q}^{-i}\right) - \dots$$

This shows that, even if each a_i has static dependence, the noise model $H(\mathrm{q},\theta)$ has dynamic dependence, i.e. v is dependent on the entire past of the scheduling signal p. Thus the assumed noise structure of an ARX model is rather artificial, implying much more conservatism than in the LTI case. It must be noted, that in the cited works [226, 225, 12, 11], only the estimation problem (9.29) has been solved in the LS setting, while the assumed noise model of this model structure and its effects have not been investigated. Recently, preliminary results have also appeared on the possible use of more realistic OE or BJ noise models in the estimation of LPV-IO models [34, 94]. Contrary to the previous results, the role of the noise model is taken into consideration in these studies and an *instrumental variable* (IV) approach is used to provide consistent estimation of the parameters with such noise models.

9.2.4.2 Comparison to LPV-SS Models

In the SS case, the model structure of the global identification methods, like the global subspace techniques and gradient methods, is given by

$$\mathrm{q}x = (A \diamond p)x + (B \diamond p)u + (E_1 \diamond p)e, \tag{9.33a}$$

$$y = (C \diamond p)x + (D \diamond p)u + (E_2 \diamond p)e, \tag{9.33b}$$

where the matrix functions (A,B,C,D) define a DT-LPV-SS representation, $E_1, E_2 \in \mathcal{R}^{n_\mathbb{X} \times 1}$, and e is a vector of independent zero-mean white noise processes (see Chap. 1). The parameter vector θ in this case is composed from the elements of $(A,\ldots,E2)$ corresponding to functions which are commonly assumed to have static p-dependence. In the SISO case, by applying a pulse basis expansion on this model, the process model G and the noise model H trivially follow:

$$G(\mathrm{q},\theta) = D + \sum_{i=1}^{\infty} C\left(\prod_{j=1}^{i-1} A^{[j]}\right) B^{[i]} \mathrm{q}^{-i}, \tag{9.34a}$$

$$H(\mathrm{q},\theta) = E_2 + \sum_{i=1}^{\infty} C\left(\prod_{j=1}^{i-1} A^{[j]}\right) E_1^{[i]} \mathrm{q}^{-i}, \tag{9.34b}$$

where $\bullet^{[j]}$ denotes that the backward-shift operator (see Definition 3.16) is applied j-times on the matrix. Note that the noise model (9.34b) involves matrices of the process part, thus it is obvious that its not independently parameterized from G and it depends on the entire past of the scheduling signal p. Furthermore, the coefficients, i.e. the matrices, do not appear linearly in (9.34a). Thus, the model (9.33a–b) can be used in the prediction-error identification setting successfully by either applying a (complicated) nonlinear estimation procedure, like gradient search [96, 97, 215, 214] or in the unrealistic case, when the state signals x are measurable. In the latter case, prediction of the state and output signals becomes linear in the coefficients, thus in a LS setting, linear regression can be used to derive a model estimate (see [124, 109, 107]). Based on this, LPV-SS models are commonly estimated in a non-prediction-error setting like the subspace approaches of [207, 52, 219]. An exception is the PBSID approach, extended to the LPV case in [210], where the Markov coefficients of the system (in terms of finite truncation of (9.34a)), are estimated first via linear regression in a prediction-error setting (by assuming an ARX noise model). Then, the estimated Markov coefficients are used in the subspace mechanism to calculate a corresponding LPV-SS form.

Comparison of the proposed model structures with the considered model structures underlines that the introduced OBFs-based series-expansion models are attractive candidates in the prediction-error identification setting of LPV systems as with an OE type of noise model they are linear in the coefficients and the noise model is independently parameterized from the process part and independent from the scheduling.

9.2.4.3 Similarity to the Nonlinear Wiener and Hammerstein Models

By comparing NW and NH models to the introduced structures, the structural similarity is immediate. However, there are some fundamental differences:

- In the NW and NH case, a static nonlinearity is assumed on the output/input of the LTI part. In the W-LPV and H-LPV OBF case, the "nonlinearity" is entering through a dynamic dependence on p, which can be composed of external (strict LPV systems) and internal (quasi-LPV systems) variables alike. Assuming that p is equal to u or y, NW and NH models can be viewed as special cases of W-LPV and H-LPV OBF models under minor restrictions on the static nonlinearities (see Sect. 7.4.3).
- The LTI parts of W-LPV and H-LPV OBF are respectively *single-input multiple-output* (SIMO) and *multiple-input single-output* (MISO) systems as opposed to the SISO LTI part of NW and NH models[1].

9.2.5 *Identification of W-LPV and H-LPV OBF Models*

In the following, a general outline of LPV identification based on W-LPV and H-LPV OBF models is presented. The major steps of the identification cycle: model structure selection, identification criterion selection, and estimation are considered. Our aim is to set the stage for the upcoming discussion of LPV-OBF identification approaches, which are based on particular choices with respect to these steps.

9.2.5.1 OBF Selection

In the case of W-LPV and H-LPV OBF models, there is a primal emphasis on the model structure selection step of the identification cycle. This is due to the fact that selection of the finite OBF set, that defines the structure, effects the approximation capabilities in terms of the convergence rate of the series-expansion with respect to

[1] Originally both Wiener and Hammerstein proposed their models with SIMO and MISO LTI parts, but because of the complexity of the problem, the LTI part has been simplified to be SISO [25, 105].

this basis. With an adequate selection of the basis functions, i.e. with high convergence rate, negligible approximation error, i.e. bias can be achieved. Beside selection of an adequate basis it is also important to decide how many of the functions in the chosen basis function sequence are considered in the model, i.e. in which degree the expansion is truncated. This decision directly effects the efficiency of the model structure in terms of the number of coefficients to be estimated.

Tools to provide adequate selection of the basis functions have already been discussed in Chap. 7, where an algorithm has been proposed to assist basis selection based on first principle information. By using this algorithm, the set of frozen pole locations of the system can be characterized, and based on this pole set, the optimal choice of a orthonormal basis follows in the Kolmogorov n-width (KnW) sense. Additionally, in Chap. 8, a clustering algorithm has been introduced, that solves the KnW optimal basis selection problem based on measured data. The latter approach is useful in a black-box identification scenario, where no reliable structural information about the LPV system/physical plant is available. Moreover, via adaptive cluster merging, the algorithm is also capable to efficiently choose the number of OBFs required for an adequate approximation of the system. This practically applicable tool, which accomplishes the model structure selection in terms of the basis functions can be summarized in the following algorithm:

Algorithm 9.1 (OBFs based LPV identification, basis function selection)

Step 1. Estimation of samples of the pole manifest set $\Omega_{\mathbb{P}}$, associated with the LPV system $\mathcal{S}$. The estimation is accomplished by LTI identification of each $F_{\mathrm{p}_i} \in \mathscr{F}_{\mathbb{P}}$ of $\mathcal{S}$ for a given set of scheduling points $\{\mathrm{p}_i\}_{i=1}^{N_\mathrm{p}} \subset \mathbb{P}$.

Step 2. Based on FKcM clustering of the sample poles, determination of an adequate (optimal) OBF set $\Phi_{n_\mathrm{g}}^{n_\mathrm{e}} \subset \mathcal{RH}_{2-}(\mathbb{E})$ with respect to $\mathcal{S}$.

9.2.5.2 Parametrization of Coefficient Dependencies

Beside the selection of basis functions, model structure selection with respect to W-LPV and H-LPV OBF also contains an equally important part: the parametrization of the functional dependence of the coefficients w_{ij} on the scheduling signal p:

$$w_{ij} = \psi_{ij}(\theta), \tag{9.35}$$

where $\psi_{ij} \in \mathcal{R}$ is meromorphic with constant parameters $\theta \in \mathbb{R}^n$. The aim of the identification is to estimate θ based on a measured data record. To simplify the estimation problem, often a linear parametrization of the structural dependence in (9.35) is used. In fact w_{ij} is considered to be a linear combination of fixed meromorphic functions $\psi_{ijl} \in \mathcal{R}|_{n_\mathbb{P}}$:

$$w_{ij} = \sum_{l=0}^{n_{ij}} \theta_{ijl}\psi_{ijl}, \tag{9.36}$$

where $\theta_{ijl} \in \mathbb{R}$. A linear parametrization not only reduces the complexity of the associated estimation problem but also makes the problem of adequate selection of the underlaying structural dependence well-posed [197]. In terms of (9.36), the selection problem of an adequate parametrization translates to a search for a set of functions $\{\psi_{ijl}\}$ such that the true expansion coefficient w_{ij}^{o} of the system with respect to the used basis functions satisfies $w_{ij} \in \mathrm{Span}(\{\psi_{ijl}\}_{l=0}^{n_{ij}})$. Considering that the class of meromorphic functions $\mathcal{R}$ presents degrees of freedom in terms of the order of dynamic dependence and in terms of functions, any structural information about the coefficients can considerably reduce the search space for an optimal choice of $\{\psi_{ijl}\}$.

In case of a black-box scenario, the choice of $\{\psi_{ijl}\}$ can be arbitrary. One can consider all $\{\psi_{ijl}\}$ to be rational functions or polynomials with a fixed degree and a fixed order of dynamic dependence. However the possible choices are enormous. By including a too large set of functions $\{\psi_{ijl}\}$ can easily lead to over-parametrization, while restriction of $\{\psi_{ijl}\}$ to only a few basic functions can lead to serious bias. In order to assist the selection of an efficient set of functional dependencies in the parametrization of linear regression models recently practically applicable approaches have been proposed in [200] and [75]. While in [75] a *dispersion functions* based approach, originating form the machine learning field, has been developed to basically learn the underlying (possibly) static nonlinear dependence of the coefficients, in [200] a coefficient shrinkage method, originating form statistics, has been introduced for this purpose. Contrary to the dispersion approach, the coefficient shrinkage method, which is based on a so-called *non-negative garotte* approach, is uses regularization in terms of weights to penalize individual elements of the parameter vector θ. In this way, the approach starts with a relatively large set of possible functional dependencies from which those functions that do not contribute significantly to the validity of the estimated model are eliminated by decreasing their weights. In this way the nonnegative garotte approach gives a practically useful tool to decide from a set of functional dependencies, expected in the model, which are needed for an efficient parametrization in terms of (9.36).

9.2.5.3 Criterion Selection and Estimation

Based on the predictor form (9.14), many different classical identification criteria can be applied for the selected model structure. A particularly interesting choice is the *least-squared* (LS) prediction-error criterion

$$\mathcal{W}_{N_{\mathrm{d}}}\left(\theta, \mathcal{D}_{N_{\mathrm{d}}}\right) = \frac{1}{N_{\mathrm{d}}} \sum_{k=0}^{N_{\mathrm{d}}-1} \varepsilon^2(k,\theta). \tag{9.37}$$

where the residual ε is given by (9.15). If the parametrization of the coefficients is linear (see (9.36)), then with respect to (9.37), the estimation of $\{\theta_{ijl}\}$, similar to the LTI case, reduces to a linear regression problem for the W-LPV and H-LPV OBF models. In other cases, when the parametrization of the coefficients is nonlinear, then estimation corresponds to a nonlinear optimization problem.

To guarantee a unique solution of (9.37), one condition is that the set of functional dependencies $\{\psi_{ijl}\}$ are chosen such that (9.21) and (9.24) are globally identifiable:

Definition 9.1 (Identifiability, [56]). A model structure $(G(\mathrm{q},\theta),H(\mathrm{q},\theta))$ with a parameter domain $\Theta \subseteq \mathbb{R}^n$ is called locally identifiable at a parameter value $\theta_1 \in \Theta$, if $\exists \varepsilon > 0$ such that for all $\theta \in \Theta$ with $\|\theta - \theta_1\| \leq \varepsilon$ the corresponding one-step-ahead predictors (see (9.14)) are distinguishable:

$$H(\mathrm{q},\theta) = H(\mathrm{q},\theta_1) \text{ and } H^{\dagger}(\mathrm{q},\theta)G(\mathrm{q},\theta) = H^{\dagger}(\mathrm{q},\theta_1)G(\mathrm{q},\theta_1) \Rightarrow \theta = \theta_1.$$

The model structure $(G(\mathrm{q},\theta),H(\mathrm{q},\theta))$ is called globally identifiable at θ_1 if it is locally identifiable at θ_1 with $\varepsilon \to \infty$. Moreover, $(G(\mathrm{q},\theta),H(\mathrm{q},\theta))$ is called globally identifiable if it is globally identifiable at all $\theta \in \Theta$. □

In terms of the 1-step ahead predictor of the considered OBF models, (9.22) and (9.25), one way to guarantee this important condition is to assume that each set $\{\psi_{ijl}\}_{l=0}^{n_{ij}}$ contains functions ψ_{ijl} which are orthogonal with respect to each other for all possible trajectories of p, i.e. $p \in \mathfrak{B}_{\mathbb{P}}$. In case of static dependence, this means that ψ_{ijl} are orthogonal with respect to each other on $\mathbb{P}$.

9.3 Identification with Static Dependence

In the previous part, we have developed truncated OBF expansion models, as the basic ingredients of a well-posed LPV identification approach. As a next step, we show how these model structures can be efficiently identified in the prediction-error setting, such that the obtained models are directly applicable for control design. First we clarify the exact identification setting (parametrization, identification criterion, etc.) in which we aim to derive the model estimates. Then we develop our approaches using either the gain-scheduling identification strategy (local approach) or a linear regression based strategy with varying scheduling trajectory (global approach). We only treat the SISO case. The MIMO extension of the developed approaches is covered later. The methods are analyzed in terms of variance, bias, consistency and validation of the model estimates is also investigated. Finally, a simulation example is studied to visualize the performance of the approaches.

9.3.1 The Identification Setting

In the previous parts, we have seen that a common feature of all LPV model structures, either based on SS, IO, or series-expansion representations, is that to represent general LPV systems they need dynamic dependence in the parametrization of their coefficients. However, LPV control design approaches often assume only static dependence of the model estimate. On the other hand, it is hard to handle the extra degree of freedom that dynamic dependence constitutes in an estimation problem.

This gives the motivation to investigate identification in the special case, when the coefficients of W-LPV and H-LPV OBF model structures are parameterized with static dependence, i.e in (9.35) the chosen structural dependence ψ_{ij} is only dependent on the instantaneous value of p. To make a clear distinction when we talk about static and respectively dynamic dependence, we use $w(p)$ to express evaluation of a static coefficient dependence along a scheduling trajectory p, contrary to $w \diamond p$.

We have already motivated that using a linear parametrization of the expansion coefficients (see (9.36)), estimation of the parameters can be formulated as a linear regression if the LS identification criterion (9.37) is used. Based on this, we aim in this section at the identification of LPV systems by the $\mathfrak{M}_{\mathrm{W}}(\Phi_{n_\mathrm{g}}^{n_\mathrm{e}}, \theta, \mathfrak{B}_{\mathbb{P}})$ and $\mathfrak{M}_{\mathrm{H}}(\Phi_{n_\mathrm{g}}^{n_\mathrm{e}}, \theta, \mathfrak{B}_{\mathbb{P}})$ model structures, where the process part G is parameterized as:

$$G(\mathrm{q},\theta) \diamond p = \underbrace{\sum_{l=0}^{n_{00}} \theta_{00l}\psi_{00l}(p)}_{w_{00} \diamond p} + \sum_{i=0}^{n_\mathrm{e}} \sum_{j=1}^{n_\mathrm{g}} \underbrace{\sum_{l=0}^{n_{ij}} \theta_{ijl}\psi_{ijl}(p)\phi_{ij}(\mathrm{q})}_{w_{ij} \diamond p}, \tag{9.38a}$$

in the Wiener case and

$$G(\mathrm{q},\theta) \diamond p = \underbrace{\sum_{l=0}^{n_{00}} \theta_{00l}\psi_{00l}(p)}_{w_{00} \diamond p} + \sum_{i=0}^{n_\mathrm{e}} \sum_{j=1}^{n_\mathrm{g}} \phi_{ij}(\mathrm{q}) \underbrace{\sum_{l=0}^{n_{ij}} \theta_{ijl}\psi_{ijl}(p)}_{w_{ij} \diamond p}. \tag{9.38b}$$

in the Hammerstein case, where $\theta_{ijl} \in \mathbb{R}$, $\psi_{ijl} \in \mathcal{R}|_{n_\mathbb{P}}$, and $\Phi_{n_\mathrm{g}}^{n_\mathrm{e}} = \{\phi_{ij}\}_{j=1,\dots,n_\mathrm{g}}^{i=0,\dots,n_\mathrm{e}}$. In these parameterizations, the basis functions $\Phi_{n_\mathrm{g}}^{n_\mathrm{e}}$ are considered to be the result of a basis selection and hence they are fixed, while the functions ψ_{ijl} are either chosen by the user, or derived from first-principle information (like by the use of the approach given in Chap. 7). It is also assumed that in each set $\{\psi_{ijl}\}_{l=0}^{n_{ij}}$ the functions ψ_{ijl} are orthogonal on $\mathbb{P}$ to ensure global identifiability of the model. Consequently, the remaining unknowns in the model are the real parameters $\{\theta_{ijl}\}$ which appear linearly in the structures. Thus by using the LS criterion they can be identified by linear regression. Here we do not consider the identification or optimal choice of the functions ψ_{ijl}. However, there are recent approaches available which can handle efficient selection of $\{\psi_{ijl}\}$ in the prediction-error setting, see [200, 75].

9.3.2 LPV Identification with Fixed OBFs

Based on the choice of linear parametrization with static dependence and the LS criterion, two pragmatical approaches are available for the identification of the LPV system $\mathcal{S}$ via W-LPV and H-LPV OBF models:

Local approach Identify $\mathrm{w}_{ij} = \sum_{l}^{n_{ij}} \theta_{ijl}\psi_{ijl}(\mathrm{p})$ for several $\mathrm{p} \in \mathbb{P}$ and interpolate the obtained estimates to calculate $\{\theta_{ijl}\}$. This gives the freedom to choose the functions $\psi_{ijl}(\mathrm{p})$ based on the local estimates w_{ij}. However, an apparent disadvantage is that many

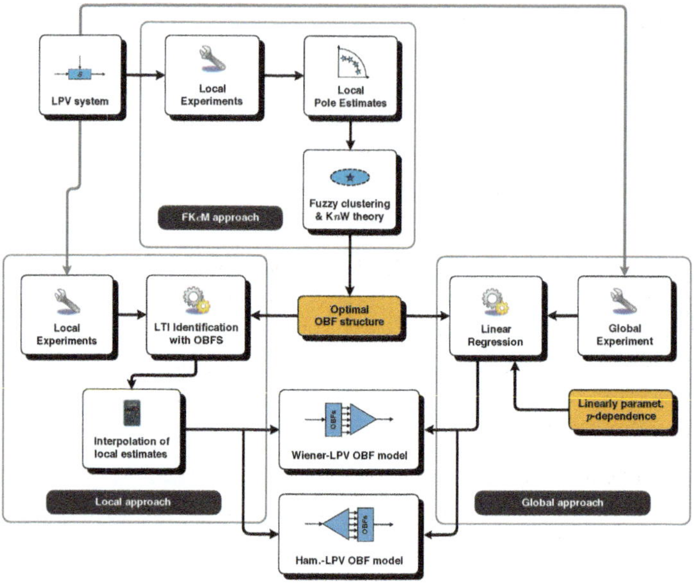

Fig. 9.3 Block diagram of local and global identification methods

experiments with different constant scheduling trajectories are needed for successful interpolation.

Global approach Using a data record with varying p, formulate a linear regression problem with respect $\{\theta_{ijl}\}$. The resulting problem has an analytic solution, giving a direct estimate of the parameters without the need of interpolation.

In the following a detailed description of these approaches is presented based on [194, 197], including an investigation in which situations one is more applicable than the other. The approaches are formulated by assuming that the basis selection phase has already been accomplished, so we pick up the line of reasoning right after Step 2 in Sect. 9.2.5.1. Fig. 9.3 illustrates the basic steps of the approaches. For reasons of simplicity, we assume that no feedthrough term is present, $w_{00} = 0$. Later, the estimation of w_{00} is also investigated for both algorithms.

9.3.3 Local Approach

Assume that a set of constant scheduling points $\mathsf{P} = \{\mathsf{p}_\tau\}_{\tau=1}^{N_{\text{loc}}} \subset \mathbb{P}$ is given for $\mathcal{S}$, where it is assumed that $\mathbb{P}$ is well covered, meaning that $\max_i \min_{j\neq i} |\mathsf{p}_i - \mathsf{p}_j|$, $i, j \in \mathbb{I}_1^{N_{\text{loc}}}$ is small enough. This is required for successful interpolation by most numerical methods. See [84] for an approach to optimally choose such scheduling points under a linear regression identification criterion. Assume also that measured data records $\mathcal{D}_{N_{\text{d}},\mathsf{p}_\tau} = \{y(k), u(k), \mathsf{p}\}_{k=0}^{N_{\text{d}}-1}$ with length $N_{\text{d}} \in \mathbb{N}$ are available. Then the identification of $\mathcal{S}$ is solved as (continuing from Step 2 in Sect. 9.2.5.1) [194]:

Algorithm 9.2 (OBFs based LPV identification, local method)

Step 3a. For a given OBF set $\Phi_{n_g}^{n_e} = \{\phi_{ij}\}_{j=1,\dots,n_g}^{i=0,\dots,n_e}$, scheduling points $\mathsf{P} = \{\mathsf{p}_\tau\}_{\tau=1}^{N_{\mathrm{loc}}}$ $\subset \mathbb{P}$, and identification criterion $\mathcal{W}$, identify each frozen system $\mathcal{F}_{\mathsf{p}_\tau} \in \mathscr{F}_{\mathsf{P}}$, $\tau \in \mathbb{I}_1^{N_{\mathrm{loc}}}$ based on the LTI-OBF model structure:

$$G_{\mathsf{p}_\tau}(\mathsf{q},\theta) = \sum_{i=0}^{n_e}\sum_{j=1}^{n_g} \mathtt{w}_{ij\tau}\phi_{ij}(\mathsf{q}), \quad H(\mathsf{q},\theta) = 1, \tag{9.39}$$

and data records $\mathcal{D}_{N_d,\mathsf{p}_\tau}$ of $\mathcal{S}$ collected with constant scheduling trajectories and with u that is persistently exciting with the required order. This results in a set of estimated coefficients $\{\hat{\mathtt{w}}_{ij\tau}\}_{i=0,j=1,\tau=1}^{n_e,n_g,N_{\mathrm{loc}}} \subset \mathbb{R}$, where $\{\hat{\mathtt{w}}_{ij\tau}\}_{i=0,j=1}^{n_e,n_g}$ describes the coefficients of $\Phi_{n_g}^{n_e}$ with respect to $\mathcal{F}_{\mathsf{p}_\tau}$.

Step 4a. Interpolation of the frozen OBF coefficients $\{\mathtt{w}_{ij\tau}\}$. For each $(i,j) \in \mathbb{I}_0^{n_e} \times \mathbb{I}_1^{n_g}$, choose a set of orthogonal interpolation functions $\{\hat{\psi}_l\}_{l=0}^{n_{ij}}$, being a set of meromorphic functions over $\mathbb{P}$, i.e. $\hat{\psi}_\tau \in \mathcal{R}|_{n_\mathbb{P}}$, and a set of constants $\{\hat{\theta}_{ijl}\}_{l=0}^{n_{ij}}$ such that

$$\hat{\mathtt{w}}_{ij\tau} = \sum_{l}^{n_{ij}} \hat{\theta}_{ijl}\,\hat{\psi}_l(\mathsf{p}_\tau), \quad \forall \tau \in \mathbb{I}_1^{N_{\mathrm{loc}}}. \tag{9.40}$$

In this way the estimate of the expansion coefficients results as

$$\hat{w}_{ij} := \sum_{l=0}^{n_{ij}} \hat{\theta}_{ijl}\,\hat{\psi}_l. \tag{9.41}$$

In general, any interpolation technique can be used to approximate the coefficient functions $\{w_{ij}\}$, however most commonly polynomial, rational, or Chebyshev interpolation provides adequate results (see [158]). Naturally, specific choices of the interpolation functions result in different estimates of (9.41). Validation of the model estimate is required to verify these choices.

9.3.4 Global Approach

Opposite to the local approach, the global approach utilizes only one data set which is collected from $\mathcal{S}$ with varying scheduling, i.e. one global experiment. Assume that measured IO data $\mathcal{D}_{N_d} = \{y(k), u(k), p(k)\}_{k=0}^{N_d-1}$ is available and it is informative for $\mathcal{S}$. Informative means in this case that with the considered parametrization (9.38a–b), a unique model in the considered model class can be found in the statistical sense such that (9.37) is minimal. For a more detailed discussion on informativity and its connection to the classical notion of persistency of excitation see Sect. 9.3.5.2. Using this data set, the global identification of $\mathcal{S}$ is solved in the W-LPV OBF case as [195]:

Algorithm 9.3 (OBFs based LPV identification, global method, Wiener case)

Step 3b. For a given OBF set $\Phi_{n_\mathrm{g}}^{n_\mathrm{e}} = \{\phi_{ij}\}_{j=1,\dots,n_\mathrm{g}}^{i=0,\dots,n_\mathrm{e}}$ and data record $\mathcal{D}_{N_\mathrm{d}}$, generate $\breve{y} = [\breve{y}_{ij}]_{j=1\dots n_\mathrm{g}}^{i=0,\dots,n_\mathrm{e}}$ with $\breve{y}_{ij} = \phi_{ij}(\mathrm{q})u$. This step can be efficiently accomplished by using a minimal balanced SS realization (A,B,C,D) of $\Phi_{n_\mathrm{g}}^{n_\mathrm{e}}$. The state evolution (9.23a) in the time interval $[0, N_\mathrm{d}-1]$ with respect to $\{u(k)\}_{k=0}^{N_\mathrm{d}-1}$ and $x(0) = 0$, gives $\breve{y}$ on $[0, N_\mathrm{d}-1]$ as $\breve{y} = x$.

Step 4b. Choose a row vector of meromorphic functions $\Psi = [\psi_l]_{l=0}^{n_\psi}$ for the parametrization of each w_{ij} in (9.21) as $w_{ij} = \sum_{l=0}^{n_\psi} \theta_{ijl}\psi_l$ where $\{\theta_{ijl}\}$ are real unknown parameters, $\psi_l \in \mathcal{R}|_{n_\mathbb{P}}$ (meromorphic functions with static dependence), $\psi_0 = 1$ and $\{\psi_l\}_{l=0}^{n_\psi}$ are orthogonal on $\mathbb{P}$.

Step 5b. Using the data set $\mathcal{D}_{N_\mathrm{d}}$, estimate the parameters $\{\theta_{ijl}\}_{i=0,j=1,l=1}^{n_\mathrm{e},n_\mathrm{g},n_\psi}$ by linear regression. Based on the predictor (9.22), define the regressors as

$$\gamma^\top(k) = \breve{y}^\top(k) \otimes \Psi(p(k)), \quad \forall k \in [0, N_\mathrm{d}-1], \tag{9.42}$$

with $\otimes$ denoting the Kronecker tensor product. Collect the data into

$$\Gamma_{N_\mathrm{d}} = \left[\gamma(0) \ \dots \ \gamma(N_\mathrm{d}-1)\right]^\top, \quad Y_{N_\mathrm{d}} = \left[y(0) \ \dots \ y(N_\mathrm{d}-1)\right]^\top.$$

Arrange the parameters to be estimated into a column vector

$$\theta = \left[\theta_{010} \dots \theta_{01n_r} \dots \theta_{n_\mathrm{e}n_\mathrm{g}n_\psi}\right]^\top. \tag{9.43}$$

Then, to minimize the LS prediction-error criterion

$$\mathcal{W}_{N_\mathrm{d}}(\theta, \mathcal{D}_{N_\mathrm{d}}) = \frac{1}{N_\mathrm{d}} \left\|Y_{N_\mathrm{d}} - \Gamma_{N_\mathrm{d}}\theta\right\|_2^2, \tag{9.44}$$

the analytic solution is obtained by

$$\hat{\theta}_{N_\mathrm{d}} = \left(\frac{1}{N_\mathrm{d}}\Gamma_{N_\mathrm{d}}^\top \Gamma_{N_\mathrm{d}}\right)^{-1} \cdot \left(\frac{1}{N_\mathrm{d}}\Gamma_{N_\mathrm{d}}^\top Y_{N_\mathrm{d}}\right). \tag{9.45}$$

In this way the estimates of the expansion coefficients result in the form of:

$$\hat{w}_{ij} := \sum_{l=0}^{n_\psi} \hat{\theta}_{ijl}\psi_l. \tag{9.46}$$

In the H-LPV OBF case, the identification procedure is similar. However, the formulation of the regressor is more complicated as the coefficients appear linearly at the input side. Next, it is described how Γ_{N_d} can be formulated for the H-LPV OBF case. The first step is the calculation of a parameter-varying transition matrix, associated with the predictor (9.25–b), such that:

$$\hat{Y}_{N_\mathrm{d}} = \begin{bmatrix} 0 & 0 & \cdots \\ CW(p(0)) & 0 & \cdots \\ CAW(p(0)) & CW(p(1)) & \\ \vdots & \vdots & \ddots \end{bmatrix} U_{N_\mathrm{d}}, \tag{9.47}$$

where $\hat{Y}_{N_\mathrm{d}} = [\,\hat{y}_\theta(0) \ldots \hat{y}_\theta(N_\mathrm{d}-1)\,]^\top$ is the stacked predicted output vector of the H-LPV OBF structure. By simple rearrangement it follows that

$$\hat{Y}_{N_\mathrm{d}} = \underbrace{\begin{bmatrix} 0 & 0 & \cdots \\ CIu(0) & 0 & \cdots \\ CAIu(0) & CIu(1) & \\ \vdots & \vdots & \ddots \end{bmatrix}}_{\hat{\mathsf{H}}_{N_\mathrm{d}}} \begin{bmatrix} W(p(0)) \\ W(p(1)) \\ \vdots \end{bmatrix}. \tag{9.48}$$

Let δ be the delta function, i.e. pulse input at $k=0$. Now define $h = [h_{ij}]_{j=1\ldots n_\mathrm{g}}^{i=0,\ldots,n_\mathrm{e}}$ as the state evolution of

$$\mathrm{q}h = A^\top h + C^\top \delta, \tag{9.49}$$

on the interval $[0, N_\mathrm{d}-1]$ with $h(0)=0$. Then h is used to calculate the columns of the transition matrix $\hat{\mathsf{H}}_{N_\mathrm{d}}$. By combining each column of $\hat{\mathsf{H}}_{N_\mathrm{d}}$ in a Kronecker product with Ψ, the parameters $\{\theta_{ijl}\}$ to be estimated are separated, giving the regressor matrix as $\Gamma_{N_\mathrm{d}} = [\,\hat{\mathsf{H}}_0 \;\ldots\; \hat{\mathsf{H}}_{n_\mathbb{P}}\,]$ where

$$\hat{\mathsf{H}}_l = \sum_{k=0}^{N_\mathrm{d}-1} \mathsf{H}_k u(k)\, \psi_l(p(k)), \tag{9.50}$$

and $\mathsf{H}_k = \begin{bmatrix} 0 & \ldots & 0 & h^\top(0) & \ldots & h^\top(N_\mathrm{d}-k-1) \end{bmatrix}^\top \in \mathbb{R}^{N_\mathrm{d} \times n_\mathrm{g}(n_\mathrm{e}+1)}$.

Algorithm 9.3 can also be extended for both model structures to estimate a direct feedthrough term. The extension is obtained by defining $w_{00} = \sum_{l=0}^{n_\psi} \theta_{00l}\psi_l$ and formulating the regressor matrix as

$$\Gamma'_{N_\mathrm{d}} = \left[\begin{array}{c|c} u(0) \otimes \psi(p(0)) & \\ \vdots & \Gamma_{N_\mathrm{d}} \\ u(N_\mathrm{d}-1) \otimes \psi(p(N_\mathrm{d}-1)) & \end{array}\right]. \tag{9.51}$$

Including $\{\theta_{00l}\}$ into θ implies that the estimate follows via (9.45).

9.3.5 *Properties*

In this part the properties of the introduced identification approaches are investigated. First we analyze the restriction of the used linear parametrization with static dependence in the identification of general LPV systems and we characterize that subclass of LPV systems where no bias results. Next we motivate when the use of a global approach or a local approach is more fruitful in practice. Then we show convergence and consistency of the parameter estimates. By using similarity of the identification methods with respect to LTI prediction-error identification, we characterize basic results about variance and bias of the estimates. Finally, (in)validation issues are discussed in the introduced framework together with the estimation of initial conditions.

9.3.5.1 Representation Capabilities via Static Dependence

It is important to investigate the effects of the linear parametrization with static dependence on the representation capabilities of the models. From the SS representation form of the models (see (9.23a–b) and (9.26a–b)) it is obvious that the assumption of static dependence restricts the class of representable LPV systems to systems which have a SS representation with static p-dependence only in the C (see (9.23a)–b) or in the B (see (9.26a)–b) system matrices. This means that systems that have no representation that satisfies these conditions are not in the model class. Hence, due to parametrization, a modeling error inevitably occurs during the identification process in these cases. Additionally, the linear parametrization of the expansion coefficients (see (9.38a–b)) implies that systems that are identifiable in the considered way are further restricted to have a SS representation where only the C (or B) matrix depends on p and the functional dependence is the linear combination of the used functions ψ_{ij}. This underlines that selection of these functions based on prior information is an important part of the model structure selection process.

A second consequence of the parametrization (9.38a–b) rises with respect to the general approximator property of the W-LPV and H-LPV OBF models. We have seen in the general case that increasing n_e lowers the achievable approximation error with these models. However, in case of static dependence, increasing n_e enlarges $\text{Span}\{\Phi_{n_g}^{n_e}\}$ which means that the worst-case representation error in the KnW sense is lowered with respect to the frozen system set of the data generating system. However, an increase in n_e may not lower the representation error of the LPV system in a global sense as the modeling error can be significantly dominated by the missing non-static p-dependence of the expansion coefficients. This results in the loss of the general approximator property, meaning that the approximation capability of the model is restricted by the absence of dynamic coefficient dependence. Thus increasing n_e in the hope of better accuracy can easily result in over-parametrization in this case.

We will see that despite the theoretically presented restrictions of the applied parametrization, commonly in a practical situation, adequate approximation of the data generating system can be achieved by the resulting model estimates. It is an important remark that the global method is also applicable in the situations where the coefficients of the W-LPV and H-LPV OBF model structures are not assumed to be static. As the estimation algorithm is independent from the choice of the ψ_{ij} functions, these functions can be considered with dynamic dependence. Then with a data record containing sufficiently exciting p, estimation of the parameters is similarly available as in the static case. Unfortunately, this property does not apply for the local method, as in that case, interpolation with dynamic dependence based on frozen estimates is an ill-conditioned problem.

9.3.5.2 Informativity of Data Sets and Persistency of Excitation

In order to estimate an adequate model in a given model set, most prediction-error algorithms require *persistency of excitation* in terms of $\mathcal{D}_{N_\mathrm{d}}$ collected from the system. This condition is required to guarantee consistency and convergence of the estimates (see Sect. 2.4.3). In order to analyze the estimation in the defined LPV-OBFs model structures, a characterization of the data sets satisfying this condition is required.

In the LTI case persistency of excitation is associated with the notion of an *informative data set*. Let $\mathcal{D}_{N_\mathrm{d}} = \{y(k), u(k)\}_{k=0}^{N_\mathrm{d}-1}$ be a data set of quasi-stationary u and y collected from the data generating system, and let $\mathcal{W}_{N_\mathrm{d}}(\theta, \mathcal{D}_{N_\mathrm{d}})$ be an identification criterion. $\mathcal{D}_{N_\mathrm{d}}$ is called informative with respect to a parametric model set with parameters θ and a given $\mathcal{W}_{N_\mathrm{d}}$ if any two models in the model set can be distinguished under $\mathcal{W}_{N_\mathrm{d}}(\theta, \mathcal{D}_{N_\mathrm{d}})$ [56]. Basically this means that if the model set is globally identifiable (no two different parameters θ_1 and θ_2 give rise to the same predictor) and the data set $\mathcal{D}_{N_\mathrm{d}}$ is informative, then $\mathcal{W}_{N_\mathrm{d}}(\theta, \mathcal{D}_{N_\mathrm{d}})$ has a global optimum in the statistical sense. The latter is the essential requirement for consistency of any minimization method. The notion of persistency of excitation of order n relates in the LTI case to an informative $\mathcal{D}_{N_\mathrm{d}}$ with respect to a model structure with n parameters. The latter is equivalent to the possibility of statistically uniquely estimating a nth order FIR filter based on $\mathcal{D}_{N_\mathrm{d}}$.

In the LPV case there are numerous differences. First of all, the requirements for identifiability imply that the linear combinations of the used functions $\{\psi_l\}_{l=0}^{n_\psi}$ in the coefficient parametrization provide inequivalent dynamical behaviors of the model structure for each θ. Moreover, the notion of an informative data set in the LPV case is not equivalent to the condition of persistency of excitation with a given order. First of all, the model parameters θ are related to signals $\psi_l(p)\phi_{ij}(\mathrm{q})u$ and not only the time-shifted versions of u and y, thus the functions $\{\psi_l\}_{l=0}^{n_\psi}$ and the scheduling trajectory p together also influence the estimation of θ. Moreover, the estimation of the parameters of a LPV-FIR filter, irrelevant to the coefficient parametrization, is not equivalent with the estimation of LPV-ARX models or LPV-OBF models due to the non-commutativity of multiplication by q. This means that the terminology of persistency of excitation with order n is ill-defined in the LPV case. Instead, the informativity of the data sets with respect to the assumed coefficient parametrization and model order must be satisfied in order to ensure consistency and convergence of the estimation. However, conditions for data sets to be informative have not been investigated directly in the LPV literature.

For the case of LPV-ARX models with polynomial dependence of the coefficients on the parameters, conditions for persistency of excitation have been studied in [57] and [225], unknowingly also addressing the question of informativity of the data set. In these works it is assumed that a LPV system is a family of LTI systems, each associated with a point in $\mathbb{P}$. If the dependence on p is polynomial of order n_ψ, then the LPV system identification can be realized by identifying at least n_ψ LTI models operating at distinct values of $\mathbb{P}$, and using these to determine the coefficients of the polynomial dependence based on the interpolation principle. This means that a data set $\mathcal{D}_{N_\mathrm{d}}$ is informative if

- p visits at least n_ψ different points $\mathsf{p}_1, \ldots, \mathsf{p}_{n_\psi} \in \mathbb{P}$.
- Each sequence of u associated with a p_l must be persistently exciting with respect to the LTI model corresponding to p_l.
- The number of revisits of each p_l must be large enough to approximate the ergodicity condition in a given neighborhood around the considered points of $\mathbb{P}$.

These conditions are rather conservative from a number of viewpoints. A LPV system can be considered as a set of LTI systems associated with points of $\mathbb{P}$, but these systems share a common memory so they can describe the continuation of the signal trajectories when p changes. This means that in terms of the above given conditions the variation of p must be infinitely slow in order to consider these systems to be independent LTI systems. However, ergodicity requires basically that the number of revisits of the chosen points must be infinite in the general case, which means that the p should vary as fast as possible to revisit these points more often. This shows that the above given conditions are too conservative for practical use. In [225], an improved version of this approach has been developed which tries to overcome the problem of conservativeness, but since it is based on the same principle of independent LTI system estimation, the question whether a data set is informative in the LPV case remains open. However, in practice, the absence of a solid criterion restricts the user to the paradigm to excite the system as much as possible in order to guarantee consistency and convergence of the estimation.

9.3.5.3 Consistency and Convergence

Similar to the classical LTI identification framework, it is possible to show that under minor conditions, the parameter estimates of local and global W-LPV and H-LPV OBF approaches are convergent and consistent. Convergence means that for $N_\mathrm{d} \to \infty$ the parameter estimate $\hat{\theta}_{N_\mathrm{d}}$ converges, i.e. $\hat{\theta}_{N_\mathrm{d}} \to \theta^*$ with probability 1, while consistency means that the convergence point θ^* is equal to the parameters of the data generating system (9.1). Obviously, the latter property requires that the data generating system is in the model class of the $\mathfrak{M}_\mathrm{W}(\Phi_{n_\mathrm{g}}^{n_\mathrm{e}}, \theta, \mathfrak{B}_\mathbb{P})$ and $\mathfrak{M}_\mathrm{W}(\Phi_{n_\mathrm{g}}^{n_\mathrm{e}}, \theta, \mathfrak{B}_\mathbb{P})$ structures. For a given OBF set $\Phi_{n_\mathrm{g}}^{n_\mathrm{e}}$ and LPV system $\mathcal{S}$, this means that for the considered IO partition, $\mathcal{S}$ has a series-expansion in terms of $\Phi_{n_\mathrm{g}}^{\infty}$ where only the first $(n_\mathrm{e}+1)n_\mathrm{g}$ expansion coefficients are not zero. Furthermore these expansion coefficients $\{w^{o}_{ij}\}^{i=0,\dots,n_e}_{j=1,\dots,n_g}$, appearing on the left (or the right) side of the basis functions, are the linear combination of the functions $\{\psi_{ijl}\}$ used in the parametrization (9.38a–b):

$$w^{o}_{ij} = \sum_{l=0}^{n_{ij}} \theta^{o}_{ijl}\psi_{ijl}. \tag{9.52}$$

Collect these true parameters $\{\theta^{o}_{ijl}\}$ into the parameter vector θ_0. Then, consistency means that $\hat{\theta}_{N_d} \to \theta_0$ with probability 1.

In the local case, assume that in the measurements $\mathfrak{D}^{\mathrm{P}}_{N_d}$, the noise is uncorrelated with u and $\mathfrak{D}^{\mathrm{P}}_{N_d}$ is informative with respect to the considered model set. Then, convergence and consistency of the estimated LTI models with parameters $\{\mathrm{w}_{ij\tau}\}$ is well-known [73, 105]. This implies the following theorem:

Theorem 9.1 (Convergence and consistency, local method). *Given a model structure $\mathfrak{M}_{\mathrm{W}}(\Phi^{n_e}_{n_g},\theta,\mathfrak{B}_{\mathbb{P}})$, with OBFs $\Phi^{n_e}_{n_g}$ and coefficient parametrization* (9.38a), *where $\theta_{ijl}\in\mathbb{R}$, each $\{\psi_{ijl}\}^{n_{ij}}_{l=0}$ is a set of orthogonal functions on $\mathbb{P}$, $\mathbb{P}$ is compact, and each $\psi_{ijl}:\mathbb{P}\to\mathbb{R}$ is Lipschitz continuous. Consider the estimate $\hat{\theta}_{N_d}$ determined by the local method for N_d-long informative data records, gathered for N_{loc} frozen scheduling signals. If $N_{loc}\to\infty$ and $N_d\to\infty$ then $\hat{\theta}_{N_d}\to\theta^*$ where θ^* is the minimizing argument of the expected value of the squared residual error, $\theta^* = \arg\min_{\theta\in\Theta}\bar{\mathcal{E}}\{\varepsilon^2(\theta)\}$. Furthermore, if S is in the model class with parameters θ_0, then $\theta^*=\theta_0$.*

Theorem (9.1) similarly holds in the H-LPV OBF case. This theorem implies that the asymptotic parameter estimate is independent from the particular noise realization in the data sequence and identification of the true system is possible if it is in the model class. Proof of the theorem follows from the consistency and convergence of the local model estimates together with the convergence of the interpolation in case of Lipschitz continuous functions (for the latter property see [8]).

Consider the global approach. Assume that the available data record is informative with respect to the considered model. Then based on the OE structure of the W-LPV and H-LPV OBF models and the applied linear parametrization of the coefficient functions, it is well-known in the nonlinear case (see [105]), that under these conditions, the least-squares estimate of θ is strongly convergent and consistent. So the following theorem obviously holds:

Theorem 9.2 (Convergence and consistency, global method). *Consider the estimate $\hat{\theta}_{N_d}$ determined by* (9.45) *with respect to the model structure $\mathfrak{M}_{\mathrm{W}}(\Phi^{n_e}_{n_g},\theta,\mathfrak{B}_{\mathbb{P}})$ or $\mathfrak{M}_{\mathrm{H}}(\Phi^{n_e}_{n_g},\theta,\mathfrak{B}_{\mathbb{P}})$. Assume that the data record $\mathfrak{D}_\infty$ is informative and the model structure is globally identifiable. If $N_d\to\infty$, then $\hat{\theta}_{N_d}\to\theta^*$ with probability 1 where θ^* is the minimizing argument of the expected value of the squared residual error, $\theta^*=\arg\min_{\theta\in\Theta}\bar{\mathcal{E}}\{\varepsilon^2(\theta)\}$. Furthermore, if S is in the model class with parameters θ_0, then $\theta^*=\theta_0$.*

9.3.5.4 The Concept of Variance and Bias

Estimation errors of the resulting model estimates can be decomposed into variance and bias parts:

$$G_0(\mathrm{q}) - G(\mathrm{q}, \hat{\theta}_{N_\mathrm{d}}) = \underbrace{G_0(\mathrm{q}) - G(\mathrm{q}, \theta^*)}_{\text{bias}} + \underbrace{G(\mathrm{q}, \theta^*) - G(\mathrm{q}, \hat{\theta}_{N_\mathrm{d}})}_{\text{variance}}. \tag{9.53}$$

where $G_0(\mathrm{q})$ is the pulse basis expansion of the process part of the data generating system (9.1), while $G(\mathrm{q}, \hat{\theta}_{N_\mathrm{d}})$ corresponds to the truncated OBF expansion form of the estimated model $\mathfrak{M}_\mathrm{W}(\Phi_{n_\mathrm{g}}^{n_\mathrm{e}}, \hat{\theta}_{N_\mathrm{d}}, \mathfrak{B}_\mathbb{P})$ or $\mathfrak{M}_\mathrm{H}(\Phi_{n_\mathrm{g}}^{n_\mathrm{e}}, \hat{\theta}_{N_\mathrm{d}}, \mathfrak{B}_\mathbb{P})$. Similar to the LTI case, the *bias* part corresponds to the structural error, i.e. the modeling error introduced by the finite truncation of the expansion and the applied coefficient parametrization, while the *variance* corresponds to the error which is due to the noise contribution on the data.

9.3.5.5 Variance

In case of the local approach, the concepts of variance and bias can be formulated in the frozen sense. By viewing the result of each frozen identification as a LTI model estimate in terms of the considered basis functions, all results of the LTI framework apply in terms of variance and bias (see Sect. 2.4.5). However, due to interpolation of these local model estimates through their expansion coefficients, there is a little hope to characterize the variance and the bias of the resulting LPV model estimate. In terms of the bias, the main difficulty is that the number of frozen models, i.e. the number of interpolation points has a significant, but not well understood effect on the bias. For the variance, the problem is that, by knowing the distribution of $\{\hat{\mathrm{w}}_{ij\tau}\}$, it is a difficult problem in general to deduce the distribution of $\{\hat{\theta}_{ijl}\}$ in (9.40). Based on these, variance and bias are only characterized in the frozen sense with respect to the estimates by the local method.

In the global case, we face a different situation. We have already shown that the parameter estimates in this case are consistent. Let $\mathcal{D}_\infty$ be informative with respect to $\mathcal{S}$. Then due to the prediction-error setting of the estimation, the classical result of the LTI framework holds:

Theorem 9.3 (Asymptotic variance, global method). *Consider the estimate* $\hat{\theta}_{N_\mathrm{d}}$ *determined by* (9.45) *with respect to the model structure* $\mathfrak{M}_\mathrm{W}(\Phi_{n_\mathrm{g}}^{n_\mathrm{e}}, \theta, \mathfrak{B}_\mathbb{P})$ *or* $\mathfrak{M}_\mathrm{H}(\Phi_{n_\mathrm{g}}^{n_\mathrm{e}}, \theta, \mathfrak{B}_\mathbb{P})$. *Assume that the data record* $\mathcal{D}_\infty$ *is informative and the used model structure is globally identifiable. Due to the convergence of* $\hat{\theta}_{N_\mathrm{d}}$, *there exists a* $\theta^* \in \Theta$ *such that* $\hat{\theta}_{N_\mathrm{d}} \to \theta^*$ *with probability 1 if* $N_\mathrm{d} \to \infty$. *Then*

$$\sqrt{N_\mathrm{d}}(\hat{\theta}_{N_\mathrm{d}} - \theta^*) \to \mathcal{N}(0, \mathcal{Q}_\theta) \quad \text{as } N_\mathrm{d} \to \infty, \tag{9.54}$$

where

$$\mathfrak{Q}_\theta = \lim_{N_\mathrm{d}\to\infty} N_\mathrm{d}\cdot\mathcal{E}\left\{\left[\frac{\partial}{\partial\theta}\mathcal{W}_{N_\mathrm{d}}(\theta^*,\mathcal{D}_{N_\mathrm{d}})\right]\left[\frac{\partial}{\partial\theta}\mathcal{W}_{N_\mathrm{d}}(\theta^*,\mathcal{D}_{N_\mathrm{d}})^\top\right]\right\}.$$

Note that the proof follows similarly as in the LTI case [105], due to the fact that the considered model sets correspond to asymptotically stable models with respect to all $\theta\in\Theta$, the estimates of θ are convergent, and the considered prediction-error framework is equivalent with the classical formulation. This basic result on the variance of the parameter estimates provides some insights, however further properties in terms of asymptotic model order or frequency characterization of the variance are hard to derive due to the parameter variation. Developing more informative expressions for the asymptotic variance is an aim of future research.

9.3.5.6 Bias

Consider the estimated models $\mathfrak{M}_\mathrm{W}(\Phi_{n_\mathrm{g}}^{n_\mathrm{e}},\hat{\theta}_{N_\mathrm{d}},\mathfrak{B}_\mathbb{P})$ and $\mathfrak{M}_\mathrm{H}(\Phi_{n_\mathrm{g}}^{n_\mathrm{e}},\hat{\theta}_{N_\mathrm{d}},\mathfrak{B}_\mathbb{P})$. In the global case it holds that, if the data record $\mathcal{D}_{N_\mathrm{d}}$ is informative and the data generating system is in the model class, then the classical results of prediction-error identification imply that the estimate $\hat{\theta}_{N_\mathrm{d}}$ of the parameters θ is unbiased. In case the process part G_O, i.e. the LPV system $\mathcal{S}$, in (9.1) is not in the model class, i.e. either in the series expansion of $\mathcal{S}$ by $\Phi_{n_\mathrm{g}}^{\infty}$ there are non-zero coefficients $\{w_{ij}^\mathrm{o}\}$ in the truncated part ($w_{ij}\neq 0$ for $i>n_\mathrm{e}$), or the coefficients have a different dependence than the used parametrization (9.38a–b), then for the asymptotic estimate θ^* it holds that for a given $p\in\mathfrak{B}_\mathbb{P}$

$$(G_\mathrm{O}(\mathrm{q})-G(\mathrm{q},\theta^*))\diamond p = \underbrace{\sum_{i=0}^{n_\mathrm{e}}\sum_{j=1}^{n_\mathrm{g}}\left(\left(w_{ij}^\mathrm{o}-\sum_{l=0}^{n_{ij}}\theta_{ijl}^*\psi_{ijl}\right)\diamond p\right)\phi_{ij}(\mathrm{q})}_{\text{parametrization bias}} + \underbrace{\sum_{i=n_\mathrm{e}+1}^{\infty}\sum_{j=1}^{n_\mathrm{g}}(w_{ij}\diamond p)\,\phi_{ij}(\mathrm{q})}_{\text{truncation bias}}, \tag{9.55}$$

where $\theta^*=[\theta_{ijl}^*]_{i=0,j=1,l=0}^{n_\mathrm{e},n_\mathrm{g},n_{ij}}$ and $\{w_{ij}^\mathrm{o}\}$ are the expansion coefficients of $\mathcal{S}$ in terms of $\Phi_{n_\mathrm{e}}^{\infty}$. The first part in expression (9.55) describes bias due to improper assumptions on the scheduling dependence while the second part describes bias due to the non-considered tail of the series expansion. Based on (9.55), it can be concluded that by extending the number of basis functions, i.e. increasing n_e, the truncation bias can be arbitrary decreased. However, increasing n_e means that more coefficients appear in the parametrization bias, which results in an eventual increase of this term. This underlines that choosing a good coefficient dependence in terms of $\{\psi_{ijl}\}$ is equally important as the choice of a OBF set with fast convergence rate. Fortunately, due to the model transformation approach of Chap. 7, first principle information can be used to assist the adequate choice of ψ_{ijl} with respect to a given basis $\Phi_{n_\mathrm{g}}^{n_\mathrm{e}}$. If reliable first principle information is not available, the recently developed statistical tool based on a *non-negative garotte* approach can accommodate adequate selection of $\{\psi_{ijl}\}$, see [200].

9.3.5.7 (In)validation

In general, reliable (in)validation of LPV model estimates is a theoretically hard task. One problem is that uncertainty of the model estimates has not been investigated in the LPV framework and, on the other hand, it is difficult to judge how the estimated model relates or fits to the available first principle knowledge.

In general, only (in)validation through model simulation is available, comparing the model output with respect to measured data records which are assumed to be informative. Here the richness of p in terms of excitation has a prime importance, as with slowly varying scheduling trajectories, model estimates with significantly different transient behavior can seem to be both valid models of the plant. Error measures like MSE, BFR and VAF, introduced in Sect. 2.4.7, can be successfully used to accommodate comparison of the simulated and measured signals and to decide on the validity of the model estimates. However, in case of an invalidated model estimate, it is hard to give any indication how to identify the system with a better end result (reconfiguration of the model structure, more exciting (u, p), other type of identification method, etc.).

Additionally, for the introduced models, (in)validation can also be accomplished based on the residual signal ε in the one-step-ahead prediction error (9.15). The residual can be easily computed for the proposed models as the inverse of the noise model is 1. Applying residual analysis, similar to the LTI case, the hypothesis that ε is white noise or ε is uncorrelated with the past inputs can be tested. If these hypothesis tests result in rejection, then the deficiency of the applied model structure is implied. The only problem is that due to the approximative nature of the applied model structure (both in terms of the finite series expansion and in terms of the assumption of static dependence), some unmodeled dynamics of the system always contribute to the residual term. This implies that, based on residual analysis, the models are most likely to be rejected. Thus, in the following, we only consider (in)validation in terms of simulations and error measures.

9.3.5.8 Initial Conditions

In practice, slow dynamics of the system, safety considerations, or high costs of long measurements often result in data records which are collected with non-zero initial conditions of the plant. For these data sets, the initial conditions need to be estimated during the identification process. In the LTI case the identification framework of OBFs supports estimation of initial conditions (see [73]), which can be efficiently applied in the local algorithm to estimate initial conditions of the frozen systems of $\mathcal{S}$. Therefore, further investigation of initial condition estimation is only interesting for the global identification approach.

In the global method the SS form of the H-LPV OBF model structure can be used to formulate an extended data matrix which involves the initial condition $x(0)$ of the model. The extended data matrix $\hat{Y}'_{N_\mathrm{d}}$ is introduced as

$$\hat{Y}'_{N_\mathrm{d}} = \begin{bmatrix} Cx(0) \\ \vdots \\ CA^{N_\mathrm{d}-1}x(0) \end{bmatrix} + \hat{Y}_{N_\mathrm{d}}. \tag{9.56}$$

where $\hat{Y}_{N_\mathrm{d}}$ satisfies (9.47), i.e. it is the predicted output of $\mathfrak{M}_\mathrm{H}(\Phi^{n_\mathrm{e}}_{n_\mathrm{g}}, \theta, \mathfrak{B}_\mathbb{P})$ with zero initial condition. By extending θ with $x(0)$ as parameters and including $[C^\top\ A^\top C^\top\ \ldots]^\top$ into Γ_{N_d}, estimation of $x(0)$ becomes available through linear regression. In the W-LPV OBF case, the extended data matrix is

$$\hat{Y}'_{N_\mathrm{d}} = \begin{bmatrix} W(p(0))x(0) \\ \vdots \\ W(p(N_\mathrm{d}))A^{N_\mathrm{d}-1}x(0) \end{bmatrix} + \hat{Y}_{N_\mathrm{d}}. \tag{9.57}$$

Based on (9.57), simultaneous estimation of θ and $x(0)$ is a bilinear optimization problem for the LS criterion. This optimization is solvable by the application of a separable least-squares strategy (see [62]), however the obtained solutions are only local minima or saddle points of the involved cost function.

9.3.5.9 (Quasi) LPV System Identification

There is an important aspect of the proposed identification methods if the data generating system is a quasi LPV system. For quasi LPV systems, generally, p cannot be held constant, as the scheduling is an internal signal of the system like elements of the state or output variables. Thus, for this case, only the global method is applicable, as the local approach needs identification of the system with respect to constant scheduling trajectories. Violation of the freedom of p, how this affects the previous results and what happens in case p is influenced by noise are not well understood phenomena and these questions are in the focus of current research.

9.3.5.10 Global versus Local Approach

As demonstrated both the global and the local approaches provide attractive ways of identifying an LPV system an obvious question is when to use which approach. In most situations the global approach is considered to be a more attractive candidate as it provides estimation of the system with a varying trajectory of p, giving a better possibility to approximate the global dynamic behavior of the system instead of just the frozen aspects. As shown, estimation in the global case can be formulated in a simple least-squares setting and cumbersome problems of interpolation are avoided due to the fixed functional dependencies of the parametrization. The theoretically better understood behavior of the stochastic nature of estimation in the global setting also suggests that it is a theoretically more sound approach than the local method if informativity of the measured data set can be guaranteed in the application.

However, practical use of LPV identification on industrial problems often turns out to favor different properties. In most practical applications, identification is must be accomplished in a closed-loop setting due to instability of the plant or because the current production can not be disturbed in the favor of identification. In terms of the local approach, the well worked out methods of the LTI framework can be fully used to solve the identification problem in a divide-and-conquer manner. This implies that closed-loop identification is practically feasible via this approach. For the global case, closed-loop identification in the general sense is not well worked out even though some recent research directions are promising [210, 94, 28].

The use of frequency-based identification is also supported in the local setting. The latter is important in mechatronic applications where the often tight modeling specifications with respect to the frozen behaviors are only available in the frequency domain. Such specifications can often not be addressed in the global setting. On the other hand interpolation can result in unexpected global behavior as the local identification approach only focuses on the frozen behaviors. However, such a drawback can be avoided by using data with varying p to assist the interpolation.

As a general receipt the use of the global approach is advised whenever there is enough possibility to perturb the system for an informative data record and if the model specifications are not given in the frequency domain. In other situations the use of a local approach is advised due to its higher capability to meet the target performance under the given information content of available data sets.

9.3.5.11 Similarity to Nonlinear Identification Methods

In Chap. 1 it has been discussed that the attractive properties of truncated OBF expansion representations inspired some identification approaches in the NL and the fuzzy field. In order to position the developed approaches with respect to those methods, i.e. to clarify the connection or dissimilarities, the following properties are important:

- In the nonlinear case, the method of [63] uses Wiener and Hammerstein type of models where the LTI part is the linear combination of a filter bank of OBFs. We have already shown that these models result as "special cases"of W-LPV and H-LPV OBF models in the quasi-LPV case. In [159], the used model structure contains fuzzy membership functions associated with a (Laguerre) basis function in the filter bank.
- In our approach, the OBFs, the backbone of the model structures, are optimized for $\mathcal{S}$, while in the methods of [159] and [63] they are assumed to be chosen by the user.
- In [63], the LTI part is parameterized and estimated as a linear combination of the chosen basis functions simultaneously with the estimation of the static nonlinearity. Similarly in [159], both the LTI part and the fuzzy part are identified

together. In our approach, the parametrization is focused on the expansion coefficients while the LTI part is fixed, thus the complicated estimation structure of the fuzzy and NL methods is not required in this case.
- Both the proposed and the NL/fuzzy approaches use the least-squares criterion for the identification of the system, however, the proposed LPV approach is simpler as it solves the estimation problem via an analytic solution (linear regression).

9.3.6 Examples

In this section, the applicability of the introduced identification methods is shown through three different examples.

9.3.6.1 SS Example with Invariant (A, B)

As a first example, consider an asymptotically stable LPV system $\mathcal{S}_1$, with an $\mathfrak{R}_{SS}(\mathcal{S}_1)$ equal to

$$\left[\begin{array}{c|c} A_1 & B_1 \\ \hline C_1 \diamond p & 0 \end{array}\right] = \left[\begin{array}{ccc|c} 0.3 & 0.2 & 0.4 & 1 \\ -0.1 & 0.2 & 0.2 & 1 \\ 0.4 & -0.1 & 0.5 & 1 \\ \hline 2p & -p^2 & \sin(p) & 0 \end{array}\right],$$

and $\mathbb{P} = [-1, 1]$. Note that in this representation, only C depends on p and the underlying dependence is static. Using the poles of A to generate Hambo basis functions Φ_3^0, the resulting OBFs satisfy with respect to $\mathfrak{F}_{\mathcal{S}_1}$ of $\mathfrak{R}_{SS}(\mathcal{S}_1)$ that $\mathfrak{F}_{\mathcal{S}_1} \subset \mathrm{Span}(\Phi_3^0)$. As only C depends on p and $\mathfrak{F}_{\mathcal{S}_1} \subset \mathrm{Span}(\Phi_3^0)$, the system lies in the model class of the W-LPV OBF structure with these basis functions. Furthermore, the expansion coefficients of $\mathfrak{R}_{SS}(\mathcal{S}_1)$ in terms of the basis functions Φ_3^0 have static dependence, thus choosing the parametrization of the W-LPV OBF model with static dependence still implies that the system lies in the model class. Based on the OBF set Φ_3^0 and 100 experiments with varying p, global identification of $\mathcal{S}$ with the W-LPV OBF structure has been carried out in 100 Monte Carlo runs. In each experiment, a $N_d = 500$ sample long data record of the system has been generated, based on white u and p with uniform distribution $\mathcal{U}(-1, 1)$. For each data record, a white output noise e with distribution $\mathcal{N}(0, 0.5)$ has been added, which matches with the noise concept of the prediction-error setting (see Sect. 9.2.1). The resulting *signal-to-noise ratio* (SNR) has been 29.5 dB, while the relative signal-to-noise amplitude has been 26%. Using the same conditions in the local case, $\mathcal{D}_{N_d,\mathrm{P}}$ data records have been collected 100 times with constant scheduling points $\mathrm{P}_{11} = \{-1 + 0.2\tau\}_{\tau=0}^{10}$. Based on these data records, 11 local estimates of $\mathfrak{R}_{SS}(\mathcal{S})$ have been produced using an LTI-OBF model structure with Φ_3^0. The resulting "frozen" basis coefficients have been interpolated in each of the 100 cases. In both the global and local methods, a 2nd-order polynomial-based parametrization has been used in the estimation of W (see (9.38a) with $\psi_{ijl}(p) = p^l$ and $n_{ij} = 2$).

Table 9.1 Validation results of 100 identification experiments by the global and local methods using the W-LPV and H-LPV OBF model structures in the considered examples. The results are given in terms of the average MSE, BFR and VAF of the simulated output signals of the model estimates.

Model	Case	MSE (dB)	BFR (%)	VAF (%)
W-LPV $\mathcal{S}_1$	local	−94.99	99.84%	99.99%
	global	−62.35	98.54%	98.98%
H-LPV $\mathcal{S}_2$	local	−95.90	99.71%	99.99%
	global	−62.78	98.39%	99.97%
W-LPV $\mathcal{S}_3$	local	−18.01	75.82%	94.12%
	global	−31.03	86.34%	98.23%
H-LPV $\mathcal{S}_3$	local	−10.16	62.13%	85.69%
	global	−26.41	82.11%	96.18%

In the first row of Table 9.1, the (in)validation results of the resulting 100 model estimates are shown in both the local and global cases. The (in)validation results are given in terms of *average* MSE, BFR and VAF of the simulated output signals of the models (see Sect. 2.4) for realizations of u and p that are different from the ones used during the identification. In Fig. 9.4, a typical plot of the simulated output signals and the resulting output-error of the models are presented. As expected, both approaches identified the system with adequate validation results. This underlines, that with respect to systems that are in the considered W-LPV OBF model class, both approaches provide reliable estimates even in case of significant output noise. By further investigating the results it is obvious that the global approach has produced a slightly worse result than the local approach. Explanation of this phenomenon lies in the much larger amount of data $(11 \cdot N_\mathrm{d})$ available in the local case.

9.3.6.2 SS Example with Invariant (A,C)

As a second example, consider an asymptotically stable LPV system $\mathcal{S}_2$, with an $\mathfrak{R}_\mathrm{SS}(\mathcal{S}_2)$ equal to the transpose of $\mathfrak{R}_\mathrm{SS}(\mathcal{S}_1)$:

$$\left[\begin{array}{c|c} A_2 & B_2 \\ \hline C_2 & 0 \end{array}\right] = \left[\begin{array}{c|c} A_1^\top & C_1^\top \\ \hline B_1^\top & 0 \end{array}\right].$$

Note that $\mathfrak{R}_\mathrm{SS}(\mathcal{S}_1)$ and $\mathfrak{R}_\mathrm{SS}(\mathcal{S}_2)$ are not equivalent. However, the OBF set Φ_3^0 of the previous example still satisfy that $\mathfrak{F}_{\mathcal{S}_2} \subset \mathrm{Span}(\Phi_3^0)$ with respect to the frozen transfer function set $\mathfrak{F}_{\mathcal{S}_2}$ of $\mathfrak{R}_\mathrm{SS}(\mathcal{S}_2)$. In this case, the true system lies in the model class of the H-LPV OBF structure with the basis functions Φ_3^0 and with static coefficient dependence. By using the same setting of data sequences and local model estimates as in the previous example, identification of $\mathcal{S}_2$ has been accomplished by

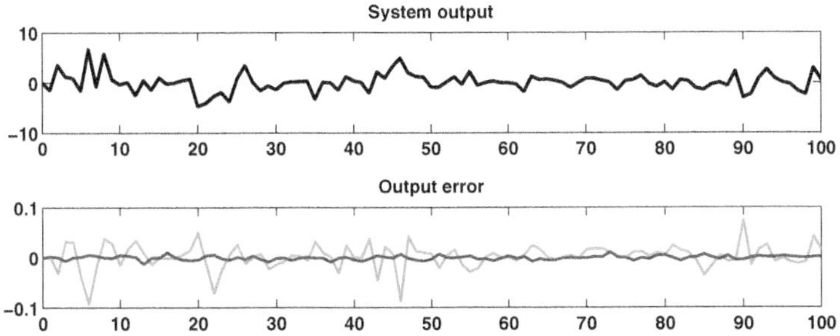

Fig. 9.4 Comparison of the identified models of $\mathcal{S}_1$ by their responses for white u, p with distribution $\mathcal{U}(-1,1)$: $\mathfrak{R}_{\mathrm{SS}}(\mathcal{S})$ (black), W-LPV OBF local (grey), W-LPV OBF global (light grey).

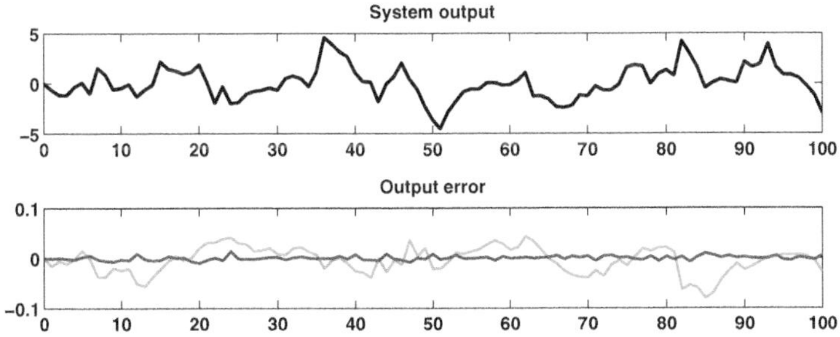

Fig. 9.5 Comparison of the identified models of $\mathcal{S}_2$ by their responses for white u, p with distribution $\mathcal{U}(-1,1)$: $\mathfrak{R}_{\mathrm{SS}}(\mathcal{S}_2)$ (black), H-LPV OBF local (grey), H-LPV OBF global (light grey).

using the H-LPV OBF model structure with Φ_3^0. The (in)validation results in terms of simulation are shown in Fig. 9.5 and in the second row of Table 9.1. As expected, both approaches identified the system adequately just like in the previous example.

9.3.6.3 IO Example

As a third example, the asymptotically stable LPV system $\mathcal{S}_3$ for which OBF selection has been extensively studied in Sect. 8.3.3 is identified. Consider the IO representation $\mathfrak{R}_{\mathrm{IO}}(\mathcal{S}_3)$ defined by (8.17) and with $\mathbb{P} = [0.6, 0.8]$. Using the poles obtained via the FKcM algorithm with fuzzyness $m = 25$ and $n_c = 8$, a OBF set Φ_8^0 has resulted, which has been found adequate for the truncated series-expansion based approximation of the system. Using this basis function set and the proposed identification algorithms, we show that quite accurate estimated models of the system can be derived.

Based on Φ_8^0 and a 2nd-order polynomial-based parametrization of the coefficients, identification of $\mathcal{S}_3$ with both methods and structures has been accomplished in 100 Monte Carlo runs. The used data sequences have been based on the same

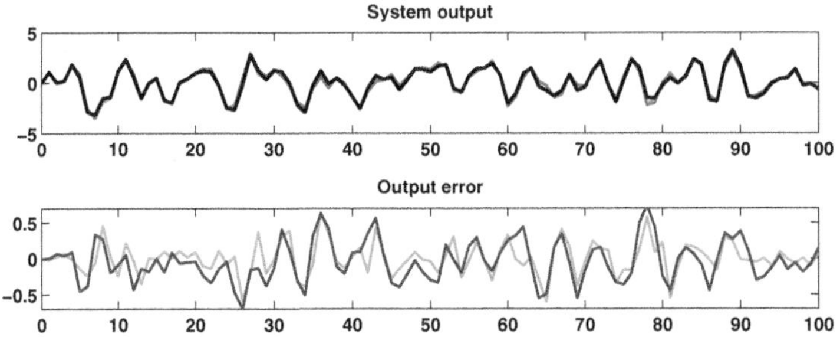

Fig. 9.6 Comparison of the identified models of $\mathcal{S}_3$ by their responses for white (u, p) with distribution $(\mathcal{U}(-1,1), \mathcal{U}(0.6,0.8))$: $\mathcal{S}_3$ (black), W-LPV OBF global (light grey), H-LPV OBF global (grey).

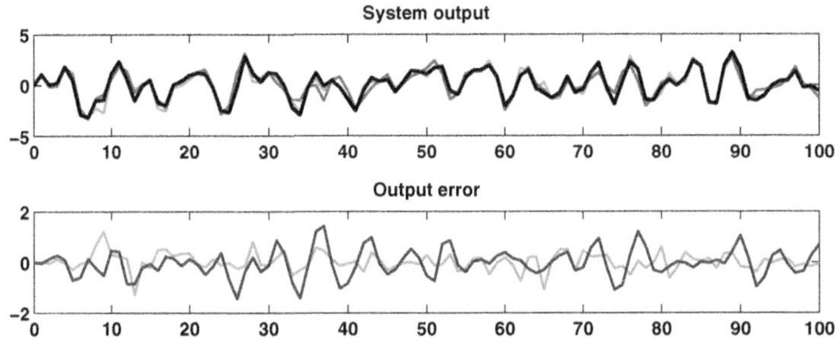

Fig. 9.7 Comparison of the identified models of $\mathcal{S}_3$ by their responses for white (u, p) with distribution $(\mathcal{U}(-1,1), \mathcal{U}(0.6,0.8))$: $\mathcal{S}_3$ (black), W-LPV OBF local (light grey), H-LPV OBF local (grey).

setting and conditions as in the previous examples, only now p has been generated with a distribution $\mathcal{U}(0.6,0.8)$. The SNR was 20 dB in the resulting data records with a relative signal-to-noise amplitude of 25 %. Calculation time of the algorithms was a few seconds on a Pentium 4, 2.8 GHz PC. In Fig. 9.6 and 9.7 and in the last two rows of Table 9.1, the (in)validation results are shown for different realizations of u, p than used during the identification. As expected, the W-LPV and H-LPV OBF structures based on coefficients with static dependence could not fully cope with the variations in the $\{a_l\}_{l=0}^{5}$ parameters. However, the global W-LPV OBF identification provided quite acceptable results for such a heavily nonlinear system. The explanation why the H-LPV OBF structure gave a worse result lies in the different approximation capabilities of these models. By computing the left-side expansion coefficients of (8.17) in terms of the used basis, which corresponds to the true coefficients of the system with respect to the W-LPV OBF model structure, the resulting expansion coefficients have a dominant part with static dependence. This means that a good approximation of the system can be found among the used W-LPV OBF models with static dependence (both the parametrization and truncation bias are small). On the other hand, the right-side expansion coefficients of (8.17) in terms of Φ_8^∞ have a dominant part with dynamic dependence. This means that the parametrization bias of H-LPV OBF models with static dependence must be larger than in the previous case. This implies that with the considered parametrization, W-LPV OBF model structures are generally better for systems with a IO representation where the dynamics are dominated by the variation of the $\{a_i\}$ coefficients, while H-LPV OBF model structures are better for the cases, where $\{b_j\}$ are dominant.

In this example global methods prevail, because they have been able to capture the transient dynamics of the system between frozen scheduling points of $\mathbb{P}$, while in the local case the number of 11 local model estimates was not enough for correct interpolation. By using $N_{\mathrm{loc}} > 11$, the local method quickly improves. Note that in the asymptotic sense (in N_{loc}) the local and global method converge to the same optimal model in the utilized model class. Extension of Φ_8 with $n_{\mathrm{e}} = 1, 2, \ldots$ has not improved the results as Φ_8 is well chosen with respect to $\mathcal{S}_3$, i.e. the local modeling error is negligible due to the optimal choice by the FKcM. Therefore, the error in Table 9.1 is mainly governed by the modeling error of the used parametrization (9.55). Using higher order polynomials in Ψ produces a 2-5% percentage improvement in the results of Table 9.1, but in order to achieve full representation with the proposed models, incorporation of dynamic dependence on p is required.

9.4 Approximation of Dynamic Dependence

In the previous part, OBFs-based model structures have been introduced for the identification of LPV systems, with the main intention to give flexible models that are able to describe general LPV systems, simplify identification, are useful for control, and are unaffected by the difficulties present in the identification of LPV-SS or IO models. It has been shown that to describe any LPV system, the coefficients of the W-LPV and H-LPV OBF models, similar to the coefficients of other LPV models, need to have dynamic dependence on p. We have motivated that dynamic dependence presents an extra freedom of the parametrization and is not supported by the existing control approaches. To overcome this problem, in Sect. 9.3 the coefficients have been restricted to static dependence. This assumption led to a novel and efficient identification approach of W-LPV and H-LPV OBF model structures based on the LS criterion. However, the drawback of the assumption has also been pointed out: it limits the class of representable LPV systems. In this section an alternative of LPV truncated OBF expansion models and its identification approach is introduced with the intention to improve the representation capabilities of the previously considered model structures and parametrization, but without the use of dynamic dependence. In fact the idea that we will apply is the introduction of an additional feedback-loop around each basis component of the W-LPV and H-LPV OBF model structures with a gain incorporating also static dependence (see Figures 9.8 and 9.9). In this way, the filter bank of OBFs as a dynamical LTI system is "reused"to provide dynamic expansion coefficients. This implies that these modified structures can approximate a much wider class of LPV systems than W-LPV and H-LPV OBF models with static dependence. The introduction of feedback-based weighting leads to

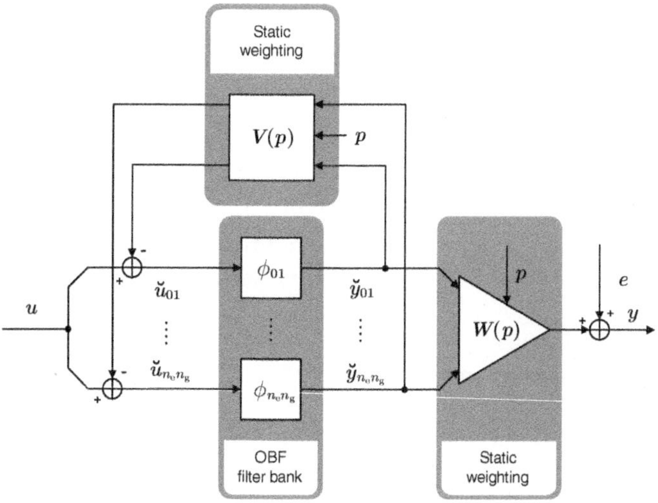

Fig. 9.8 IO signal flow graph of WF-LPV OBF models with feedback coefficients $V = [v_{01} \; \dots \; v_{n_e n_g}]$ and output-side coefficients $W = [w_{01} \; \dots \; w_{n_e n_g}]$ without a feedthrough term. All coefficients are considered with static dependence.

two new model structures given in Fig. 9.8 and 9.9, which we call *Wiener feedback* (WF) and *Hammerstein feedback* (HF) LPV OBF models.

9.4.1 Feedback-Based OBF Model Structures

Again we consider the prediction-error setting of Sect. 9.2.1. Let $\Phi_{n_g}^{n_e}$ be a set of Hambo basis functions in $\mathcal{RH}_{2-}(\mathbb{E})$. Denote the input and output of each basis function in $\Phi_{n_g}^{n_e}$ by $\breve{u}_{ij}$ and $\breve{y}_{ij}$ satisfying:

$$\breve{y}_{ij} = \phi_j(\mathrm{q}) G_\mathrm{b}^i(\mathrm{q}) \breve{u}_{ij}. \tag{9.58}$$

Additionally, let $(A_{ij}, B_{ij}, C_{ij}, D_{ij})$ be a minimal balanced SS realization of each basis function $\phi_j G_\mathrm{b}^i$ and introduce $A = \mathrm{Diag}(A_{01}, \dots, A_{n_e n_g})$. Similarly define B and C. Denote $\breve{u} = [\, \breve{u}_{01} \; \dots \; \breve{u}_{n_e n_g} \,]^\top$ and $\breve{y} = [\, \breve{y}_{01} \; \dots \; \breve{y}_{n_e n_g} \,]^\top$. Then, the SS form of the IO relation (9.58) is

$$\mathrm{q}x = Ax + B\breve{u}, \tag{9.59a}$$
$$\breve{y} = Cx. \tag{9.59b}$$

Note that this is a non-minimal SS representation, but it is needed to introduce the feedback loops around each basis function separately (see Figures 9.8 and 9.9). Let $\mathcal{S}$ be an asymptotically stable SISO LPV system with scheduling space $\mathbb{P} \subseteq \mathbb{R}^{n_\mathbb{P}}$, scheduling signal p, and IO partition (u, y). Then the feedback model structures of Figures 9.8 and 9.9 are formulated as follows [195]:

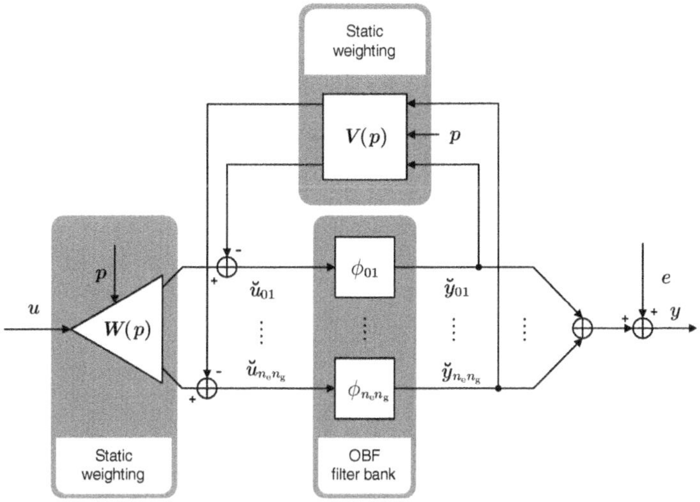

Fig. 9.9 IO signal flow graph of HF-LPV OBF models with feedback coefficients $V = [v_{01} \ \dots \ v_{n_e n_g}]$ and input-side coefficients $W = [w_{01} \ \dots \ w_{n_e n_g}]$ without a feedthrough term. All coefficients are considered with static dependence.

- Wiener feedback LPV OBF model (WF-LPV OBF)

$$\breve{y}_{ij} = \phi_j(q) G_b^i(q) u - \phi_j(q) G_b^i(q) v_{ij}(p) \breve{y}_{ij}, \tag{9.60a}$$

$$y = e + w_{00}(p) u + \sum_{i=0}^{n_e} \sum_{j=1}^{n_g} w_{ij}(p) \breve{y}_{ij}, \tag{9.60b}$$

with $p \in \mathfrak{B}_{\mathbb{P}}$, where $\mathfrak{B}_{\mathbb{P}}$ is considered to be known and with

$$\theta = \begin{bmatrix} w_{00} & w_{01} & \dots & w_{n_e n_g} & v_{01} & \dots & v_{n_e n_g} \end{bmatrix}^\top \in (\mathcal{R}|_{n_\mathbb{P}})^{1+2(n_e+1)n_g}.$$

This model, given in Fig. 9.8, is called the WF-LPV OBF model and denoted by $\mathfrak{M}_{\mathrm{WF}}(\Phi_{n_g}^{n_e}, \theta, \mathfrak{B}_{\mathbb{P}})$. As $\breve{y}_{ij}$ is independent from e (u and p are assumed to be deterministic), the one-step-ahead predictor in this case reads as

$$\breve{y}_{ij} = \phi_j(q) G_b^i(q) u - \phi_j(q) G_b^i(q) v_{ij}(p) \breve{y}_{ij}, \tag{9.61a}$$

$$\hat{y}_\theta = w_{00}(p) u + \sum_{i=0}^{n_e} \sum_{j=1}^{n_g} w_{ij}(p) \breve{y}_{ij}. \tag{9.61b}$$

Denote $W = \mathrm{Diag}(w_{01}, \dots, w_{n_e n_g})$ and V accordingly and let $E = [1 \ \dots \ 1]$. Then, the SS equivalent of (9.61a–b) is given by

$$qx = (A - BV(p)C)x + BE^\top u, \tag{9.62a}$$

$$\hat{y}_\theta = W(p)Cx + w_{00}(p)u. \tag{9.62b}$$

- Hammerstein feedback LPV OBF model (HF-LPV OBF)

$$\breve{u}_{ij} = w_{ij}(p)u - v_{ij}(p)\phi_j(\mathrm{q})G_{\mathrm{b}}^i(\mathrm{q})\breve{u}_{ij}, \tag{9.63a}$$

$$y = e + w_{00}(p)u + \sum_{i=0}^{n_e}\sum_{j=1}^{n_g} \phi_j(\mathrm{q})G_{\mathrm{b}}^i(\mathrm{q})\breve{u}_{ij}. \tag{9.63b}$$

with $p \in \mathfrak{B}_{\mathbb{P}}$, where $\mathfrak{B}_{\mathbb{P}}$ is considered to be known, and with

$$\theta = \begin{bmatrix} w_{00} & w_{01} & \dots & w_{n_e n_g} & v_{01} & \dots & v_{n_e n_g} \end{bmatrix}^\top \in (\mathcal{R}|_{n_\mathbb{P}})^{1+2(n_e+1)n_g}.$$

This model, given in Fig. 9.9, is called the HF-LPV OBF model and denoted by $\mathfrak{M}_{\mathrm{HF}}(\Phi_{n_g}^{n_e}, \theta, \mathfrak{B}_{\mathbb{P}})$. As $\breve{u}_{ij}$ is independent from e, the one-step-ahead predictor reads as

$$\breve{u}_{ij} = w_{ij}(p)u - v_{ij}(p)\phi_j(\mathrm{q})G_{\mathrm{b}}^i(\mathrm{q})\breve{u}_{ij}, \tag{9.64a}$$

$$\hat{y}_\theta = w_{00}(p)u + \sum_{i=0}^{n_e}\sum_{j=1}^{n_g} \phi_j(\mathrm{q})G_{\mathrm{b}}^i(\mathrm{q})\breve{u}_{ij}. \tag{9.64b}$$

Following a similar formulation as in the Wiener case, the SS equivalent of (9.64a–b) is given by

$$\mathrm{q}x = (A - BV(p)C)x + BW(p)u, \tag{9.65a}$$

$$\hat{y}_\theta = ECx + w_{00}(p)u, \tag{9.65b}$$

Note that the weighting functions $\{w_{ij}\}$, $\{v_{ij}\}$ are not necessarily equivalent in (9.60a–b) and (9.63a–b). Thus similar to the previous case, these model structures are distinguished in the sequel.

9.4.2 Properties of Wiener and Hammerstein Feedback Models

In order to show the importance of the introduced feedback based model structures, their key properties are investigated in this section.

9.4.2.1 Representation of Dynamic Dependence

First, the representation capabilities of the feedback based model structures are discussed:

Property 9.1 (Representation of dynamic dependence). Let $\mathcal{S}$ be an asymptotically stable LPV system and $\Phi_{n_g}^{\infty} \subset \mathcal{RH}_{2-}(\mathbb{E})$ be a Hambo basis. Consider a Wiener-feedback model $\mathfrak{M}_{\mathrm{WF}}(\Phi_{n_g}^{n_e}, \theta, \mathfrak{B}_{\mathbb{P}})$ of $\mathcal{S}$ with coefficients w_{ij} and v_{ij} having static dependence. If $\mathfrak{M}_{\mathrm{WF}}(\Phi_{n_g}^{n_e}, \theta, \mathfrak{B}_{\mathbb{P}})$ is asymptotically stable, then its deterministic part has a convergent series-expansion in terms of $\Phi_{n_g}^{\infty}$. If there is a feedback coefficient v_{ij} of $\mathfrak{M}_{\mathrm{WF}}(\Phi_{n_g}^{n_e}, \theta, \mathfrak{B}_{\mathbb{P}})$ which is dependent on p, i.e. not constant, then the coefficients of the series-expansion have dynamic dependence, otherwise all expansion coefficients have static dependence.

Obviously the same property holds for the Hammerstein case. For a proof see Appendix A.4. Note that in terms of Property 9.1, a WF-LPV or a HF-LPV OBF model can be considered to be equivalent with a W-LPV or a H-LPV OBF model with dynamic coefficient dependence. However, the converse does not hold in general. Thus, through static weighting functions $w_{ij}, v_{ij} \in \mathcal{R}|_{n_{\mathbb{P}}}$ the feedback based OBF models can approximate general LPV systems. It is trivial that such an approximation is more adequate than using W-LPV and H-LPV OBF models with static coefficient dependence. In fact, those models are special cases of the WF-LPV OBF and HF-LPV OBF structures. Equations (9.62a–b) and (9.65a–b) imply that the WF-LPV and HF-LPV OBF structures can approximate static scheduling dependence in the A matrix as well as static scheduling dependence in the autoregressive part, see (9.60a–b) and (9.63a–b). However, this improved representation capability comes at a price, namely that due to the feedback, stability of the model is not internally guaranteed like in the W-LPV or the H-LPV OBF case.

9.4.2.2 General Approximation Property

Based on the previous part, a WF-LPV or a HF-LPV OBF model can only approximate and not represent an arbitrary LPV system $\mathcal{S}$. So the general approximator property does not hold in this case as it would require dynamic dependence of W and V. However, these models have a series-expansion representation with coefficients incorporating dynamic dependence, thus they give better approximations of general LPV systems than W-LPV and H-LPV OBF models restricted to have static dependence.

9.4.2.3 Loss of Linearity in the Coefficients

The second price to be paid is that, by the introduction of the feedback loop, the linear-in-the-coefficients property of the original series-expansion structure is lost. The resulting predictors (9.61a–b) and (9.64a–b) are bilinear in the coefficients.

9.4.2.4 McMillan Degree Property

Due to the fact that the introduced feedback based structures are still based on truncated series expansions, just like the W-LPV and H-LPV OBF models, they are well structured against changes of the McMillan degree in the frozen system set $\mathscr{F}_{\mathcal{S}}$.

9.4.2.5 Models for Control

The existing approaches of LPV control theory are also directly applicable for estimates with the introduced model structures. Through (9.62a–b) and (9.65a–b), immediate SS realizations of estimated models are available where all dependencies are static. Opposite to the previous case, in (9.62a) and (9.65a) the resulting A matrix is dependent on p, thus control design does not simplify for these models. However, the specific structure of the dependence may be exploited during control design. The resulting SS forms also have a trivial LFR realization if the coefficients have polynomial or rational dependence on p.

9.4.3 *Identification by Dynamic Dependence Approximation*

In the following, an approach is proposed based on [195] for the identification with the introduced WF-LPV and HF-LPV OBF model structures using the LS criterion. This identification approach is the extension of the global method of Sect. 9.3. Again we consider the parametrization of each $\mathfrak{w}_{ij}$ and $\mathfrak{v}_{ij}$ as

$$\mathfrak{w}_{ij} = \sum_{l=0}^{n_\mathrm{w}} \theta_{ijl}^{\mathrm{w}} \psi_l^{\mathrm{w}}, \quad \mathfrak{v}_{ij} = \sum_{l=0}^{n_\mathrm{v}} \theta_{ijl}^{\mathrm{v}} \psi_l^{\mathrm{v}}, \tag{9.66}$$

where $\{\theta_{ijl}^{\mathrm{w}}\}$ and $\{\theta_{ijl}^{\mathrm{v}}\}$ are real-valued unknown coefficients and $\psi_l^{\mathrm{w}}, \psi_l^{\mathrm{v}} \in \mathcal{R}|_{n_\mathbb{P}}$, with $\psi_0^{\mathrm{w}} = \psi_0^{\mathrm{v}} = 1$, are given orthogonal meromorphic functions with static dependence. We also assume that no feedthrough term is present, $\mathfrak{w}_{00} = 0$, however later the estimation of $\mathfrak{w}_{00}$ is also investigated for the algorithm. Similar to the previous case, the approach is formulated by assuming that the basis selection phase has already been accomplished, so we pick up the line of reasoning right after Step 2 in Sect. 9.2.5.1.

Similar to the global approach of Sect. 9.3, the estimation phase of the identification approach based on WF-LPV and HF-LPV OBF model structures uses only one data set $\mathcal{D}_{N_\mathrm{d}} = \{y(k), u(k), p(k)\}_{k=0}^{N_\mathrm{d}-1}$ which is collected from $\mathcal{S}$ with varying scheduling and it is assumed to be informative with respect to the considered model structure. Note that due to the linear parametrization of the coefficients, the unknown parameters $\{\theta_{ijl}^{\mathrm{v}}\}$ and $\{\theta_{ijl}^{\mathrm{v}}\}$ appear in a bilinear relationship in the predictors (9.61a–b) and (9.64a–b). This implies that by using the LS criterion (9.37) as an identification criterion with residual (9.15), the minimization of (9.37) can be tackled by a separable least-squares algorithm. In each step of this iterative solution, either $\{\theta_{ijl}^{\mathrm{v}}\}$ or $\{\theta_{ijl}^{\mathrm{v}}\}$ is fixed, while the other parameter set is estimated by linear regression. This iterative scheme is repeated, till (9.37) converges. The procedure for the Wiener case is given in detail as follows:

Algorithm 9.4 (OBFs based LPV identification, Wiener-Feedback case)

Step 3c. Given an OBF set $\Phi_{n_\mathrm{g}}^{n_\mathrm{e}} = \{\phi_{ij}\}_{j=1,\dots,n_\mathrm{g}}^{i=0,\dots,n_\mathrm{e}}$ and data record $\mathcal{D}_{N_\mathrm{d}}$ of $\mathcal{S}$. Parameterize each $\mathfrak{w}_{ij}$ and $\mathfrak{v}_{ij}$ of (9.60a–b) according to (9.66) where $\psi_l^{\mathrm{w}}, \psi_l^{\mathrm{v}} \in \mathcal{R}|_{n_\mathbb{P}}$ are meromorphic functional dependencies chosen by the user with $\psi_0^{\mathrm{w}} = \psi_0^{\mathrm{v}} = 1$. Collect these functions as

$$\Psi^{\mathrm{w}} = \begin{bmatrix} \psi_0^{\mathrm{w}} & \dots & \psi_{n_{\mathrm{w}}}^{\mathrm{w}} \end{bmatrix}, \quad \Psi^{\mathrm{v}} = \begin{bmatrix} \psi_0^{\mathrm{v}} & \dots & \psi_{n_{\mathrm{w}}}^{\mathrm{v}} \end{bmatrix},$$

and also collect the real parameters, associated with the parametrization (9.66), into the vectors:

$$\theta^{\mathrm{w}} = \begin{bmatrix} \theta_{010}^{\mathrm{w}} & \theta_{011}^{\mathrm{w}} & \dots & \theta_{n_{\mathrm{e}}n_{\mathrm{g}}n_{\mathrm{w}}}^{\mathrm{w}} \end{bmatrix}^{\top}, \quad \theta^{\mathrm{v}} = \begin{bmatrix} \theta_{010}^{\mathrm{v}} & \theta_{011}^{\mathrm{v}} & \dots & \theta_{n_{\mathrm{e}}n_{\mathrm{g}}n_{\mathrm{v}}}^{\mathrm{v}} \end{bmatrix}^{\top}.$$

Step 4c. Choose a set of initial values for the parameters $\{\theta_{ijl}^{\mathrm{v}}\}$, like $\theta_{ijl}^{\mathrm{v}} = 0$.

Step 5c. Based on $\mathcal{D}_{N_{\mathrm{d}}}$, compute $\breve{y} = [\breve{y}_{ij}]_{j=1\dots n_{\mathrm{g}}}^{i=0,\dots,n_{\mathrm{e}}}$ via (9.60a) with respect to the OBF set $\Phi_{n_{\mathrm{g}}}^{n_{\mathrm{e}}}$.

Step 6c. Estimate the parameter set $\{\theta_{ijl}^{\mathrm{w}}\}_{i=0,j=1,l=0}^{n_{\mathrm{e}},n_{\mathrm{g}},n_{\psi}}$ by linear regression with respect to fixed $\{\theta_{ijl}^{\mathrm{v}}\}_{i=0,j=1,l=0}^{n_{\mathrm{e}},n_{\mathrm{g}},n_{\psi}}$. This is done by defining the regressors as

$$\gamma^{\top}(k) = \breve{y}(k) \otimes \Psi^{\mathrm{w}}(p(k)), \quad k \in [0, N_{\mathrm{d}} - 1]. \tag{9.67}$$

Collect the data into

$$\Gamma_{N_{\mathrm{d}}} = \begin{bmatrix} \gamma(0) & \dots & \gamma(N_{\mathrm{d}} - 1) \end{bmatrix}^{\top}, \quad Y_{N_{\mathrm{d}}} = \begin{bmatrix} y(0) & \dots & y(N_{\mathrm{d}} - 1) \end{bmatrix}^{\top}.$$

Then, to minimize the prediction-error criterion (9.44), the analytic solution θ is obtained via (9.45).

Step 7c. Fix $\{\theta_{ijl}^{\mathrm{w}}\}$ to $\hat{\theta}_{N_{\mathrm{d}}}^{\mathrm{w}}$ and estimate the parameters $\{\theta_{ijl}^{\mathrm{v}}\}$ in the following iterative way. In each iteration step we calculate an update for each $\theta_{ij}^{\mathrm{v}} = [\theta_{ij0}^{\mathrm{v}} \dots \theta_{ijn_{\mathrm{v}}}^{\mathrm{v}}]$ and choose that estimate which gives the best improvement on the prediction error of the model. This is formalized in the following steps:

1. For each basis function $\phi_j G_{\mathrm{b}}^i$, compute

$$\tilde{y}_{ij} = \frac{1}{w_{ij}(p)} \left(y - \sum_{\substack{k=0 \\ k \neq i}}^{n_{\mathrm{e}}} \sum_{\substack{l=1 \\ l \neq j}}^{n_{\mathrm{g}}} w_{kl}(p) \breve{y}_{kl} \right). \tag{9.68}$$

If $w_{ij}(p(k)) = 0$ for some k, then do not consider those time instants in the further procedure.

2. Collect each $\tilde{y}_{ij}$ into $\tilde{Y}_{N_{\mathrm{d}}}^{(ij)}$ and u into $U_{N_{\mathrm{d}}}$ similar to $Y_{N_{\mathrm{d}}}$. Let H_{ij} be the lower triangular Toeplitz matrix of the Markov parameters associated with (A_{ij}, B_{ij}, C_{ij}):

$$\mathsf{H}_{ij} = \begin{bmatrix} 0 & 0 & \dots & \dots \\ C_{ij}B_{ij} & 0 & \dots & \dots \\ C_{ij}A_{ij}B_{ij} & C_{ij}B_{ij} & 0 & \dots \\ \vdots & \vdots & \vdots & \ddots \end{bmatrix}. \tag{9.69}$$

Define

$$\gamma_{ij}^{\top}(k) = \breve{y}_{ij}(k) \otimes \psi^{\mathrm{v}}(p(k)), \quad k \in [0, N_{\mathrm{d}}-1], \tag{9.70}$$

and collect it into $\Gamma_{N_{\mathrm{d}}}^{(ij)}$. Then, based on (9.62a–b), it holds that

$$\hat{Y}_{N_{\mathrm{d}}}^{(ij)} = \mathsf{H}_{ij} U_{N_{\mathrm{d}}} - \mathsf{H}_{ij} \Gamma_{N_{\mathrm{d}}}^{(ij)} \theta_{ij}^{\mathrm{v}}, \tag{9.71}$$

where

$$\hat{Y}_{N_{\mathrm{d}}}^{(ij)} = \left[\hat{y}_{\theta}^{(ij)}(0) \;\; \hat{y}_{\theta}^{(ij)}(1) \;\; \dots \;\; \hat{y}_{\theta}^{(ij)}(N_{\mathrm{d}}-1) \right]^{\top}, \tag{9.72}$$

is the predicted output of the basis function $\phi_j(\mathrm{q})G_{\mathrm{b}}^{i}(\mathrm{q})$. Estimation of θ_{ij}^{v} can be formulated as a linear regression, similarly as in Step 6.c, to minimize the residual of $\tilde{Y}_{N_{\mathrm{d}}}^{(ij)} - \hat{Y}_{N_{\mathrm{d}}}^{(ij)}$. The regressor in this case is $\mathsf{H}_{ij}\Gamma_{N_{\mathrm{d}}}^{(ij)}$ and the data matrix is $\mathsf{H}_{ij}U_{N_{\mathrm{d}}} - \tilde{Y}_{N_{\mathrm{d}}}^{(ij)}$.

3. For each $\hat{\theta}_{ij}^{\mathrm{v}}$, compute the prediction error with only this element updated in θ^{v}. Choose the $\hat{\theta}_{ij}^{\mathrm{v}}$ which renders the smallest error and only update the value of θ^{v} with this element.
4. If the overall prediction error did not converge, then goto Step 7.c.1.

Step 8c. If the prediction error converged with respect to both $\{\theta_{ijl}^{\mathrm{w}}\}$ and $\{\theta_{ijl}^{\mathrm{v}}\}$, then stop, else goto 5.c.

In the HF-LPV OBF case, the identification procedure is similar. However, the formulation of the regressor is accomplished differently just like in Sect. 9.3. Based on (9.65a–b), for $x(0) = 0$ and $\mathfrak{w}_0 = 0$, the predicted output of the HF-LPV OBF model $\hat{Y}_{N_{\mathrm{d}}} = [\,\hat{y}_{\theta}(0) \;\; \hat{y}_{\theta}(1) \;\; \dots \;\; \hat{y}_{\theta}(N_{\mathrm{d}}-1)\,]^{\top}$ satisfies

$$\hat{Y}_{N_{\mathrm{d}}} = \begin{bmatrix} 0 & 0 & \cdots \\ ECBW(p(0)) & 0 & \cdots \\ EC[A - BV(p(1))C]BW(p(0)) & ECBW(p(1)) & \\ \vdots & \vdots & \ddots \end{bmatrix} U_{N_{\mathrm{d}}}.$$

By simple rearrangement it follows that

$$\hat{Y}_{N_{\mathrm{d}}} = \underbrace{\begin{bmatrix} 0 & 0 & \cdots \\ ECBIu(0) & 0 & \cdots \\ EC[A - BV(p(1))C]BIu(0) & ECBIu(1) & \\ \vdots & \vdots & \ddots \end{bmatrix}}_{\hat{\mathsf{H}}_{N_{\mathrm{d}}}} \begin{bmatrix} W(p(0)) \\ W(p(1)) \\ \vdots \end{bmatrix}$$

Now similar to the global method of Sect. 9.3, the formulation of the regressor can be accomplished based on $\hat{\mathsf{H}}_{N_\mathrm{d}}$ and the estimate of $\{\theta^{\mathrm{w}}_{ijl}\}$ follows via linear regression.

The other difference occurs in Step 7.c.1 where, due to the Hammerstein structure of the model, (9.68) simplifies to

$$\tilde{y}_{ij} = y - \sum_{\substack{k=0 \\ k\neq i}}^{n_\mathrm{e}} \sum_{\substack{l=1 \\ l\neq j}}^{n_\mathrm{g}} \breve{y}_{kl}. \tag{9.73}$$

Furthermore, (9.71) is translated to

$$\hat{Y}^{(ij)}_{N_\mathrm{d}} = \mathsf{H}_{ij}\hat{U}^{(ij)}_{N_\mathrm{d}} - \mathsf{H}_{ij}\Gamma^{(ij)}_{N_\mathrm{d}}\theta^{\mathrm{v}}_{ij}, \tag{9.74}$$

where

$$\hat{U}^{(ij)}_{N_\mathrm{d}} = \left[\, w_{ij}(p(0))u(0) \quad w_{ij}(p(1))u(1) \quad \dots \quad w_{ij}(p(N_\mathrm{d}-1))u(N_\mathrm{d}-1) \,\right]^\top.$$

The procedure can also be extended to estimate a direct feedthrough term in the same way as discussed in Sect. 9.3.

9.4.4 Properties

In the following important properties of the introduced identification approach of WF-LPV and HF-LPV OBF models are investigated. It is motivated that the separable least-squares estimation scheme is convergent and the obtained parameter estimates are local minima or saddle points of the LS criterion.

9.4.4.1 Convergence of the Iterative Estimation Scheme

In the proposed identification scheme, the coefficients w_{ij} and v_{ij} are parameterized linearly via (9.66), resulting in a set of unknown parameters $\{\theta^{\mathrm{w}}_{ijl}\}$ and $\{\theta^{\mathrm{v}}_{ijl}\}$. Due to the bilinear relationship in which these parameters appear in the one-step-ahead predictors (9.61a) and (9.64a), a separable least-squares algorithm is applied to tackle the minimization of the LS criterion. It is a cardinal question if this iterative scheme is convergent, i.e. is it guaranteed that the value of the LS criterion decreases in each consecutive iteration step. In each iteration cycle of this scheme, one set of the parameters is fixed, $\{\theta^{\mathrm{w}}_{ijl}\}$ or $\{\theta^{\mathrm{v}}_{ijl}\}$, in order to form a linear regression based estimation of the other set by minimizing the mean squared error of the residual. This results in a steepest descend type of iterative optimization in the search for the optimal LS prediction error. For such a separable least squares strategy, it is well known that it is convergent and the convergence point is a saddle point or a local minimum of the cost function [62]. The exact convergence point is characterized by the initial choice of $\{\theta^{\mathrm{v}}_{ijl}\}$ in Step 3c. The estimates converge to that point of the parameter space Θ whose associated region of attraction, based on the given data set, contains the initial choice of $\{\theta^{\mathrm{v}}_{ijl}\}$. Similar to the numerical optimization schemes of LTI OE or Box-Jenkins models, the global optimum of (9.37) can only be obtained by starting the iterative search from different initial values and comparing the results [105].

9.4.4.2 Unstable Model Estimates

A further problem may arise in cases where the resulting model estimate is unstable, even if $\mathcal{S}$ is asymptotically stable. This phenomenon is due to the fact that the feedback weighting is tuned on a particular, finite scheduling trajectory. As this feedback tuning can be thought of as the reoptimization of the basis with respect to $\mathcal{D}_{N_{\mathrm{d}}}$, the finite data length and the excitation capabilities of the input and scheduling signals directly effect the estimation. Thus, even if the resulting model is stable with respect to the scheduling trajectory in $\mathcal{D}_{N_{\mathrm{d}}}$ it is not guaranteed that it is stable for any other $p \in \mathfrak{B}_{\mathbb{P}}$.

9.4.4.3 Practical Use and (In)validation

Local minima and the possibility of unstable model estimates do not necessarily create problems in practice. If the resulting model passes the validation test it should be an acceptable model [105]. The problem is that the only theoretically sound (in)validation approach of the model estimates is based on simulation. By computing error measures of the difference of simulated and measured outputs, the qualities of the model estimates can be compared. As is shown in the example of Sect. 9.4.5, the proposed method quickly converges in practice and provides a reliable estimate of LPV systems.

9.4.4.4 Consistency, Variance, and Bias of the Estimates

In the previous section, the linear-in-the-parameter property of the used model structures in the one-step-ahead predictor enabled results on the consistency, variance, and bias of the resulting model estimates. In the feedback case, the linear-in-the-parameter property is lost, due to the bilinear relationship of the coefficients. Thus the previously developed results do not hold in this case. In nonlinear system identification, there are many results on the consistency, variance, and bias of model estimates obtained via different types of separable least-squares strategies. However, none of these results seem to apply to the considered model structures and estimation strategies used here. Investigation of consistency, variance, and bias of the estimation mechanism remains the subject of future research for the feedback based model structures.

Table 9.2 Validation results of 100 identification experiments with the Wiener (W) and the Wiener Feedback (WF) model structures. The results are given in terms of the average MSE, BFR and VAF of the simulated output signals of the model estimates. Subscript 1 denotes the values of these error measures for simulation with multisine u and p and subscript 2 denotes the case when uniform noise superimposed on multisines is used as an excitation.

LPV Wiener model

SNR	MSE_1 (dB)	BFR_1 (%)	VAF_1 (%)	MSE_2 (dB)	BFR_2 (%)	VAF_2 (%)
no noise	−18.23	85.39	97.84	−34.96	90.04	99.00
35 dB	−18.21	85.38	97.82	−34.77	89.92	98.99
20 dB	−17.81	85.17	97.75	−32.75	88.69	98.71
10 dB	−20.68	86.27	98.47	−31.81	87.73	98.19

LPV Wiener Feedback model

SNR	MSE_1 (dB)	BFR_1 (%)	VAF_1 (%)	MSE_2 (dB)	BFR_2 (%)	VAF_2 (%)
no noise	−31.32	94.04	99.55	−39.75	92.40	99.42
35 dB	−31.30	93.64	99.53	−39.17	92.15	99.39
20 dB	−21.60	89.30	98.52	−35.01	90.06	99.00
10 dB	−22.60	88.44	98.98	−32.38	88.15	98.59

9.4.5 Example

In this section applicability of the WF-LPV OBF model structure and its identification approach for the approximation of general LPV systems is shown through an example. Comparison is made with the static-coefficient-function based Wiener OBF model structure to show that by using the proposed feedback structure, better performance of the model estimates can be achieved.

As in the previous case, the asymptotically stable LPV system $\mathcal{S}_3$ is considered which has been studied in Sect. 8.3.3. The IO representation of $\mathcal{S}_3$, $\mathfrak{R}_{\text{IO}}(\mathcal{S}_3)$ is defined by (8.17). This system has also been identified in Sect. 9.3.6 by using the basis functions set Φ_8^0 obtained via the FKcM algorithm with fuzzyness $m = 25$ and $n_c = 8$.

Here we aim at the identification of $\mathcal{S}_3$ with the WF-LPV and W-LPV OBF model structures. To ensure fair comparison of the results, both model structures are used with basis functions Φ_8^0 and the coefficients in W are parameterized as 2nd-order polynomials in p. In the feedback case, the coefficients in V are parameterized as 3rd-order polynomials. These orders have been found optimal after several trial and error experiments. Identification of $\mathcal{S}_3$ with the global approach has been accomplished 100 times in 4 different noise settings with both the Wiener and the Wiener-feedback model structures. The data records for each identification have been generated by a white (u, p) with uniform distribution $(\mathcal{U}(-0.5, 0.5), \mathcal{U}(0.65, 0.75))$ superimposed on random multisines (3 sines with random phase in $[0, \pi]$ and frequency in $[0, \pi/5]$ and with overall amplitude of 0.5 and 0.05). The reason why this

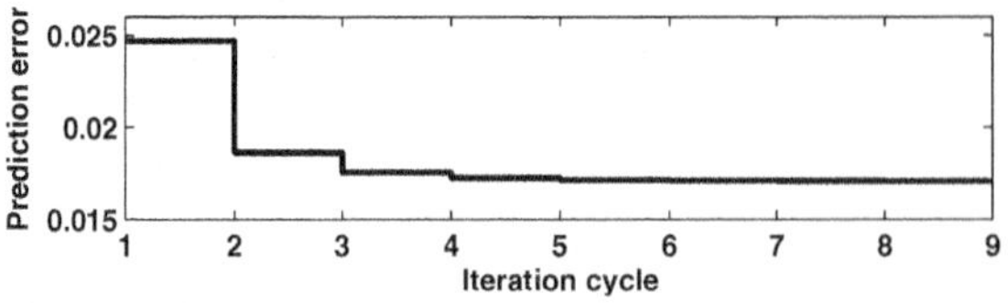

Fig. 9.10 Typical convergence plot of the prediction error in terms of (9.37) for the iterative WF-LPV OBF identification algorithm.

particular excitation has been used is explained later. For each data record the identification has been accomplished in a noiseless setting and also with additive zero-mean white output noise with normal distribution and variance $\sigma_e^2 = 0.01, 0.1$, and 0.5. The resulting SNR has been 35 dB, 20 dB, and 10 dB, while the relative noise amplitudes have been 7%, 25%, and 54% in average for the three noise cases. For the $\sigma^2 = 0.5$ noise case $N_d = 1000$, and in the other cases $N_d = 500$ samples long data records have been used. In the iterative identification method, the feedback weights have been initialized at zero. The iterative identification method converged in an average of 14 iterations for the 4×100 runs. A typical convergence plot is given in Fig. 9.10 and the typical output trajectories of the resulting estimates are shown in Fig. 9.11. Calculation time for each data set has been approximately 2 minutes with the WF-LPV OBF structure and only a few seconds with the W-LPV OBF structure on a Pentium 4, 2.8 GHz PC. In Table 9.2, the (in)validation results are shown for multisine (u, p) with random frequencies and phases and also for uniform noise superimposed on random multisine, similarly generated like the identification data. As expected, both approaches identified the system with adequate average MSE, BFR and VAF even in case of a heavy output noise, which underlines the effectiveness of the proposed OBF identification philosophy. For all measures, validation signals, and noise cases, the WF-LPV OBF model provided better estimates than the pure static dependence based W-LPV OBF model estimate. This clearly shows the improvement in the approximation capability due to the approximation of dynamic dependence with feedback-based weighting. Additional extension of Φ_8^0 with $n_e = 1, 2, \ldots$ has not improved the results as Φ_8^0 is well chosen with respect to $\mathcal{S}$, i.e. the local modeling error is negligible due to the FKcM (see Sect. 8.3.3). Even in the SNR= 10dB case, the model estimates proved to be accurate, showing that the proposed identification scheme is applicable even in the presence of significant measurement noise.

It has been observed that by using multisines superimposed on the realization of a noise sequence for the excitation of the system, the results have been better with the feedback-based structures in the presence of output additive noise. Explanation lies in the presence of the feedback gain of the model structure. In case significant noise is present in short data records, it is possible that the feedback weights are fitted to the noise process during the iterative optimization. By using multisines that emphasizes certain frequencies in the input and the scheduling signal, this effect can be attenuated. This underlines that much more understanding of feedback-based

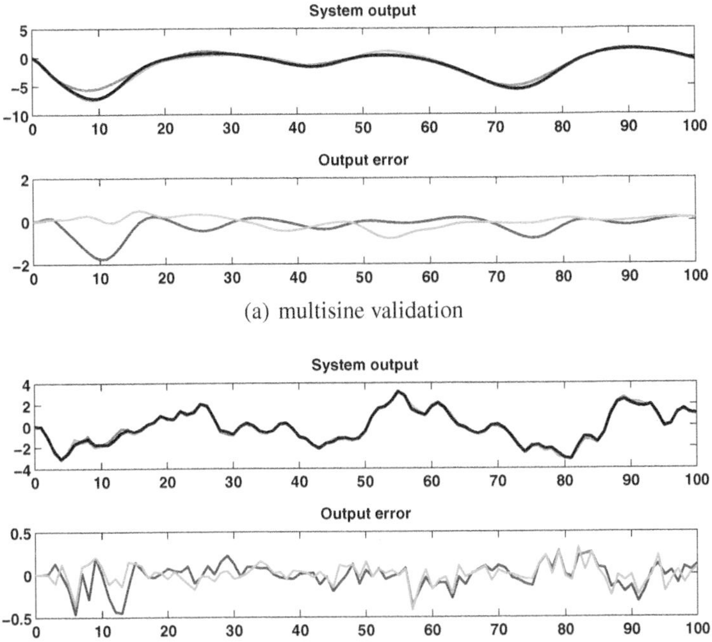

(a) multisine validation

(b) uniform noise + multisine validation

Fig. 9.11 Typical validation results of identified WF-LPV OBF (light grey) and W-LPV OBF (grey) models in the SNR = 20dB case. The response of the true system is given in black.

OBF models, especially in terms of sufficient excitation, is needed to achieve high quality model estimates.

9.5 Extension towards MIMO Systems

Next, we investigate how the developed approaches and model structures can be used for identification in the MIMO case. We show that by different choices of the basis (scalar or multivariable) different extensions of the previously studied approaches are available.

9.5.1 Scalar Basis Functions

The OBF expansion representation of asymptotically stable LPV systems has been developed in Sect. 5.3 in a general sense, using a set of scalar basis functions. This implies that such a series expansion is also available in the MIMO case, resulting in a sequence of multidimensional expansion coefficients $w_{ij} \in \mathcal{R}^{n_{\mathbb{Y}} \times n_{\mathbb{U}}}$. As a consequence, the formulation of the W-LPV and H-LPV OBF model structures for asymptotically stable MIMO LPV systems follows similarly as in Sect. 9.2 and the

MIMO W-LPV OBF structure like (9.21) or a MIMO H-LPV OBF model like (9.24) can be formulated with $w_{ij} \in \mathcal{R}^{n_Y \times n_U}$. By assuming static dependence of these coefficients, identification follows by applying the algorithms of Sect. 9.3 with some extra book keeping due to the multidimensional input and output signals. Additionally, the basis selection algorithm of Chap. 8 is also applicable in this case, by selecting the scalar basis with respect to the pole manifest set of the frozen MIMO systems of $\mathcal{S}$.

Note that a multidimensional p results in a multidimensional interpolation in the local case or the use of multidimensional $\psi(p)$ in the global case. Multidimensional interpolation problems are hard to solve and they require many interpolation points, i.e. many local experiments. Therefore, the global method is practically better-applicable for large scheduling dimensions. Note that the global approach is even applicable for significantly large scheduling dimensions which is an advantageous property with respect to LPV subspace identification methods due to the serious increase of block dimensions in such cases.

9.5.2 Multivariable Basis Functions

Due to the scalar basis sequence, it can happen that a MIMO series expansion converges fast with respect to specific input-output channels, but quite slow for others just like in the LTI case (see Sect. 2.3). As a result, some elements of the resulting coefficient matrices can have negligibly small amplitudes over all trajectories of the scheduling behavior $\mathfrak{B}_{\mathbb{P}}$. This implies that their estimation can result in a considerable variance. To overcome this effect, series expansion of asymptotically stable multivariable LPV systems can be introduced using a MIMO basis of $\mathcal{RH}_{2-}^{n_Y \times n_U}(\mathbb{E})$. Again, the key idea is to use MIMO basis functions that are composed from scalar basis sequences (see Sect. 2.3):

$$\breve{\phi}_l(\mathrm{q}) := \begin{bmatrix} \phi_{l11}(\mathrm{q}) & \cdots & \phi_{l1n_U}(\mathrm{q}) \\ \vdots & \ddots & \vdots \\ \phi_{ln_Y1}(\mathrm{q}) & \cdots & \phi_{ln_Yn_U}(\mathrm{q}) \end{bmatrix}, \tag{9.75}$$

where each $\{\phi_{lij}\}_{l=1}^{\infty}$ separately corresponds to a basis of $\mathcal{RH}_{2-}(\mathbb{E})$. Then, in terms of Th. 5.2, an asymptotically stable LPV system, with respect to a given IO partition, can be represented as

$$y = (W_0 \diamond p)u + \sum_{i=0}^{\infty} \left((W_i \diamond p) \odot \breve{\phi}_i(\mathrm{q})\right) u, \tag{9.76}$$

where $W_i \in \mathcal{R}^{n_Y \times n_U}$ and $\odot$ denotes the element-by-element matrix product. By choosing the basis sequences appropriately, a fast convergence rate can be achieved for each IO channel. The basis selection can follow by the OBF selection procedure of Chap. 8 with respect to each input-output channel separately. This means that the method is applied for a set of pole samples that are associated with the set of SISO systems describing the frozen behavior of $\mathcal{S}$ restricted to the considered IO channel.

As a next step, we formulate these model structures for a given MIMO basis sequence $\Phi_n = \{\breve{\phi}_i\}_{i=1}^{n}$ composed from these scalar elements. Note that, in this case, each basis sequence $\{\phi_{lij}\}_{l=1}^{\infty}$ is generated with a different inner function, thus Φ_n denotes the first n functions in each sequence.

- MIMO Wiener LPV-OBF model, $\mathfrak{M}_{\mathrm{W}}(\Phi_n, \theta, \mathfrak{B}_{\mathbb{P}})$

$$G(\mathrm{q}, \theta) \diamond p = W_0 \diamond p + \sum_{i=0}^{n} (W_i \diamond p) \odot \breve{\phi}_i(\mathrm{q}), \quad H(\mathrm{q}, \theta) = I, \tag{9.77}$$

where $W_i \in \mathcal{R}^{n_{\mathrm{Y}} \times n_{\mathrm{U}}}$ and e in this case is a vector of independent zero-mean white noise processes. The parameter vector of the coefficients is

$$\theta = \left[W_0 \,\vdots\, W_1 \,\vdots\, \dots \,\vdots\, W_n \right]^{\top}. \tag{9.78}$$

In this case, the one-step-ahead predictor is formulated as

$$\hat{y}_\theta = (W_0 \diamond p) u + \sum_{i=0}^{n} \left((W_i \diamond p) \odot \breve{\phi}_j(\mathrm{q}) \right) u. \tag{9.79}$$

Let $E = [1 \ \dots \ 1]^{\top}$ and introduce $\breve{y}_i : \mathbb{Z} \to \mathbb{R}^{n_{\mathrm{Y}} \times n_{\mathrm{U}}}$ defined as $\breve{y}_i = \breve{\phi}_j(\mathrm{q}) \odot (Eu^{\top})$. These signals are the multidimensional counterparts of the output signals of the filter bank part of the model structures in the SISO case (see Fig. 9.2).

- MIMO Hammerstein LPV-OBF model, $\mathfrak{M}_{\mathrm{H}}(\Phi_n, \theta, \mathfrak{B}_{\mathbb{P}})$

$$G(\mathrm{q}, \theta) \diamond p = W_0 \diamond p + \sum_{i=0}^{n} \breve{\phi}_i(\mathrm{q}) \odot (W_i \diamond p), \quad H(\mathrm{q}, \theta) = I, \tag{9.80}$$

where $W_i \in \mathcal{R}^{n_{\mathrm{Y}} \times n_{\mathrm{U}}}$. The one-step-ahead predictor is formulated in this case as

$$\hat{y}_\theta = (W_0 \diamond p) u + \sum_{i=0}^{n} \left(\breve{\phi}_j(\mathrm{q}) \odot (W_i \diamond p) \right) u. \tag{9.81}$$

Additionally, the multidimensional $\breve{u}_i : \mathbb{Z} \to \mathbb{R}^{n_{\mathrm{Y}} \times n_{\mathrm{U}}}$ (see Fig. 9.2) are introduced as $\breve{u}_i = (W_i \diamond p) \odot (Eu^{\top})$.

By assuming static dependence of the matrix coefficients $\{W_i\}$ and introducing a parametrization of $\{W_i\}$ in terms of linear combination of predefined orthogonal functions:

$$W_i = \sum_{l=0}^{n_i} \theta_l \odot \psi_l \tag{9.82}$$

where $\theta_l \in \mathbb{R}^{n_{\mathrm{Y}} \times n_{\mathrm{U}}}$ and $\psi_l \in (\mathcal{R}|_{n_{\mathbb{P}}})^{n_{\mathrm{Y}} \times n_{\mathrm{U}}}$, identification follows by applying the algorithms of Sect. 9.3 with an extended matrix regressor γ and a stacked output record $Y_{N_{\mathrm{d}}}$ formulated as $Y_{N_{\mathrm{d}}} = [y_1(0) \dots y_{n_{\mathrm{Y}}}(0) \ y_1(1) \dots y_{n_{\mathrm{Y}}}(N_{\mathrm{d}} - 1)]^{\top}$. All results in this case about consistency, bias, variance, etc. obviously extend to the MIMO case.

9.5.3 Multivariable Basis Functions in the Feedback Case

In case of the feedback structures of Sect. 9.4, the MIMO version of the model structures based on a given MIMO basis sequence Φ_n composed from scalar elements can also be formulated.

- MIMO WF-LPV OBF model, $\mathfrak{M}_{\mathrm{WF}}(\Phi_n, \theta, \mathfrak{B}_{\mathbb{P}})$

$$\breve{y}_i = \breve{\phi}_i(\mathrm{q}) \odot (E_{\mathrm{y}} u^\top) - \breve{\phi}_i(\mathrm{q}) \odot (V_i \diamond p) \odot \breve{y}_i \tag{9.83a}$$

$$y = e + (W_0 \diamond p)u + \sum_{i=0}^{n} \big((W_i \diamond p) \odot \breve{y}_i\big) E_{\mathrm{u}}, \tag{9.83b}$$

where $E_{\mathrm{y}} \in \mathbb{R}^{n_{\mathbb{Y}}}$, $E_{\mathrm{y}} = [1\ \dots\ 1]^\top$ and $E_{\mathrm{u}} \in \mathbb{R}^{n_{\mathbb{U}}}$, $E_{\mathrm{u}} = [1\ \dots\ 1]^\top$. Furthermore, each $\breve{y}_i : \mathbb{Z} \to \mathbb{R}^{n_{\mathbb{Y}} \times n_{\mathbb{U}}}$ is a multidimensional signal, $W_i, V_j \in \mathcal{R}^{n_{\mathbb{Y}} \times n_{\mathbb{U}}}$ and e is a vector of independent zero-mean white noise processes. Again as each $\breve{y}_i$ is independent from e, the one-step-ahead predictor reads as

$$\hat{y}_\theta = (W_0 \diamond p)u + \sum_{i=0}^{n} \big((W_i \diamond p) \odot \breve{y}_i\big) E_{\mathrm{u}}, \tag{9.84}$$

- MIMO HF-LPV OBF model, $\mathfrak{M}_{\mathrm{HF}}(\Phi_n, \theta, \mathfrak{B}_{\mathbb{P}})$

$$\breve{u}_i = (W_i \diamond p) \odot (E_{\mathrm{y}} u^\top) - (V_i \diamond p) \odot \breve{\phi}_i(\mathrm{q}) \odot \breve{u}_i, \tag{9.85a}$$

$$y = e + (W_0 \diamond p)u + \sum_{i=0}^{n} \big(\breve{\phi}_i(\mathrm{q}) \odot \breve{u}_i\big) E_{\mathrm{u}}. \tag{9.85b}$$

where each $\breve{u}_i : \mathbb{Z} \to \mathbb{R}^{n_{\mathbb{Y}} \times n_{\mathbb{U}}}$ is a multidimensional signal, $W_i, V_j \in \mathcal{R}^{n_{\mathbb{Y}} \times n_{\mathbb{U}}}$. The one-step-ahead predictor reads as

$$\hat{y}_\theta = (W_0 \diamond p)u + \sum_{i=0}^{n} \big(\breve{\phi}_i(\mathrm{q}) \odot \breve{u}_i\big) E_{\mathrm{u}}. \tag{9.86}$$

Assume that each element of the coefficient matrix functions of these structures is parameterized as a linear combination of functions with static dependence. Then the specific structure of the used MIMO forms enables the application of the separable least-squares algorithm for the estimation of the coefficient functions. Note that the crucial point of the algorithm is the propagation of the error backwards to the output of a basis function (see (9.68) or (9.73)). The propagated error can be computed for each element of the basis-function matrix separately. Each OBF has a separate feedback via (9.85a) or (9.83a), thus re-optimization of each feedback weight on an output channel can be done in parallel.

9.5.4 General Remarks on the MIMO Extension

Similar to the LTI case, OBFs based identification of MIMO LPV systems has a much larger freedom than the LTI case. There are different approaches that can be applied to derive a MIMO basis expansion, to formulate MIMO version of W-LPV and H-LPV OBF models, and to develop identification algorithms that can deliver the estimate. We have taken here specific choices in order to be able to apply the introduced estimation mechanisms of Sect. 9.3 and 9.4 without any modification or reformulation. However, there are numerous possibilities that may be interesting for further investigation. For example, by considering left or right side placement of the expansion coefficients for specific IO channels, like mixing the structures of H-LPV OBF and W-LPV OBF models, or using other formulations of MIMO orthonormal basis functions, like presented in Sect. 2.3.

In conclusion, it can be stated that the introduced identification approaches can directly be applied to MIMO LPV systems without the need of any modification or further assumptions. This implies that the proposed identification scheme in terms of model structure selection, parametrization, and estimation offers an effective and easily applicable identification method for a wide variety of LPV systems.

9.6 Conclusions

In this final chapter we have extended the classical prediction-error setting to the LPV framework. Using this framework and the concept of LPV series expansion we have introduced discrete-time OBF based model structures and identification methods of LPV systems. Our purpose was to establish a well understood framework of LPV identification and to introduce an efficient LPV system identification approach which benefits from the attractive properties of truncated expansion model structures.

Due to the absence of a transfer function type of description in the LPV case, we have seen that the behavioral theory and the series expansion representation of LPV systems were crucial ingredients in order to formulate the extension of the prediction-error framework to the LPV case. The process and noise models were formulated based on their impulse response presentation which made it possible to establish the one-step-ahead predictor in the classical sense, under the assumption that the scheduling variable is a deterministic signal.

We have shown that using truncated OBF-expansion models it is possible to benefit from a number of attractive properties both in terms of local and global type of LPV system identification. Beside the simple linear regression based estimation schemes it has been also shown that the introduced structures can approximate any LPV system with arbitrary accuracy if the coefficients have dynamic dependence. Furthermore, these models are not effected by changing McMillan degree of the frozen aspects of the LPV system, which represents a significant difficulty for the identification of other type of LPV model structures. We have motivated that the introduced models are effectively applicable for control as they have a direct state-space realization, however to satisfy assumptions of the control approaches, their coefficient dependencies must be restricted to be static. The introduced approaches have also been analyzed in terms of variance, bias, and consistency.

To enable the estimation of truncated expansion models with dynamic coefficient dependencies, but at the same time still provide applicable models for control, feedback-based OBF model structures have been proposed in Sect. 9.4. It has been shown, that by using static dependence, the introduced model structure is equivalent to OBF expansion models with dynamic coefficient dependencies. For the identification with such feedback-based model structures, the previously developed global approach is extended, formulating the parameter estimation through a separable least-squares approach. Finally, in Sect. 9.5, the extension of the introduced approaches to MIMO systems is investigated. It has been shown that for OBF models, based on MIMO functions composed from scalar basis sequences, the developed approaches and results are applicable.

We can conclude that we achieved our primary objective, giving an alternative, practically applicable approach for identification of LPV systems as models of an underlying physical process. On the other hand we left many open problems in terms of the whole identification cycle in the LPV context. These provide ample food for thought and objectives of future research.

Appendix A
Proofs

Abstract. In this appendix, the proofs of the theories and lemmas of this book are presented. The proofs relay heavily on the notation and concepts introduced in the previous chapters.

A.1 Proofs of Chapter 3

A.1.1 The Injective Cogenerator Property

Proof. The concept of the proof is based on [239]. Let $\mathbb{R}_\infty = \mathbb{R} \cup \{-\infty, \infty\}$ and denote by $\mathcal{Q}_n$ all maps from $(\mathbb{R}, \mathbb{R}^n)$ to $\mathbb{R}_\infty$ bounded almost everywhere on $\mathbb{R}^n$, i.e. for each $w \in \mathcal{Q}_n$ there exists a set $\mathbb{S}(w) \subset \mathbb{R}^n$ such that $w \in \mathbb{R}^{\mathbb{R} \times (\mathbb{R}^n \setminus \mathbb{S}(w))}$ and $\mathbb{S}(w)$ has a measure 0. The set $\mathcal{Q}_n$ is a real vector space for each $n \in \mathbb{N}$. Denote by $\bar{\mathcal{Q}}_n \subset \mathcal{Q}_n$ the set of all $w \in \mathcal{Q}_n$ for which there exist a $t \in \mathbb{R}$ and $\mathrm{x}_1, \ldots \mathrm{x}_n \in \mathbb{R}$ such that $w(t, \mathrm{x}_1, \ldots, \mathrm{x}_n) \neq w(t, \mathrm{x}_1, \ldots, \mathrm{x}_{n-1}, 0)$. Denote $\mathcal{Q} = \bigcup_{n \in \mathbb{N}} \bar{\mathcal{Q}}_n$. $\mathcal{Q}$ is an (additive) Abelian group.

Consider a $R \in \mathcal{R}[\xi]^{n_\mathrm{r} \times n_\mathrm{w}}$ with $\mathbb{P} = \mathbb{R}^{n_\mathbb{P}}$. For a $w \in \mathcal{Q}$, $R \circledast w = 0$ means that any $(w, p) \in \mathcal{L}_1^{\mathrm{loc}}(\mathbb{R}, \mathbb{R}^{n_\mathrm{w}} \times \mathbb{R}^{n_\mathbb{P}})$ satisfying

$$w(t) = w(t, [\, p(t) \;\; \tfrac{\mathrm{d}}{\mathrm{d}t} p(t) \;\; \tfrac{\mathrm{d}^2}{\mathrm{d}t^2} p(t) \;\; \ldots]), \tag{A.1}$$

for all $k \in \mathbb{Z}$ with p being smooth enough, also satisfies $(R(\frac{\mathrm{d}}{\mathrm{d}t}) \diamond p)(t) w(t) = 0$ weakly for all $t \in \mathbb{R} \setminus \mathbb{K}(w, p)$, where $\mathbb{K}(w, p) = \{t \in \mathbb{Z} \mid [\, p(t) \;\; \frac{\mathrm{d}}{\mathrm{d}t} p(t) \;\; \frac{\mathrm{d}^2}{\mathrm{d}t^2} p(t) \;\; \ldots] \in \mathbb{S}(w)\}$. As $\mathbb{S}(w)$ has zero measure, this means that there exists a $(w, p) \in \mathcal{L}_1^{\mathrm{loc}}(\mathbb{R}, \mathbb{R}^{n_\mathrm{w}} \times \mathbb{R}^{n_\mathbb{P}})$ satisfying (A.1) such that $(R(\frac{\mathrm{d}}{\mathrm{d}t}) \diamond p)(t) w(t) = 0$ holds weakly for all $t \in \mathbb{R}$. The set $\mathfrak{B}_*$ given as $\mathfrak{B}_* = \{w \in \mathcal{Q}^{n_\mathrm{w}} \mid R \circledast w = 0\}$, is called the complete solution space of the linear system of PV differential equations (KR-representation) $(R(\frac{\mathrm{d}}{\mathrm{d}t}) \diamond p) w = 0$. Note that the behavior $\mathfrak{B}$ of R (see Def. 3.10), contains the set of trajectories (w, p) that satisfy $w \in \mathfrak{B}_*$, while $\mathfrak{B}_*$ describes the relationship of the trajectories containing the descriptions of possible solutions that are excluded from $\mathfrak{B}$ due to the singularity of the coefficients in R.

Let $M_1 \in \mathcal{R}[\xi]^{n_r \times n_r}$ and $M_2 \in \mathcal{R}[\xi]^{n_w \times n_w}$ be unimodular matrices such that (3.35) is the Jacobson form of R with $Q = \mathrm{Diag}(1,\ldots,1,r) \in \mathcal{R}[\xi]^{n \times n}$ and $0 \neq r \in \mathcal{R}[\xi]$. It can be shown (see [41, 188]), that $(R(\frac{d}{dt}) \diamond p)w = 0$ has the same solutions as

$$(M_1(\frac{d}{dt})R(\frac{d}{dt}) \diamond p)w = (Q(\frac{d}{dt})M_2^{\dagger}(\frac{d}{dt}) \diamond p)w = 0, \tag{A.2}$$

so there is an isomorphism between solution spaces

$$\mathfrak{B}_* \cong \tilde{\mathfrak{B}}_* := \{\tilde{w} \in \mathcal{Q}^{n_w} \mid [Q \;\; 0] \circledast \tilde{w} = 0\}, \tag{A.3a}$$

$$w \to \tilde{w} := M_2^{\dagger}(q)w, \tag{A.3b}$$

where $\tilde{w}_i = 0$ for $i \in \{1,\ldots,n-1\}$ and $r_n \circledast \tilde{w}_n = 0$. Introduce $\mathcal{M}_R = \mathrm{Module}_{\mathcal{R}[\xi]}(R)$ as the left module in $\mathcal{R}[\xi]^{n_r \times n_w}$ generated by the rows of $R \in \mathcal{R}[\xi]^{n_r \times n_w}$. Then

$$\mathfrak{B}_* \cong \mathrm{Hom}_{\mathcal{R}[\xi]}(\mathcal{M}_R, \mathcal{Q}^{n_w}), \tag{A.4}$$

which corresponds to the so-called Malgrange isomorphism. Explicitly, (A.4) assigns to each $w \in \mathfrak{B}_*$ the linear map $\phi_w : \mathcal{M}_R \to \mathcal{Q}$ defined by $\phi_w([r]) := r(q)w$ where $[r]$ denotes the residue class of $r \in \mathcal{R}[\xi]^{1 \times n_w}$ in $\mathcal{M}_R$, and the well definedness of ϕ_w follows from

$$[r_1] = [r_2] \;\to\; r_1 - r_2 \in \mathrm{Span}_{\mathcal{R}}^{\mathrm{row}}(R) \;\to\; r_1(q)w = r_2(q)w,$$

for all $w \in \mathfrak{B}_*$ which also implies that $\mathcal{Q}^{n_w} \cong \mathrm{Hom}_{\mathcal{R}[\xi]}(\mathcal{R}[\xi]^{1 \times n_w}, \mathcal{Q}^{n_w})$. Conversely, for a linear map $\phi : \mathcal{M}_R \to \mathcal{Q}$ one defines $w_i := \phi([e_i])$, where e_i is the i-th natural basis vector of $\mathcal{R}[\xi]^{1 \times n_w}$. Then we have

$$\phi([r]) = \phi\left(\left[\textstyle\sum_{i=1}^{n_w} r_i e_i\right]\right) = \textstyle\sum_{i=1}^{n_w} r_i(q)\phi([e_i]) = \textstyle\sum_{i=1}^{n_w} r_i(q)w_i = r(q)w.$$

Due to (A.1), the above equation implies an isomorphism of left modules:

$$\mathrm{Module}_{\mathcal{R}[\xi]}(R) \cong \mathrm{Module}_{\mathcal{R}[\xi]}([Q \;\; 0]), \tag{A.5a}$$

$$[r] \to [rM_2]. \tag{A.5b}$$

Let $\mathcal{M}_1, \mathcal{M}_2, \mathcal{M}_3$ be left modules in $\mathcal{R}[\xi]^{n_r \times n_w}$ and let $\phi_{12} : \mathcal{M}_1 \to \mathcal{M}_2$ and $\phi_{23} : \mathcal{M}_2 \to \mathcal{M}_3$ be linear maps, i.e. left module homomorphisms. Then

$$\mathcal{M}_1 \overset{\phi_{12}}{\to} \mathcal{M}_2 \overset{\phi_{23}}{\to} \mathcal{M}_3 \tag{A.6}$$

is exact if $\mathrm{im}(\phi_{12}) = \ker(\phi_{23})$. The same notion can be used if $\mathcal{M}_1, \mathcal{M}_2, \mathcal{M}_3$ are Abelian groups and ϕ_{12}, ϕ_{23} are group homomorphisms. Then $\mathcal{Q}$ is called an injective cogenerator if the sequence

$$\mathcal{M}_1 \to \mathcal{M}_2 \to \mathcal{M}_3, \tag{A.7}$$

is exact iff the sequence

$$\mathrm{Hom}_{\mathcal{R}[\xi]}(\mathcal{M}_1, \mathcal{Q}^{n_w}) \leftarrow \mathrm{Hom}_{\mathcal{R}[\xi]}(\mathcal{M}_2, \mathcal{Q}^{n_w}) \leftarrow \mathrm{Hom}_{\mathcal{R}[\xi]}(\mathcal{M}_3, \mathcal{Q}^{n_w})$$

of Abelian groups is exact.

For injectivity, one needs to prove according to Corollary 3.17 of [92]: For every $0 \neq R \in \mathcal{R}[\xi]$ and every $w_u \in \mathcal{Q}$, there exists a $w_y \in \mathcal{Q}$ such that $R \circledast w_y = w_u$. Let $R(\xi) = \sum_{i=0}^{n_\xi} r_i \xi^i$ be given with $r_{n_\xi} \neq 0$. If $n_\xi = 0$, there is nothing to prove. Since $\mathcal{R}$ is a field, assume that $r_{n_\xi} = 1$. Then $R \circledast w_y = w_u$ can be rewritten as a first-order system

$$(r_{x,1}\frac{d}{dt} + r_{x,0}) \circledast w_x = r_u \circledast w_u, \tag{A.8}$$

where $w_x = [\, w_y \ \ \dots \ \ \frac{d^{n_\xi - 1}}{dt^{n_\xi - 1}} w_y \,]^\top$, $r_{x,1} = I$, $r_u = [0 \ \ \dots \ \ 0 \ \ 1]^\top \in \mathbb{R}^{n_\xi}$ and

$$r_{x,0} = \begin{bmatrix} 0 & -I \\ r_0 & r_* \end{bmatrix} \in \mathcal{R}^{n_\xi \times n_\xi}, \quad \text{with } r_* = [\, r_1 \ \ \dots \ \ r_n \,]. \tag{A.9}$$

Let $\mathbb{S}(R)$ denote the discrete set of singularities of the meromorphic coefficients r_i in R. Let $\mathbb{S}(w_y) := \mathbb{S}(w_u) \cup \mathbb{S}(R)$ which is still discrete. Hence, $\mathbb{R}^{\cdot} \setminus \mathbb{S}(w_y)$ is a countable union of open intervals $I_i \in \mathbb{R}^{\cdot}$ and on each I_i it holds that R_0 and w_u are bounded. Therefore there exists a bounded solution $w_x : (\mathbb{Z} \times I_i) \rightarrow \mathbb{R}^{n_\xi}$ to (A.8) on each I_i. By concatenating them, one gets a solution $w_x \in \mathcal{Q}^{n_\xi}$ and thus $w_y \in \mathcal{Q}$.

For the cogenerator property, it has to be shown that if for some $R \in \mathcal{R}[\xi]$, $R \circledast w_y = 0$ has only the zero solution, then this implies that $R \in \mathcal{R}$ and $R \neq 0$. Assume the contrary and let $\mathrm{Deg}(R) = n_\xi \geq 1$. Then one can rewrite $R \circledast w_y = 0$ as $\frac{d}{dt} w_x = -r_{x,0} \circledast w_x$ like in the previous part. Let $\mathbb{S}(w_y) = \mathbb{S}(R)$, then on each of the intervals I_i, the solution set of this homogenous equation is an n_ξ-dimensional subspace of $(\mathbb{R}^{n_\xi})^{\mathbb{R} \times I_i}$, in particular there exist non-zero solutions. By concatenating them, we get a non-zero solution $w_x \in \mathcal{Q}^n$. If $w_y = w_{x,1}$ was identically zero, then $w_x = [\, w_y \ \ \dots \ \ \frac{d^{n_\xi - 1}}{dt^{n_\xi - 1}} w_y \,]^\top$ would be identically zero which leads to a contradiction. The same proof can be given in discrete-time with minor changes due to the difference in the solution spaces and the Jacobson form (see [201] for the details).□

A.1.2 *Existence of Full Row Rank KR Representation*

Proof (Theorem 3.2). Consider a continuous-time $\mathfrak{R}_K(\mathcal{S})$ with $R \in \mathcal{R}[\xi]^{n_r \times n_w}$, $\mathbb{P} = \mathbb{R}^{n_p}$, and represented behavior $\mathfrak{B}$. Without loss of generality, let $R \neq 0$ as the behavior $\mathfrak{B} = \mathcal{L}_1^{\mathrm{loc}}(\mathbb{R}, \mathbb{R}^{n_w} \times \mathbb{R}^{n_p})$ for $R = 0$ can be represented by the empty matrix which is full rank by convention. Let $M_1 \in \mathcal{R}[\xi]^{n_r \times n_r}$ and $M_2 \in \mathcal{R}[\xi]^{n_w \times n_w}$ be unimodular matrices such that (3.35) is the Jacobson form of R in terms of Theorem 3.1 with $Q = \mathrm{Diag}(1, \dots, 1, r) \in \mathcal{R}[\xi]^{n \times n}$ and $0 \neq r \in \mathcal{R}[\xi]$. Partition $M_2^\dagger = [W_1 \ W_2]^\top$ according to the partition of the Jacobson form. Since M_1 is unimodular, the solution space of $(R(\frac{d}{dt}) \diamond p)w = 0$ is equal to the solution space of $(M_1(\frac{d}{dt})R(\frac{d}{dt}) \diamond p)w = 0$ (see the previous proof in Sect. A.1.1). Thus $R'(\xi) := Q(\xi)W_1(\xi)$ also represents $\mathfrak{B}$ in an almost everywhere sense, i.e. for all trajectories of $p \in \mathfrak{B}_p$ for which the coefficients of R' are bounded, and $\mathrm{Rank}(R') = n$. For the DT case, the proof analogously follows only the formulation of the Jacobson form and the definition of solution spaces are different. □

A.1.3 Elimination Property

Proof (Theorem 3.3). Based on the proof of the injective cogenerator property, consider

$$\mathfrak{B}_* = \{w \in \mathcal{Q}^{n_\mathrm{W}} \mid \exists w_\mathrm{L} \in \mathcal{Q}^{n_\mathrm{L}} \text{ s.t. } R_\mathrm{W} \circledast w = R_\mathrm{L} \circledast w_\mathrm{L}\}, \tag{A.10}$$

where $R_\mathrm{W} \in \mathcal{R}[\xi]^{n_\mathrm{r} \times n_\mathrm{W}}$ and $R_\mathrm{L} \in \mathcal{R}[\xi]^{n_\mathrm{r} \times n_\mathrm{L}}$ define an LPV latent variable representation in the form of (3.39) with $\mathbb{P} = \mathbb{R}^{n_\mathbb{P}}$. Then showing that $\mathfrak{B}_*$ has a kernel representation is equivalent with showing that the manifest behavior of (3.39) has a kernel representation in an almost everywhere sense. Define the left kernel of R_L as

$$\mathrm{Ker}_{\mathcal{R}[\xi]}(R_\mathrm{L}) = \{r \in \mathcal{R}[\xi]^{1 \times n_\mathrm{r}} \mid r(\xi) R_\mathrm{L}(\xi) = 0\}, \tag{A.11}$$

which is a left submodule of $\mathcal{R}[\xi]^{1 \times n_\mathrm{r}}$. Thus, it is finitely generated, i.e. there exists a $Q \in \mathcal{R}[\xi]^{n \times n_\mathrm{r}}$ such that $\mathrm{Img}_{\mathcal{R}[\xi]}(Q) := \{r(\xi) Q(\xi) \mid r(\xi) \in \mathcal{R}[\xi]^{1 \times n_\mathrm{r}}\}$ is equal to $\mathrm{Ker}_{\mathcal{R}[\xi]}(R_\mathrm{L})$. Then we have an exact sequence

$$\mathcal{R}[\xi]^{1 \times n} \overset{\cdot Q}{\to} \mathcal{R}[\xi]^{1 \times n_\mathrm{r}} \overset{\cdot R_\mathrm{L}}{\to} \mathcal{R}[\xi]^{1 \times n_\mathrm{L}} \tag{A.12}$$

and therefore the sequence $\mathcal{Q}^n \overset{Q(q)}{\leftarrow} \mathcal{Q}^{n_\mathrm{r}} \overset{R_\mathrm{L}(q)}{\leftarrow} \mathcal{Q}^{n_\mathrm{L}}$ is also exact. This signifies that $R_\mathrm{W}(q) w \in \mathrm{Img}_\mathcal{Q}(R_\mathrm{L}) := \{R_\mathrm{L}(q) w_\mathrm{L} \mid w_\mathrm{L} \in \mathcal{Q}^{n_\mathrm{L}}\}$ iff $R_\mathrm{W}(q) w \in \mathrm{Ker}_\mathcal{Q}(Q)$, i.e. $\mathfrak{B}_* = \{w \in \mathcal{Q}^{n_\mathrm{W}} \mid Q R_\mathrm{W} \circledast w = 0\}$. □

A.1.4 State-Kernel Form

Proof (Theorem 3.4). The concept of the proof is based on [153]. To simplify the discussion, we prove only the so-called *Markovian case* as the state case follows trivially from this concept due to the linearity and time-invariance of LPV systems. We call the continuous-time LPV system $\mathcal{S} = (\mathbb{R}, \mathbb{P}, \mathbb{W}, \mathfrak{B})$ Markovian, if for all $p \in \mathfrak{B}_\mathbb{P}$

$$(w_1, w_2 \in \mathfrak{B}_p) \wedge (w_1, w_2 \text{ continuous at } 0) \wedge (w_1(0) = w_2(0)) \Rightarrow (w_1 \underset{0}{\wedge} w_2) \in \mathfrak{B}_p.$$

In case of a discrete-time $\mathcal{S}$, i.e. $\mathbb{T} = \mathbb{Z}$, the definition is similar except continuity of w_1 and w_2 is not required at 0. In the following we prove that $\mathcal{S}$ is Markovian iff there exist matrices $r_0, r_1 \in \mathcal{R}^{n_\mathrm{r} \times n_\mathrm{W}}$ such that $\mathfrak{B}$ has the kernel representation:

$$r_0 w + r_1 \xi w = 0. \tag{A.13}$$

For the sake of simplicity we consider only the continuous-time case as the DT case follows similarly. For $\mathbb{T}=\mathbb{R}$, $\xi=\frac{\mathrm{d}}{\mathrm{d}t}$ and the "if" part is trivial. To show the "only if" case, assume that a KR representation of $\mathcal{S}$ is given with $R\in\mathcal{R}[\xi]^{n_\mathrm{r}\times n_\mathbb{W}}$ for which the solutions of (3.24) satisfy the above given connectability condition. Without loss of generality it can be assumed that R is full row rank. Also, there exists a unimodular $M\in\mathcal{R}[\xi]^{\cdot\times n_\mathrm{r}}$ such that $R'(\xi)=M(\xi)R(\xi)$ is in a row reduced form, meaning that the matrix formed by the coefficient functions of the highest powers in ξ of the rows $R'(\xi)$ has full row rank. Due to the fact that M is a left-side unimodular transformation, the behaviors of R and R' are equivalent.

We now show that $\mathrm{Deg}(R')=1$. Assume the contrary and write R' in the IO form:

$$(R_1(\frac{\mathrm{d}}{\mathrm{d}t})\diamond p)w_1=(R_2(\frac{\mathrm{d}}{\mathrm{d}t})\diamond p)w_2, \tag{A.14}$$

where $\mathrm{Col}(w_1,w_2)=w$ corresponds to an IO partition and $\mathrm{Deg}(R_1)\geq\mathrm{Deg}(R_2)$. The assumption that $\mathrm{Deg}(R')>1$ implies that $\mathrm{Deg}(R_1)>1$. Similarly, the assumption that $(R'(\frac{\mathrm{d}}{\mathrm{d}t})\diamond p)w=0$ is Markovian implies that $(R_1(\frac{\mathrm{d}}{\mathrm{d}t})\diamond p)w_1=0$ is Markovian.

Now let w_1',w_1'' be the solutions of $(R_1(\frac{\mathrm{d}}{\mathrm{d}t})\diamond p)w_1=0$ for a $p\in\mathfrak{B}_\mathbb{P}$ with $w_1'(0)=w_1''(0)$. Since (w_1,w_2) is an IO partition of $\mathcal{S}$, thus $\mathrm{Col}(w_1',0)$ and $\mathrm{Col}(w_1'',0)$ are also solutions of $(R'(\frac{\mathrm{d}}{\mathrm{d}t})\diamond p)w=0$ and due to Markovian property, they are connectable. This implies that in order to obtain contradiction it suffices to prove contradiction for autonomous systems. Let $n_\xi=\mathrm{Deg}(R_1)$ and by assumption $n_\xi>1$. Introduce auxiliary variables $\breve{w}_{ij}$ defined as

$$\breve{w}_{ij}:=\frac{\mathrm{d}^i}{\mathrm{d}t^i}w_j,\quad (i,j)\in\mathbb{I}_0^{n_\xi}\times\mathbb{I}_1^{n_\mathbb{W}}, \tag{A.15}$$

where $w=\begin{bmatrix} w_1 & \dots & w_{n_\mathbb{W}}\end{bmatrix}^\top$. Collect these variables in a column vector

$$\breve{w}=\begin{bmatrix}\breve{w}_{01} & \breve{w}_{02} & \dots & \breve{w}_{0n_\mathbb{W}} & \breve{w}_{11} & \dots & \breve{w}_{n_\xi n_\mathbb{W}}\end{bmatrix}^\top. \tag{A.16}$$

Now consider the system with latent variable $\breve{w}$ as

$$\frac{\mathrm{d}}{\mathrm{d}t}\breve{w}=(r\diamond p)\breve{w}, \tag{A.17a}$$

$$w_j=\breve{w}_{0j},\quad \forall j\in\mathbb{I}_1^{n_\mathbb{W}}. \tag{A.17b}$$

where the coefficient $r\in\mathcal{R}^{(n_\xi n_\mathbb{W})\times(n_\xi n_\mathbb{W})}$ is determined from the coefficients of $R_1(\xi)$ and the definition (A.15). The manifest behavior of (A.17a) is equivalent with the manifest behavior of $R_1(\xi)$, which can be checked by elimination of the latent variables of (A.17a–b). However, the manifest behavior can not be Markovian as (A.17a–b) has exactly one solution $(w,\breve{w})$ for each initial condition $\breve{w}(0)$ and scheduling trajectory $p\in\mathfrak{B}_\mathbb{P}$. This contradicts Markovianity, since two solutions $(w,\breve{w})$ and $(w',\breve{w}')$ with $\breve{w}_{0j}(0)=\breve{w}_{0j}'(0)$, $\forall j\in\mathbb{I}_1^{n_\mathbb{W}}$ cannot be connected unless also $\breve{w}_{ij}(0)=\breve{w}_{ij}'(0)$, $\forall(i,j)\in\mathbb{I}_1^{n_\xi-1}\times\mathbb{I}_1^{n_\mathbb{W}}$. □

A.1.5 Left/Right-Side Unimodular Transformation

Proof (Theorem 3.7 & 3.8). First consider the left side transformation. Let $R \in \mathcal{R}[\xi]^{n_r \times n_W}$ and $R' \in \mathcal{R}[\xi]^{n \times n_r}$ and $\mathbb{P} = \mathbb{R}^{n_P}$. Based on the proof of the injective cogenerator property, consider $\mathfrak{B}_*$ and $\mathfrak{B}'_*$ as the complete behaviors of R and R'. Then the inclusion $\mathfrak{B}'_* \subseteq \mathfrak{B}_*$ can be expressed as an exact sequence

$$0 \longrightarrow \mathfrak{B}'_* \longrightarrow \mathfrak{B}_*, \tag{A.18}$$

which is equivalent to the exact sequence

$$0 \longleftarrow \text{Module}_{\mathcal{R}[\xi]}(R') \longleftarrow \text{Module}_{\mathcal{R}[\xi]}(R). \tag{A.19}$$

Equivalently, we have $\text{Span}^{\text{row}}_{\mathcal{R}}(R') \supseteq \text{Span}^{\text{row}}_{\mathcal{R}}(R)$ or $R'(\xi) = Q(\xi)R(\xi)$ for some $Q \in \mathcal{R}[\xi]^{n \times n_r}$. If $\mathfrak{B}_* = \mathfrak{B}'_*$, then $R'(\xi) = Q_1(\xi)R(\xi)$ and $R(\xi) = Q_2(\xi)R'(\xi)$, which shows that R and R' has the same rank. If additionally, R and R' are full rank, then this implies that $Q_1 = Q_2^\dagger$, ergo Q_1 and Q_2 are unimodular. As the complete behaviors are equal therefore this implies that the behaviors of R and R' for each commonly valid trajectory of p are equal.

Consider the right side transformation. Based on the proof of the injective cogenerator property, there is a homomorphism between the the complete behaviors of $R(\xi)$ and $R'(\xi) = R(\xi)Q_1(\xi)$ and also between $R(\xi) = R'(\xi)Q_2(\xi)$ and $R'(\xi)$. This implies that if $Q_1 = Q_2^\dagger$, ergo Q_1 and Q_2 are unimodular, then there exists a isomorphism between the behaviors. □

A.2 Proofs of Chapter 5

A.2.1 LPV Series Expansion, Pulse Basis

Proof (Theorem 5.1). Let $\mathcal{S} = (\mathbb{Z}, \mathbb{P} \subseteq \mathbb{R}^{n_P}, \mathbb{R}^{n_W}, \mathfrak{B})$ be a discrete-time LPV system. Assume that $\mathcal{S}$ is not autonomous, i.e., in terms of Def. 3.11, $\mathcal{S}$ has an IO partition which can be non-unique. Denote the scheduling variable of $\mathcal{S}$ as p. For a given IO partition of $\mathcal{S}$ with output dimension n_Y and input dimension n_U, an IO representation of $\mathcal{S}$ is characterized by the polynomials $R_u \in \mathcal{R}[\xi]^{n_Y \times n_U}$ and full rank $R_y \in \mathcal{R}[\xi]^{n_Y \times n_Y}$. Among the IO representations that belong to the equivalence class of $\mathcal{S}$ for this IO partition, there exists a subset of representations where R_y and R_u are coprime. Those representations are called minimal (see Def. 3.28). Among these minimal representations the representation with monic R_y is unique. Consider this unique, minimal IO representation of $\mathcal{S}$ and denote its polynomials with R_y and R_u. Then the dynamic relation reads as

$$\mathrm{q}^{n_a} y + \sum_{i=0}^{n_a-1} (a_i \diamond p)\mathrm{q}^i y = \sum_{j=0}^{n_b} (b_j \diamond p)\mathrm{q}^j u, \tag{A.20}$$

where $a_i \in \mathcal{R}^{n_Y \times n_Y}$ and $b_j \in \mathcal{R}^{n_Y \times n_U}$ are the coefficients of R_y and R_u. Due to the maximum freedom of u in the IO partition, it holds that $\mathrm{Deg}(R_y) = n_a \geq n_b = \mathrm{Deg}(R_u)$.

Assume that $\mathcal{S}$ is asymptotically stable. Multiply the expression (A.20) with q^{-n_a} which according to the non-commutative multiplication rules in discrete time (see Def. 3.16) results in

$$y = -\sum_{i=0}^{n_a-1} \left(\overleftarrow{a_i}^{(n_a)} \diamond p\right) q^{i-n_a} y + \sum_{j=0}^{n_b} \left(\overleftarrow{b_j}^{(n_a)} \diamond p\right) q^{j-n_a} u, \tag{A.21}$$

where $\overleftarrow{\cdot}^{(n_a)}$ denotes the backward shift operator applied on the coefficient function for n_a times. We call (A.21) the filter form of (A.20). Now substitute $q^{-1}y$ in (A.21) by the left-hand side of (A.21) multiplied by q^{-1}. This results in the following expression:

$$\begin{aligned} y = &-\overleftarrow{a_{n_a-1}}^{(n_a)} \left(-\sum_{i=0}^{n_a-1} \left(\overleftarrow{a_i}^{(n_a+1)} \diamond p\right) q^{i-1-n_a} y + \sum_{j=0}^{n_b} \left(\overleftarrow{b_j}^{(n_a+1)} \diamond p\right) q^{j-1-n_a} u \right) \\ &- \sum_{i=0}^{n_a-2} \left(\overleftarrow{a_i}^{(n_a)} \diamond p\right) q^{i-n_a} y + \sum_{j=0}^{n_b} \left(\overleftarrow{b_j}^{(n_a)} \diamond p\right) q^{j-n_a} u, \end{aligned} \tag{A.22}$$

Notice that in (A.22), the smallest time-shift of y has the order of -2 and that all coefficient relations with (A.21) are uniquely determined. Note that (A.20) represents $\mathfrak{B}$ restricted to signals y and u with left compact support. Thus, by applying this procedure recursively on $q^{-2}y$, $q^{-3}y$, etc., there exists a $n \in \mathbb{N}$ for every $(u, y, p) \in \mathfrak{B}$ such that y vanishes after substitution of $q^{-n}y$. This yields that the recursive procedure results in a Laurent-like series expansion of (A.20).

Denote by g_i the resulting expressions of the coefficients of R_y and R_u associated with each $q^{-i}u$ in (A.22). It is obvious that $g_i \in \mathcal{R}^{n_Y \times n_U}$. However, it is not obvious how this coefficient sequence behaves for increasing i and if it is convergent or not. As a next step, we investigate this property.

Due to the asymptotic stability of $\mathcal{S}$, it holds that for all $p \in \mathfrak{B}_{\mathbb{P}}$ and $k \in \mathbb{Z}$

$$\lim_{i \to \infty} (g_i \diamond p)(k) = 0. \tag{A.23}$$

Otherwise there exists a $k_0 \in \mathbb{Z}$, and an input signal

$$u(k) = \begin{cases} 1, & \text{if } k = k_0; \\ 0, & \text{else.} \end{cases} \tag{A.24}$$

such that the associated output trajectory y in terms of the IO representation does not converge to the origin when $k \to \infty$, which contradicts with the asymptotic stability. This implies that

$$\lim_{i \to \infty} (g_i \diamond p) = 0 \quad \forall p \in \mathfrak{B}_{\mathbb{P}}, \tag{A.25}$$

meaning that the coefficient sequence of the expansion converges to the zero function on $\mathfrak{B}_{\mathbb{P}}$. In this way, the limit of the function sequence $\{\mathrm{g}_0, \mathrm{g}_1, \mathrm{g}_2, \ldots\}$ can be considered zero.

Asymptotic stability of $\mathcal{S}$ also implies global BIBO stability in the ℓ_∞ norm:

$$\sup_{k\geq 0} \|u(k)\| < \infty \quad \Rightarrow \quad \sup_{k\geq 0} \|y(k)\| < \infty.$$

In the view of global ℓ_∞ BIBO stability, the resulting series expansion form satisfies

$$\sup_{k\geq 0} \|y(k)\| = \sup_{k\geq 0} \|(\mathrm{g}_0 \diamond p)(k)u(k) + (\mathrm{g}_1 \diamond p)(k)u(k-1) + \ldots\|,$$

for all $(u,y,p) \in \mathfrak{B}$. As $\sup_{k\geq 0} \|u(k)\| < \infty$ and $\sup_{k\geq 0} \|y(k)\| < \infty$, the above equation implies that

$$\sup_{k\geq 0} \sum_{i=0}^{\infty} \|(\mathrm{g}_i \diamond p)(k)\| < \infty, \tag{A.26}$$

for all $p \in \mathfrak{B}_{\mathbb{P}}$. These properties prove that

$$y = \sum_{i=0}^{\infty} (\mathrm{g}_i \diamond p)\mathrm{q}^{-i} u, \tag{A.27}$$

exists and it is satisfied for all $(u,y,p) \in \mathfrak{B}$ with left compact support. In this way, an asymptotically stable discrete-time LPV system has a unique, convergent series expansion in terms of the LTI pulse basis $\{1, \mathrm{q}^{-1}, \mathrm{q}^{-2}, \ldots\}$ with expansion coefficients in $\mathcal{R}^{n_{\mathbb{Y}} \times n_{\mathbb{U}}}$. □

A.2.2 LPV Series Expansion, OBFs

Proof (Theorem 5.2). To simplify the discussion consider the SISO case as the MIMO case analogously follows. In the LTI theory it is proven that any pulse basis function q^{-i}, $i > 0$, has a unique series expansion in terms of an arbitrary basis sequence $\Phi_\infty = \{\phi_i\}_{i=1}^{\infty}$ in $\mathcal{RH}_{2-}(\mathbb{E})$ [73]. This implies that

$$\begin{bmatrix} \mathrm{q}^{-1} \\ \mathrm{q}^{-2} \\ \mathrm{q}^{-3} \\ \vdots \end{bmatrix} = \begin{bmatrix} \mathrm{w}_{11} & \mathrm{w}_{12} & \mathrm{w}_{13} & \cdots \\ \mathrm{w}_{21} & \mathrm{w}_{22} & \mathrm{w}_{23} & \cdots \\ \mathrm{w}_{31} & \mathrm{w}_{32} & \mathrm{w}_{33} & \cdots \\ \vdots & \vdots & \vdots & \ddots \end{bmatrix} \begin{bmatrix} \phi_1(\mathrm{q}) \\ \phi_2(\mathrm{q}) \\ \phi_3(\mathrm{q}) \\ \vdots \end{bmatrix}, \tag{A.28}$$

where $\mathrm{w}_{ij} \in \mathbb{R}$, $i,j \in \mathbb{I}_1^\infty$, are the expansion coefficients. It holds for all $i \in \mathbb{I}_1^\infty$, that the sequence $\{\mathrm{w}_{ij}\}_{j=1}^{\infty}$ is an ℓ_2 sequence [73]. The same property also holds for each sequence $\{\mathrm{w}_{ij}\}_{i=1}^{\infty}$.

Based on Th. 5.1, there exists a pulse basis series expansion of any discrete-time asymptotically stable LPV system $\mathcal{S} = (\mathbb{Z}, \mathbb{P}, \mathbb{W}, \mathfrak{B})$ with IO partition (u,y). This series expansion can be written in the form of

$$y = \mathrm{g}_0 u + \left(\begin{bmatrix} \mathrm{g}_1 & \mathrm{g}_2 & \mathrm{g}_3 & \dots \end{bmatrix} \diamond p\right) \begin{bmatrix} \mathrm{q}^{-1} \\ \mathrm{q}^{-2} \\ \mathrm{q}^{-3} \\ \vdots \end{bmatrix} u, \tag{A.29}$$

where $\mathrm{g}_i \in \mathcal{R}$, $i \in \mathbb{I}_0^{\infty}$. Now by substituting (A.28) into (A.29) it follows that

$$y = \mathrm{g}_0 u + \left(\begin{bmatrix} w_1 & w_2 & w_3 & \dots \end{bmatrix} \diamond p\right) \begin{bmatrix} \phi_1 \\ \phi_2 \\ \phi_3 \\ \vdots \end{bmatrix} u, \tag{A.30}$$

where

$$\begin{bmatrix} w_1 & w_2 & w_3 & \dots \end{bmatrix} = \begin{bmatrix} \mathrm{g}_1 & \mathrm{g}_2 & \mathrm{g}_3 & \dots \end{bmatrix} \begin{bmatrix} \mathrm{w}_{11} & \mathrm{w}_{12} & \mathrm{w}_{13} & \dots \\ \mathrm{w}_{21} & \mathrm{w}_{22} & \mathrm{w}_{23} & \dots \\ \mathrm{w}_{31} & \mathrm{w}_{32} & \mathrm{w}_{33} & \dots \\ \vdots & \vdots & \vdots & \ddots \end{bmatrix}. \tag{A.31}$$

Due to the fact that $\{\mathrm{g}_i\}_{i=1}^{\infty}$ converges to zero and it satisfies (A.26) for all $p \in \mathfrak{B}_{\mathbb{P}}$ and each $\{\mathrm{w}_{ij}\}_{i=1}^{\infty}$ is an ℓ_2 sequence, each $w_i \in \mathcal{R}^{n_{\mathbb{Y}} \times n_{\mathbb{U}}}$ is unique and well defined. On the other hand, convergence of each $\{\mathrm{w}_{ij}\}_{j=1}^{\infty}$ to zero implies that

$$\lim_{i \to \infty} (w_i \diamond p) = 0 \quad \forall p \in \mathfrak{B}_{\mathbb{P}}, \tag{A.32}$$

meaning that the new coefficient sequence converges to the zero function on $\mathfrak{B}_{\mathbb{P}}$. This provides that (A.30) is a unique, convergent series expansion of the asymptotically stable LPV system in terms of $\Phi_\infty \subset \mathcal{RH}_{2-}(\mathbb{E})$. As in the considered system class any system has a convergent series expansion in terms of the pulse basis, therefore any of these systems has a convergent series expansion in terms of an arbitrary $\Phi_\infty \subset \mathcal{RH}_{2-}(\mathbb{E})$ basis. □

A.3 Proofs of Chapter 8

A.3.1 Optimal Partition

Proof (Theorem 8.1). The proof is given in an alternating minimization sense. First, fix V and define $\hat{J}_m(U) = J_m(U,V)$, for $U \in \mathcal{U}_{n_c}^{N_z}$. Since the membership values $[\mu_{ik}]_{i=1}^{n_c}$ of z_k to the fixed clusters are not depending on the memberships of other data points, the columns of U are degenerate to each other (decoupled) in the minimization of $\hat{J}_m(U)$, therefore:

$$\min_{U \in \mathcal{U}_{n_c}^{N_z}} \hat{J}_m(U) = \min_{U \in \mathcal{U}_{n_c}^{N_z}} \max_{k \in \mathbb{I}_1^{N_z}} \sum_{i=1}^{n_c} \mu_{ik}^m d_{ik} = \max_{k \in \mathbb{I}_1^{n_c}} \min_{U \in \mathcal{U}_{n_c}^{N_z}} \sum_{i=1}^{N_z} \mu_{ik}^m d_{ik}. \tag{A.33}$$

Denote $\hat{J}_m^{(k)}(U) = \sum_{i=1}^{n_c} \mu_{ik}^m d_{ik}$. To introduce the constraints $\mathcal{U}_{n_c}^{N_z}$, the Lagrangian $\Delta_k(\delta_k, U)$ of $\hat{J}_m^{(k)}(U)$ is defined for each $k \in \mathbb{I}_1^{N_z}$ as

$$\Delta_k(\delta_k, U) = \sum_{i=1}^{n_c} \mu_{ik}^m d_{ik} - \delta_k \left(\left(\sum_{i=1}^{n_c} \mu_{ik} \right) - 1 \right). \tag{A.34}$$

Assume that $\mathbb{S}(V, \mathrm{z}_k) = \emptyset$, then (δ_k, U) is a stationary point for Δ_k, only if

$$\frac{\partial}{\partial(\delta, U)} \Delta_k(\delta_k, U) = \left(0^{N_z}, 0^{n_c \times N_z}\right), \tag{A.35}$$

for all $k \in \mathbb{I}_1^{N_z}$. Setting all of these gradients equal to zero yields that

$$\frac{\partial \Delta_k(\delta_k, U)}{\partial \delta_k} = \sum_{i=1}^{n_c} \mu_{ik} - 1 = 0, \tag{A.36a}$$

$$\frac{\partial \Delta_k(\delta_k, U)}{\partial \mu_{ik}} = m\mu_{ik}^{m-1} d_{ik} - \delta_k = 0, \tag{A.36b}$$

for every $k \in \mathbb{I}_1^{N_z}$ and $i \in \mathbb{I}_1^{n_c}$. From (A.36b), it follows that

$$\mu_{ik} = \left(\frac{\delta_k}{m d_{ik}} \right)^{\frac{1}{m-1}}. \tag{A.37}$$

Moreover, by substitution of (A.37) into (A.36a):

$$0 = \sum_{l=1}^{n_c} \left(\frac{\delta_k}{m} \right)^{\frac{1}{m-1}} \left(\frac{1}{d_{lk}} \right)^{\frac{1}{m-1}} - 1 \tag{A.38a}$$

$$\left(\frac{\delta_k}{m} \right)^{\frac{1}{m-1}} = \left(\sum_{l=1}^{n_c} \left(\frac{1}{d_{lk}} \right)^{\frac{1}{m-1}} \right)^{-1}. \tag{A.38b}$$

If (A.38b) is substituted back into (A.37), then

$$\mu_{ik} = \frac{\left(\frac{1}{d_{ik}} \right)^{\frac{1}{m-1}}}{\sum_{l=1}^{n_c} \left(\frac{1}{d_{lk}} \right)^{\frac{1}{m-1}}} = \frac{1}{\sum_{l=1}^{n_c} \left(\frac{d_{ik}}{d_{lk}} \right)^{\frac{1}{m-1}}}. \tag{A.39}$$

In this way we have proved that in a local minima of $J_m(U, V)$, all μ_{ik} have to satisfy (8.11a). If $\mathbb{S}(V, \mathrm{z}_k) \neq \emptyset$, then (A.39) is singular. In this situation, choosing μ_{ik} as given by (8.11a) results in $\hat{J}_m^{(k)}(U) = 0$, because the non-zero weights are placed on zero distances, while positive distances with nonzero weights would increase

$\check{J}_m^{(k)}(U)$, contradicting minimality. As the zero-distances can have arbitrary weights, for the shake of simplicity equal weights are considered fulfilling (8.11a). Note, that such a singularity hardly occurs in reality, since machine round-off prevents its encounter.

To establish (8.11b), fix $U \in \mathcal{U}_{n_c}^{N_z}$ and define $\check{J}_m(V) = J_m(U,V)$. Minimization of $\check{J}_m(V)$ is unconstrained on $\mathbb{D}^{n_c}$ and it is decoupled for each υ_i. Therefore

$$\min_{V\in\mathbb{D}^{n_c}} \check{J}_m(V) = \min_{V\in\mathbb{D}^{n_c}} \max_{k\in\mathbb{I}_1^{N_z}} \sum_{i=1}^{n_c} \mu_{ik}^m d_{ik} = \sum_{i=1}^{n_c} \min_{V\in\mathbb{D}^{n_c}} \check{J}_m^{(i)}(V), \tag{A.40}$$

where $\check{J}_m^{(i)}(V) = \max_{k\in\mathbb{I}_1^{N_z}} \mu_{ik}^m d_{ik}$, depending only on υ_i. This means that

$$\upsilon_i = \arg\min_{V\in\mathbb{D}^{n_c}} \check{J}_m^{(i)}(V) = \arg\min_{\upsilon_i\in\mathbb{D}} \max_{k\in\mathbb{I}_1^{N_z}} \mu_{ik}^m d_{ik}. \tag{A.41}$$

Optimization (A.41) can be formulated as a matrix inequalities constrained minimization problem. Denote

$$\gamma_i = \check{J}_m^{(i)}(V) = \max_{k\in\mathbb{I}_1^{N_z}} \mu_{ik}^m d_{ik}, \tag{A.42}$$

then the solution of (A.41) can be obtained by solving

$$\begin{aligned} \text{minimize} \quad & \gamma_i \geq 0, \\ \text{subject to} \quad & \mu_{ik}^m \left| \tfrac{\mathrm{z}_k-\upsilon}{1-\mathrm{z}_k^*\upsilon} \right| \leq \gamma_i, \ \forall k \in \mathbb{I}_1^{N_z}, \\ & \upsilon \in \mathbb{D}. \end{aligned}$$

The constraints of this minimization can be written for each k as

$$\mu_{ik}^m \left| \frac{\mathrm{z}_k-\upsilon}{1-\mathrm{z}_k^*\upsilon} \right| \leq \gamma_i, \tag{A.43a}$$

$$\mu_{ik}^{2m} \left|\mathrm{z}_k-\upsilon\right|^2 \left|1-\mathrm{z}_k^*\upsilon\right|^{-2} \leq \gamma_i^2. \tag{A.43b}$$

From the Schur-complement of (A.43b) it follows that (A.43a) holds iff

$$\begin{bmatrix} |1-\mathrm{z}_k^*\upsilon|^2 & \mu_{ik}^m(\mathrm{z}_k-\upsilon) \\ \mu_{ik}^m(\mathrm{z}_k-\upsilon)^* & \gamma_i^2 \end{bmatrix} \succeq 0, \quad \forall k \in \mathbb{I}_1^{N_z}, \tag{A.44}$$

where $\upsilon \in \mathbb{D}$. Then a necessary but not sufficient condition for (U,V) being local minima of J_m is to satisfy (A.39) and (A.41). This concludes the proof. It is important to remark that $J_m(U,V)$ has more stationary points than what can be reached through alternating minimization, however all points fulfilling Th. 8.1 are stationary points of $J_m(U,V)$. □

A.3.2 Asymptotic Property of J_m

Proof (Theorem 8.2). As the cluster centers of V are assumed to be "nonsingular" with respect to Z, i.e. $d_{ik} > 0$ for all $(i,k) \in \mathbb{I}_1^{n_c} \times \mathbb{I}_1^{N_z}$, thus based on the optimality of U_m, substitution of (A.39) into (8.9) implies, that for $m > 1$:

$$J_m(U_m,V) = \max_{k\in\mathbb{I}_1^{N_z}} \sum_{i=1}^{n_c} \mu_{ik}^m d_{ik} = \max_{k\in\mathbb{I}_1^{N_z}} \sum_{i=1}^{n_c} \mu_{ik}\, \mu_{ik}^{m-1} d_{ik} =$$

$$= \max_{k\in\mathbb{I}_1^{N_z}} \sum_{i=1}^{n_c} \mu_{ik} \frac{d_{ik}}{d_{ik}\left(\sum\limits_{l=1}^{n_c}\left(\frac{1}{d_{lk}}\right)^{\frac{1}{m-1}}\right)^{m-1}} =$$

$$= \max_{k\in\mathbb{I}_1^{N_z}} \left(\sum_{l=1}^{n_c} (d_{lk})^{\frac{1}{1-m}}\right)^{1-m},$$

holds as $\sum_{i=1}^{n_c} \mu_{ik} = 1$. Now introduce

$$\bar{J}_\tau^{(k)}(V) = \left(\sum_{i=1}^{n_c} \frac{1}{n_c} (d_{ik})^\tau\right)^{1/\tau}, \tag{A.45}$$

with $\tau = \frac{1}{1-m}$. Then

$$J_m(U_m,V) = J_{\frac{\tau-1}{\tau}}(U_{\frac{\tau-1}{\tau}},V) = n_c^{1/\tau} \max_{k\in\mathbb{I}_1^{N_z}} \bar{J}_\tau^{(k)}(V). \tag{A.46}$$

Equation (A.45) is called the *Hölder* or *generalized mean* [33] of d_{ik}. Based on the properties of the generalized mean in terms of τ, the following hold:

- Case $m \to 1 \Leftrightarrow \tau \to -\infty \Rightarrow \bar{J}_\tau^{(k)}(V) \to \min_{i\in\mathbb{I}_1^{n_c}}\{d_{ik}\}$ for all $k \in \mathbb{I}_1^{N_z}$. Since $n_c^{1-m} \to 1$, the minimum over $\mathbb{I}_1^{n_c}$ is unique for each k:

$$\lim_{m\to 1} J_m(U_m,V) = \max_{k\in\mathbb{I}_1^{N_z}} \min_{i\in\mathbb{I}_1^{n_c}} \{d_{ik}\}. \tag{A.47}$$

- Case $m = 2 \Leftrightarrow \tau = -1$. Then $\bar{J}_{-1}^{(k)}(V)$ is the harmonic mean of $\{d_{ik}\}_{i=1}^{n_c}$ for each $k \in \mathbb{I}_1^{N_z}$, so

$$J_2(U_2,V) = \frac{1}{n_c} \max_{k\in\mathbb{I}_1^{N_z}} \frac{n_c}{\sum_{i=1}^{n_c} \frac{1}{d_{ik}}}. \tag{A.48}$$

- Case $m \to \infty \Leftrightarrow \tau \to 0$. Then, the asymptotic convergence of the generalized mean to the geometric mean yields: $\bar{J}_\tau^{(k)}(V) = \left(\prod_{i=1}^{n_c} d_{ik}\right)^{1/n_c} + \mathcal{O}(e^{\frac{1}{\tau}})$, which gives

$$J_m(U_m,V) = n_c^{1-m} \max_{k\in\mathbb{I}_1^{N_z}} \left(\prod_{i=1}^{n_c} d_{ik}\right)^{\frac{1}{n_c}} + \mathcal{O}(e^{-m}), \tag{A.49}$$

and since $n_c^{1-m} \to 0$, therefore

$$\lim_{m \to \infty} J_m(U_m, V) = 0. \tag{A.50}$$

□

A.3.3 h-Center Relation

Proof (Lemma 8.3). From Lemma 8.2 we obtain that for any $K_h(\mathtt{e}_h, \mathtt{r}_h)$, there exists a $K(\mathtt{e}, \mathtt{r})$, such that all $z \in K_h(\mathtt{e}_h, \mathtt{r}_h)$ satisfies both

$$\left| \frac{z - \mathtt{e}_h}{1 - z^* \mathtt{e}_h} \right| = \mathtt{r}_h, \quad |z - \mathtt{e}| = \mathtt{r}. \tag{A.51}$$

Straightforward calculus leads to

$$\mathtt{e} = \frac{1 - \mathtt{r}_h^2}{1 - \mathtt{r}_h^2 |\mathtt{e}_h|^2} \mathtt{e}_h, \quad \mathtt{r} = \frac{1 - |\mathtt{e}_h|^2}{1 - \mathtt{r}_h^2 |\mathtt{e}_h|^2} \mathtt{r}_h, \tag{A.52}$$

concluding that

$$\mathtt{e} = \varphi_h \mathtt{e}_h \quad \text{where} \quad \varphi_h = \frac{1 - \mathtt{r}_h^2}{1 - \mathtt{r}_h^2 |\mathtt{e}_h|^2} \in \mathbb{R}. \tag{A.53}$$

□

A.3.4 κ_l-Metric

Proof (Theorem 8.4). In order to prove that KM is a metric on $\mathbb{D}$, the following 3 properties have to be verified for all $\mathtt{x}, \mathtt{y}, \mathtt{z} \in \mathbb{D}$:

(i) Zero metricity: $\kappa_l(\mathtt{x}, \mathtt{x}) = 0$. By substitution:

$$\kappa_l(\mathtt{x}, \mathtt{x}) = \left| \frac{\mathtt{x} - \mathtt{x}}{1 - \mathtt{x}^* \mathtt{x}} \right| = \left| \frac{0}{1 - |\mathtt{x}|^2} \right| = 0, \quad \forall \mathtt{x} \in \mathbb{D}. \tag{A.54}$$

(ii) Symmetry: $\kappa_l(\mathtt{x}, \mathtt{y}) = \kappa_l(\mathtt{y}, \mathtt{x})$. Let $\mathtt{x}, \mathtt{y} \in \mathbb{D}$ be arbitrary, then:

$$\kappa_l(\mathtt{x}, \mathtt{y}) = \frac{|\mathtt{x} - \mathtt{y}|}{|1 - \mathtt{x}^* \mathtt{y}|} = \frac{|-(\mathtt{y} - \mathtt{x})|}{|(1 - (\mathtt{x}^* \mathtt{y}))^*|} = \kappa_l(\mathtt{y}, \mathtt{x}). \tag{A.55}$$

(iii) Triangular inequality: $\kappa_l(\mathtt{x}, \mathtt{y}) \leq \kappa_l(\mathtt{z}, \mathtt{x}) + \kappa_l(\mathtt{z}, \mathtt{y})$. Assume that $\mathtt{x}, \mathtt{y} \in \mathbb{D}$ and $\mathtt{z} \in H_{\mathtt{xy}}$ (see Fig. A.1.a). If $\mathtt{x} = \mathtt{y}$, then *(iii)* holds trivially as $\kappa_l(\mathtt{x}, \mathtt{x}) = 0 \leq 2\kappa_l(\mathtt{z}, \mathtt{x})$. Moreover, if $\mathtt{z} = \mathtt{x}$ or $\mathtt{z} = \mathtt{y}$ then *(iii)* holds with equality as $\kappa_l(\mathtt{x}, \mathtt{y}) = \kappa_l(\mathtt{x}, \mathtt{x}) + \kappa_l(\mathtt{x}, \mathtt{y}) = \kappa_l(\mathtt{x}, \mathtt{y})$ or $\kappa_l(\mathtt{x}, \mathtt{y}) = \kappa_l(\mathtt{y}, \mathtt{y}) + \kappa_l(\mathtt{x}, \mathtt{y}) = \kappa_l(\mathtt{x}, \mathtt{y})$. Now assume that $\mathtt{x}, \mathtt{y}, \mathtt{z}$ are distinct points. Define $\hat{\mathtt{x}} = \mathfrak{h}_{\mathtt{z}}(\mathtt{x})$ and $\hat{\mathtt{y}} = \mathfrak{h}_{\mathtt{z}}(\mathtt{y})$, where $\mathfrak{h}_{\mathtt{z}}$ is the h-inversion that maps $\mathtt{z} \to 0$. Then, $\mathfrak{h}_{\mathtt{z}}(H_{\mathtt{xy}}) = H_{\hat{\mathtt{x}}\hat{\mathtt{y}}}$ is a segment of an Euclidian line, a diameter of $\mathbb{J}$ (dark grey line in

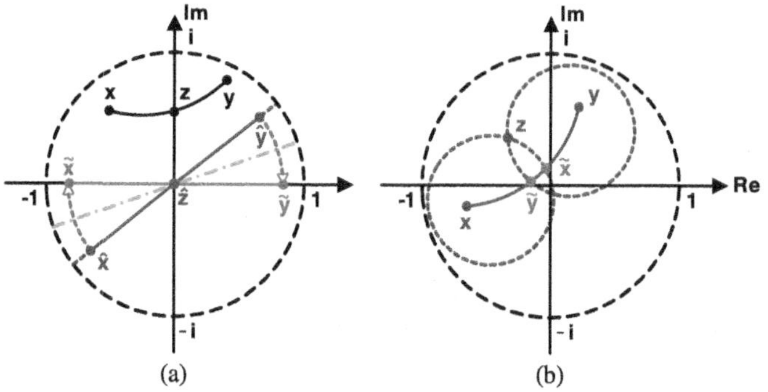

Fig. A.1 Inversion of h-lines and points proving the Triangular equality.

Fig. A.1.a). Let $\tilde{H}$ be the diameter line (also an h-line) bisecting the angle between $H_{0\hat{x}}$ and the real axis. Then $\mathfrak{h}_{\tilde{H}}(\hat{x}) = \tilde{x}$ and $\mathfrak{h}_{\tilde{H}}(\hat{y}) = \tilde{y}$ are points of the real axis and $\mathfrak{h}_{\tilde{H}}(H_{\hat{x}\hat{y}}) = H_{\tilde{x}\tilde{y}} \subset (-1,1)$ with $0 \in H_{\tilde{x}\tilde{y}}$ (light grey line in Fig. A.1.a). As $\mathfrak{h}_{\tilde{H}} \circ \mathfrak{h}_z \in \mathcal{D}$, then by Corollary 8.1, *(iii)* can be written as $\kappa_1(\tilde{x},\tilde{y}) \leq \kappa_1(0,\tilde{x}) + \kappa_1(0,\tilde{y})$. Assuming that $\tilde{x}$ and $\tilde{y}$ are ordered as $\tilde{x} \in (-1,0)$ and $\tilde{y} \in (0,1)$, then

$$\begin{aligned}\frac{|\tilde{x}-\tilde{y}|}{|1-\tilde{x}^*\tilde{y}|} &\leq -\tilde{x}+\tilde{y},\\ \frac{\tilde{y}-\tilde{x}}{1+|\tilde{x}|\,\tilde{y}} &\leq \tilde{y}-\tilde{x},\\ 1 &\leq 1+|\tilde{x}|\,\tilde{y},\\ 0 &\leq |\tilde{x}|\,\tilde{y}.\end{aligned}$$

In the second part of the proof assume that $z \notin H_{xy}$. Define h-circles $K_{h1}(x, r_{h1})$ and $K_{h2}(y, r_{h2})$ (dark grey circles in Fig. A.1.b) such that $z \in K_{h1}, K_{h2}$. Note that $\kappa_1(z,x) = r_{h1}$ and $\kappa_1(z,y) = r_{h2}$. If $\max(r_{h1}, r_{h2}) \leq \kappa_1(x,y)$, then $K_{h1} \cap H_{xy} = \tilde{x}$ and $K_{h2} \cap H_{xy} = \tilde{y}$ exist (light grey points in Fig. A.1.b). Since $\tilde{x}, z \in K_{h1}$, therefore $\kappa_1(\tilde{x},x) = \kappa_1(z,x)$ and since $\tilde{y}, z \in K_{h2}$, therefore $\kappa_1(\tilde{y},y) = \kappa_1(z,y)$. Moreover, $\kappa_1(\tilde{x},x) > \kappa_1(\tilde{y},x)$ and $\kappa_1(\tilde{y},y) > \kappa_1(\tilde{x},y)$, otherwise $K_{h1} \cap K_{h2} = \emptyset$. Then, $\kappa_1(x,y) \leq \kappa_1(\tilde{x},x) + \kappa_1(\tilde{x},y)$, which gives $\kappa_1(x,y) \leq \kappa_1(z,x) + \kappa_1(\tilde{y},y)$ implying *(iii)*. If $\max(r_{h1}, r_{h2}) > \kappa_1(x,y)$, then either $\tilde{x} \notin H_{xy}$ or $\tilde{y} \notin H_{xy}$. $\tilde{x} \notin H_{xy}$ means that the circle K_{h1} is containing the circle $K_h(x, \kappa_1(x,y))$, so $\kappa_1(x,y) < r_{h1} = \kappa_1(z,x)$. This implies that *(iii)* holds. The case of $\tilde{y} \notin H_{xy}$ analogously follows. □

A.3.5 h-Segment Worst-Case Distance

Proof (Theorem 8.5). Define $\hat{x} = \mathfrak{h}_v(x)$, $\hat{y} = \mathfrak{h}_v(y)$ and $H_{\hat{x}\hat{y}} = \mathfrak{h}_v(H_{xy})$. As $\mathfrak{h}_v(v) = 0$, $\kappa_1(0,\hat{z}) = |\hat{z}|$ for all $\hat{z} \in H_{\hat{x}\hat{y}}$. Moreover, $\mathfrak{h}_v$ keeps distances in the Kolmogorov sense (Corollary 8.1), therefore

$$\max_{z \in H_{\mathrm{xy}}} \kappa_1(\mathrm{v}, z) = \max_{\hat{z} \in H_{\hat{\mathrm{x}}\hat{\mathrm{y}}}} |\hat{z}| . \tag{A.56}$$

In accordance to this, the most distant point of $H_{\hat{\mathrm{x}}\hat{\mathrm{y}}}$ in the Euclidian sense from the origin represents the furthest point of H_{xy} in the Kolmogorov sense from v. Based on the Euclidian geometry:

$$\max_{\hat{z} \in H_{\hat{\mathrm{x}}\hat{\mathrm{y}}}} |\hat{z}| = \max(|\hat{\mathrm{x}}|, |\hat{\mathrm{y}}|) = \max(\kappa_1(\mathrm{x}, \mathrm{v}), \kappa_1(\mathrm{y}, \mathrm{v})) . \tag{A.57}$$

□

A.3.6 h-Disc Worst-Case Distance

Proof (Theorem 8.6). Define $\hat{\mathrm{e}}_{\mathrm{h}} = \mathfrak{h}_{\mathrm{v}}(\mathrm{e}_{\mathrm{h}})$, then due to Lemma 8.6, $\mathfrak{h}_{\mathrm{v}}(D_{\mathrm{h}}(\mathrm{e}_{\mathrm{h}}, \mathrm{r}_{\mathrm{h}})) = D_{\mathrm{h}}(\hat{\mathrm{e}}_{\mathrm{h}}, \mathrm{r}_{\mathrm{h}})$. As $\mathfrak{h}_{\mathrm{v}}(\mathrm{v}) = 0$, thus $\kappa_1(0, \hat{z}) = |\hat{z}|$ for every $\hat{z} \in D_{\mathrm{h}}(\hat{\mathrm{e}}_{\mathrm{h}}, \mathrm{r}_{\mathrm{h}})$. Due to Corollary 8.1:

$$\breve{z} = \arg \max_{z \in D_{\mathrm{h}}(\mathrm{e}_{\mathrm{h}}, \mathrm{r}_{\mathrm{h}})} \kappa_1(\mathrm{v}, z) = \mathfrak{h}_{\mathrm{v}}(\arg \max_{\hat{z} \in D_{\mathrm{h}}(\hat{\mathrm{e}}_{\mathrm{h}}, \mathrm{r}_{\mathrm{h}})} |\hat{z}|). \tag{A.58}$$

Assume that $\mathrm{v} = \mathrm{e}_{\mathrm{h}}$, then $\hat{\mathrm{e}}_{\mathrm{h}} = 0$ and $\mathfrak{h}_{\mathrm{v}}(D_{\mathrm{h}}(\mathrm{e}_{\mathrm{h}}, \mathrm{r}_{\mathrm{h}}))$ is equal to the Euclidean disc $D(0, \mathrm{r}_{\mathrm{h}})$. Then

$$\arg \max_{\hat{z} \in D(0, \mathrm{r}_{\mathrm{h}})} |\hat{z}| = K(0, \mathrm{r}_{\mathrm{h}}), \tag{A.59}$$

and $\breve{z} = \mathfrak{h}_{\mathrm{v}}(K(0, \mathrm{r}_{\mathrm{h}})) = K_{\mathrm{h}}(\mathrm{e}_{\mathrm{h}}, \mathrm{r}_{\mathrm{h}})$, i.e. all the perimeter points of $D_{\mathrm{h}}(\mathrm{e}_{\mathrm{h}}, \mathrm{r}_{\mathrm{h}})$, with

$$\max_{z \in D_{\mathrm{h}}(\mathrm{e}_{\mathrm{h}}, \mathrm{r}_{\mathrm{h}})} \kappa_1(\mathrm{v}, z) = \mathrm{r}_{\mathrm{h}}. \tag{A.60}$$

If $\mathrm{v} \neq \mathrm{e}_{\mathrm{h}}$, then due to Lemma 8.3, $\hat{\mathrm{e}}_{\mathrm{h}}$ and the Euclidian center of $D_{\mathrm{h}}(\hat{\mathrm{e}}_{\mathrm{h}}, \mathrm{r}_{\mathrm{h}})$ are on the same Euclidian line from the origin. Therefore, based on the Euclidian geometry:

$$\breve{z} = \mathfrak{h}_{\mathrm{v}}(\arg \max_{\hat{z} \in \{\hat{\mathrm{z}}_1, \hat{\mathrm{z}}_2\}} |\hat{z}|) \tag{A.61}$$

where $\{\hat{\mathrm{z}}_1, \hat{\mathrm{z}}_2\} = E(0, \hat{\mathrm{r}}) \cap K_{\mathrm{h}}(\hat{\mathrm{e}}_{\mathrm{h}}, \mathrm{r}_{\mathrm{h}})$ with $E(0, \hat{\mathrm{r}})$, the Euclidian line connecting the origin with $\hat{\mathrm{e}}_{\mathrm{h}}$. Furthermore,

$$\max_{z \in D_{\mathrm{h}}(\mathrm{e}_{\mathrm{h}}, \mathrm{r}_{\mathrm{h}})} \kappa_1(\mathrm{v}, z) = \max(|\hat{\mathrm{z}}_1|, |\hat{\mathrm{z}}_2|) . \tag{A.62}$$

As $\mathfrak{h}_{\mathrm{v}}(E(0, \hat{\mathrm{r}})) = H_{\mathrm{v}\mathrm{e}_{\mathrm{h}}}$ the h-line connecting v with e_{h}, therefore $H_{\mathrm{v}\mathrm{e}_{\mathrm{h}}} \cap K_{\mathrm{h}}(\mathrm{e}_{\mathrm{h}}, \mathrm{r}_{\mathrm{h}}) = \{\mathfrak{h}_{\mathrm{v}}(\hat{\mathrm{z}}_1), \mathfrak{h}_{\mathrm{v}}(\hat{\mathrm{z}}_2)\}$. □

A.3.7 Convexity

Proof (Theorem 8.7). Any conformal mapping in $\mathbb{D}$ preserves convexity of subsets of $\mathbb{D}$ [26]. As any $\mathfrak{h}_{\mathrm{e}}$, with $\mathrm{e} \in \mathbb{D}$, is a conformal mapping, therefore $\mathfrak{h}_{\mathrm{e}}$ preserves convexity of sets in $\mathbb{D}$ (convexity preservation in the KM sense). Let an arbitrary h-segment $H_{\mathtt{xy}}$ be given. Then $\mathtt{e} = H_{\mathtt{xy}} \cap H^{\perp(\mathtt{xy})}$ is the midpoint of $H_{\mathtt{xy}}$ and $H_{\hat{\mathtt{x}}\hat{\mathtt{y}}} = \mathfrak{h}_{\mathtt{e}}(H_{\mathtt{xy}})$ is the part of a diameter of $\mathbb{D}$. Then based on the convexity of Euclidian lines, for any $\mathtt{z}_1, \mathtt{z}_2 \in H_{\hat{\mathtt{x}}\hat{\mathtt{y}}}$ and $\mathtt{r} \in [0,1]$ it holds that $\mathtt{r}\mathtt{z}_1 + (1-\mathtt{r})\mathtt{z}_2 \in H_{\hat{\mathtt{x}}\hat{\mathtt{y}}}$. Therefore convexity holds for $H_{\mathtt{xy}}$ in the KM sense. Furthermore, for any $D_{\mathrm{h}}(\mathtt{e}_{\mathrm{h}}, \mathtt{r}_{\mathrm{h}})$, there exists an equivalent Euclidian disc $D(\mathtt{e}, \mathtt{r})$. Based on the convexity of Euclidian discs, the convexity of $D_{\mathrm{h}}(\mathtt{e}_{\mathrm{h}}, \mathtt{r}_{\mathrm{h}})$ is straightforward. □

A.3.8 *Optimal Robust Partition*

Proof (Theorem 8.8). Similar to the proof of Th. 8.1 an alternating minimization approach is utilized. First, fix V and define $\hat{J}_m(U) = J_m(U,V)$, for $U \in \mathcal{U}_{n_c}^{N_z}$. Since the minimization of $\hat{J}_m(U)$ does not depend on the actual computation of $\{d_{ik}\}$, therefore the first part of the proof deriving the condition (8.30a) is the same as given for the proof of Th. 8.1 with the exception, that in this case no singularity can occur as $d_{ik} > 0$ for all $(i,j) \in \mathbb{I}_1^{n_c} \times \mathbb{I}_1^{N_z}$. See Sect. 8.3 for details. To establish (8.30b), fix $U \in \mathcal{U}_{n_c}^{N_z}$ and define $\check{J}_m(V) = J_m(U,V)$. Minimization of $\check{J}_m(V)$ is decoupled for each υ_i, therefore

$$\min_{V \in \mathbb{D}^c} \check{J}_m(V) = \min_{V \in \mathbb{D}^{n_c}} \max_{k \in \mathbb{I}_1^{N_z}} \sum_{i=1}^{n_c} \mu_{ik}^m d_{ik} = \sum_{i=1}^{n_c} \min_{V \in \mathbb{D}^{n_c}} \check{J}_m^{(i)}(V), \tag{A.63}$$

where $\check{J}_m^{(i)}(V) = \max_{k \in \mathbb{I}_1^{N_z}} \mu_{ik}^m d_{ik}$, depending only on υ_i. As the elements of V are degenerate to each other (decoupled) in the minimization of $\check{J}_m^{(i)}(V)$, the minimizer of $\check{J}_m^{(i)}$ is obtained as

$$\upsilon_i = \arg \min_{\upsilon_i \in \mathbb{D}} \max_{k \in \mathbb{I}_1^{N_z}} \max_{z_k \in \mathcal{Z}_k} \mu_{ik}^m \kappa_1(\upsilon_i, z_k). \tag{A.64}$$

Optimization (A.64) can be formulated as a QCs' constrained minimization problem. Denote

$$\gamma_i = \check{J}_m^{(i)}(V) = \max_{k \in \mathbb{I}_1^{N_z}} \mu_{ik}^m d_{ik}, \tag{A.65}$$

then the solution of (A.64) can be obtained by solving

$$\begin{aligned} \text{minimize} \quad & \gamma_i \geq 0, \\ \text{subject to} \quad & \mu_{ik}^m \left| \tfrac{z - \mathtt{v}}{1 - z^* \mathtt{v}} \right| \leq \gamma_i, \ \forall k \in \mathbb{I}_1^{n_c}, \ \forall z \in \mathcal{Z}_k \\ & \mathtt{v} \in \mathbb{D}. \end{aligned}$$

Moreover, the constraints of the minimization can be rewritten for each k as (A.43a–b). The same holds for the h-discs induced constraints. From the Schur-complement it follows that (A.43a) holds iff (8.29) is fulfilled and the h-disc coverage induced constraints hold iff (8.35) is fulfilled. Then a necessary but not sufficient condition for (U,V) being a local minimum of J_m is to satisfy (8.30a) and (8.30b). □

A.4 Proofs of Chapter 9

A.4.1 *Representation of Dynamic Dependence*

Proof (Property 9.1). Let $\mathcal{S}$ be an asymptotically stable SISO LPV system and $\Phi_{n_g}^{\infty} \subset \mathcal{RH}_{2-}(\mathbb{E})$ be a Hambo basis. Consider an asymptotically stable Wiener-feedback model $\mathfrak{M}_{\mathrm{WF}}(\Phi_{n_g}^{n_e}, \theta, \mathfrak{B}_{\mathbb{P}})$ of $\mathcal{S}$. The deterministic part $F(\mathrm{q}, \theta)$ of $\mathfrak{M}_{\mathrm{WF}}(\Phi_{n_g}^{n_e}, \theta, \mathfrak{B}_{\mathbb{P}})$ has the representation:

$$\breve{y}_{ij} = \phi_j(\mathrm{q})G_{\mathrm{b}}^i(\mathrm{q})u - \phi_j(\mathrm{q})G_{\mathrm{b}}^i(\mathrm{q})v_{ij}(p)\breve{y}_{ij}, \tag{A.66a}$$

$$y = w_{00}(p)u + \sum_{i=0}^{n_e}\sum_{j=1}^{n_g} w_{ij}(p)\breve{y}_{ij}, \tag{A.66b}$$

where $w_{ij}, v_{ij} \in \mathcal{R}|_{n_{\mathbb{P}}}$ are meromorphic coefficients with static dependence. As $\mathfrak{M}_{\mathrm{WF}}(\Phi_{n_g}^{n_e}, \theta, \mathfrak{B}_{\mathbb{P}})$ is asymptotically stable, (A.66a–b) has a convergent series expansion in terms of $\Phi_{n_g}^{\infty}$ (see Th. 5.2). By substituting (A.66a) into (A.66b) recursively, this series expansion can be derived as follows:

$$\begin{aligned} y = {} & w_{00}(p)u + \sum_{i=0}^{n_e}\sum_{j=1}^{n_g} w_{ij}(p)\phi_j(\mathrm{q})G_{\mathrm{b}}^i(\mathrm{q})u \\ & - \sum_{i=0}^{n_e}\sum_{j=1}^{n_g} w_{ij}(p)\phi_j(\mathrm{q})G_{\mathrm{b}}^i(\mathrm{q})v_{ij}(p)\phi_j(\mathrm{q})G_{\mathrm{b}}^i(\mathrm{q})u + \ldots \end{aligned} \tag{A.67}$$

Consider $\phi_j(\mathrm{q})G_{\mathrm{b}}^i(\mathrm{q})v_{ij}(p)$. In this term, $\phi_j(\mathrm{q})G_{\mathrm{b}}^i(\mathrm{q})$ is a stable LTI filter, hence there exist polynomials $R_1, R_2 \in \mathbb{R}[\xi]$ such that

$$\phi_j(\mathrm{q})G_{\mathrm{b}}^i(\mathrm{q}) = \frac{R_1(q^{-1})}{R_2(q^{-1})}, \tag{A.68}$$

where the polynomial $q^n R_2(q^{-1})$ in q is monic for $n = \mathrm{Deg}(R_2(q))$. Then due the stability of (A.68), there exists a $\hat{v}_{ij} \in \mathcal{R}$ such that

$$\begin{aligned} \frac{R_1(\mathrm{q}^{-1})}{R_2(\mathrm{q}^{-1})} v_{ij}(p) &= \sum_{l=0}^{\infty} \left(1 - R_2(\mathrm{q}^{-1})\right)^l R_1(\mathrm{q}^{-1}) v_{ij}(p) \\ &= \sum_{l=0}^{\infty} (\hat{v}_{ij} \diamond p)\left(1 - R_2(\mathrm{q}^{-1})\right)^l R_1(\mathrm{q}^{-1}) = (\hat{v}_{ij} \diamond p)\frac{R_1(\mathrm{q}^{-1})}{R_2(\mathrm{q}^{-1})}. \end{aligned}$$

If $v_{ij} \in \mathbb{R}$, i.e. v_{ij} is a constant, then $\hat{v}_{ij} = v_{ij}$, otherwise $\hat{v}_{ij}$ has dynamic dependence on p due to the shift operation (see Def. 3.16). Then in (A.67), the terms in the second summation reads as

$$w_{ij}(p)\phi_j(\mathrm{q})G_{\mathrm{b}}^i(\mathrm{q})v_{ij}(p)\phi_j(\mathrm{q})G_{\mathrm{b}}^i(\mathrm{q})u = w_{ij}(p)(\hat{v}_{ij} \diamond p)\phi_j^2(\mathrm{q})G_{\mathrm{b}}^{2i}(\mathrm{q})u. \tag{A.69}$$

For this expression it holds true that $\phi_j^2(\mathrm{q})G_\mathrm{b}^{2i}$ can be written a finite sum of $\{\phi_\tau(\mathrm{q})G_\mathrm{b}^l\}_{\tau=1,\dots,n_\mathrm{g}}^{l=0,\dots,2i+1}$. Then by considering all summation terms in (A.67) in analogy with the previous manipulations, (A.67) reads as

$$y = w_{00}(p)u + \sum_{i=0}^{\infty}\sum_{j=1}^{n_\mathrm{g}} (\hat{w}_{ij} \diamond p)\phi_j(\mathrm{q})G_\mathrm{b}^i(\mathrm{q})u, \tag{A.70}$$

where $\hat{w}_{ij} \in \mathcal{R}$. It is obvious that in (A.70) $\hat{w}_{ij} = w_{ij}$ for $i \leq n_\mathrm{e}$ and $\hat{w}_{ij}$ for $i > n_\mathrm{e}$ has static dependence on p iff each v_{ij} is constant in $\mathfrak{M}_{\mathrm{WF}}(\Phi_{n_\mathrm{g}}^{n_\mathrm{e}}, \theta, \mathfrak{B}_\mathbb{P})$. This concludes the proof in the WF-LPV OBF case. With respect to HF-LPV OBF models, the proof follows similarly. □

References

1. Abbas, H., Tóth, R., Werner, H.: State-space realization of LPV input-output models: practical methods for the user. Accepted for the American Control Conf. Baltimore, Maryland, USA (2010)
2. Anderson, B.D.O.: Internal and external stability of linear time-varying systems. SIAM Journal on Control and Optimization 20, 408–413 (1982)
3. Angelis, G.Z.: System analysis, modeling and control with polytopic linear models. Ph.D. thesis, Eindhoven University of Technology (2001)
4. Apkarian, P.: On the discretization of LMI-synthesized linear parameter-varying controllers. Automatica 33(4), 655–661 (1997)
5. Apkarian, P., Adams, R.J.: Advanced gain-scheduling techniques for uncertain systems. IEEE Trans. on Control Systems Technology 9(1), 21–32 (1988)
6. Apkarian, P., Gahinet, P.: A convex characterization of gain-scheduled $\mathcal{H}_\infty$ controllers. IEEE Trans. on Automatic Control 40(5), 853–864 (1995)
7. Åström, K.J., Wittenmark, B.: Adaptive Control. Addison-Wesley, Reading (1989)
8. Atkinson, K.E.: An Introduction to Numerical Analysis. John Wiley and Sons, Chichester (1989)
9. Backer, E., Jain, A.K.: A clustering performance measure based on fuzzy set decomposition. IEEE Trans. on Pattern Analysis and Machine Intelligence 3, 66–75 (1981)
10. Balas, G., Bokor, J., Szabó, Z.: Invariant subspaces for LPV systems and their applications. IEEE Trans. on Automatic Control 48(11), 2065–2069 (2003)
11. Bamieh, B., Giarré, L.: Identification of linear parameter varying models. In: Proc. of the 38th IEEE Conf. on Decision and Control, Phoenix, Arizona, USA, pp. 1505–1510 (1999)
12. Bamieh, B., Giarré, L.: Identification for a general class of LPV models. In: Proc. of the 11th IFAC Symposium on System Identification, Santa Barbara, California, USA (2000)
13. Bamieh, B., Giarré, L.: Identification of linear parameter varying models. Int. Journal of Robust and Nonlinear Control 12, 841–853 (2002)
14. Banks, S.P., Al-Jurani, S.K.: Pseudo-linear systems, Lie's algebra, and stability. IMA Journal of Mathematical Control and Information 13, 385–401 (1996)
15. Baslamisli, S., Köse, I.E., Anlaş, G.: Gain-scheduled integrated active steering and differential control for vehicle handling improvement. Vehicle System Dynamics 47(1), 99–119 (2009)
16. Beauzamy, B.: Introduction to operator theory and invariant subspaces. North-Holland, Amsterdam (1988)

17. Becker, G., Packard, A., Philbrick, D., Balas, G.: Control of parametrically-dependent linear systems: A single quadratic Lyapunov function approach. In: Proc. of the American Control Conf., San Francisco, California, USA, pp. 2795–2799 (1993)
18. Belforte, G., Dabbene, F., Gay, P.: LPV approximation of distributed parameter systems in environmental modeling. Environmental Modeling & Software 20, 1063–1070 (2005)
19. Belforte, G., Gay, P.: Optimal worst case estimation for LPV-FIR models with bounded errors. Systems & Control Letters 53, 259–268 (2004)
20. Bemporad, A., Morari, M.: Control of systems integrating logics, dynamics, and constraints. Automatica 35(5), 407–427 (1999)
21. Besselmann, T., Löfberg, J., Morari, M.: Explicit model predictive control for linear parameter-varying systems. In: Proc. of the 47th IEEE Conf. on Decision and Control, Cancun, Mexico, pp. 3848–3853 (2008)
22. Bezdek, J.C.: Pattern Recognition with Fuzzy Objective Function Algorithms. Plenum Press, New York (1981)
23. Bianchi, F.D., De Battista, H., Mantz, R.J.: Wind Turbine Control Systems; Principles, modeling and gain scheduling design. Springer, Heidelberg (2007)
24. Bianchi, F.D., Sánchez-Peña, R.S.: Robust identification/invalidation in an LPV framework. International Journal of Robust and Nonlinear Control (2009)
25. Billings, S., Fakhouri, S.: Identification of systems containing linear dynamic and static nonlinear elements. Automatica 18(1), 15–26 (1982)
26. Blair, D.E.: Inversion theory and conformal mapping. Student Mathematical Library, vol. 9. American Mathematical Society (2000)
27. Bolea, Y., Blesa, J., Puig, V.: LPV modeling and identification of an open canal for control. In: Proc. of the European Control Conf., Kos, Greece, pp. 3584–3591 (2007)
28. Boonto, S., Werner, H.: Closed-loop system identification of LPV input-output models - application to an arm-driven pendulum. In: Proc. of the 47th IEEE Conf. on Decision and Control, Cancun, Mexico, pp. 2606–2611 (2008)
29. Bos, R., Bombois, X., Van den Hof, P.M.J.: Accelerating simulations of first principle models of complex industrial systems using quasi linear parameter varying models. In: Proc. of the 44th IEEE Conf. on Decision and Control, Seville, Spain, pp. 4146–4151 (2005)
30. Boyd, S., Chua, L.O.: Fading memory and the problem of approximating nonlinear operators with Volterra series. IEEE Trans. on Circuits and Systems 32(11), 1150–1161 (1985)
31. Boyd, S., Ghaoui, L.E., Feron, E., Balakrishnan, V.: Linear Matrix Inequalities in System and Control Theory. SIAM Press, Philadelphia (1994)
32. Brannan, D.A., Esplen, M.F., Gray, J.J.: Geometry. Cambridge University Press, Cambridge (1999)
33. Bullen, P.S.: Handbook of Means and Their Inequalities. Kluwer Academic Publishers, Dordrecht (2003)
34. Butcher, M., Karimi, A., Longchamp, R.: On the consistency of certain identification methods for linear parameter varying systems. In: Proc. of the 17th IFAC World Congress, Seoul, Korea, pp. 4018–4023 (2008)
35. Callier, F.M., Desoer, C.A.: Linear system theory. Springer, Heidelberg (1991)
36. Carter, L.H., Shamma, J.S.: Gain-scheduled bank-to-turn autopilot design using linear parameter varying transformations. Int. Journal of Control 19(5), 1056–1063 (1996)
37. Casella, F., Donida, F., Lovera, M.: Automatic generation of LFTs from object-oriented non-linear models with uncertain parameters. In: Proc. of the 6th Vienna Conference on Mathematical Modeling, Vienna, Austria, pp. 1359–1367 (2009)

38. Cerone, V., Regruto, D.: Set-membership identification of LPV models with uncertain measurements of the time-varying parameter. In: Proc. of the 47th IEEE Conf. on Decision and Control, Cancun, Mexico, pp. 4491–4496 (2008)
39. Chiuso, A.: The role of vector autoregressive modeling in predictor-based subspace identification. Automatica 43, 1034–1048 (2007)
40. Chiuso, A., Picci, G.: Consistency analysis of some closed-loop subspace identification methods. Automatica 41, 377–391 (2005)
41. Cohn, P.M.: Free rings and their relations. Academic Press, London (1971)
42. Cook, M.V.: Flight dynamics principles. Springer, Heidelberg (1997)
43. Davies, D.I., Bouldin, D.W.: A cluster separation measure. IEEE Trans. on Pattern Analysis and Machine Intelligence 1(2), 95–104 (1979)
44. DeCarlo, R.A.: Linear systems: a state variable approach with numerical implementation. Prentice-Hall, Englewood Cliffs (1989)
45. Dettori, M., Scherer, C.W.: LPV design for a CD player: An experimental evaluation of performance. In: Proc. of the 40th IEEE Conf. on Decision and Control, Orlando, Florida, USA, pp. 4711–4716 (2001)
46. Donida, F., Romani, C., Casella, F., Lovera, M.: Towards integrated modeling and parameter estimation: an LFT-Modelica approach. In: Proc. of the 15th IFAC Symposium on System Identification, Saint-Malo, France, pp. 1286–1291 (2009)
47. dos Santos, P.L., Ramos, J.A., de Carvalho, J.L.M.: Identification of linear parameter varying systems using an iterative deterministic-stochastic subspace approach. In: Proc. of the European Control Conf., Kos, Greece, pp. 4867–4873 (2007)
48. dos Santos, P.L., Ramos, J.A., de Carvalho, J.L.M.: Identification of LPV systems using successive approximations. In: Proc. of the 47th IEEE Conf. on Decision and Control, Cancun, Mexico, pp. 4509–4515 (2008)
49. Douma, S.G., Van den Hof, P.M.J.: Relations between uncertainty structures in identification for robust control. Automatica 41(3), 439–457 (2005)
50. Douwe, K., Van den Hof, P.M.J.: Frequency domain identification with generalized orthonormal basis functions. IEEE Trans. on Automatic Control 43(5), 656–669 (1998)
51. Evans, L.C.: Partial differential equations. American Mathematical Society, Providence (1998)
52. Felici, F., van Wingerden, J.W., Verhaegen, M.: Subspace identification of MIMO LPV systems using a periodic scheduling sequence. Automatica 43(10), 1684–1697 (2006)
53. Felici, F., van Wingerden, J.W., Verhaegen, M.: Dedicated periodic scheduling sequences for LPV system identification. In: Proc. of the European Control Conf., Kos, Greece, pp. 4896–4902 (2007)
54. Gáspár, P., Szabó, Z., Bokor, J.: A grey-box continuous-time parameter identification for LPV models with vehicle dynamics applications. In: Proc. of the 13th Mediterranean Conf. on Control and Automation, Limassol, Cyprus, pp. 393–397 (2005)
55. Gáspár, P., Szabó, Z., Bokor, J.: A grey-box identification of an LPV vehicle model for observer-based side-slip angle estimation. In: Proc. of the American Control Conf., New York City, USA, pp. 2961–2965 (2007)
56. Gevers, M., Bazanella, A., Mišković, L.: Informative data: how to get just sufficiently rich? In: Proc. of the 47th IEEE Conf. on Decision and Control, Cancun, Mexico, pp. 4522–4527 (2008)
57. Giarré, L., Bauso, D., Falugi, P., Bamieh, B.: LPV model identification for gain scheduling control: An application to rotating stall and surge control problem. Control Engineering Practice 14, 351–361 (2006)
58. Gil, P., Henriques, J., Dourado, A., Duarte-Ramos, H.: On state-space neural networks for systems identification: Stability and complexity. In: Proc. of the IEEE Conf. on Cybernetics and Intelligent Systems, pp. 1–5 (2006)

59. Ginnakis, G.B., Serpedin, E.: A bibliography on nonlinear system identification. Signal Processing 81(3), 533–580 (2001)
60. Glover, K.: All optimal Hankel norm approximations of linear multivariable systems and their L_∞-error bound. Int. Journal of Control 39(6), 1115–1193 (1984)
61. Gohberg, I., Kaashoek, M.A., Lerer, L.: Minimality and realization of discrete time-varying systems. In: Gohberg, I. (ed.) Time-variant systems and interpolation, pp. 261–296. Birkhäuser, Basel (1992)
62. Golub, G.H., Pereyra, V.: The differentiation of pseudo-inverses and nonlinear least squares problems whose variables are separate. SIAM Journal on Numerical Analysis 10(2), 413–432 (1973)
63. Gómez, J., Baeyens, E.: Identification of block-oriented nonlinear systems using orthonormal bases. Journal of Process Control 14, 685–697 (2004)
64. Gooderal, K.R., Warfield, R.B.: An introduction to noncommutative Noetherian rings. Cambridge University Press, Cambridge (1989)
65. Guidorzi, R., Diversi, R.: Minimal representations of MIMO time-varying systems and realization of cyclostationary models. Automatica 39(11), 1903–1914 (2003)
66. Guillaume, P., Schoukens, J., Pintelon, R.: Sensitivity of roots to errors in the coefficient of polynomials obtained by frequency-domain estimation methods. IEEE Trans. on Instrumentation and Measurement 38(6), 1050–1056 (1989)
67. Hakvoort, R.G., Van den Hof, P.M.J.: Identification of probabilistic system uncertainty regions by explicit evaluation of bias and variance errors. IEEE Trans. on Automatic Control 42(11), 1516–1528 (1997)
68. Halanay, A., Ionescu, V.: Time-varying discrete linear systems: input-output operators, Riccati equations, disturbance attenuation. Birkhäuser Verlag, Basel (1994)
69. Hallouzi, R., Verhaegen, M., Kanev, S.: Model weight estimation for FDI using convex fault models. In: Proc. of the 6th IFAC Symposium on Fault Detection, Supervision and Safety of Technical Processes, Beijing, China, pp. 847–852 (2006)
70. Hanselmann, H.: Implementation of digital controllers - A survey. Automatica 23(1), 7–32 (1987)
71. Hecker, S., Varga, A.: Symbolic manipulation techniques for low order LFR-based parametric uncertainty modeling. International Journal of Control 79(11), 1485–1494 (2006)
72. Heuberger, P.S.C.: On approximate system identification with system based orthonormal functions. Ph.D. thesis, Delft University of Technology (1990)
73. Heuberger, P.S.C., Van den Hof, P.M.J., Wahlberg, B.: Modeling and Identification with Rational Orthonormal Basis Functions. Springer, Heidelberg (2005)
74. Heuberger, P.S.C., Van den Hof, P.M.J., Bosgra, O.H.: A generalized orthonormal basis for linear dynamical systems. IEEE Trans. on Automatic Control 40(3), 451–465 (1995)
75. Hsu, K., Vincent, T.L., Poolla, K.: Nonparametric methods for the identification of linear parameter varying systems. In: Proc. of the Int. Symposium on Computer-Aided Control System Design, San Antonio, Texas, USA, pp. 846–851 (2008)
76. Hunt, K.J., Haas, R., Murray-Smith, R.: Design analysis of gain-scheduled local controller networks. Int. Journal of Control 66(5), 619–651 (1997)
77. Hyde, R.A., Glover, K.: The application of scheduled $\mathcal{H}_\infty$ controllers to a VSTOL aircraft. IEEE Trans. on Automatic Control 38, 1021–1039 (1993)
78. Ilchmann, A., Mehrmann, V.: A behavioral approach to time-varying linear systems. Part 1: General theory. SIAM Journal of Control Optimization 44(5), 1725–1747 (2005)
79. Jain, A.K., Dubes, R.C.: Algorithms for clustering data. Prentice-Hall, Englewood Cliffs (1988)

80. Kalman, R.E.: Mathematical description of linear dynamical systems. SIAM Journal on Control 1, 152–192 (1963)
81. Kamen, E.W., Khargonekar, P.P., Poolla, K.R.: A transfer-function approach to linear time-varying discrete-time systems. SIAM Journal of Control and Optimization 23(4), 550–565 (1985)
82. Kanev, S.: Polytopic model set generation for fault detection and diagnosis. Tech. Rep. 06-016, Delft University of Technology, Delft Center for Systems and Control (2006)
83. Kaymak, U., Setnes, M.: Fuzzy clustering with volume prototypes and adaptive cluster merging. IEEE Trans. on Fuzzy Systems 10(6), 705–711 (2002)
84. Khalate, A.A., Bombois, X., Tóth, R., Babuška, R.: Optimal experimental design for LPV identification using a local approach. In: Proc. of the 15th IFAC Symposium on System Identification, Saint-Malo, France, pp. 162–167 (2009)
85. Kiriakidis, K.: Nonlinear modeling by interpolation between linear dynamics and its application in control. Journal of Dynamic Systems, Measurement, and Control 129(6), 813–824 (2007)
86. Korba, P.: A gain-scheduling approach to model-based fuzzy control. Ph.D. thesis, Universität Dusiburg (2000)
87. Korba, P., Babuška, R., Verbruggen, H.B., Frank, P.M.: Fuzzy gain scheduling: controller and observer design based on Lyapunov method and convex optimization. IEEE Trans. on Fuzzy Systems 11, 285–298 (2003)
88. Krantz, S.G.: Handbook of complex variables. Birkhäuser Verlag, Basel (1999)
89. Kulcsár, B., Bokor, J., Shinar, J.: Unknown input reconstruction for LPV systems. International Journal of Robust and Nonlinear Control 20(5), 579–595 (2009)
90. Kwiatkowski, A., Werner, H., Boll, M.T.: Automated generation and assessment of affine LPV models. In: Proc. of the 45th IEEE Conf. on Decision and Control, San Diego, California, USA, pp. 6690–6695 (2006)
91. Lachhab, N., Abbas, H., Werner, H.: A neural-network based technique for modeling and LPV control of an arm-driven inverted pendulum. In: Proc. of the 47th IEEE Conf. on Decision and Control, Cancun, Mexico, pp. 3860–3865 (2008)
92. Lam, T.Y.: On the quality of row rank and column rank. Expositiones Mathematicae 18, 161–164 (2000)
93. Landau, I.D., Karimi, A.: Recursive algorithms for identification in closed loop: A unified approach and evaluation. Automatica 33, 1499–1523 (1997)
94. Laurain, V., Gilson, M., Tóth, R., Garnier, H.: Refined instrumental variable methods for identification of LPV Box-Jenkins models. Automatica, in print, (2010)
95. Lawrence, D.A., Rugh, W.J.: Gain scheduling dynamic linear controllers for a nonlinear plant. Automatica 3(31), 381–390 (1995)
96. Lee, L.H.: Identification and robust control of linear parameter-varying systems. Ph.D. thesis, University of California, Berkeley (1997)
97. Lee, L.H., Poolla, K.R.: Identification of linear parameter-varying systems via LFTs. In: Proc. of the 35th IEEE Conf. on Decision and Control, Kobe, Japan, pp. 1545–1550 (1996)
98. Lee, L.H., Poolla, K.R.: Identifiability issues for parameter-varying and multidimensional linear systems. In: Proc. of the ASME Design, Engineering Technical Conf. Sacramento, California (1997)
99. Leith, D.J., Leithhead, W.E.: Appropriate realization of gain-scheduled controllers with application to wind turbine regulation. International Journal of Control 65(2), 223–248 (1996)
100. Leith, D.J., Leithhead, W.E.: Gain-scheduled controller design: An analytic framework directly incorporating non-equilibrium plant dynamics. Int. Journal of Control 70, 249–269 (1998)

101. Leith, D.J., Leithhead, W.E.: Gain-scheduled nonlinear systems: Dynamic analysis by velocity based linearization families. Int. Journal of Control 70, 289–317 (1998)
102. Leith, D.J., Leithhead, W.E.: Comments on the prevalence of linear parameter varying systems. Tech. rep., Department of Electronic and Electrical Engineering, University of Strathclyde, Glasgow, Scotland (1999)
103. Lescher, F., Zhao, J.Y., Martinez, A.: Multiobjective $\mathcal{H}_2/\mathcal{H}_\infty$ control of a pitch regulated wind turbine for mechanical load reduction. In: Proc. of the European Wind Energy Conf. Athens, Greece (2006)
104. Liu, K.: Identification of linear time varying systems. Journal of Sound and Vibration 206(4), 487–505 (1997)
105. Ljung, L.: System Identification, theory for the user. Prentice-Hall, Englewood Cliffs (1999)
106. Ljung, L.: System Identification Toolbox, for use with Matlab. The Mathworks Inc. (2006)
107. Lovera, M.: Subspace identification methods: theory and applications. Ph.D. thesis, Department of Electronics and Information, Politechnico di Milano (1997)
108. Lovera, M., Mercère, G.: Identification for gain-scheduling: a balanced subspace approach. In: Proc. of the American Control Conf., New York City, USA, pp. 858–863 (2007)
109. Lovera, M., Verhaegen, M., Chou, C.T.: State space identification of MIMO linear parameter varying models. In: Proc. of the Int. Symposium on the Mathematical Theory of Networks and Systems, Padova, Italy, pp. 839–842 (1998)
110. Luenberger, D.G.: Canonical forms of linear multivariable systems. IEEE Trans. on Automatic Control 12(3), 290–293 (1967)
111. Mäkilä, P.M., Partington, J.R.: Robust approximate modeling of stable linear systems. Int. Journal of Control 58(3), 665–683 (1993)
112. Marcos, A., Balas, G.J.: Development of linear-parameter-varying models for aircraft. Journal of Guidance, Control and Dynamics 27(2), 218–228 (2004)
113. Marcovitz, A.B.: Linear time-varying discrete time systems. The Franklin Institute, Lancaster (1964)
114. Mårtensson, J., Hjalmarsson, H.: Variance error quantification for identified poles and zeros. Automatica 45(11), 2512–2525 (2009)
115. Martínez-González, R., Bolea, Y., Grau, A., Martínez-García, H.: An LPV fractional model for canal control. Mathematical Problems in Engineering, 1–18 (2009)
116. Mason, S.J.: Feedback theory, further properties of signal flow graphs. Proc. of the IRE 44(8), 920–926 (1956)
117. Matz, G., Hlawatsch, F.: Extending the transfer function calculus of time-varying linear systems: a generalized underspread theory. In: Proc. of the 23rd IEEE Int. Conf. on Acoustics, Speech, and Signal Processing, Seattle, Washington, USA, pp. 2189–2192 (1998)
118. Mazzaro, M.C., Movsichoff, E.A., Pena, R.S.S.: Robust identification of linear parameter varying systems. In: Proc. of the American Control Conf., San Diego, California, USA, pp. 2282–2284 (1999)
119. Middleton, R.H., Goodwin, G.C.: Digital control and estimation - A unified approach. Prentice-Hall, Englewood Cliffs (1990)
120. Milanese, M., Vicino, A.: Optimal estimation theory for dynamic systems with set membership uncertainty: An overview. Automatica 27(6), 997–1009 (1991)
121. Murray-Smith, R., Johansen, T.A.: Multiple model approaches to modeling and control. Taylor and Francis, Abington (1997)

122. Murray-Smith, R., Johansen, T.A., Shorten, R.: On the interpretation of local models in blended multiple model structures. Int. Journal of Control 72(7-8), 620–628 (1999)
123. Murray-Smith, R., Johansen, T.A., Shorten, R.: On transient dynamics, off-equilibrium behavior and identification in blended multiple model structures. In: Proc. of the European Control Conf., Karlsruhe, Germany (1999)
124. Nemani, M., Ravikanth, R., Bamieh, B.A.: Identification of linear parametrically varying systems. In: Proc. of the 34th IEEE Conf. on Decision and Control, New Orleans, Louisiana, USA, pp. 2990–2995 (1995)
125. Nichols, R.A., Reichert, R.T., Rugh, W.J.: Gain scheduling for $\mathcal{H}$-infinity controllers: A flight control example. IEEE Trans. on Control Systems Technology 1(2), 69–79 (1993)
126. Niedźwiecki, M.: Identification of time-varying processes. John Wiley and Sons, Chichester (2000)
127. Ninness, B., Goodwin, G.C.: Estimation of model quality. Automatica 31(12), 1771–1797 (1995)
128. Ninness, B.M., Gómez, J.C.: Asymptotic analysis of MIMO systems estimates by the use of orthonormal basis. In: Proc. of the 13th IFAC World Congress, San Francisco, California, USA, pp. 363–368 (1996)
129. Ninness, B.M., Gustafsson, F.: A unifying construction of orthonormal bases for system identification. IEEE Trans. on Automatic Control 42(4), 515–521 (1997)
130. Nobakhti, A., Munro, N.: Eigenfitting: An alternative LPV modeling technique. In: Proc. of the 10th Mediterranean Conf. on Control and Automation, Lisbon, Portugal (2002)
131. Oliveira e Silva, T.: A n-width result for the generalized orthonormal basis function model. In: Proc. of the 13th IFAC World Congress, Sydney, Australia, pp. 375–380 (1996)
132. Ore, O.: Theory of non-commutative polynomials. Annals of Mathematics 34(2), 480–508 (1933)
133. Packard, A.: Gain scheduling via linear fractional transformations. System & Control Letters 22(2), 79–92 (1994)
134. Packard, A., Becker, G.: Quadratic stabilization of parametrically-dependent linear system using parametrically-dependent linear, dynamic feedback. Advances in Robust and Nonlinear Control Systems DSC (43), 29–36 (1992)
135. Paijmans, B., Symens, W., Van Brussel, H., Swevers, J.: A gain-scheduling-control technique for mechatronic systems with position dependent dynamics. In: Proc. of the American Control Conf., Minneapolis, Minnesota, USA, pp. 2933–2938 (2006)
136. Paijmans, B., Symens, W., Van Brussel, H., Swevers, J.: Interpolating affine LPV identification for mechatronic systems with one varying parameter. European Journal of Control 14(1) (2008)
137. Papageorgiou, G.: Robust control system design, $\mathcal{H}_\infty$ loop shaping and aerospace application. Ph.D. thesis, University of Cambridge (1998)
138. Papageorgiou, G., Glover, K., D'Mello, G., Patel, Y.: Taking robust LPV control into flight on the VAAC Harrier. In: Proc. of the 39th IEEE Conf. on Decision and Control, Sydney, Australia, pp. 4558–4564 (2000)
139. Park, B.P., Verriest, E.I.: Canonical forms for linear time-varying multivariable discrete systems. In: Proc. of the 28th Annual Allerton Conf. on Communication, Control, and Computers, Urbana, Illinois, USA (1990)
140. Peters, M.A., Iglesias, P.A.: A spectral test for observability and reachability of time-varying systems and the Ricatti difference equation. SIAM Journal on Control and Optimization 37(5), 1330–1344 (1999)

141. Petersson, D., Löfberg, J.: Optimization based LPV-approximation of multiple model systems. In: Proc. of the European Control Conf., Budapest, Hungary, pp. 3172–3177 (2009)
142. Pfifer, H., Hecker, S.: Generation of optimal linear parametric models for LFT-based robust stability analysis and control design. In: Proc. of the 47th IEEE Conf. on Decision and Control, Cancun, Mexico, pp. 3866–3871 (2008)
143. Pinkus, A.: n-Widths in Approximation Theory. Springer, Heidelberg (1985)
144. Pintelon, R., Schoukens, J.: System identification, a frequency domain approach. IEEE press, Los Alamitos (2001)
145. Polderman, J.W.: Proper elimination of latent variables. In: Proceedings of the 12th IFAC Symposium on System Identification, Sydney, Australia, pp. 73–76 (1992)
146. Polderman, J.W., Willems, J.C.: Introduction to Mathematical Systems Theory, A Behavioral Approach. Springer, Heidelberg (1991)
147. Prempain, E., Postlethwaite, I., Benchaib, A.: Linear parameter variant H_∞ control design for an induction motor. Control Engineering Practice 10(6), 663–644 (2002)
148. Previdi, F., Lovera, M.: Identification of a class of linear models with nonlinearly varying parameters. In: Proc. of the European Control Conf., Karlsruhe, Germany (1999)
149. Previdi, F., Lovera, M.: Identification of a class of nonlinear parametrically varying models. In: Proc. of the European Control Conf., Porto, Portugal, pp. 3086–3091 (2001)
150. Previdi, F., Lovera, M.: Identification of a class of non-linear parametrically varying models. Int. Journal on Adaptive Control and Signal Processing 17, 33–50 (2003)
151. Previdi, F., Lovera, M.: Identification of nonlinear parametrically varying models using separable least squares. Int. Journal of Control 77(12), 1382–1392 (2004)
152. Qin, W., Wang, Q.: An LPV approximation for admission control of an internet web server: Identification and control. Control Engineering Practice 15, 1457–1467 (2007)
153. Rapisarda, P., Willems, J.C.: State maps for linear systems. SIAM Journal on Control and Optimization 35(3), 1053–1091 (1997)
154. Ravit, R., Nagpal, K.M., Khargonekar, P.P.: $\mathcal{H}_\infty$ control of linear time-varying systems: A state-space approach. SIAM Journal on Control and Optimization 29(6), 1394–1413 (1991)
155. Rosenbrock, H.H.: The stability of linear time-dependent control systems. Journal of Electronic Control 15, 73–80 (1963)
156. Rugh, W., Shamma, J.S.: Research on gain scheduling. Automatica 36(10), 1401–1425 (2000)
157. Rugh, W.J.: Analytical framework for gain scheduling. IEEE Control Systems Magazine 11(1), 79–84 (1991)
158. Sakhnovich, L.A.: Interpolation theory and its applications. Kluwer Academic, Dordrecht (1997)
159. Sbárbaro, D., Johansen, T.: Multiple local Laguerre models for modeling nonlinear dynamic systems of the Wiener class. IEEE Process Control Theory Applications 144(5), 375–380 (1997)
160. Scherer, C.W.: Mixed $\mathcal{H}_2/\mathcal{H}_\infty$ control for time-varying and linear parametrically-varying systems. Int. Journal of Robust and Nonlinear Control 6(9-10), 929–952 (1996)
161. Scherer, C.W.: LPV control and full block multipliers. Automatica 37(3), 361–375 (2001)
162. Scherer, C.W., Hol, C.W.J.: Matrix sum-of-squares relaxations for robust semi-definit programs. Mathematical Programming 107(1), 189–211 (2006)
163. Setnes, M., Babuška, R.: Rule based reduction: some comments on the use of orthogonal transforms. IEEE Trans. on Systems, Man and Cybernetics 31(2), 199–206 (2001)

164. Shamma, J.S., Athans, M.: Analysis of gain scheduled control for nonlinear plants. IEEE Trans. on Automatic Control 35(8), 898–907 (1990)
165. Shamma, J.S., Athans, M.: Guaranteed properties of gain-scheduled control for nonlinear-plants. Automatica 27, 559–564 (1991)
166. Shamma, J.S., Athans, M.: Gain scheduling: potential hazards and possible remedies. IEEE Control Systems Magazine 12(3), 101–107 (1992)
167. Shamma, J.S., Cloutier, J.R.: Gain-scheduled missile autopilot design using linear-parameter varying transformations. AIAA Journal of Guidance, Control and Dynamics 16(2), 256–263 (1993)
168. Shin, J.Y.: Worst-case analysis and linear parameter-varying gain-scheduled control of aerospace systems. Ph.D. thesis, University of Minnesota, Minneapolis (2000)
169. Shin, J.Y.: Quasi-linear parameter varying representation of general aircraft dynamics over non-trim region. Tech. rep., NASA: Langley Research Center (2007)
170. Shin, J.Y., Balas, G., Kaya, M.A.: Blending methodology of linear parameter varying control synthesis of F-16 aircraft system. Journal of Guidance, Control, and Dynamics 25(6), 1040–1048 (2002)
171. Silverman, L.M.: Transformation of time-variable systems to canonical (phase-variable) form. IEEE Trans. on Automatic Control 11(2), 300–303 (1966)
172. Silverman, L.M.: Realization of linear dynamical systems. IEEE Trans. on Automatic Control 16(6), 554–567 (1971)
173. Silverman, L.M., Meadows, H.E.: Degrees of controllability in time-variable systems. In: Proc. of the National Electronics Conf., Chicago, vol. 21, pp. 689–693 (1965)
174. Silverman, L.M., Meadows, H.E.: Controllability and observability in time-variable linear systems. SIAM Journal on Control 5, 64–73 (1967)
175. Silverman, L.M., Meadows, H.E.: Equivalent realizations of linear systems. SIAM Journal on Applied Mathematics 17(2), 393–408 (1969)
176. Skoog, R.A., Lau, C.G.: Instability of slowly varying systems. IEEE Trans. on Automatic Control 17(1), 86–92 (1972)
177. Spong, M.: Modeling and control of elastic joint robots. Journal of dynamic systems, measurement, and control 109(4), 310–319 (1987)
178. Sreedhar, N., Rao, S.N.: Stability of linear time-varying systems. Int. Journal of Control 7(6), 591–594 (1968)
179. Stein, G.: Adaptive flight control - A pragmatic view. In: Applications of Adaptive Control. Academic, New York (1980)
180. Steinbuch, M., van de Molengraft, R., van der Voort, A.: Experimental modeling and LPV control of a motion system. In: Proc. of the American Control Conf., Denver, Colorado, USA, pp. 1374–1379 (2003)
181. Stevens, B.L., Lewis, F.L.: Aircraft Control and Simulation. John Wiley and Sons, Chichester (2003)
182. Stilwell, D.J., Rugh, W.J.: Stability preserving interpolation methods for synthesis of gain scheduled controllers. Automatica 36(5), 665–671 (2000)
183. Sturm, J.: Using SeDuMi 1.02, a Matlab toolbox for optimization over symmetric cones. Optimization Methods and Software 11-12, 625–653 (1999)
184. Sznaier, M., Mazzaro, C., Inanc, T.: An LMI approach to control oriented identification of LPV systems. In: Proc. of the American Control Conf., Chicago, Illinois, USA, pp. 3682–3686 (2000)
185. Tan, W.: Application of linear parameter-varying control theory. Master's thesis, University of California, Berkeley (1997)
186. Tan, W., Packard, A., Balas, G.: Quasi-LPV modeling and LPV control of a generic missile. In: Proc. of the American Control Conf., Chicago, Illinois, USA, pp. 3692–3696 (2000)

187. Tanelli, M., Ardagna, D., Lovera, M.: LPV model identification in virtualized service center environments. In: Proc. of the 15th IFAC Symposium on System Identification, Saint-Malo, France, pp. 862–867 (2009)
188. Tóth, R.: Modeling and identification of linear parameter-varying systems, an orthonormal basis function approach. Ph.D. thesis, Delft University of Technology (2008)
189. Tóth, R., Felici, F., Heuberger, P.S.C., Van den Hof, P.M.J.: Discrete time LPV I/O and state space representations, differences of behavior and pitfalls of interpolation. In: Proc. of the European Control Conf., Kos, Greece, pp. 5418–5425 (2007)
190. Tóth, R., Felici, F., Heuberger, P.S.C., Van den Hof, P.M.J.: Crucial aspects of zero-order-hold LPV state-space system discretization. In: Proc. of the 17th IFAC World Congress, Seoul, Korea, pp. 3246–3251 (2008)
191. Tóth, R., Fodor, D.: Speed sensorless mixed sensitivity linear parameter variant $\mathcal{H}_\infty$ control of the induction motor. Journal of Electrical Engineering 6(4), 12/1–6 (2006)
192. Tóth, R., Heuberger, P.S.C., Van den Hof, P.M.J.: Optimal pole selection for LPV system identification with OBFs, a clustering approach. In: Proc. of the 14th IFAC Symposium on System Identification, Newcastle, Australia, pp. 356–361 (2006)
193. Tóth, R., Heuberger, P.S.C., Van den Hof, P.M.J.: Orthonormal basis selection for LPV system identification, the Fuzzy-Kolmogorov c-Max approach. In: Proc. of the 45th IEEE Conf. on Decision and Control, San Diego, California, USA, pp. 2529–2534 (2006)
194. Tóth, R., Heuberger, P.S.C., Van den Hof, P.M.J.: LPV system identification with globally fixed orthonormal basis functions. In: Proc. of the 46th IEEE Conf. on Decision and Control, New Orleans, Louisiana, USA, pp. 3646–3653 (2007)
195. Tóth, R., Heuberger, P.S.C., Van den Hof, P.M.J.: Flexible model structures for LPV identification with static scheduling dependency. In: Proc. of the 47th IEEE Conf. on Decision and Control, Cancun, Mexico, pp. 4522–4527 (2008)
196. Tóth, R., Heuberger, P.S.C., Van den Hof, P.M.J.: Asymptotically optimal orthonormal basis functions for LPV system identification. Automatica 45(6), 1359–1370 (2009)
197. Tóth, R., Heuberger, P.S.C., Van den Hof, P.M.J.: An LPV identification framework based on orthonormal basis functions. In: Proc. of the 15th IFAC Symposium on System Identification, Saint-Malo, France, pp. 1328–1333 (2009)
198. Tóth, R., Heuberger, P.S.C., Van den Hof, P.M.J.: On the calculation of accurate uncertainty regions for estimated system poles and zeros. Tech. Rep. 10-010, Delft University of Technology (2010)
199. Tóth, R., Lovera, M., Heuberger, P.S.C., Van den Hof, P.M.J.: Discretization of linear fractional representations of LPV systems. In: Proc. of the 48th IEEE Conf. on Decision and Control, Shanghai, China, pp. 7424–7429 (2009)
200. Tóth, R., Lyzell, C., Enqvist, M., Heuberger, P.S.C., Van den Hof, P.M.J.: Order and structural dependence selection of LPV-ARX models using a nonnegative garrote approach. In: Proc. of the 48th IEEE Conf. on Decision and Control, Shanghai, China, pp. 7406–7411 (2009)
201. Tóth, R., Willems, J.C., Heuberger, P.S.C., Van den Hof, P.M.J.: The behavioral approach to linear parameter-varying systems. Accepted to the IEEE Trans. on Automatic Control (2010)
202. Tóth, R., Willems, J.C., Heuberger, P.S.C., Van den Hof, P.M.J.: A behavioral approach to LPV systems. In: Proc. of the European Control Conf., Budapest, Hungary, pp. 2015–2020 (2009)
203. Trangbæk, K.: Linear parameter varying control of induction motors. Ph.D. thesis, Aalborg University (2001)

204. van der Voort, A.: LPV control based on a pick-and-place unit. Ph.D. thesis, Delft University of Technology (2002)
205. van Donkelaar, E.T.: Improvement of efficiency in identification and model predictive control of industrial processes, a flexible linear parametrization approach. Ph.D. thesis, Delft University of Technology (2000)
206. van Wingerden, J.W.: Control of wind turbines with "smart" rotors: Proof of concept & LPV subspace identification. Ph.D. thesis, Delft University of Technology (2008)
207. van Wingerden, J.W., Felici, F., Verhaegen, M.: Subspace identification of MIMO LPV systems using a piecewise constant scheduling sequence with hard/soft switching. In: Proc. of the European Control Conf., Kos, Greece, pp. 927–934 (2007)
208. van Wingerden, J.W., Verhaegen, M.: Subspace identification of MIMO LPV systems: The PBSID approach. In: Proc. of the 47th IEEE Conf. on Decision and Control, Cancun, Mexico, pp. 4516–4521 (2008)
209. van Wingerden, J.W., Verhaegen, M.: Subspace identification of multivariable LPV systems: a novel approach. In: Proc. of the Int. Symposium on Computer-Aided Control System Design, San Antonio, Texas, USA, pp. 840–845 (2008)
210. van Wingerden, J.W., Verhaegen, M.: Subspace identification of bilinear and LPV systems for open- and closed-loop data. Automatica 45, 372–381 (2009)
211. Varga, A., Looye, G., Moorman, D., Grübel, G.: Automated generation of LFR-based parametric uncertainty descriptions from generic aircraft models. Mathematical and Computer Modeling of Dynamical Systems 4, 249–274 (1998)
212. Veenman, J., Scherer, C.W., Köroglu, H.: IQC-based LPV controller synthesis for the NASA HL20 atmospheric re-entry vehicle. In: Proc. of the AIAA Guidance, Navigation, and Control Conference, Chicago, Illinois, USA, pp. 1–18 (2009)
213. Verdult, V.: Nonlinear system identification: a state-space approach. Ph.D. thesis, University of Twente (2002)
214. Verdult, V., Bergoer, N., Verhaegen, M.: Identification of fully parameterized linear and non-linear state-space systems by projected gradient search. In: Proc. of the 13th IFAC Symposium on System Identification, Rotterdam, The Netherlands, pp. 737–742 (2003)
215. Verdult, V., Ljung, L., Verhaegen, M.: Identification of composite local linear state-space models using a projected gradient search. Int. Journal of Control 75(16-17), 1125–1153 (2002)
216. Verdult, V., Lovera, M., Verhaegen, M.: Identification of linear parameter-varying state space models with application to helicopter rotor dynamics. Int. Journal of Control 77(13), 1149–1159 (2004)
217. Verdult, V., Verhaegen, M.: Subspace identification of multivariable linear parameter-varying systems. Automatica 38(5), 805–814 (2002)
218. Verdult, V., Verhaegen, M.: Subspace identification of piecewise linear systems. In: Proc. of the 43rd IEEE Conf. on Decision and Control, Atlantis, Paradise Island, Bahamas, pp. 3838–3843 (2004)
219. Verdult, V., Verhaegen, M.: Kernel methods for subspace identification of multivariable LPV and bilinear systems. Automatica 41(9), 1557–1565 (2005)
220. Verhaegen, M.: Identification of the deterministic part of MIMO state space models given in innovations form from input-output data. Automatica 30(1), 61–74 (1994)
221. Verhaegen, M., Dewilde, P.: Subspace model identification Part 1: The output-error state-space model identification class of algorithms. Int. Journal of Control 56(5), 1187–1210 (1992)
222. Verhaegen, M., Yu, X.: A class of subspace model identification algorithms to identify periodically and arbitrary time-varying systems. Automatica 31(2), 201–216 (1995)

223. Vuerinckx, R., Pintelon, R., Schoukens, J., Rolain, Y.: Obtaining accurate confidence regions for the estimated zeros and poles in system identification problems. IEEE Trans. on Automatic Control 46(4), 656–659 (2001)
224. Wassink, M.G., van de Wal, M., Scherer, C.W., Bosgra, O.: LPV control for a wafer stage: Beyond the theoretical solution. Control Engineering Practice 13, 231–245 (2004)
225. Wei, X.: Advanced LPV techniques for diesel engines. Ph.D. thesis, Johannes Kepler University, Linz (2006)
226. Wei, X., Del Re, L.: On persistent excitation for parameter estimation of quasi-LPV systems and its application in modeling of diesel engine torque. In: Proc. of the 14th IFAC Symposium on System Identification, Newcastle, Australia, pp. 517–522 (2006)
227. Weiss, G.: Memoryless output nullification and canonical forms, for time varying-systems. Int. Journal of Control 78(15), 1174–1181 (2005)
228. Whatley, M.J., Pot, D.C.: Adaptive gain improves reactor control. Hydrocarbon Processing, pp. 75–78 (1984)
229. Wijnheijmer, F., Naus, G., Post, W., Steinbuch, M., Teerhuis, P.: Modeling and LPV control of an electro-hydraulic servo system. In: Proc. of the IEEE International Conf. on Control Applications, Munich, Germany, pp. 3116–3120 (2006)
230. Willems, J.C.: Paradigms and puzzles in the theory of dynamical systems. IEEE Tran. on Automatic Control 36(3), 259–294 (1991)
231. Willems, J.C.: The behavioral approach to open and interconnected systems. IEEE Control Systems Magazine 27(6), 46–99 (2007)
232. Wood, G.D., Goddard, P.J., Glover, I.: Approximation of linear parameter-varying systems. In: Proc. of the 35th IEEE Conf. on Decision and Control, Kobe, Japan, pp. 406–411 (1996)
233. Wu, F., Dong, K.: Gain-scheduling control of LFT systems using parameter-dependent Lyapunov functions. Automatica 42, 39–50 (2006)
234. Xie, X.L.X., Beni, G.: A validity measure for fuzzy clustering. IEEE Trans. on Pattern Analysis and Machine Intelligence 13(8), 841–847 (1991)
235. Yaz, E., Niu, X.: Stability robustness of linear discrete-time systems in the presence of uncertainty. Int. Journal of Control 50(1), 173–182 (1989)
236. Yen, J., Wang, L.: Simplifying fuzzy rule-based models using orthogonal transformation methods. IEEE Trans. on Systems, Man and Cybernetics 29(1), 13–24 (1999)
237. Young, J.: Gain scheduling for geometrically nonlinear flexible space structures. Ph.D. thesis, Massachusetts Institute of Technology (2002)
238. Zenger, K., Ylinen, R.: Pole placement of time-varying state space representations. In: Proc. of the 44th IEEE Conf. on Decision and Control, Seville, Spain, pp. 6527–6532 (2005)
239. Zerz, E.: An algebraic analysis approach to linear time-varying systems. IMA Journal of Mathematical Control and Information 23, 113–126 (2006)
240. Zhou, K., Doyle, J.C.: Essentials of Robust Control. Prentice-Hall, Englewood Cliffs (1998)
241. Zhu, Y., Ji, G.: LPV model identification using blended linear models with given weightings. In: Proc. of the 15th IFAC Symposium on System Identification, Saint-Malo, France, pp. 1674–1679 (2009)
242. Zhu, Y., Xu, X.: A method of LPV model identification for control. In: Proc. of the 17th IFAC World Congress, Seoul, Korea, pp. 5018–5024 (2008)

Index

Lecture Notes in Control and Information Sciences

Edited by M. Thoma, F. Allgöwer, M. Morari

Further volumes of this series can be found on our homepage:
springer.com

Vol. 403: Tóth, R.;
Modeling and Identification of
Linear Parameter-Varying Systems
319 p. 2010 [978-3-642-13811-9]

Vol. 402: del Re, L.; Allgöwer, F.;
Glielmo, L.; Guardiola, C.;
Kolmanovsky, I. (Eds.):
Automotive Model Predictive Control
284 p. 2010 [978-1-84996-070-0]

Vol. 401: Chesi, G.; Hashimoto, K. (Eds.):
Visual Servoing via Advanced
Numerical Methods
393 p. 2010 [978-1-84996-088-5]

Vol. 400: Tomás-Rodríguez, M.;
Banks, S.P.:
Linear, Time-varying Approximations
to Nonlinear Dynamical Systems
298 p. 2010 [978-1-84996-100-4]

Vol. 399: Edwards, C.; Lombaerts, T.;
Smaili, H. (Eds.):
Fault Tolerant Flight Control
appro. 350 p. 2010 [978-3-642-11689-6]

Vol. 398: Hara, S.; Ohta, Y.;
Willems, J.C.; Hisaya, F. (Eds.):
Perspectives in Mathematical System
Theory, Control, and Signal Processing
appro. 370 p. 2010 [978-3-540-93917-7]

Vol. 397: Yang, H.; Jiang, B.;
Cocquempot, V.:
Fault Tolerant Control Design for
Hybrid Systems
191 p. 2010 [978-3-642-10680-4]

Vol. 396: Kozlowski, K. (Ed.):
Robot Motion and Control 2009
475 p. 2009 [978-1-84882-984-8]

Vol. 395: Talebi, H.A.; Abdollahi, F.;
Patel, R.V.; Khorasani, K.:
Neural Network-Based State
Estimation of Nonlinear Systems
appro. 175 p. 2010 [978-1-4419-1437-8]

Vol. 394: Pipeleers, G.; Demeulenaere, B.;
Swevers, J.:
Optimal Linear Controller Design for
Periodic Inputs
177 p. 2009 [978-1-84882-974-9]

Vol. 393: Ghosh, B.K.; Martin, C.F.;
Zhou, Y.:
Emergent Problems in Nonlinear
Systems and Control
285 p. 2009 [978-3-642-03626-2]

Vol. 392: Bandyopadhyay, B.;
Deepak, F.; Kim, K.-S.:
Sliding Mode Control Using Novel Sliding
Surfaces
137 p. 2009 [978-3-642-03447-3]

Vol. 391: Khaki-Sedigh, A.; Moaveni, B.:
Control Configuration Selection for
Multivariable Plants
232 p. 2009 [978-3-642-03192-2]

Vol. 390: Chesi, G.; Garulli, A.;
Tesi, A.; Vicino, A.:
Homogeneous Polynomial Forms for
Robustness Analysis of Uncertain Systems
197 p. 2009 [978-1-84882-780-6]

Vol. 389: Bru, R.; Romero-Vivó, S. (Eds.):
Positive Systems
398 p. 2009 [978-3-642-02893-9]

Vol. 388: Jacques Loiseau, J.; Michiels, W.;
Niculescu, S-I.; Sipahi, R. (Eds.):
Topics in Time Delay Systems
418 p. 2009 [978-3-642-02896-0]

Vol. 387: Xia, Y.;
Fu, M.; Shi, P.:
Analysis and Synthesis of
Dynamical Systems with Time-Delays
283 p. 2009 [978-3-642-02695-9]

Vol. 386: Huang, D.;
Nguang, S.K.:
Robust Control for Uncertain
Networked Control Systems with
Random Delays
159 p. 2009 [978-1-84882-677-9]

Vol. 385: Jungers, R.:
The Joint Spectral Radius
144 p. 2009 [978-3-540-95979-3]

Vol. 384: Magni, L.; Raimondo, D.M.;
Allgöwer, F. (Eds.):
Nonlinear Model Predictive Control
572 p. 2009 [978-3-642-01093-4]

Vol. 383: Sobhani-Tehrani E.:
Khorasani K.;
Fault Diagnosis of Nonlinear Systems
Using a Hybrid Approach
360 p. 2009 [978-0-387-92906-4]

Vol. 382: Bartoszewicz A.;
Nowacka-Leverton A.:
Time-Varying Sliding Modes for Second
and Third Order Systems
192 p. 2009 [978-3-540-92216-2]

Vol. 381: Hirsch M.J.; Commander C.W.;
Pardalos P.M.; Murphey R. (Eds.):
Optimization and Cooperative
Control Strategies: Proceedings of the 8th
International Conference on Cooperative
Control and Optimization
459 p. 2009 [978-3-540-88062-2]

Vol. 380: Basin M.:
New Trends in Optimal Filtering and Control for
Polynomial and Time-Delay Systems
206 p. 2008 [978-3-540-70802-5]

Vol. 379: Mellodge P.; Kachroo P.:
Model Abstraction in Dynamical Systems:
Application to Mobile Robot Control
116 p. 2008 [978-3-540-70792-9]

Vol. 378: Femat R.; Solis-Perales G.:
Robust Synchronization of Chaotic Systems
Via Feedback
199 p. 2008 [978-3-540-69306-2]

Vol. 377: Patan K.:
Artificial Neural Networks for
the Modelling and Fault
Diagnosis of Technical Processes
206 p. 2008 [978-3-540-79871-2]

Vol. 376: Hasegawa Y.:
Approximate and Noisy Realization of
Discrete-Time Dynamical Systems
245 p. 2008 [978-3-540-79433-2]

Vol. 375: Bartolini G.; Fridman L.;
Pisano A.; Usai E. (Eds.):
Modern Sliding Mode Control Theory
465 p. 2008 [978-3-540-79015-0]

Vol. 374: Huang B.; Kadali R.:
Dynamic Modeling, Predictive Control
and Performance Monitoring
240 p. 2008 [978-1-84800-232-6]

Vol. 373: Wang Q.-G.; Ye Z.; Cai W.-J.;
Hang C.-C.:
PID Control for Multivariable Processes
264 p. 2008 [978-3-540-78481-4]

Vol. 372: Zhou J.; Wen C.:
Adaptive Backstepping Control of Uncertain
Systems
241 p. 2008 [978-3-540-77806-6]

Vol. 371: Blondel V.D.; Boyd S.P.;
Kimura H. (Eds.):
Recent Advances in Learning and Control
279 p. 2008 [978-1-84800-154-1]

Vol. 370: Lee S.; Suh I.H.;
Kim M.S. (Eds.):
Recent Progress in Robotics:
Viable Robotic Service to Human
410 p. 2008 [978-3-540-76728-2]

Vol. 369: Hirsch M.J.; Pardalos P.M.;
Murphey R.; Grundel D.:
Advances in Cooperative Control and
Optimization
423 p. 2007 [978-3-540-74354-5]

Vol. 368: Chee F.; Fernando T.
Closed-Loop Control of Blood Glucose
157 p. 2007 [978-3-540-74030-8]

Vol. 367: Turner M.C.; Bates D.G. (Eds.):
Mathematical Methods for Robust and
Nonlinear Control
444 p. 2007 [978-1-84800-024-7]

Vol. 366: Bullo F.; Fujimoto K. (Eds.):
Lagrangian and Hamiltonian Methods for
Nonlinear Control 2006
398 p. 2007 [978-3-540-73889-3]

Vol. 365: Bates D.; Hagström M. (Eds.):
Nonlinear Analysis and Synthesis
Techniques for Aircraft Control
360 p. 2007 [978-3-540-73718-6]

Vol. 364: Chiuso A.; Ferrante A.;
Pinzoni S. (Eds.):
Modeling, Estimation and Control
356 p. 2007 [978-3-540-73569-4]

Vol. 363: Besançon G. (Ed.):
Nonlinear Observers and Applications
224 p. 2007 [978-3-540-73502-1]

Vol. 362: Tarn T.-J.; Chen S.-B.;
Zhou C. (Eds.):
Robotic Welding, Intelligence and Automation
562 p. 2007 [978-3-540-73373-7]

Vol. 361: Méndez-Acosta H.O.; Femat R.;
González-Álvarez V. (Eds.):
Selected Topics in Dynamics and Control of
Chemical and Biological Processes
320 p. 2007 [978-3-540-73187-0]

Vol. 360: Kozlowski K. (Ed.):
Robot Motion and Control 2007
452 p. 2007 [978-1-84628-973-6]

Vol. 359: Christophersen F.J.:
Optimal Control of Constrained
Piecewise Affine Systems
190 p. 2007 [978-3-540-72700-2]